Regenerated Cellulose

Scrivener Publishing
100 Cummings Center, Suite 541J
Beverly, MA 01915-6106

Publishers at Scrivener
Martin Scrivener (martin@scrivenerpublishing.com)
Phillip Carmical (pcarmical@scrivenerpublishing.com)

Regenerated Cellulose

A Renewable Alternative to Petroleum-Based Plastics

Edited by

Mohd Shabbir
School of Engineering and Technology (SET), CGC University, Mohali, Punjab, India

Scrivener Publishing

WILEY

This edition first published 2026 by John Wiley & Sons, Inc., 111 River Street, Hoboken, NJ 07030, USA and Scrivener Publishing LLC, 100 Cummings Center, Suite 541J, Beverly, MA 01915, USA

For more information about Scrivener publications please visit www.scrivenerpublishing.com.

Wiley Global Headquarters
111 River Street, Hoboken, NJ 07030, USA

For details of our global editorial offices, customer services, and more information about Wiley products visit us at www.wiley.com.

The manufacturer's authorized representative according to the EU General Product Safety Regulation is Wiley-VCH GmbH, Boschstr. 12, 69469 Weinheim, Germany, e-mail: Product_Safety@wiley.com.

Library of Congress Cataloging-in-Publication Data

ISBN 978-1-394-34313-3

Cover image: Generated by AI using Adobe Firefly
Cover design by Russell Richardson

Set in size of 11pt and Minion Pro by Manila Typesetting Company, Makati, Philippines

Printed in the USA

10 9 8 7 6 5 4 3 2 1

Contents

Preface

Today, we prefer natural resources or products derived from nature for our daily use, since over-utilization of petroleum resources and synthetic materials has contributed to climate change and a rise in global temperature. Cellulose, one of the most abundant natural polymers on earth, plays an important role in meeting increased demands. The question is how can we use it in more applications?

Significant problems arise from the insolubility of cellulose in common solvents. This book focuses on the dissolution and regeneration for various applications. Different morphologies enable new applications of regenerated cellulose materials, which broadens the range of applications. The accumulation of material herein and the examination of specific applications and processes will be useful for researchers, scholars, engineers, and industrial workers.

This book is composed of fifteen chapters from various research areas dealing with regeneration, morphological aspects of cellulose, and characteristic applications. The first part of the book includes chapters discussing cellulose procurement, its derivatives, modifications, and regeneration after dissolution, along with the economic perspectives. Chapter One outlines the dissolution of cellulose and regeneration into various morphologies. Chapters Two and Three explore cellulose modifications and derivatization of cellulose for various applications. Chapter Four discusses the economic perspectives of cellulose and its utilization in several industries or as a better alternative to petroleum products.

The next part of the book is focused on environmental remediation applications *via* different mechanisms of action, including how various pollutants in water or the environment are classified and removed with the help of regenerated cellulose materials. Sensing metal ions and other pollutants, followed by the adsorptive or catalytic removal of pollutants from wastewater, are discussed first, and then alternate solutions for packaging and storage using cellulose are elaborated.

The next part of chapters reviews pharmaceutical and biomedical applications. Drug delivery, wound dressing, bone tissue engineering, etc., are discussed throughout these chapters. The last two chapters are based on industrial applications: the textile industry and construction materials. The role of regenerated cellulose in textiles is discussed in Chapter Fourteen, while concrete materials designed with the help of cellulose are explored in the last chapter.

The book contains informative chapters from authors from all over the world and from various fields, addressing cellulose and its applications in different sectors. Students, researchers, and academicians in materials science, chemical engineering, environmental science, textiles, the food packaging industry, civil engineering, and biomedical research will find this book of great interest and useful in their curriculum. This book will definitely assist in generating emerging new ideas in cellulose research, leading to interdisciplinary research collaborations.

I would like to express my heartfelt thanks to all those who supported the creation of this book. I am deeply grateful to the eminent authors whose valuable contributions made this work possible, as well as to my family, colleagues, and students for their unwavering encouragement. Finally, we wish to thank Martin Scrivener and the team at Scrivener Publishing for their efforts in publishing this work.

Mohd Shabbir
August 2026

Part I

CELLULOSE PRODUCTION AND ECONOMIC PERSPECTIVE

1

Cellulose: An Introduction to Sources, Regeneration, and Solvents

Perneta, Manish Sharma and Mohd Shabbir*

School of Engineering and Technology (SET), CGC University, Mohali, Punjab, India

Abstract

Cellulose is the earth's most abundant naturally occurring polymer and primary structural component of plant cells. Despite being lightweight, it exhibits exceptional strength and stiffness. Owing to these properties, cellulose finds its applications as a reinforcing component in biocomposites. Being an ecofriendly biomaterial, processing of cellulose and regeneration for various applications needs special attention for keeping all sustainable and compatible towards ecosystem. The choice of solvents for dissolving cellulose in the first step of regeneration is crucial, as cellulose dissolution is challenging. These solvents are key for transforming cellulose into several forms including fibers, membranes/films/sheets, beads/microspheres, and hydrogels/aerogels, among others. In recent times, several green solvents are evolved to counter the application of harmful chemicals and the conventional processes of cellulose regeneration. This chapter is all about the introduction of the various aspects of cellulose, solvents, and regeneration to different morphologies.

Keywords: Regenerated cellulose, cellulose microspheres, hydrogels, cellulose membranes, agricultural waste, ionic liquids, deep eutectic solvents, polymer composites

1.1 Introduction

In recent times, since the petroleum-based resources are depleting, there is a need for renewable resources, energy, and environment-friendly materials,

**Corresponding author*: shabbirmeo@gmail.com

Mohd Shabbir (ed.) Regenerated Cellulose: A Renewable Alternative to Petroleum-Based Plastics, (3–14)

as well as sustainable techniques and processes. Green and sustainability has been in trend for the themes of research and scientific gatherings in previous few decades. Products from natural resources fulfil most of the criterion for green and sustainable materials. Natural polymers received much attention of the scientific community for the development of ecofriendly and biocompatible fuels and products. Cellulose stands as the most abundant renewable polymer resource currently available and has attracted significant interest due to its numerous beneficial properties. Its renewability, wide availability, nontoxicity, low cost, and environmental friendliness make it highly appealing. Additionally, cellulose is biocompatible, biodegradable, and exhibits excellent thermal and chemical stability, along with versatile derivatization potential [1]. Plants produce cellulose in abundance in their cell walls and it provides strength to the plants. Cellulose is a linear polymer consisting of repeating anhydroglucose units $(C_6H_{10}O_5)_n$. These units are connected by covalent bonds through an oxygen atom linking the C1 carbon of one glucose ring to the C4 carbon of the next, forming a 1→4 glycosidic linkage. This β 1→4 glycosidic bond results in a flat and ribbon-like structure [2].

Cellulose had been difficult to utilize due to the high insolubility in common solvents and subsequently regeneration to functional materials.

Unmodified cellulose fibers, chemically modified cellulose, such as cellulose derivatives, and nanocellulose are processed to give many types of biodegradable polymeric materials such as films, composites, nanocomposite systems, and composites films for various applications.

1.2 Cellulose Sources

Cellulose is primarily obtained from agricultural waste, bacterial and algal sources, wood, plants etc. (Figure 1.1). Agricultural waste comprises biomass of pineapple peel, sugarcane bagasse, oat husk, sisal, rice and wheat stalks, straw, leaves, etc., which can be treated by various chemical processes to obtain cellulose such as alkali treatment, acid hydrolysis and bleaching. Agricultural waste generally contains other components apart from cellulose such as fiber, lignin, hemicelluloses, sugar, etc. The structure of cellulose obtained may vary as per the sources and method of procurement. Forest wood is another primary source for cellulose and exists in abundance on earth. Cellulose content from this source contains high quality cellulose of great mechanical strength and ductile properties. Oxidation or acid treatment and some biological methods by using fungi on softwoods are the methods to procure cellulose from forest wood. Bacteria and the algal species are another source of cellulose. Cellulose is extracted from *Cladophora*,

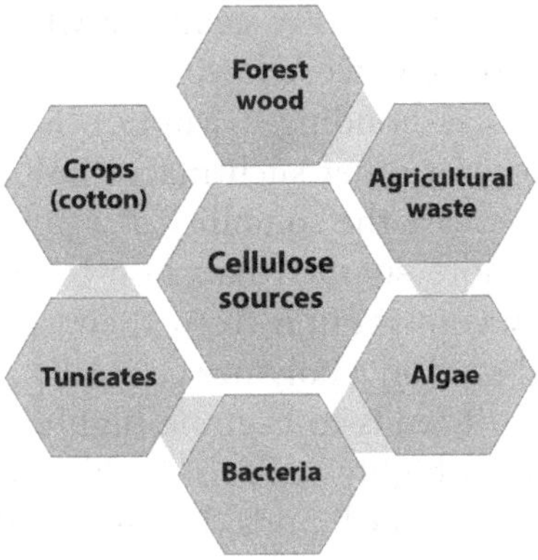

Figure 1.1 Various resources of cellulose.

which is also known as green algae in algal species. Bacterial species such as *Sarcina*, *Gluconacetobacter*, *Agrobacterium*, and *Rhizobium* produce nanocellulose as an extracellular product through the process of fermentation. A significant portion of wastepaper, including newspapers and newsprints, serves as a source for cellulose extraction. In addition to paper waste, cellulose can also be sourced from marine invertebrate organisms known as tunicates, such as *Halocynthia papillosa*, *Metandroxarpa uedai*, and *Halocynthia roretzi*. These organisms possess an outer tissue layer called the tunic, which contains highly pure crystalline cellulose [3, 4].

1.3 Cellulose Dissolution and Solvents

Cellulose has been used for various applications, but some of them involve cellulose dissolution and it is a challenging part. Finding novel solvents for cellulose based on a critical analysis of the intermolecular interactions involved and mechanisms of dissolution had been an area of research. Cellulose solvents are classified into two categories, which involve a chemical modification, are derivatizing solvents, while those who do not are nonderivatizing solvents. Since cellulose, having several –OH groups, is a polar molecule, and thus expected having a good hydrogen bonding ability, there is no issue to understand the insolubility of cellulose in organic solvents of nonpolar nature. Besides this, the aqueous insolubility had been highly challenging to understand and has inspired researchers. Dissolution also depends on thermodynamic and kinetic parameters, or we can say favourable free energy conditions affect the dissolution. There is interchain H-bonding between cellulose chains which provides the stability and insolubility to the cellulose, hence the solvents which are expected to break this

H-bonding are supposed to dissolve cellulose which is contradictory to the dissolution behaviour of glucose having –OH groups and H-bonding (Figure 1.2). This shows H-bonding disruption is not the sole reason of the solubility of cellulose or other such macromolecules. Introduction of charges also helps to increase the solubility of a polymer and polyelectrolytes are significantly more soluble than non-ionic polymers, primarily because of the entropy contribution from their small counterions. As a result, increasing the charge on a polymer generally enhances its solubility. In the case of cellulose, it tends to become highly soluble or more easily penetrated by water under highly acidic or highly basic conditions [5].

In recent years, utilizing ionic liquids (ILs) to dissolve cellulose in industrial applications has become a widely adopted and established approach. Initially, research focused on ILs with relatively high melting points. However, the focus has shifted toward using ionic liquids with lower melting points as nonderivatizing solvents for cellulose. Eventually, room temperature ILs, with melting points at or near 20°C, became the preferred choice [7]. Another emerging class of sustainable solvents that have gained significant attention in the field of cellulose chemistry is of deep eutectic solvents (DESs). With their low cost and physicochemical properties like those of ionic liquids, DESs are considered promising alternatives for dissolving cellulose. However, the solubility of cellulose in DESs currently

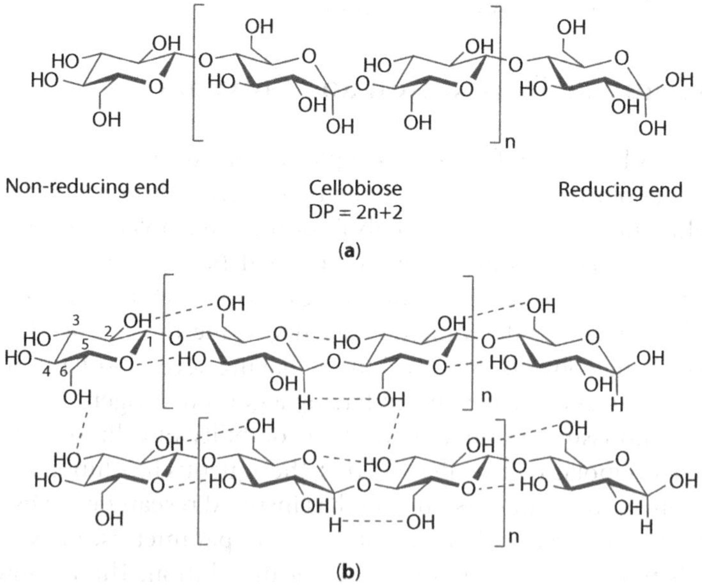

Figure 1.2 (a) Molecular structure of cellulose. (b) Hydrogen bonding (intrachain and interchain) in cellulose chains [6].

remains lower than in many ionic liquids. Compounds with strong hydrogen bond accepting capabilities like Cl^-, OAc^-, $HCOO^-$, imidazole, etc. are promising components for creating deep eutectic solvents (DESs) that can effectively dissolve cellulose [8].

Solvent systems based on aqueous sodium hydroxide (H_2O–NaOH)-based have become widely used in recent years due to their biocompatibility and eco-friendly properties. The addition of urea also aligns in the lines of green chemistry principles. Although the precise mechanism of the cellulose–NaOH interaction is still not fully understood, at low temperatures, NaOH molecules in aqueous solution are thought to form hydrates. These NaOH hydrates are believed to disrupt the tightly packed structure of cellulose by forming hydrogen bonds with one or two hydroxyl groups on each anhydroglucose unit of the cellulose chains. This interaction weakens the extensive hydrogen bonding network within the cellulose chains, leading to swelling and partial dissolution [6]. The improved dissolution of cellulose in the NaOH–water system, particularly with the inclusion of additives such as urea, is attributed to the formation of an inclusion complex. This complex is believed to originate from the self-assembly of urea hydrates at the interface between NaOH and cellulose. Both urea and thiourea have been shown to act as hydrogen bond donors and acceptors, which helps stabilize the cellulose chains in solution and prevents them from reassociating. This interaction helps improve the stability of the solution [9–12].

N,N-dimethylacetamide/lithium chloride (DMAc/LiCl) System and N-methylmorpholine-N-oxide (NMMO) are the solvent systems used earlier for cellulose dissolution (Figure 1.3). Several studies carried out to understand the mechanism of dissolution and the strategies to improve the dissolution.

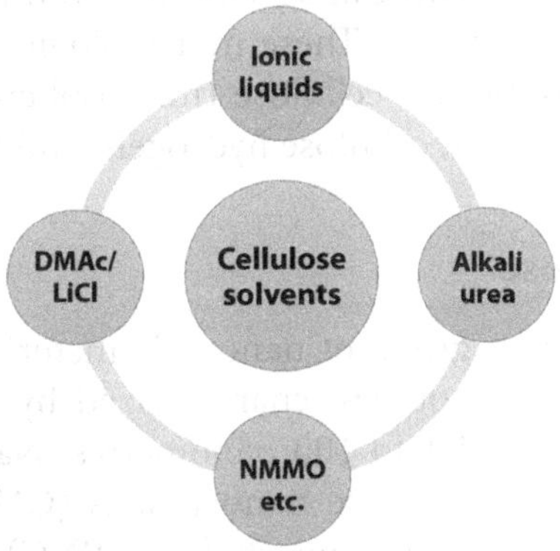

Figure 1.3 Solvents for cellulose dissolution.

To improve the cellulose solubility in the DMAc/LiCl system, various pretreatment or activation methods have been employed. These processes are designed to induce structural and morphological changes in the cellulose, such as disrupting its crystalline regions or increasing its surface area. Such modifications create a more favorable environment for solvent interaction and allow for easier diffusion of DMAc into the cellulose macromolecules. As a result, these pretreatment steps significantly enhance the efficiency and extent of cellulose dissolution in the DMAc/LiCl system. Cellulose activation can be performed using the solvent-exchange method, where cellulose is initially swollen in a polar medium-typically water-followed by activation in liquid ammonia and a low-concentration acetic acid solution. Additionally, a dioxane-based activation technique is employed for processing larger quantities of cellulose [13–15]. NMMO provides an efficient, nonderivatizing method for cellulose dissolution, valued for its simplicity of preparation and broad range of applications. It plays a key role in the lyocell process, which is the most widely used industrial method after the viscose process for cellulose dissolution. Unlike the viscose process, which requires activation or derivatization, NMMO allows for the direct dissolution of cellulose, streamlining the production process. Furthermore, because the viscose process involves the release of the toxic CS2, the lyocell process is seen as a more environmentally friendly alternative for producing industrial textile fibers [16, 17]. Since dissolution of cellulose is a challenging part in cellulose processing, more attention to this part still needed and more ecofriendly and green solvents are yet to discover for broadening the application area of cellulose.

1.4 Regeneration of Cellulose to Various Morphologies

Dissolved cellulose in solvents can be solidified into various morphologies depending on the applications. These morphologies (Figure 1.4) may be nanocellulose, microspherical cellulose (microspheres and beads), cellulose membranes and sheets, cellulose hydrogels, and fibers.

1.4.1 Nanocellulose

Nanocellulose refers to a group of nanoscale materials derived from cellulose or cellulose-based sources, characterized by having one or more dimension in the range of 1 to 100 nanometres. Nanocellulose exists in several forms, including cellulose nanocrystals (CNCs), cellulose nanofibrils (CNFs), and bacterial nanocellulose (BNC), each distinguished by its own structural features and functional properties. Renowned for

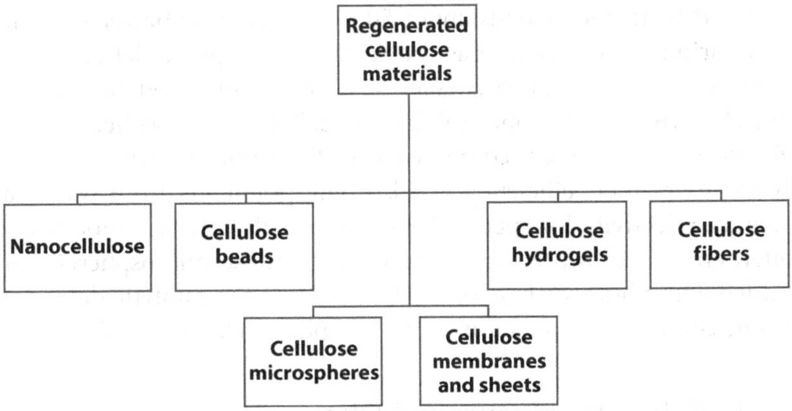

Figure 1.4 Various morphologies of regenerated cellulose.

its outstanding mechanical strength, low density, large surface area, and biodegradability, nanocellulose has gained considerable attention across a wide range of industries. Its versatility allows it to be utilized in diverse applications, including biomedical devices, wound dressings, flexible electronics, sustainable packaging, and eco-friendly composite materials. Due to its nontoxic and biocompatible nature, nanocellulose is particularly well suited for advancing sustainable technologies and promoting environmentally conscious innovation.

Nanocellulose is obtained through mechanical, chemical, or biological methods that reduce cellulose fibers to nanoscale dimensions. The production technique influences the properties of the final material, which can range from highly crystalline to highly flexible forms. Thanks to its exceptional strength-to-weight ratio, nanocellulose serves as an excellent reinforcement material in polymers and construction composites. Additionally, its capacity to form transparent and flexible films makes it highly suitable for applications in electronic displays, sensors, and other advanced technologies.

1.4.2 Spherical Cellulose (Microspheres and Beads)

Cellulose microspheres are valued for their spherical morphology, which provides a large specific surface area. Their porous structure further enhances this surface area, expanding the range of practical applications for regenerated cellulose microspheres. Depending on their size and shape, these spherical cellulose structures are categorized into different types, such as microspheres, beads, and nanocellulose. Cellulose-based spherical materials are typically classified as microspheres when their size is up to 500 μm and beads if their size extends above 500 μm. Cellulose beads are synthesized by various methods such as dropping (e.g. dropping, jet cutting etc.) and mostly, microspheres in smaller sizes are synthesized by dispersion methods [10, 18, 19]. The spherical shape of cellulose microspheres provides a high surface area, which, combined with their porous structure, significantly improves their efficiency in adsorbing pollutants from wastewater. To enhance selectivity for specific ions or molecules, surface modifications are often employed. Moreover, regenerated cellulose microspheres can be embedded with photocatalytic materials, enabling the catalytic degradation of organic contaminants including dyes and pharmaceuticals [20, 21].

1.4.3 Cellulose Membranes and Films

Membranes, sheets, and films can also be fabricated from regeneration of cellulose, typically using the casting method in which cellulose is dissolved in an appropriate solvent first and then casted onto a glass like flat surface. The thickness of the membrane is easily controlled during this stage, allowing for tailored properties depending on the application. Various studies have shown that cellulose films can be prepared by casting cellulose solutions onto glass surfaces, then undergoing a gelation step, and finally immersing them in antisolvents to form hydrogels. These hydrogels are then dried, either by air or under vacuum, to produce solid films. The final properties of the cellulose films are greatly affected by variables such as the choice of coagulant and the coagulation bath temperature [22, 23]. Regenerated cellulose membranes find applications across various fields, such as food and pharmaceutical packaging, water purification, cosmetics, sensing devices etc. A variety of organic and inorganic additives are integrated into the cellulose structure to enhance its functionality for specific purposes. These additives include metal nanoparticles, fluorescent dyes, organic colorants, and antigen–antibody systems. By incorporating these components, the cellulose materials can be tailored for specialized applications [24].

1.4.4 Cellulose Hydrogels

Hydrogels are three-dimensional networks made of polymers. They have the ability to absorb and retain large amounts of water. Additionally, they can reversibly release the absorbed water when needed. They are often produced by gelating cellulose solutions. However, cellulose-based hydrogels sometimes exhibit limited mechanical strength, which can be improved by incorporating crosslinking agents such as epichlorohydrin (ECH), aldehydes, or urea-based compounds to reinforce the network structure [25].

Physical and chemical interactions between polymer chains play a crucial role in restricting dissolution in solvents and minimizing excessive swelling after water absorption, thereby helping maintain the structural integrity of the hydrogel. The ability of hydrogels to absorb water is mainly due to the presence of hydrophilic functional groups within their structure. These groups, such as -OH and -COOH, are distributed throughout the three-dimensional polymer network. Their interaction with water molecules enables efficient water absorption [19]. The surface polarity of cellulose gels is significantly influenced through the coagulant selection during their preparation. Cellulose-based hydrogels are widely utilized across various fields thanks to their hydrophilic nature, biocompatibility, responsiveness to external stimuli, and excellent adsorption capacity, characterized by a high swelling ratio. Cellulose can be blended with other biodegradable polymers, such as chitin or alginate. This approach allows the combined material to exhibit the unique properties of each individual polymer. As a result, the overall performance of the material can be enhanced for specific applications. This approach brings the development of multifunctional composite hydrogels with enhanced performance. These hydrogels can be specifically tailored to meet the requirements of various applications [6].

1.4.5 Cellulose Fibers

Natural cellulose fibers derived from various biorenewable sources have garnered significant interest from the global research community due to their distinctive inherent properties, including biodegradability, abundant availability, environmental sustainability, flexibility, ease of processing, and remarkable physicomechanical characteristics. Materials based on these natural fibers are being applied across diverse industries. Moreover, cellulose fibers are commonly used as reinforcing agents in polymer matrices to impart specific functional and mechanical properties to the final products [26–28]. Common regenerated cellulose fibers used in clothing, such as rayon, are typically produced through the viscose process. In this method, cellulose is first treated with NaOH and then reacted with carbon disulfide (CS_2) to form cellulose xanthogenate. This viscose solution is then extruded through a spinneret into an acidic coagulation bath, where it solidifies into regenerated cellulose filaments [19].

AI disclosure statement: The author acknowledges the use of AI-based tools for minor language editing or grammatical correction. However, all intellectual contributions, analyses, and conclusions presented in this chapter are entirely the author's own.

References

1. Wang, S., Lu, A., Zhang, L., Recent advances in regenerated cellulose materials. *Prog. Polym. Sci.*, 53, 169–206, 2016.
2. Shaghaleh, H., Xu, X., Wang, S., Current progress in production of biopolymeric materials based on cellulose, cellulose nanofibers, and cellulose derivatives. *RSC Adv.*, 8, 2, 825–842, 2018.
3. Kirtania, M.D., *et al.*, Sources and Applications of Nano Cellulose Tailored Materials, in: *Nanocellulose-Sources, Preparations, and Applications*, IntechOpen, United Kingdom, 2024.
4. Shak, K.P.Y., Pang, Y.L., Mah, S.K., Nanocellulose: Recent advances and its prospects in environmental remediation. *Beilstein J. Nanotechnol.*, 9, 1, 2479–2498, 2018.
5. Medronho, B. and Lindman, B., Brief overview on cellulose dissolution/regeneration interactions and mechanisms. *Adv. Colloid Interface Sci.*, 222, 502–508, 2015.
6. Acharya, S., *et al.*, Utilization of cellulose to its full potential: A review on cellulose dissolution, regeneration, and applications. *Polymers*, 13, 24, 4344, 2021.
7. Wu, J., *et al.*, Homogeneous acetylation of cellulose in a new ionic liquid. *Biomacromolecules*, 5, 2, 266–268, 2004.
8. Chen, Y.-L., *et al.*, Deep eutectic solvents (DESs) for cellulose dissolution: A mini-review. *Cellulose*, 26, 205–213, 2019.
9. Lue, A., *et al.*, Light scattering study on the dynamic behaviour of cellulose inclusion complex in LiOH/urea aqueous solution. *Polymer*, 52, 17, 3857–3864, 2011.
10. Gong, Y., *et al.*, pH-Responsive Cellulose-Based Microspheres Designed as an Effective Oral Delivery System for Insulin. *ACS Omega*, 6, 4, 2734–2741, 2021.
11. Xia, J., *et al.*, Cobalt oxyhydroxide nanoflakes/cellulose composite membranes with enhanced detection of ascorbic acid. *ACS Appl. Polym. Mater.*, 4, 1, 469–478, 2021.
12. Zhang, H., *et al.*, Early-warning and semi-quantitative colorimetric detection of Hg (II) with lysine-bis-Schiff base cellulose membranes designed by simple interfacial covalent bonding. *Sens. Actuators, B*, 346, 130435, 2021.
13. Raus, V., *et al.*, Activation of cellulose by 1, 4-dioxane for dissolution in N, N-dimethylacetamide/LiCl. *Cellulose*, 19, 1893–1906, 2012.
14. Röder, T., *et al.*, Solutions of cellulose in N, N-dimethylacetamide/lithium chloride studied by light scattering methods. *Polymer*, 42, 16, 6765–6773, 2001.
15. Hu, Y., *et al.*, Role of low-concentration acetic acid in promoting cellulose dissolution. *Cellulose*, 25, 4389–4405, 2018.
16. Gao, Q., Shen, X., Lu, X., Regenerated bacterial cellulose fibers prepared by the NMMO· H2O process. *Carbohydr. Polym.*, 83, 3, 1253–1256, 2011.

17. Zhang, S., *et al.*, Regenerated cellulose by the lyocell process, a brief review of the process and properties. *BioResources*, 13, 2, 4577–4592, 2018.
18. Gericke, M., Trygg, J., Fardim, P., Functional cellulose beads: preparation, characterization, and applications. *Chem. Rev.*, 113, 7, 4812–36, 2013.
19. Khatoon, F., Shabbir, M., Annu, An Introduction to Regenerated Cellulose: Morphologies and Applications, in: *Regenerated Cellulose and Composites: Morphology-Property Relationship*, pp. 1–7, Springer, Singapore, 2023.
20. Luo, X. and Zhang, L., High effective adsorption of organic dyes on magnetic cellulose beads entrapping activated carbon. *J. Hazard. Mater.*, 171, 1, 340–347, 2009.
21. Luo, X., *et al.*, Removal of Heavy Metal Ions from Water by Magnetic Cellulose-Based Beads with Embedded Chemically Modified Magnetite Nanoparticles and Activated Carbon. *ACS Sustainable Chem. Eng.*, 4, 7, 3960–3969, 2016.
22. Cai, J., Wang, L., Zhang, L., Influence of coagulation temperature on pore size and properties of cellulose membranes prepared from NaOH–urea aqueous solution. *Cellulose*, 14, 205–215, 2007.
23. Yang, Q., *et al.*, Transparent cellulose films with high gas barrier properties fabricated from aqueous alkali/urea solutions. *Biomacromolecules*, 12, 7, 2766–2771, 2011.
24. Fan, J., *et al.*, Recent advances in cellulose-based membranes for their sensing applications. *Cellulose*, 27, 16, 9157–9179, 2020.
25. Sikdar, P., *et al.*, Recent advances in the synthesis of smart hydrogels. *Mater. Adv.*, 2, 14, 4532–4573, 2021.
26. Thakur, V.K. and Thakur, M.K., Processing and characterization of natural cellulose fibers/thermoset polymer composites. *Carbohydr. Polym.*, 109, 102–117, 2014.
27. Shabbir, M. and Mohammad, F., Natural textile fibers: Polymeric base materials for textile industry, in: *Natural Polymers: Derivatives, Blends and Composites*, vol. 2, pp. 89–102, 2017.
28. Shabbir, M. and Mohammad, F., Sustainable production of regenerated cellulosic fibres, in: *Sustainable Fibres and Textiles*, pp. 171–189, Elsevier, United Kingdom, 2017.

2

Cellulose and Cellulose Derivatives

Tanvir Ahmad Yamin[1], Sha Md. Mosaddak Shehab[1], A.S.M. Rowson Jamil[1] and Md. Arif Mahmud[2]*

[1]Ahsanullah University of Science & Technology, Department of Textile Engineering, Dhaka, Bangladesh
[2]The University of Manchester, Department of Materials, Manchester, England, United Kingdom

Abstract
Cellulose is a highly popular organic material in the world and it is a naturally occurring polymer which can be found inside plant cell walls. It is obviously a crucial structural element in plants because of its great tensile strength and chemical durability. It is made up of linear chains of $\beta(1\rightarrow4)$ connected D-glucose units. Beyond its native state, cellulose can undergo chemical modification to yield a vast range of derivatives with various characteristics and uses. These cellulose derivatives, which have distinct uses in the food, paper, textile, and pharmaceutical industries, include cellulose ethers, esters, and regenerated cellulose. Cellulose ethers, such as methylcellulose and carboxymethylcellulose, are effective thickeners and stabilizers because they increase viscosity and water solubility. Photofilms and coatings use cellulose esters, such as cellulose acetate, which are prized for their ability to create films. Materials, such as viscose rayon and cellulose nitrate, are examples of regenerated cellulose, which is versatile in its ability to produce fibers and polymers. Cellulose chemistry is a field that is constantly expanding in use due to research and development, which advances material science and promotes sustainability. This chapter will discuss the chemical processes for the abovementioned cellulose derivatives that have been deployed in place of synthetic polymers in various applications.

***Keywords*:** Cellulose, regenerated cellulose, biodegradable polymers, sustainability, bio-compatibility, cellulose acetate (CA), green chemistry, food packaging

**Corresponding author*: arif.shawon@hotmail.com

Mohd Shabbir (ed.) Regenerated Cellulose: A Renewable Alternative to Petroleum-Based Plastics, (15–58) © 2026 Scrivener Publishing LLC

2.1 Introduction

The subject of climate change has become a serious concern for humanity and the world after COVID-19, alarming people and highlighting the need for sustainability in all facets of existence. As climate change becomes a more serious threat to humanity, there is no other way to address it in a sustainable way so the most can be extracted out of all the natural resources instead of using synthetic materials [1]. In particular, the use of oil as well as other fossil fuels in order to create plastic cannot be a reasonable practice; instead, sustainable materials and production techniques are required. In order to enable a bio-based circular economy, we must expedite the shift to materials that are renewable, biodegradable, sustainable, and ability to be recycled in a closed-loop system, at the very least. These are a few of the factors that strongly support the use of cellulose. At the end of its useful life, cellulose, a substance formed from plants, may be recycled back into the environment. In order to support biological development, it may also take up carbon dioxide from the environment [2]. The development of procedures and methods is the focus of recent efforts. The depletion of petroleum resources, the scarcity of landfill space, the persistence of plastics in the environment, the disposal of plastic waste, and the consumption of plastic packaging by birds and marine life are some of the problems caused by plastic. As a result, there is a lot of interest in using biodegradable or natural ingredients in a variety of ways in place of part or all synthetic plastics. It is evident that the challenge of gathering solid waste composed of artificially inert polymers can be partially and alternatively solved by using biodegradable polymers for packaging [3].

Several techniques are put forth to functionalize the surface in order to change cellulose from a low-value traditional material to a valuable, clever, and useful stuff. By adding nanoparticles or other functional elements, cellulose-based materials can be made to have tunable characteristics like photosensitivity, electrical conductivity, catalytic activity, magnetic qualities, and other special qualities [4]. It is highly appropriate to review the most recent advancements in this field so that the study in this area might inspire the next generation of scholars to conduct even more studies to expand this field. First, we go over the fundamental anatomy of cellulose in this section.

2.2 Cellulose: Formation and Structure

With a global production and degradation of approximately 1.5 trillion tons annually, the topmost prevalent polymer found in nature is cellulose in the biosphere and truly equivalent to planetary reserves about the primary fossil, as well as mineral reserves [5]. Also, the material's long-standing scientific enthusiasm, the use of cellulose in the form of a renewable as well as biodegradable raw material within countless implementations always is an incredible solution for contemporary structural complication to completely minimize environmental along with recycling challenges [6].

The work of French scientist Anselme Payen, who demonstrated that different plant ingredients produced a fibrous product following extraction using water, alcohol, and ether, and purification using acid-ammonia treatment, marked the beginning of the earliest organized trials that led to the revelation of cellulose back in 1837. Those resultant carbohydrate was finally given the name "cellulose" by the French Academy [7]. Figure 2.1 shows the structure of a cellulose unit. These days, cellulose is isolated using a variety of techniques, such as the alkaline, bisulfite, and sulfate (kraft-) procedures combined with temperature and mechanical treatments. Pulp's fiber strengths fluctuate as a result of the various treatments [8].

The lowest energy conformation, the 4C1-chair structure, is exhibited by D-glucopyranose ring units, which compose cellulose [9]. These components are connected by β-1,4-glycosidic linkages, which causes the cellulose chain axis to alternately turn by 180 degrees. The repeating unit that made cellulose is known as cellobiose, and it contains a measure in length of 1.3 nm [13]. Each and every anhydroglucose unit (AGU) in the cellulose chain has three reactive hydroxyl groups: two secondary groups at C2 and C3, which are located in the ring's plane, and a primary group at C6. The cellulose molecule has chemically distinct chain ends, as is common for a polymer created *via* "polycondensation" [14]. A glucopyranose unit in equilibrium along with the aldehyde function is present at one end, while an anomeric C atom linked by glycosidic bondage is present at the another (nonreducing end).

Reactions causing the cellulose chain to hydrolyze or oxidize are the source of molecular structural changes. These reactions mostly take place in amorphous areas or on the fibril surface.

Native cellulose from different sources has a degree of polymerization (DP) ranging from 1,000 to 30,000, meaning that its chain has a length range from 500 to 15,000 nm. These DP values of these cellulose samples are obtained by quarantine techniques range from 800 to 3,000 [13]. The DP is an average value since cellulose samples are polydisperse.

Numerous methods, such as size-separation chromatography, light scattering, and viscosity measurements, may provide data about the molar masses and their dispersal.

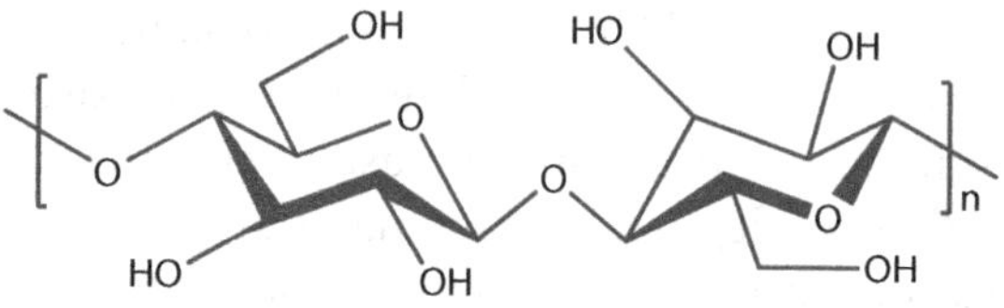

Figure 2.1 Unit of cellulose [11].

The many hydrogens bonding systems found in cellulose have a big impact on its characteristics [10]. For example, the hydroxyl groups' reactivity, the cellulose samples' low solubility in most solvents, and their crystallinity are all caused by robust hydrogen bonding systems. Hydrophobic regions around the C atoms in cellulose also got an impact on the material's general characteristics, such as solubility.

2.3 Cellulose Derivatives That Can Be Used as Packaging Materials

2.3.1 Cellulose Acetate

Paul Schützenberger synthesized cellulose acetate (CA) first time in the year of 1865 by treating wood pulp together with acetic anhydride. Using H_2SO_4 as their catalyst, the chemical reaction of cotton and acetic acid yields CA through another process. In order to properly ensure proper and repetitive acetylation with cellulose, it is at the beginning treated along with acetic acid. After that it is dehydrated and reacts solely with acetic anhydride in the presence of a suitable catalyst–solvent system, just like sulfuric acid-anhydrous acetic acid.

Final generated material is cellulose triacetate. It must be further acid-hydrolyzed to produce cellulose diacetate and, ultimately, CA to a lesser extent. This cellulose ester's degree of substitution (DS) is a crucial feature since it affects the solubility and Journal Pre-proof 11 biodegradability [12]. A number of CA production methods have been set up recently. They offer ring-opening esterification, trans-esterification, and esterification employing iminium chloride or N,N-carbonyl diimidazole. Among CA's noteworthy characteristics are its hydrolytic stability, high mechanical qualities, and insolubility in water. Its natural origin is associated with its biodegradability, biocompatibility, nontoxicity, and inexpensive cost [14].

Membranes for biomedical applications are a significant area of application for CA in the realm of biomaterials [15]. Because of its exceptional biocompatibility, it has been demonstrated CA is a suitable product for fabrication of membranes for the process of osteointegration. When it breaks down, glucose bits that are nonharmful to adjacent tissues are released along with acetyl groups that alter the pH a little bit locally. Those CA films were applied as received to conceal transplants made of magnesium alloy, and they demonstrated favorable outcomes *in vitro* as well as after intramedullary canal implantation in mice [16]. The osseointegration processes of CA membranes functionalized with resveratrol [17] or sericin [18] were successfully applied. These operative fundamentals used for membrane modification favor those proliferation ability of MC3-T3 osteoblast harbinger units, indirectly enhancing the implants' osseointegration that are protected by these membranes.

Particularly in the dentistry area, controlled release of antibiotics to those embedment area might avert any after care difficulties. For these purposes, dual-role films who promote osseointegration, as well as permit the discharge of the medicinally active ingredient doxycycline had been created [19]. Fillers like hydroxyapatite [17], silica nanowires [20], alternatively some combination of carbon nanotubes and graphene [21] were used to create composite CA membranes with hemodialysis potential. Every synthetic material demonstrated a high flow rate, strong bovine serum albumin (BSA) binding capacity, and non-cytotoxic nature. Additional uses for composite CA membranes include the creation of antibacterial membranes (such as graphene or carbon nanotube composite membranes functionalized with silver nanoparticles) [22–25] or the ability of composite membranes with layered double hydroxides to retain antibiotics from aqueous solutions [26].

The incredibly simple and efficient electrospinning method opened up new possibilities for polymers based on CA. According to Konwarh *et al.*, electro spun CA fibers have found use as protein immobilization membranes, tissue engineering scaffolds, wound healing mats, and biosensors. Silver oxide nanoparticles and CA-based chemicals, such as silver sulfadiazine, were used to create antibacterial electro spun wound healing mats. The resulting nanofibers shown significant antibacterial efficacy against microbiological threads of Gram-positive *Bacillus subtilis* and Gram-negative *Escherichia coli* as well as Gram-positive *Staphylococcus aureus* [27–29] Khoshnevisan *et al.* impregnated amalgam of CA/polycaprolactone (PCL) nanofibrous mats in a solution containing propolis to provide them antibacterial and antioxidant qualities. The results of the 2,2-diphenyl-1-picrylhydrazyl (DPPH) experiment indicated about these propolis-CA/

PCL mats had a high level of antioxidant activity. Furthermore, according to Khoshnevisan *et al.*, the biocomposites displayed robust bactericidal operation counter to each Gram-positive and Gram-negative microbes.

The microbicidal, antiparasitic, virucidal, fungicidal, insecticidal, analgesic, as well as anti-swelling qualities of herbal extracts have drawn many interests in study [30].

The essential oils of *Mentha piperita* (Peppermint), Cinnamomum (Cinnamon), and Cymbopogon (Lemongrass) were effectively wrapped in CA nano-sized particles made using solvent/antisolvent methods.

Both in planktonic cultures and biofilms, *in vitro* biological experiments demonstrated very minimal cytotoxicity as well as strong antimicrobial effectiveness contrary to the highest common parasitic infections. Furthermore, the encapsulated essential oils' luminous qualities suggest using the nanoparticles for imaging therapeutic applications [31]. Peng *et al.* reported creating radiant luminous nano-size particles by encasing different microscopic fluorophores (methylene blue, rhodamine, stilbene, and tracer yellow) in spherical nanoarchitectures (CA-dots) made of CA/pluronic F-127 orb-shaped nanostructures, which were produced with nano-sized precipitation. Some anionic, amphiphilic polymer with PEG blocks within their structure is called Pluronic F-127. PEG-functionalized nanoparticles were produced by co-assembling the two polymers. In order to target certain types of cancer, those PEG exterior clusters improved the emulsified firmness in addition to that these enabled enhancement of nanoparticles by physiologically significant compounds, which is folates. According to Peng *et al.*, the nanoparticles were deemed completely appropriate for *in vivo* tumor imaging and targeted implementations.

Because they may slow the spread of infections, functional textiles are also very interesting to the healthcare industry. Jia *et al.* treated cotton textiles which are electrospun complex materials dependent on CA, TiO_2 nanofibers, and graphene oxide (GO) panels to generate antibacterial nanofibers/cotton hybrid yarns. The TiO_2 and GO were distributed uniformly throughout the CA nanofibers, as seen by these SEM pictures of CA/TiO_2/GO. Also, the augmentation of GO as well as TiO_2 to CA significantly increased its hydrophilicity. *Bacillus cereus* and *B. subtilis* underwent antibacterial testing. According to Jia *et al.*, these cotton textiles treated along with TiO_2/GO/CA showed outstanding antibacterial endeavor, with the inhibition percentage much greater than the obvious 95% against other bacterial strains. Utilizing acetoxy propyl cellulose (APC) and CA as printing materials, surface tailoring of cotton fabrics was also accomplished by 3D printing. CA's linear structure gave it a stronger affinity for the cotton substrates, which gave the printed structures outstanding adhesion qualities.

APC was less firmly attached because its branching molecular structure had not match properly with substrates keeping cellulose molecules inside. Both materials, however, were thought to be appropriate for functionalization and textile modification; this work brings up new avenues as a reason for the creation of innovative, all-cellulose textile materials with a variety of uses, especially in the biomedical industry [32]. According to research, CA electrospun fibers are suitable delivery systems for transdermal retinoic acid along with alpha tocopherol as well as nonsteroidal anti-inflammatory medications (NSAIDs) which includes ibuprofen, naproxen, and indomethacin [33]. However, because of the wide outer layer of the fibers as well as the high concentration with medicine close to their outer area in the case of standard monoaxial blend electrospinning, a sudden expulsion concerning these therapeutic agents was noticed within the early hours. As a result, using a modified triaxial electrospinning technique, unique core-shell nanostructures were created, consisting of some drug-protein nanocomposite covered with a few light and slim CA shell. The external fluid found simply was acetone/acetic acid solvent, the inner fluid was an electro spinnable ibuprofen-gliadin mixture, and the outer fluid was a non-electrospinnable CA. The resulting solid dispersion of ibuprofen with gliadin was encapsulated in a panel within CA, which avoided these initial burst release seen in traditional composites and ensured a regulated ejection of these therapeutic medicine for high up to 43 hours [34].

Increasing the drug carriers' surface hydrophobicity is another way to achieve sustained drug release. Using the co-electrospinning method, Chen *et al.* produced thymol-loaded CA layers with the ellipse-shaped nanopores that were created *in-situ* with the fiber outer layer as a result of a quick solvent evaporation process that was thermally driven. The perforations made the membranes more hydrophobic by trapping air at the contact. Porous membranes were found to be more effective than non-porous ones in terms of both starting stage and sustained drug excretion, according to *in vitro* drug release experiments. Additionally, according to Chen *et al.*, they had a larger specific surface area, which led to a greater drug use ratio [35].

According to several studies, the extracellular matrix (ECM)-like structure of CA based nanofibrous scaffolds, which may be created by electrospinning techniques, promotes cellular adhesion and improvement. Polyethylene glycol (PEG) and CA butyrate (CAB) were combined to create composite scaffolds that had improved mechanical and hydrophilic qualities. According to Tan *et al.* [139], the constructions showed minimal cytotoxicity about natural individuals' dermal fibroblasts, indicating that those are potentially useful ingredients about tissue technological implementations. By deacetylating electrospun CA in a NaOH/EtOH substance, regenerable cellulose scaffolds with fiber diameters between 300 and 500 nm were produced. Such small fiber diameters were shown to produce a moderate degradation rate, a higher porosity, and a higher swelling ability. Using the MC3T3-E1 osteoblast cell line, the materials were tested for cell viability and proliferation; the positive results demonstrated the materials' viability as possible surfaces for bone-tissue engineering [36]. After two processes, biomimetic scaffolds made of collagen or chitosan blends and modeled after the architecture of native decellularized tissues were produced. The electrospun membrane was first immersed in a collagen or chitosan solution to create a fibrous CA basis. This allowed for the incorporation of the sponge component. Depending on their composition, the resulting composites displayed an ECM-like feature with varying multi-hierarchical structures [37].

Matos *et al.* investigated the use of magnetic nanoparticles (NP) in conjunction with non-woven CA membranes created using electrospinning to cure cancer through magnetic hyperthermia. Either *via* adding to the electrospinning fluid or by accumulation at the fiber sheen, the nanoparticles were integrated. Because there were more nanoparticles on the CA fibers, the composite layers made *via* retention of nanoparticles within the fiber exterior were more heat-resistant than those made by incorporation, but they also displayed cytotoxicity. To make sure about the biocompatibility about the composite interfaces, it is important to carefully control the density along with the type of incorporation as well as discharge rate of nanoparticles.

Phase inversion is another method that is commonly used in the manufacturing of CA. This technique has been applied in several studies to construct biologically compatible along with hydrodynamically smooth composite membranes for hemodialysis along with protein sustenance to its great versatility [21, 38]. Supercritical phase inversion-prepared composite interfaces dependent on CA and Laponite® were studied for use in tissue engineering used as structures in the purpose of cellular proliferation. A profoundly permeable composition (~85% porosity) composed of a normal size of pore in the range between 09 to 13 μm was observed in the materials.

It was discovered that the initial CA/Laponite® ratio determined about cellular fastening %, hole diameters, temp control as well as mechanical information. The clay had a beneficial impact on all of the aforementioned qualities [39]. Phase-inverted CA membranes were also functionalized with resveratrol through covalent immobilization or modified with hydroxyapatite as well as impact on the osseointegration capacity for the metallic transplant was examined on them. During the *in vitro* studies, good osteoblast viability as well as differentiation potential were noted, along with the notable rise within alkaline phosphatase performance and extracellular matrix crystallization. According to these acquired results [17, 40], these products contain a huge deal of opportunity for applications involving directed bone regeneration.

For use in biosensors and bioanalysis, composite materials with some CA matrix plus some different kinds of effective stuffings were crafted. Yu *et al.*, as example, created the novel optical fiber biosensor which is used for the detection of glucose by dip-coating about 600 μm optical fiber's end face by taking help from a fragile membrane of glucose oxide (GOD) and carbon quantum dots (CQDs) that were soaked or dipped in CA. These enzyme-based and fluorescent fiber-optic biosensor demonstrated about a high degree of targeted approach with addition to persistence and anti-interference performance in addition to a quick fluorescence response to glucose. These characteristics allowed for non-stop live glucose spotting at low concentrations too [41].

A glassy carbon (GC) electrode was used to drop-cast the resulting aqueous dispersion, which included carbon nanotubes (MWCNTs) into CA who were previously multi-walled. This created a sensing probe that could be used within mentioned electrochemical spotting of many significant catecholamines, including dopamine and epinephrine. This random orientation of the carbonaceous fibers and their formation of interconnected structures with the organic matrix were shown with some scanning electron microscopy (SEM) pictures. Also, to strengthen the MWCNTs' mechanical durability and adherence to these GC electrode, CA served as the permanent-exclusive wall which can prevent catalyst poisoning along with boosting analytical abilities in the process. For the analytical measurement of dopamine along with adrenaline inside medical vials and urine of human, the GC/MWCNT/CA biosensor was productively utilized inside processes like differential pulse voltammetry (DPV) and solid phase removal which are connected with the liquid chromatography (SPE-LC) [42]. Another kind which is carbon-based extenders that finally was utilized in order to improve the bending result as well as reaction time of electroactive polymer (EAP) activators depended over CA is graphene nanopowder (GN). The CA/GN composite filter was made with a biocompatible electrode made of poly(3,4-ethylene-dioxythiophene)-poly styrene sulfonate (PEDOT: PSS), a plasticizer made of ionic liquid (IL). Compared to the tidy CA actuator, the insertion of GN to the CA matrix resulted with a 4.8 times rapid reaction along with a 2.9 times higher bending deformation. The improved mechanical, specific capacitance, and electrical conductivity that came about as a result of the powerful ionic contacts, as well as physical crosslinking among CA, IL, and GN, were credited with the performance gain. The created CA/IL/GN activator o starter was regarded as some viable contender for human-friendly gadgets such as artificial muscles as well as flexible haptic devices because of its affordability, quick reaction, and enhanced functionality [43].

2.3.2 Cellulose Sulfate

Cellulose sulfate (CS) is an ester which is created with cellulose produced by heterogeneous, homogeneous or quasi-homogeneous sulfation processes is called CS) In comparison to pure cellulose, CS contains several advantages which are water solubility as well as strong antibacterial qualities [44, 45]. Isopropyl alcohol is commonly employed in the reaction between pure cellulose and a sulfating agent (such as H_2SO_4) within a suitable repercussion vessel to produce heterogeneous sulfation. After this reaction, sulfate groups take the place of mentioned hydroxyl groups left top of the outer layer. Mentioned process which is a homogeneous sulfation was initially described by R.G. Schweiger back in 1979. It involved solubilizing cellulose within a SO3-pyridine complex with N,N-dimethylformamide acting as a co-solvent. These results were CS containing lot of substitution degree as well as some intermediate reaction material which is cellulose nitrite. According to Rohowsky *et al.* and Zhang *et al.*, mentioning quasi-homogeneous process, which highly is depended on acetosulfation, is characterized by a revolutionizing dissolution of cellulose in N,N-dimethylformamide combined with the solution of sulfating along with acetylating agent (e.g., chlorosulfuric acid/acetic anhydride). This is trailed by the shearing with acetyl groups upon precipitation, which results in the changing of CA-sulfate finally to CS. One can produce a regioselective sulfation, depending on the method employed. For instance, acetosulfation protects the hydroxyl teams in the C2 as well as C3 stage of mentioned anhydroglucose item along with results in the topmost degree of substitution in mentioned C6 step; homogenous sulfation also affects the C2 and C3 locations [46]. Also, the mentioned biological properties of CS are highly impacted through mentioned molecular weight and degree of substitution. For instance, a greater substitution degree was demonstrated in order to stimulate 3T3 fibroblast cell forming *in vivo*, and greater molecular mass CS demonstrated the improved microbicidal action in opposition to the human papilloma virus (HPV) [47]. When CS was injected intravenously into rabbit models, it mainly discovered to contain some dose-dependent anticoagulant action akin to heparin (HEP) [48]. Similar to normally happening sulfated polysaccharides (also known as glycosaminoglycans, or GAGs) similar to dermatan sulfate along with chondroitin sulfate as well as heparan sulfate, the biological action of CS can be likened. The creation of electrospun CS/gelatin-based GAG mimetics which mimic this function about the nerve tissue extracellular matrix (ECM) was recently described by Menezes *et al.* NGF bondage as well as permissive neurite outgrowth were found to be influenced by some degree and location about mentioned sulfate circle in incomplete (pCelS) and completely (fCelS) sulfated cellulose. In comparison to mentioned scaffolds which have chondroitin sulfate A and C (CSA/Gel and CSC/Gel), the partial GAGs found inside neural extracellular matrix, the CS/gelatin constructions (pCelS/Gel and fCelS/Gel) were highly significant about neurite enlargement. Higher levels of sulfation were found to positively affect NGF binding; the fCelS/Gel scaffolds produced the greatest neural growth after the pCelS/Gel ones [49]. When added to gelatin (Gel) scaffolds at low concentrations (0.1 wt%), fully sulfated cellulose also aided in the promotion of chondrogenesis [50].

Moreover, regioselectively sulfated cellulose increased the development of endothelial tubules caused by vascular endothelial growth factor (VEGF) when added in collagen/hydroxyapatite nanocomposite scaffolds within use in structural tissue engineering [51].

Some production of influenza vaccines using cell cultures explicitly used the advantageous molecular binding characteristics offered within the sulfate ligands within the CS backbone to purify the viral ingredients using capturing chromatography methods. At-first, a pseudo-affinity CS membrane diffuser was employed to catch the viral ingredients and lower these mentioned levels of protein and DNA contamination. It was made using a technique outlined by Wolff *et al.* [52]. Secondly, as a polishing step, some salt-tolerant membrane (STIC®PA) with a polyallylamine ligand which are used commercially was implemented to further eliminate any remaining contaminants. Influenza virus clarity levels attained for this simple double-step variation method met those requirements within some European Pharmacopeia as a result of vaccines generated from cell cultures intended to normal use of common people. For these kinds of applications, using CS film adsorbers rather than commercial bead-based resins (like Cellufine™ Sulfate, Capto™ DeVirS) was also linked to increased productivity. This is because the membranes' superior mechanical qualities allow for higher flow rates and also guarantee that the virus is more accessible to the binding sites [53].

Sodium salt (NaCS), the polyanionic complex item having subtle film-forming properties, is profoundly more stable version of CS. It is really active in the encapsulation of growth factors [54, 55] pharmaceutical compounds [56], and human as well as mammalian cells for fixed delivery implementations [57]. Under the trade named Cell-in-a-box®, manual encapsulation kits depended on conventional NaCS capsules which are normally offered for sale. Research revealed that the NaCS capsules offer a farfetched milieu for cell survival in addition to shielding cells from immune system rejection [57]. Salmons *et al.*'s rigorous analysis of mentioned sizes of particles which may be removed within Cell-in-a-Box® systems revealed that micro biomolecules, including cytokines as well as proteins along with some molecular mass which are about to reach 70 kDa, were effectively released through the encapsulated cells [58].

NaCS/chitosan hydrochloride (CHC) polyelectrolyte complexes (PEC) were investigated for being possible solvent jar in the shape of microcapsules which are crosslinked along with polyphosphates (sodium pyrophosphate, or SPPP, sodium hexametaphosphate, or SHPP) to deliver small molecular drugs such as 5-aminosalicylic acid. The mentioned polyphosphates contain a major effect on the interactions among NaCS as well as CHC. Also, the formation of microstructures. The medication was released *via* a diffusion approach for SPPP-crosslinked capsules and a non-Fickian transport mechanism for STPP and SHPP. Their results showed about their PEC microcapsules might be developed as drug transporting vessels with tunable characteristics, according to Su *et al.* Primary human T cells were cultivated in PEC capsules made of polydiallyl dimethyl ammonium chloride (PDADMAC) and NaCS. It was observed that the capsules achieved some of the *in vivo* conditions for expansion and differentiation, creating a biomimetic microenvironment that the T cells could self-condition [59, 60].

Semi-synthetic antibiotics, likely ceftriaxone along with cefotaxime were mixed with the sodium and potassium as well as lithium salts of CS acetate. These immobilized on activated carbon in a trial to develop an edible dosage formation except the compromising therapeutic efficacy. Completely nonstoichiometric complexes were formed by 0.4 to 0.6 moles of antibiotic for each cellobiose unit. The medication was effectively transferred to the gastrointestinal system except undergoing any alterations. Additionally, a less amount of antibiotic was used in the CS/AC tablet shape (0.075 g) to achieve the same biological activity as parenteral injections (0.1 g) [61].

2.3.3 Cellulose Nitrate (CN)

The primary component of smokeless gunpowder is nitrocellulose (CN), commonly referred to as gun cotton, due to its propensity for explosive breakdown [62]. Cellulose made from wood pulp or cotton linters is nitrated to produce CN using a strong nitrating agent, such as nitric acid. Researches show that H2SO4 is a good product to stabilize the nitration mixture and prevent unwanted explosions [63]. The H2SO4/HNO3 combination is frequently used in industry to nitrate a range of substrates, such as glycerol and cellulose. During the process of nitration, the mentioned hydroxyl groups on the outer layer of cellulose are replaced by nitrate esters as a result of the electrophilic assault of NO2+ ions on its OH moieties.

Although mentioned process was initially investigated at some context of nitration of alcohols along with amines, later researches showed about it was also applied to cellulose. A total nitration of cellulose is defined as a final nitrogen content of 13.5% or above [64]. A number of factors, including the substrate's source, the makeup of the nitrating mixture, and the earlier treatments used, might affect how well the reaction proceeds. For instance, cellulose extracted from *Stipa tenacissima* grass demonstrated a higher nitration yield in comparison to other sources because of its structure's numerous crystalline α regions, which affected the nitrating agents' diffusion [65].

The implementations of Cellulose Nitrate are heavily influenced by some nitrogen content. Li and Frey [63] state that while Cellulose Nitrate with nitrogen ratios below 12.6% has strong biocompatibility as well as physicochemical stability which makes it appropriate in case of use in biomedical implementations. Cellulose Nitrate with nitrogen percentages between 12.6 to 13.3% is classified as explosive. It is used as cannon propellant. It has been shown that CN membranes are very effective in retaining proteins as it is one of the non-covalent interactions that take place among some nitro groups of mentioned polymers as well as some amine functional teams in the protein property.

This feature suggests that they be used in biosensors [23, 66–68] and hemodialysis [69]. Osmosis as well as ultrafiltration membranes along with electrophoresis films, are more applications for CN [70].

Commercial lateral flow assay (LFA) kits, which are depended on antibodies or nucleic acids immobilized on a Cellulose Nitrate membrane, enable their identification of certain biomolecules from the examined materials. This creates a platform for the detection of infectious diseases, microbe identification, oncogene mutation testing, and prenatal testing [71]. Yew *et al.* reported on improved LFA biosensors that electrospun-coated NC membranes with the hydrophobic chemical polycaprolactone (PCL) to decrease the sample flow rate and lengthen the biomolecule contact time. For the NC/PCL membranes, synthetic Zika virus oligonucleotides showed an up to ten-fold increase in detection sensitivity. As a result, they could be utilized as a quicker and less expensive substitute for traditional detection platforms like the enzyme-linked immunosorbent assay (ELISA) and quantitative polymerase chain reaction (qPCR) [72]. Transparency, which CN membranes acquire after 100% ethanol and xylene treatments, is an underutilized characteristic. After staining, cells sown on such substrates may be seen under an optical microscope. By putting eight distinct cell lines on CN membranes and assessing their adhesion, growth, spread, and viability %, the potential of CN membranes as cell culture scaffolds was examined. The CN scaffolds exhibited superior handling qualities, remained intact after autoclaving sterilization processes, and showed superior cytocompatibility as compared to standard glass coverslips. The examined cells' microscope pictures showed a variety of pseudopod-like features that suggested a strong interaction with the cellulose substrate.

In the case of coverslip-grown cells, no such structures were seen [70].

2.3.4 Carboxymethyl Cellulose (CMC)

As the highest significant cellulose ethers, carboxymethyl cellulose (CMC) is made by reacting raw cellulose along with an aqueous solution of sodium hydroxide. After removing the resulting alkali cellulose and then reacting with either sodium monochloro-acetate or monochloroacetic acid in an alcohol-supporting medium [73, 74]. A variety of cellulose sources, including wood residue, paper sludge, mesquite tree, grains or maize husks, were described for the synthesis of carboxymethyl cellulose [75, 76]. Carboxymethyl cellulose is categorized as technical, semi-purified, and purified depending on how pure it is (cellulose gum, for example, is a high purity grade carboxymethyl cellulose used as an additive in the food sector) [77, 78]. A linear, lengthy polysaccharide backbone with both carboxyl and hydroxyl surface groups is produced during the etherification process. Carboxymethyl cellulose chemical reactivity and water solubility are provided by the distinctive chemical structure [78, 79]. One significant factor affecting the characteristics of the polymer is the degree of substitution (DS). Capanema *et al.* synthesized carboxymethyl cellulose/doxorubicin (DOX) prodrug bioconjugates with the help of sodium carboxymethyl cellulose (NaCMC). By the help of two different carboxymethylation degrees (0.77 and 1.22) which are crosslinked with citric acid. They investigated the influence of DS on the parts of carboxymethyl cellulose hydrogels. It was shown that the concentration of DOX and the degree of NaCMC substitution had an impact on the swelling behavior and gel fraction. Higher degree of substitution hydrogels from NaCMC completely dissolved in water, whereas hydrogels with smaller DS showed more stiffness. It was thought that more COO- groups at greater carboxymethylation degrees restricted the development of crosslinks by causing repulsion between neighboring negatively charged polymeric chains. Conversely, a smaller DS made it normal for covalent connections about development among mentioned active teams. For both varieties of NaCMC, the swelling was significantly reduced following some incorporation of DOX into the hydrogel structure. This alteration was ascribed into the existence of hydrophobic areas in the drug property as well as the decrease in system polarity following covalent coupling along with DOX [80]. The same research group's subsequent paper, which documented the ecologically non-hostile which are one-pot synthesis of superabsorbent carboxymethyl cellulose/polyethylene glycol (PEG) hydrogels crosslinked along with citric acid, supported those findings. By altering the NaCMC replacement degree, crosslinking time, and PEG concentration, the swelling degree of the composites was adjusted between 100% and 5000% [81].

Now highest widely made cellulose ethers on a factory level are carboxymethyl cellulose as well as its sodium salt. They have FDA approval for parenteral usage in medicinal items [74]. NaCMC was utilized to create amorphous drug-polyelectrolyte nanoparticle complexes (nanoplexes), which are an efficient way to increase the solubility of lower basic and less soluble medications like ciprofloxacin (CIP). This is because NaCMC inhibits the formation of crystals.

NaCMC/CIP nanoplex demonstrated better solubility enhancement, longer-term physical stability, and a larger drug loading capacity as compared to traditional dextran sulfate (DXT)-based systems. In light of these findings, NaCMC was determined to be a better polyanion than DXT in the place of the complexation of medications that are lowly soluble [82]. Using the nanoprecipitation technique, nanoparticulated fixed transportation vessels depended on rosin gum (RG) as well as carboxymethyl cellulose were created. These vehicles effectively released 5-aminosalycilic acid (5-ASA) at the level of the upper gastrointestinal tract. 5-ASA was removed outside slowly over the course of 12 hours *via* a non-Fickian diffusion mechanism. Also, a zero-order release kinetics. The structural network formed by grafting hydrophobic RG on the hydrophilic CMC backbone was found to be responsible for the drug delivery system's efficiency [78]. CMC and NaCMC have been studied for a range of biomedical applications, especially about those creation of developed stimuli-responsive hydrogels utilized in controlled drug infusion as a result of their strong gelation ability and pH-sensitive swelling abilities. The synthesis of CMC/quaternized chitosan (HACC) hydrogels crosslinked by cysteamine hydrochloride demonstrated by Wang *et al.* included increased mechanical and swelling capabilities, as well as behavior responsive to redox and pH.

5-fluorouracil (5-FU), a medication with broad-spectrum antitumoral effects, was added to the composite hydrogels. Under normal circumstances (pH 7.4), the model drug's release was maintained, but was accelerated under weakly acidic or reducing environments. HepG2 cells were used in cytotoxicity experiments, which demonstrated the 5-FU/CMC/HACC composites' good biocompatibility and outstanding capacity to inhibit malignancy. Furthermore, those hydrogels prevented the growth of S. aureus and E. coli bacterial strains as a result of the antibacterial activity of HACC [83]. On the other hand, collagen-based hydrogels made in a new biphasic acetic acid/1-ethyl-3-methylimidazolium acetate (AA/[EMIM][Ac]) solvent system were crosslinked using CMC dialdehyde. According to Yang, Shen, Bu, and Li, the ECM-like network architectures demonstrated remarkable stability and mechanical strength, making them suitable for three-dimensional cell culture. Blending CMC with polyethylene oxide (PEO) also showed improvements in swelling ability and thermal stability. In comparison to pure CMC hydrogels, the resulting composites' extremely porous structure enabled for the trapping of a higher drug content [84].

An extra method for the betterment of the physicochemical items of the carboxymethyl cellulose constructions is the use of different natural fillers, in addition to blending with other polymers [85, 86]. For instance, Oun *et al.* used chitin nanocrystals as a reinforcing ingredient to create highly capable carboxymethyl cellulose solution-casted layers with enhanced mechanical as well as hydro vapor barrier qualities. Sadeghi *et al.*'s grafting of human hair keratin on CMC/haloysite nanotube hydrogels produced comparable results [87, 88]. CMC hydrogels crosslinked with graphene quantum dots (GQDs) were studied for pH-sensitive oral medication administration of the anticancer drug doxorubicin (DOX). At pH 7.4 as opposed to pH 2, the nanocomposite hydrogels swelled noticeably more, which was associated with a greater release of drugs. Their sensitivity to pH and capacity to shield the medication from the stomach acid were thus verified. When compared to free DOX, the cellular viability tests demonstrated that CMC/GQD composites were less hazardous. In addition to facilitating drug delivery, the GQD-induced fluorescence of the composites suggested them for possible imaging applications [89].

As a result of their inexpensive pricing and biodegradability along with their biocompatibility, lack of toxicity also little immunogenicity carboxymethyl cellulose was frequently utilized in tissue engineering applications. Creating tissue adhesives that match the specifications for biomedical applications while having bonding qualities akin to general purpose glues is one of the problems in this field. Biomimetic CMC/

dopamine tissue adhesive hydrogels were created, drawing inspiration from the adhesive proteins found in mussels. In the presence of H2O2, the hydrogels underwent either self-crosslinking or an enzymatic crosslinking process using horseradish peroxidase (HRP). Changing the proportion of the crosslinking agent or polymer allowed for easy customization of the rheological characteristics, swelling ratio, and gelation time. The materials' interfacial interactions with mentioned biomolecules within the non-dry tissue outer layer and the catechol moieties of the hydrogel produced a six-fold increase in wet tissue adhesion force compared to a suture agent (fibrin glue). According to Chen *et al.* and Zhong *et al.*, the gels have shown proper biodegradation behavior as well as biocompatibility *in vitro*. In order to create antiadhesive composite materials with topmost cytocompatibility and minimal immunogenicity that could be utilized as some harsh barrier for post-operative tissue joining, enzymatic crosslinking technology was also utilized to modify collagen/pullulan or chitosan/CMC/collagen [35, 90, 91, 141].

Numerous studies on CMC's applications in bone tissue engineering have been carried out in response to recent reports that it can induce osteogenic differentiation [92–95]. The ability of calcium phosphate (CaP)-loaded NaCMC nonwoven layers to promote *in vitro* osteoblast differentiation and *in vivo* bone regeneration was examined. These layers were made by alternating immersion of protonated- carboxymethyl cellulose in CaCl2 and NaHPO4 solutions. Due to the increased quantity of carboxymethyl active groups on top of the surface of the cellulosic fibers, which facilitate the nucleation of those calcium phosphates, the carboxymethyl cellulose layers with a higher degree of substitution were able to load phosphate more efficiently. While CaP-loaded carboxymethyl cellulose sheets enhanced marker gene expressions relevant to later stages of cellular differentiation as well as calcification, neat CMC sheets accelerated early osteoblast differentiation. In a mouse model of cranial bone deficiency, new bone development transpired 4 weeks after the CaP/CMC sheet was implanted [96]. Porous CMC/chitosan scaffolds that were freeze-dried and had mechanical strengths similar to cancellous bone were created by adding AgNP-CCNW (cellulose nanoparticle-modified nanohiskers) to the polymeric matrix. The resulting materials enhanced the adhesion as well as proliferation of MG63 cells to displaying protein adsorption and biomineralization capabilities. The addition of CCNWs enhanced the mechanical and thermal properties of mentioned nanocomposite scaffolds, and the combined antimicrobial activity of AgNPs and chitosan eliminated microbial strains associated with bone infections (*Escherichia coli*, *Enterococcus hirae*), including osteomyelitis [97]. In a study by Maver *et al.*, AISI 316 LVM stainless steel alloys were spin-coated alternatively with CMC and diclofenac (DCF) solutions, to create a multilayered system, taking advantage of the outstanding film forming capacity of CMC. The coatings were intended to improve the compatibility between metal and tissue, minimize long-term corrosion, and lessen the possibility of metallic particles releasing into the body. Although a significant portion of DCF was released during the first six hours, this feature is adjustable by changing the quantity and sequence in which DCF and CMC layers are deposited. Following exposure to the coated surfaces, *in vitro* experiments showed elevated expressions of genes encoding for osteomodulin (OMD) and bone morphogenetic protein 2 (BMP-2) in a primary culture of osteoblasts. These findings supported the idea that using CMC-based materials to change the surface of metallic implants can increase osseointegration rates [98].

Carboxymethyl cellulose was discovered to be a suitable item in replace of the manufacture of high fluorescence carbon dots by a straightforward hydrothermal treatment with regard to the biosensing applications. Sarkar *et al.* created proper system based on hydroxyapatite and carbon dots decorated- carboxymethyl cellulose, with good outcomes in the controlled distribution of doxorubicin, metal ion detection, osteogenesis enhancement, and bioimaging. The carbon dots that were formed exhibited a strong emission of blue fluorescence at around 440 nm. This emission was considerably suppressed when $Fe3+$ ions were present, suggesting that ferric ions are more sensitive than other metal ions [99]. Additionally, carboxymethyl cellulose effectively supports the immobilization of proteins and enzymes, guaranteeing storage stability and reusability even after extended room temperature storage. For instance, compared to the free enzyme, Jack bean urease immobilized onto amino-functionalized, epichlorohydrin-cross linked carboxymethylcellulose (ECH-CMC) beads showed superior thermal stability at comparatively high temperatures. Even after ten reuse cycles, 88% of the initial enzyme activity was still present in the ECH-CMC beads [77]. In addition, compared to alternative supports, the carboxymethyl cellulose/chitosan multilayered magnetic fluorescent systems, which were made *via* high gravity reactive precipitation, showed a comparatively high capacity for immobilizing bovine serum albumin (BSA). The process of protein immobilization was significantly influenced by the intraparticle diffusion along with the chemical chelating reaction among BSA as well as the active sites of the multilayered systems. These results create new opportunities for the application of materials obtained from biomass in the field of biosensors and bioanalysis [100].

2.3.5 Ethyl Cellulose (EC) and Methyl Cellulose (MC)

Two cellulose derivatives with comparatively likely chemical structures are methyl cellulose (MC) and ethyl cellulose (EC). Alkali cellulose and ethyl chloride react to produce Ethyl Cellulose, a partly ethylated cellulosic ether. Under the brand name Ethocel™, various EC kinds are sold. These types are often identified by their viscosity, molecular weight, and degree of substitution [101]. Commercially utilized N-type EC is produced by carefully replacing the hydroxyl groups on the surface. When the ethoxy concentration hits 47.5–49%, or a substitution degree of 2.5, the process is typically terminated. While EC is insoluble in water, it can be dissolved in a number of organic solvents, including toluene, ethanol, methanol, chloroform, and ethyl acetate. EC becomes extremely intractable if the degree of substitution is greater than 2.8 [102, 103].

When employed as a tablet matrix, Ethyl cellulose creates a very porous formation that is simple to decorate with the right size of particles. Ksenofontov *et al.* took advantage of this property to create a novel, inexpensive ethanol fluorescence investigating sensor based on Ethyl Cellulose doped with zinc (II)bis(diphyrromethenate). The sensors' ethanol vapor investigation limit in air mixes began at 0.56 parts per trillion (ppt) [104]. For coating and microencapsulating soluble and poorly soluble pharmaceutical compounds, such as acetylsalicylic acid, tacrolimus, theophylline, and propranolol in hydrochloride form, EC is recommended due to its non-toxicity, stability, hydrophobicity, and biodegradability [105, 106].

After being isolated from the olive oil tree and its leaves, hydroxytyrosol (HT), a highly hydrophilic phenolic chemical, was successfully loaded into EC microcapsules made using a water/oil/water double emulsion technique. In simulated stomach contents, the antioxidant activity of HT was kept above 50%, and the microcapsules exhibited a fair degree of stability throughout a broad temperature range. Further, the amount of HT released in the artificial intestinal fluid was greater than that typically found in olive oil; as a result, the microencapsulation of HT in EC-based structures may be a practical way to develop pharmaceuticals, nutraceuticals, and functional foods with improved antioxidant qualities [107].

A unique EC-based medication delivery system included 3D-printed tablets with an interior scaffold structure for ibuprofen's prolonged release. Hot melt extrusion was employed to convert the therapeutic agent and polymer into filaments, and fused deposition molding was then used to print the tablets. Release modifiers, such as hydroxypropyl methyl cellulose (HPMC), poly (vinyl alcohol) (PVA), sodium alginate, and xanthan gum, were added to the tablets to reduce the hydrophobicity of EC and increase its biodegradability. This resulted in an optimal sustained drug delivery that lasted for up to 24 hours. The best addition was HPMC, which increased the EC-based tablet's water uptake, swelling, and erosion, all of which helped to stimulate the release of ibuprofen. Other factors influencing the release behavior were the drug concentration, the internal structure architecture, and the thickness of the EC shell [108].The benefit of topical drug delivery is that it can prolong the therapeutic effect at the target site while avoiding any negative effects linked to oral administration.

The regulated distribution of tolmetin sodium using matrix-type transdermal patches enhanced by eucalyptus oil for penetration was reported using EC/HPMC blends. The formulations demonstrated a non-Fickian diffusion mechanism and zero order release kinetics. Studies on *in vitro* drug permeation employing cellophane membranes as penetration barriers showed that adding eucalyptus oil to the mixture enhanced the drug's permeation and flux while a higher percentage of HPMC increased the total amount of drug release. According to Prasad *et al.* [140], the medication did not interact with the polymeric matrix while being stored, and the drug remained stable. To create EC/polyvinylpyrrolidone (PVP) composites containing the anti-inflammatory medication naproxen for use in the initial phases of wound care, monoaxial electrospinning was utilized. With a controlled medication release for more than 80 hours, the nanofibrous mats showed tremendous promise in the long-term treatment of pain and inflammation associated with injuries [109]. Micro sponges have garnered significant interest as topical delivery systems recently due to their unique particle size and porous structure, which sets them apart from other carrier types [110].

In comparison to the tidy copper nanoparticles, the composites demonstrated a strong development inhibition of *Candida albicans* and *B. subtilis* as well as an increased antioxidant activity. With potentially fewer adverse effects than traditional drug-based therapy, encapsulated metallic nanoparticle systems have demonstrated antibacterial and antioxidant qualities that make them a viable substitute [111].

MC is the methyl ester of cellulose that is produced when methyl chloride and alkali cellulose combine. Twenty-seven percent to 31% of the hydroxyl moieties in the anhydroglucose units are replaced by methoxy groups after the esterification process. It is available commercially across a broad range of molecular weights, albeit its low molecular weight (~10000 Da) MC is most commonly utilized as an excipient in pharmaceutical formulations, either as a matrix for regulated drug administration [112] or as a coating agent [113, 114]. To enhance flow performance, MC was used as a viscosity modifier to injectable pastes containing porous calcium phosphate glass microspheres. Nevertheless, additional testing revealed that MC-based pastes only behaved as solid materials at 20°C; above body temperature (37°C), they lost their consistency [115].

MC may generate thermo-sensitive hydrogels with a substitution degree of 1.7, where a rise in temperature triggers a reversible sol-gel transition. Kim *et al.* reported the fabrication of MC hydrogels that exhibit thermoresponsive activity when exposed to body temperature. The MC matrix was modified to include a silver acetate precursor salt (CH3COOAg) in order to produce silver oxide nanoparticles (AgNPs) *in situ* from Ag+ ions during the hydrogelation process. It was observed that the salt-out action brought on by the remaining CH3COO− ions reduced the gelation time and temperature. The creation of AgNPs was verified by XRD and XPS spectra, and their uniform distribution within the gel network was demonstrated by SEM pictures.

The hybrid materials showed a lot of promise as injectable hydrogels for skin regeneration and performed well in tests for burn wound healing. The production of thermoresponsive hydrogels with potential uses in peritoneal tissue adhesion prevention after surgery was also described by Sultana *et al.* Methyl cellulose (MC), carboxymethyl cellulose, and polyethylene glycol (PEG) were combined to create the hydrogels. Based on earlier research showing CMC/PEG/MC composites to be useful lubricant coatings for preventing undesired tissue adhesion during surgery, this particular polymeric blend was selected. Fillers called TEMPO-oxidized cellulose nanofibers (TOCN) were utilized to extend the materials' biological residence duration at the implantation site. deterioration experiments showed that after 14 days, without a discernible change in shape, the hydrogel with the highest percentage of TOCN had the lowest deterioration rate, at 40.31 ± 6.09%. According to Sultana *et al.*, the hydrogelation behavior of the composites was suitable for the intended application, as they remained in a solution form at low temperatures (4°C) and spontaneously transformed into gel when exposed to body temperature (37°C).

Although MC films are transparent, have strong mechanical qualities, and are resistant to oils and grease, their water solubility restricts the range of possible packaging uses. One of the most common techniques for altering MC hydrophilicity is crosslinking, which can be accomplished by chemical crosslinking agents such epoxy compounds or aldehydes or by radiation [116]. MC and HPMC, which are cellulose derivatives, have the ability to self-assemble when they are functionalized with particular biomolecules, resulting in the formation of water-stable nanoaggregates. In a new study, cellulosic nanoaggregates were created by grafting testosterone, vitamins D2 and E, and MC. These nanoaggregates were then utilized to encapsulate camptothecin (CPT), a potent anticancer medication that is currently prohibited from clinical usage due to severe side effects and decreased solubility. Since the cytotoxicity of CPT-loaded cellulose nanoparticles against MCF-7 tumor cell lines was comparable to that of pure CPT, the encapsulation had no effect on the therapeutic activity of the medication. The drug release process was thought to be predicated on the diffusion of CPT from the nanoaggregate core following the ester connections between the MC shell, testosterone, and vitamins being hydrolyzed. Prior studies on CPT have reported on hyaluronic acid-based carriers [117, 118]. However, this new cellulose-based encapsulation technology offers some advantages over those studies in terms of material availability and cost-effectiveness [119].

2.3.6 Nanocellulose

Nanocellulose (NC) is a naturally occurring nanomaterial that is taken from plant cell walls. It has unique properties because of its nano size, including a large surface area, a stiffness of up to 220 Gpa elastic modulus, and a strength of up to 10 Gpa TS. Furthermore, their structure has a large number of hydroxyl groups, which allow for additional alteration. Since NC is made entirely of cellulose, it is thought to be biodegradable. It also has exceptional strength, a low density (around 1.6 g/cm^3), and a low weight. There are three distinct varieties of nanocellulose that have distinct morphologies but comparable chemical compositions. Based on their size, source, function, and synthesis method, three types of nanocellulose (NC) exist: bacterial nanocellulose (BNC), cellulose nanocrystals (CNCs), and nano fibrillated cellulose (NFC). They have similar chemical compositions but distinct morphologies [120].

Other names for CNCs include cellulose nanowhiskers, cellulose nanocrystalline cellulose, and cellulose nanocrystals. Cellulose nanofibrils undergo acid hydrolysis to primarily obtain them. Their diameter and length range from 2 to 20 μm and from 100 to 500 nm, respectively, and their rod- or whisker-shaped structure indicate that they are composed entirely of cellulose. NFCs are also referred to as NC fibrils, cellulose microfibrils, cellulose nanofibers, and microfabricated cellulose. They are flexible, lengthy, and knotted.

The dimensions range from 500 to 2000 nm in length and from 1 to 100 nm in diameter. Furthermore, they are made entirely of cellulose, both crystalline and amorphous. NFCs are longer, have a larger surface area, and a higher aspect ratio than nanocrystals. Unlike the other mentioned forms of cellulose, which are taken from lignocellulosic biomasses, BNC is obtained from the low molecular weight of sugars by the bacterium gluconacetobacter xylinus. According to Phanthong *et al.*, it has a twisted ribbon-like structure and measures 20 to 100 nm [120]. Numerous research demonstrates the uses of NC and its variants in food packaging technologies.

The usage of nanofibril cellulose in the production of biodegradable composite membranes was described by Sundaram *et al.* The encapsulated S-nitroso-N-acetyl-D-penicillamine chitosan capsules were transported using nanofibrillated cellulose as a carrier [121].

2.3.7 Bacterial Cellulose

An organic substance known as "bacterial cellulose" is made by specific bacteria, particularly those belonging to the genera Acetobacter, *Sarcina ventriculi*, and Agrobacterium. Microbial or bacterial, cellulose is distinguished from plant cellulose by its superior strength, moldability, purity, and capacity to retain more water. Majority of bacteria generate extracellular polysaccharides like cellulose in their natural habitats, which surround the cells in safety layers. Though bacterial cellulose is produced naturally, there are now a number of techniques under examination to enhance cellulose advancement from cultures in labs as a grand-scale procedure. The derived microbial cellulose can be made to have particular qualities of desire by adjusting the synthesis techniques. For instance, the peculiar mechanical characteristics of its cellulose and its uses in biotechnology, microbiology, and materials science have drawn attention to the bacteria *Acetobacter xylinum*. Bacterial cellulose is being employed for broad applications in commerce, such as textiles, cosmetics, health food products, coatings, reinforcement for optically transparent films, and medical applications [122].

In 2008, Chawla *et al.* conducted research on microbial cellulose as a potential recently developed resource for some applications [123]. The inexpensive raw material is one positive aspect of adopting industrial or agricultural residual streams as feedstock for the bacterial cellulose synthesis process. The breakdown of polysaccharides and lignin in lignocellulosic feedstocks under high pressure and temperature culminates in the formation of inhibitory chemicals [124]. To allow the growth of bacteria and the synthesis of BC, over liming was used in the research by Hong and Qiu and Hong *et al.* to detoxify the hydrolysates generated from the acid hydrolysis of konjak glucomannan and wheat straw [125, 126].

Furthermore, Cavka *et al.* looked at the suitability of waste fiber sludge for the synthesis of bacterial cellulose (BC) and the potential to integrate BC manufacturing with the synthesis of hydrolytic enzymes that are helpful in the degradation of lignocellulose. The fiber sludges were from a lignocellulosic biorefinery that used a sulfite-based process and a pulp mill that used a sulfate-based method (kraft pulping). The filamentous fungus *Trichoderma reesei* and the bacteria *Gluconacetobacter xylinus* were used in successive fermentations to create the BC and enzymes.

Without first undergoing a thermochemical pretreatment, fiber sludges from the sulfate (SAFS) and sulfite (SIFS) processes were hydrolyzed enzymatically, and the resulting hydrolysates were utilized to produce BC. It is possible to use the cellulase generated in this way to hydrolyze new fiber sludge and create a medium that may be used to make BC in the very biorefinery. Fiber sludge is mostly composed of cellulose and hemicellulose, with lignin often making up less than 5% of the material. Fiber sludges' structure and composition make them generally amenable to enzymatic hydrolysis without the need for a previous thermochemical pretreatment. This could result in hydrolysates with high glucose concentrations and low inhibitory component contents. Inhibitory chemical content should be favorable for the bacterial strains utilized to produce BC. Enzymatic hydrolysis does have certain disadvantages, though, chief among them being the high price of the hydrolytic enzymes required for the procedure [127].

2.4 Crosslinking of Cellulose

The mechanical, thermal, and water stability of cellulose can all be refined by crosslinking. This method creates crosslinked covalent bonds between cellulose and its derivatives' –OHs or functional groups and crosslinking agents [128].

2.4.1 Crosslinking on Hydroxyl Groups of Cellulose

As found in a study, NaH2PO4 was used as a catalyst to dip papers in aqueous solutions containing the polyfunctional carboxylic acids 1,2,3,4-butanetetra-carboxylic acid (BTCA), tricarballylic acid (TCA), or succinic acid (SA) for 30 minutes in order to chemically crosslink the softwood cellulose papers. After that, the treated cellulose sheets were cured for five to thirty minutes at 150°C in an oven. When it came to crosslinking, BTCA outperformed TCA. As the number of crosslinking sites in cellulose papers rose, their wet σ increased in comparison to pristine paper.

The alteration of cellulose by SA led to the attachment of cellulosic units with a single pendant carboxylic group, which has poor reactivity with the cellulosic -OHs and so provides little crosslinking [129]. As a result, sheets treated with SA demonstrated very little wet strength gain. Since it is non-toxic, citric acid (CAC) is frequently utilized as a secure natural addition in the food sector. In order to make CNC more hydrophobic, He and colleagues employed CAC as a crosslinker. Using NaH2PO4 as a catalyst, an aqueous colloidal suspension of CNC and CAC was reacted for four hours at 130°C in an oven. Compared to 41° for CNC, the WCA of CAC crosslinked CNC (CAC-CNC) was only 55°.

Nevertheless, set side by side to the unmodified CNC (Td-onset = 249°C), the CAC-CNC demonstrated a superior thermal stability (Td-onset = 298°C) [130]. In a different work, filter paper coated with sorbitol plasticized nanocellulose was crosslinked in an aqueous CAC solution and then cured for five minutes at 150°C in an oven with NaH2PO4 acting as a catalyst. The plasticized, crosslinked, and coated filter paper that resulted from applying three layers of CAC crosslinked sorbitol-NC coating showed a lower WVP and OP of 0.5 g mm kPa^{-1} m^{-2} day^{-1} and 2 mL µm m^{-2} day^{-1} kPa^{-1}, compared to 4.0 g mm kPa^{-1} m^{-2} day^{-1} and 197 mL µm m^{-2} day^{-1} kPa^{-1} for the unmodified filter paper.

While there were no discernible changes in σ, the changed paper's E dropped from 570 to 310 MPa and its ε marginally rose from 2.5% to 3.7% in comparison to the control paper. Compared to the non-modified material, the modified paper's Td-onset dropped from 311°C to 288°C [131].

2.4.2 Crosslinking with Functional Groups of Cellulose Derivatives

Different functional groups, namely hydroxy, propyl, methyl, carboxymethyl, and formyl groups, are present in cellulose derivatives. Five weight percent CAC was used as a crosslinker for hydroxy propyl methyl cellulose (HPMC) and reduced the WS of CAC crosslinked HPMC (CAC-HPMC) films by 74% when compared with pristine HPMC film. The highest reduction in WVTR was observed for CAC-HPMC film, occurring at 190°C for 15 minutes with PEG400 as a plasticizer and NaH2PO4 as a catalyst [132]. Carboxymethyl cellulose (CMCS) was altered in two ways by another investigation. Using sodium benzoate (SB) as a photo-initiator, one technique employed photo-crosslinking *via* UV irradiation at room temperature for 30 to 180 minutes.

Using gelatin as a synergistic crosslinker and saturated glutaraldehyde (GLA) vapor, the other method utilized chemical crosslinking at 80°C for 30 to 180 minutes. The best crosslinking degrees were found in the photo-crosslinked film treated with 20 weight percent SB and subjected to GLA vapor for 180 minutes, as well as in the chemically crosslinked film modified by adding 0.2 g of gelatin and exposed to GLA vapor for 90 minutes. In terms of WCA, WVP, and σ, the photo-crosslinking treatment outperformed the chemical crosslinking method. The WS of the CMCS/SB/UV and CMCS/GLA/gelatin films decreased to 57.1% and 50.1%, respectively, in contrast to 78.2% for the pristine CMCS film.

The WCA (78°) of the CMCS/SB/UV film was somewhat greater than that of the CMCS/GLA/gelatin film (67°). The weight loss percentage (WVP) of the CMCS/SB/UV film was significantly reduced to 9×10^{-7} g m^{-1} s^{-1} Pa^{-1} (as opposed to 819×10^{-7} g m^{-1} s^{-1} Pa^{-1} for the neat CMCS film). In contrast, the WVP of the CMCS/GLA/gelatin film fell less significantly, to 50×10^{-7} g m^{-1} s^{-1} Pa^{-1}. The σ of the CMCS/SB/UV and CMCS/GLA/gelatin films increased significantly to 46.8 MPa and 33.7 MPa, respectively, in contrast to the unaltered CMCS film's 14.2 MPa. Nevertheless, after being crosslinked with SB/UV and GLA/gelatin, ε of the CMCS films dropped from 19.9% to 9.1% and 13.8%, respectively.

Both the chemically and photo-crosslinked films exhibited negligible cytotoxicity [133]. In a different investigation, sodium periodate with 6% of AGU on oxidized cellulose periodate-oxidized softwood CNF to dialdehyde CNF (DACNF).

At 60°C for three hours, the Maillard reaction was used to crosslink gelatin to DACNF. When submerged in DI water after one hour, the weight of the gelatin-crosslinked DACNF (G-DACNF) film (crosslinking degree = 57%) rose by only 44%, compared to a 167% increase in the weight of the neat CNF film. This indicates a significant decrease in WA. Furthermore, wet mechanical strength (wet ε: 23.2%, wet σ: 15.4 MPa) of the G-DACNF film was significantly greater than that of the unmodified CFR film (6.7%, 0.9 MPa) and DACNF (16.5%, 9.9 MPa) [134].

Chitosan was employed as the DACNF crosslinker in a different investigation. The WVP of the crosslinked films was marginally higher than that of the raw cellulose film. The OP of the pure cellulose film was over the OP analyzer's detection limit, whereas the OP of the chitosan crosslinked films was significantly lowered to 4.3-8.9 $\times 10^{-15}$ cm^3 cm/cm^2 s Pa. Furthermore, the crosslinked films' σ value was significantly greater than the original film's [135].

Additionally, it was discovered that 4-(4,6-dimethoxy-1,3,5-triazin-2-yl)-4-methyl-morpholinium chloride (DMTMM) was an effective crosslinking agent that encouraged the formation of esters through the reaction of carboxylic groups and -OHs under mild reaction conditions. Beghetto *et al.* used DMTMM as a crosslinking agent for CMCS, using gelatin as a plasticizer. After 2 hours of RT in DI water, the reaction produced a transparent DMTMM crosslinked CMCS film (DMTMM-CMCS) with an 80% to 90% optical transmittance. 2,4-dimethoxy-6-hydroxy-1,3,5-triazine (DMTOH) and N-methyl morpholinium hydrochloride were by-products of the reaction between DMTMM and water washing. Both of these compounds are harmless. One more benefit is that DMTOH can be extracted and repurposed to create DMTMM.

Additionally, the lowest WS (18.1%) was observed in the 5-weight percent DMTMM-CMCS film, and the CMCS fully dissolved in DI water in 4 hours. The WVP and oil resistance ability of the 10 wt% DMTMM-CMCS film were significantly lower (0.68×10^{-7} g m^{-1} h^{-1} Pa^{-1} and 0.29%) than those of the unmodified CMCS film (1.14×10^{-7} g m^{-1} h^{-1} Pa^{-1}, 1.89%).

While σ steadily declined, increasing DMTMM levels resulted in a much-improved σ. When compared to the neat CMCS film, the σ increased from 32 MPa to 54.9 MPa and the ε decreased from 30.1% to 25.4% with the addition of 5 weight percent DMTMM.

The 5% DMTMM-CMCS film was plasticized using 50% wt% glycerol to obtain the ideal σ and ε (52.3 MPa, 37.3%), meeting the requirements for high σ and high ε in food packaging applications. Additionally, compared to the unmodified CMCS film, the crosslinked CMCS film demonstrated greater thermal stability [136].

2.5 Application and Advantages of Various Cellulose Based Packaging Materials

Cellulose Acetate

Applications:

- Frequently utilized for food packaging in movies.
- Because CA is biodegradable, it is used to make cigarette filters.
- Applied to packaging as a coating substance to increase moisture resistance.

Advantages:

- Outstanding gloss and transparency.
- It can be used for food wraps because it is resistant to oils and grease.
- Biodegradable in the context of industrial composting [40].

Cellulose Sulfate

Applications:

- Food packaging made of edible films. It is being investigated for usage in edible or dissolvent packaging.
- Barrier coatings to increase the water resistance of cardboard or paper packing.

Advantages:

- High water absorption, which is helpful in applications where controlling moisture levels is essential.
- Possibility of being a biodegradable substitute for coatings made of synthetic plastic [61].

Cellulose Nitrate

Applications:

- Mostly employed for its good film-forming qualities in specialty packaging applications, like lacquers and films.
- Specialty packaging is less prevalent in regular packing since it is flammable; it is used for things that need transparent, lightweight barriers.

Advantages:

- Creates resilient, flexible films that can act as layers of protection [70].

Nanocellulose (NC)

Applications:

- Barrier films: Food packaging can benefit greatly from the strong oxygen, grease, and oil barriers that nanocellulose can give in coatings.
- Flexible packaging: Films made of nanocellulose can be added to flexible films that are used to wrap fresh products.
- Composite packaging materials: To make environmentally friendly packaging, nanocellulose can be combined with biopolymers or even recycled plastics.

Advantages:

- High rigidity and strength.
- Excellent oxygen-barrier qualities make it helpful for food preservation [137].

Bacterial Cellulose (BC)

Applications:

- Utilized in films for biodegradable packaging.
- Excellent because of its ability to retain moisture for packaging fresh foods like fruits and vegetables.
- Can be utilized to create packing materials that decompose.

Advantages:

- Robust mechanical characteristics.
- Excellent ability to retain water, extending the freshness of perishables.
- Fully degradable by nature [122].

Carboxymethyl Cellulose (CMC)

Applications:

- Applied to food packaging as a coating agent to enhance moisture retention and stop spoiling.
- Can be mixed with additional biopolymers to make composite packaging films.
- Utilized in adhesive layers for packaging constructions with many layers as well.

Advantages:

- Outstanding capacity to produce films.
- Both biodegradable and offering strong moisture-barrier qualities.
- To enhance their mechanical qualities, bio-based materials are frequently combined with one another [91].

2.6 Benefits of Packaging Made of Cellulose Instead of Synthetic Polymer

Because it has so many advantages over plastic, cellulose is used in many practical applications. The following are some of the many benefits that make cellulose packaging a desirable alternative to plastic:

- **Renewability**: Less dependence on finite fossil fuels is made possible by the fact that renewable resources such as algae and cotton are used to make cellulose.
- **Biodegradability**: Biodegradable cellulose packaging reduces waste in the environment because it breaks down more slowly than conventional plastic packaging.
- **Composability**: Plant growth and soil nutrition can be enhanced and cellulose packing can be composted.
- **Lightweight**: Since cellulose weighs less than conventional materials, less carbon dioxide is released during distribution and transit.
- **Strong**: Although cellulose is light, it is robust and long-lasting, offering sufficient safeguarding.
- **Temperature resistance**: Because cellulose packaging is resistant to temperature changes, it is ideal for carrying pre-made foods.
- **Versatility**: Cellulose may be shaped and tailored to match any application.
- **Low-cost**: Cellulose packaging has been used since 1912 as a byproduct of the paper industry. This, combined with the plentiful of its raw materials, results in a lower cost than other eco-friendly plastic replacement.
- **Lower carbon footprint:** The manufacturing of cellulose packaging requires minimal energy and emits lesser greenhouse emissions. Furthermore, it minimizes reliance on fossil fuels, and hence greenhouse gas emissions.
- **Transparency**: Cellulose-based and cellophane bioplastics are more convenient to create transparent and clear than traditional plastics.
- **Non-Toxic**: Cellulose is non-toxic, meaning it will not infuse poisons into your food or the environment [138].

2.7 Conclusion

The shift toward sustainability is necessary since environmental issues are posing numerous challenges to our society today. The fact that there are plentiful nonbiodegradable materials in the world means that abundant agricultural and forest resources are used. The numerous characteristics of cellulose have shown its promise as a strengthening agent. Distinct cellulose supplies have varying mechanical properties. For instance, the majority of cotton (more than 94%) and wood (more than 50%) are composed of cellulose. The two leading raw materials used to make all cellulose products, including textiles, paper, building supplies, cardboard, and cellophane, rayon, and CA, are cotton and wood. Nonetheless, cellulose, which is derived from many plant sources, provides a compelling mix of mechanical and thermal characteristics. Several uses in the packaging domain include nanocomposites, films, bio composites, and coatings to use extracted cellulose as a raw material. These items are made by a complicated process including the conversion of cellulose. In particular, wood pulp is processed into a viscous liquid by a number of chemical steps before being turned into film. This liquid is filtered to remove impurities and increase material purity in order to produce the highest quality film possible. To guarantee that the film has the correct optical and mechanical qualities, the viscose is extruded and then cast *via* a sequence of rollers and baths. Uncoated cellulose films are utilized for items like fresh bread and some candies that do not need a whole moisture or gas barrier because of their significant permeability to water vapor. Every cellulose film is printable, machinable, heat sealable, and resistant to heat. Additives like cellulose derivatives are used in a lot of thermoplastic and thermosetting coating systems. As film formers, cellulose esters offer crosslinking reactions, quick drying, early hardness buildup, and suppression of flooding and floating. Plastic is the most sought-after material for packaging because of its amazing qualities, which suit every aspect of packaging. However, the intricacy of managing the waste plastics has compelled efforts to be made in the field of innovative biosourced plastic alternatives. There are several uses for these biosourced plastics in packaging because of the wide variations in their feedstocks and manufacturing processes. The packaging business has a ton of potential applications for plastic packaging made from biosource materials. Conversely, the second iteration of biosourced plastics, or biodegradable plastic, is also effectively addressing issues related to public health and the natural environment brought on by the waste plastic that comes from packaging. Plastics derived from biological sources and those that degrade naturally will undoubtedly be important in the future. A few issues still need to be resolved, though, including its high cost, the availability of land for the production of raw materials, the degree of durability, and the material's performance in comparison to plastics based on petrochemicals. The Complete encapsulation of the biodegradable polymers' socio-economic profits require further research and developments. Future scientists will have to deal with these matters in question and generate the required resources in order to shape a promising future for plastics derived from biosources.

AI disclosure statement: The author acknowledges the use of AI-based tools for minor language editing or grammatical correction. However, all intellectual contributions, analyses, and conclusions presented in this chapter are entirely the author's own.

References

1. Subbotina, E., Ram, F., Dvinskikh, S.V., Berglund, L.A., Olsén, P., Aqueous synthesis of highly functional, hydrophobic, and chemically recyclable cellulose nanomaterials through oxime ligation. *Nat. Commun.*, 13, 1, Dec. 2022, doi: 10.1038/s41467-022-34697-5.
2. Eichhorn, S.J., *et al.*, Current international research into cellulose as a functional nanomaterial for advanced applications. *J. Mater. Sci.*, Mar. 01, 2022, Springer. doi: 10.1007/s10853-022-06903-8.
3. Jayasekara, R., Harding, I., Bowater, I., Christie, G.B.Y., Lonergan, G.T., Preparation, surface modification and characterisation of solution cast starch PVA blended films. *Polym. Test.*, 23, 1, 17–27, Feb. 2004, doi: 10.1016/S0142-9418(03)00049-7.
4. Qi, H., Cellulose-based functional and smart materials, in: *SpringerBriefs in Applied Sciences and Technology*, pp. 45–67, no. 9783319495910, 2017.
5. Klemm, D., Heublein, B., Fink, H.P., Bohn, A., Cellulose: Fascinating biopolymer and sustainable raw material. *Angewandte Chemie International Edition*, May 30, 2005, doi: 10.1002/anie.200460587.
6. Mohanty, A.K., Misra, M., Hinrichsen, G., Biofibres, biodegradable polymers and biocomposites: An overview Summary of contents 1. Introduction 2. Reinforcing biofibres 2.1 Chemical constituents and structural aspects 2.2 Properties of biofibres 2.3 Degradation properties of biofibres 2.4 Cost aspects, availability and sustainable development of biofibres. *Macromol. Mater. Eng.*, 2000.
7. Wisniak, J., Anselme Payen. *Educ. Quím.*, 16, 4, 568, Aug. 2018, doi: 10.22201/fq.18708404e.2005.4.66095.
8. Young, R.A., Comparison of the cellulose pulps properties of chemical, 1994.

9. Heinze, T., Cellulose: Structure and properties. *Adv. Polym. Sci.*, 271, 1–52, 2015, doi: 10.1007/12_2015_319.
10. Kondo, T., *The Relationship between Intramolecular Hydrogen Bonds and Certain Physical Properties of Regioselectively Substituted Cellulose Derivatives*, John Wiley & Sons, Inc, Hoboken, New Jersey, USA, 1997.
11. Willgert, M., Leijonmarck, S., Lindbergh, G., Malmström, E., Johansson, M., Cellulose nanofibril reinforced composite electrolytes for lithium ion battery applications. *J. Mater. Chem. A Mater.*, 2, 33, 13556–13564, Jul. 2014, doi: 10.1039/C4TA01139B.
12. Cheng, H.N., Dowd, M.K., Selling, G.W., Biswas, A., Synthesis of cellulose acetate from cotton byproducts. *Carbohydr. Polym.*, 80, 2, 449–452, Apr. 2010, doi: 10.1016/j.carbpol.2009.11.048.
13. Iodine-catalyzed esterification of polysaccharides | Request PDF, Accessed: Aug. 28, 2024. [Online]. Available: https://www.researchgate.net/publication/283844040_Iodine-catalyzed_esterification_of_polysaccharides.
14. Khoshnevisan, K., *et al.*, Cellulose acetate electrospun nanofibers for drug delivery systems: Applications and recent advances. *Carbohydr. Polym.*, 198, 131–141, Oct. 2018, doi: 10.1016/J.CARBPOL.2018.06.072.
15. Thakur, V.K. and Voicu, S.I., Recent advances in cellulose and chitosan based membranes for water purification: A concise review. *Carbohydr. Polym.*, Aug. 01, 2016, Elsevier Ltd. doi: 10.1016/j.carbpol.2016.03.030.
16. Neacsu, P., *et al.*, Characterization and *in vitro* and *in vivo* assessment of a novel cellulose acetate-coated Mg-based alloy for orthopedic applications. *Materials*, 10, 7, Jun. 2017, doi: 10.3390/ma10070686.
17. Pandele, A.M., *et al.*, Cellulose acetate membranes functionalized with resveratrol by covalent immobilization for improved osseointegration. *Appl. Surf. Sci.*, 438, 2–13, Apr. 2018, doi: 10.1016/j.apsusc.2017.11.102.
18. Voicu, S.I., *et al.*, Sericin Covalent Immobilization onto Cellulose Acetate Membrane for Biomedical Applications. *ACS Sustainable Chem. Eng.*, 4, 3, 1765–1774, Mar. 2016, doi: 10.1021/acssuschemeng.5b01756.
19. Corobea, M.S., *et al.*, Modification of titanium surface with collagen and doxycycline as a new approach in dental implants. *J. Adhes. Sci. Technol.*, 29, 23, 2537–2550, Dec. 2015, doi: 10.1080/01694243.2015.1073661.
20. Corobea, M.C., *et al.*, Novel nanocomposite membranes from cellulose acetate and clay-silica nanowires. *Polym. Adv. Technol.*, 27, 12, 1586–1595, Dec. 2016, doi: 10.1002/pat.3835.
21. Ioniţă, M., *et al.*, Synergistic effect of carbon nanotubes and graphene for high performance cellulose acetate membranes in biomedical applications. *Carbohydr. Polym.*, 183, 50–61, Mar. 2018, doi: 10.1016/j.carbpol.2017.10.095.
22. Esmaeili, E., *et al.*, The biomedical potential of cellulose acetate/polyurethane nanofibrous mats containing reduced graphene oxide/silver nanocomposites and curcumin: Antimicrobial performance and cutaneous wound healing. *Int. J. Biol. Macromol.*, 152, 418–427, Jun. 2020, doi: 10.1016/j.ijbiomac.2020.02.295.

23. Chen, L., Li, Z., Chen, M., Facile production of silver-reduced graphene oxide nanocomposite with highly effective antibacterial performance. *J. Environ. Chem. Eng.*, 7, 3, Jun. 2019, doi: 10.1016/j.jece.2019.103160.
24. de Faria, A.F., de Moraes, A.C.M., Andrade, P.F., da Silva, D.S., do Carmo Gonçalves, M., Alves, O.L., Cellulose acetate membrane embedded with graphene oxide-silver nanocomposites and its ability to suppress microbial proliferation. *Cellulose*, 24, 2, 781–796, Feb. 2017, doi: 10.1007/s10570-016-1140-6.
25. Rusen, E., *et al.*, Design of antimicrobial membrane based on polymer colloids/multiwall carbon nanotubes hybrid material with silver nanoparticles. *ACS Appl. Mater. Interfaces*, 6, 20, 17384–17393, Oct. 2014, doi: 10.1021/am505024p.
26. Raicopol, M.D., Andronescu, C., Voicu, S.I., Vasile, E., Pandele, A.M., Cellulose acetate/layered double hydroxide adsorptive membranes for efficient removal of pharmaceutical environmental contaminants. *Carbohydr. Polym.*, 214, 204–212, Jun. 2019, doi: 10.1016/j.carbpol.2019.03.042.
27. Bie, X., *et al.*, Fabrication and characterization of wound dressings containing gentamicin/silver for wounds in diabetes mellitus patients. *Mater. Res. Express*, 7, 4, Apr. 2020, doi: 10.1088/2053-1591/ab8337.
28. Jatoi, A.W., Kim, I.S., Ni, Q.Q., Cellulose acetate nanofibers embedded with AgNPs anchored TiO 2 nanoparticles for long term excellent antibacterial applications. *Carbohydr. Polym.*, 207, 640–649, Mar. 2019, doi: 10.1016/j.carbpol.2018.12.029.
29. Khan, M.Q., *et al.*, Fabrication of antibacterial electrospun cellulose acetate/silver-sulfadiazine nanofibers composites for wound dressings applications. *Polym. Test.*, 74, 39–44, Apr. 2019, doi: 10.1016/j.polymertesting.2018.12.015.
30. Chen, G. and Liu, B., Cellulose sulfate based film with slow-release antimicrobial properties prepared by incorporation of mustard essential oil and β-cyclodextrin. *Food Hydrocolloids*, 55, 100–107, Apr. 2016, doi: 10.1016/j.foodhyd.2015.11.009.
31. Sci-Hub, Cellulose acetate - essential oil nanocapsules with antimicrobial activity for biomedical applications. *Colloids Surf. B: Biointerfaces*, 2018, 10.1016/j.colsurfb.2018.08.069. Accessed: Aug. 27, 2024. [Online]. Available: https://sci-hub.se/10.1016/j.colsurfb.2018.08.069.
32. Tenhunen, T.M., *et al.*, Surface tailoring and design-driven prototyping of fabrics with 3D-printing: An all-cellulose approach. *Mater. Des.*, 140, 409–419, Feb. 2018, doi: 10.1016/j.matdes.2017.12.012.
33. Tungprapa, S., Jangchud, I., Supaphol, P., Release characteristics of four model drugs from drug-loaded electrospun cellulose acetate fiber mats. *Polymer (Guildf)*, 48, 17, 5030–5041, Aug. 2007, doi: 10.1016/j.polymer.2007.06.061.
34. Yang, Y., Li, W., Yu, D.G., Wang, G., Williams, G.R., Zhang, Z., Tunable drug release from nanofibers coated with blank cellulose acetate layers fabricated using tri-axial electrospinning. *Carbohydr. Polym.*, 203, 228–237, Jan. 2019, doi: 10.1016/j.carbpol.2018.09.061.

35. Chen, Y., Qiu, Y., Chen, W., Wei, Q., Electrospun thymol-loaded porous cellulose acetate fibers with potential biomedical applications. *Mater. Sci. Eng., C*, 109, Apr. 2020, doi: 10.1016/j.msec.2019.110536.
36. Chakraborty, P.K., Adhikari, J., Saha, P., Facile fabrication of electrospun regenerated cellulose nanofiber scaffold for potential bone-tissue engineering application. *Int. J. Biol. Macromol.*, 122, 644–652, Feb. 2019, doi: 10.1016/j.ijbiomac.2018.10.216.
37. Sci-Hub, Multi-hierarchical tissue-engineering ECM-like scaffolds based on cellulose acetate with collagen and chitosan fillers. *Carbohydr. Polym.*, 191, 119–126, 2018, 10.1016/j.carbpol.2018.02.061. Accessed: Aug. 27, 2024. [Online]. Available: https://sci-hub.se/10.1016/j.carbpol.2018.02.061.
38. Dumitriu, C., *et al.*, Production and characterization of cellulose acetate – titanium dioxide nanotubes membrane fraxiparinized through polydopamine for clinical applications. *Carbohydr. Polym.*, 181, 215–223, Feb. 2018, doi: 10.1016/j.carbpol.2017.10.082.
39. Tabernero, A., Cardea, S., Masa, R., Baldino, L., del Valle, E.M.M., Reverchon, E., Preparation and characterization of cellulose acetate-Laponite® composite membranes produced by supercritical phase inversion. *J. Supercrit. Fluids*, 155, Jan. 2020, doi: 10.1016/j.supflu.2019.104651.
40. Pandele, A.M., *et al.*, Synthesis and characterization of cellulose acetate-hydroxyapatite micro and nano composites membranes for water purification and biomedical applications. *Vacuum*, 146, 599–605, Dec. 2017, doi: 10.1016/j.vacuum.2017.05.008.
41. Yu, S., Ding, L., Lin, H., Wu, W., Huang, J., A novel optical fiber glucose biosensor based on carbon quantum dots-glucose oxidase/cellulose acetate complex sensitive film. *Biosens. Bioelectron.*, 146, Dec. 2019, doi: 10.1016/j.bios.2019.111760.
42. Casella, I.G., Gioia, D., Rutilo, M., A multi-walled carbon nanotubes/cellulose acetate composite electrode (MWCNT/CA) as sensing probe for the amperometric determination of some catecholamines. *Sens. Actuators B Chem.*, 255, 3533–3540, Feb. 2018, doi: 10.1016/j.snb.2017.09.188.
43. Nan, M., *et al.*, Ecofriendly high-performance ionic soft actuators based on graphene-mediated cellulose acetate. *Sens. Actuators B Chem.*, 301, Dec. 2019, doi: 10.1016/j.snb.2019.127127.
44. Aggarwal, N., Altgärde, N., Svedhem, S., Zhang, K., Fischer, S., Groth, T., Effect of molecular composition of heparin and cellulose sulfate on multilayer formation and cell response. *Langmuir*, 29, 45, 13853–13864, Nov. 2013, doi: 10.1021/la4028157.
45. Tao, W., Richards, C., Hamer, D., Short communication: Enhancement of HIV infection by cellulose sulfate. *AIDS Res. Hum. Retroviruses*, 24, 7, 925–929, Jul. 2008, doi: 10.1089/aid.2008.0043.
46. Strätz, J., Liedmann, A., Heinze, T., Fischer, S., Groth, T., Effect of Sulfation Route and Subsequent Oxidation on Derivatization Degree and

Biocompatibility of Cellulose Sulfates. *Macromol. Biosci.*, 20, 2, Feb. 2020, doi: 10.1002/mabi.201900403.

47. Gericke, M., Doliška, A., Stana, J., Liebert, T., Heinze, T., Stana-Kleinschek, K., Semi-Synthetic Polysaccharide Sulfates as Anticoagulant Coatings for PET, 1 - Cellulose Sulfate. *Macromol. Biosci.*, 11, 4, 549–556, Apr. 2011, doi: 10.1002/mabi.201000419.
48. Sci-Hub, Dose Dependence of the Anticoagulant Effect of Intravenously Administered Cellulose Sulfate. *Bull. Exp. Biol. Med.*, 160, 6, 767–770, 2016, 10.1007/s10517-016-3305-6. Accessed: Aug. 27, 2024. [Online]. Available: https://sci-hub.se/10.1007/s10517-016-3305-6.
49. Menezes, R., *et al.*, Investigation of glycosaminoglycan mimetic scaffolds for neurite growth. *Acta Biomater.*, 90, 169–178, May 2019, doi: 10.1016/j.actbio.2019.03.024.
50. Huang, G.P., Molina, A., Tran, N., Collins, G., Arinzeh, T.L., Investigating cellulose derived glycosaminoglycan mimetic scaffolds for cartilage tissue engineering applications. *J. Tissue Eng. Regener. Med.*, 12, 1, e592–e603, Jan. 2018, doi: 10.1002/term.2331.
51. Weber, D., *et al.*, Influence of Regioselectively Sulfated Cellulose on *in Vitro* Vascularization of Biomimetic Bone Matrices. *Biomacromolecules*, 19, 11, 4228–4238, Nov. 2018, doi: 10.1021/ACS.BIOMAC.8B01004.
52. 021 B2 1. Method for the preparation of sulfated cellulose membranes and sulfated cellulose membranes.
53. Fortuna, A.R., *et al.*, Use of sulfated cellulose membrane adsorbers for chromatographic purification of cell cultured-derived influenza A and B viruses. *Sep. Purif. Technol.*, 226, 350–358, Nov. 2019, doi: 10.1016/j.seppur.2019.05.101.
54. Müller, M., *et al.*, Switchable release of bone morphogenetic protein from thermoresponsive poly(NIPAM-co-DMAEMA)/cellulose sulfate particle coatings. *Polymers (Basel)*, 10, 12, Nov. 2018, doi: 10.3390/polym10121314.
55. Kauschke, V., Hessland, F.M., Vehlow, D., Müller, M., Heiss, C., Lips, K.S., High concentrations of polyelectrolyte complex nanoparticles decrease activity of osteoclasts. *Molecules*, 24, 12, 2019, doi: 10.3390/molecules24122346.
56. Vehlow, D., Schmidt, R., Gebert, A., Siebert, M., Lips, K.S., Müller, M., Polyelectrolyte complex based interfacial drug delivery system with controlled loading and improved release performance for bone therapeutics. *Nanomaterials*, 6, 3, Mar. 2016, doi: 10.3390/nano6030053.
57. Bentham Science Publishers, *Bioencapsulation of Living Cells for Diverse Medical Applications*, Bentham Science Publishers, Sharjah, United Arab Emirates, May 2013, doi: 10.2174/97816080572071130101.
58. Salmons, B. and Gunzburg, W.H., Release characteristics of cellulose sulphate capsules and production of cytokines from encapsulated cells. *Int. J. Pharm.*, 548, 1, 15–22, Sep. 2018, doi: 10.1016/J.IJPHARM.2018.06.040.

59. Kaiser, P., Werner, M., Jérôme, V., Freitag, R., Scale-up of the *ex vivo* expansion of encapsulated primary human T lymphocytes. *Biotechnol. Bioeng.*, 115, 10, 2632–2642, Oct. 2018, doi: 10.1002/bit.26786.
60. Jérôme, V., Werner, M., Kaiser, P., Freitag, R., Creating a Biomimetic Microenvironment for the *Ex Vivo* Expansion of Primary Human T Lymphocytes. *Macromol. Biosci.*, 17, 9, Sep. 2017, doi: 10.1002/mabi.201700091.
61. Savitskaya, T.A., Shakhno, E.A., Grinshpan, D.D., Ivashkevich, O.A., Soluble Polymer Complexes of Ceftriaxone and Cefotaxime with Cellulose Sulfate Acetate Salt Forms and Their Adsorption on Carbon Sorbents. *Polym. Sci. - Ser. A*, 61, 3, 274–286, May 2019, doi: 10.1134/S0965545X1903012X.
62. McKeen, L.W., *The effect of sterilization on plastics and elastomers*, third edition, William Andrew, Oxford, United Kingdom, 2012.
63. Li, L. and Frey, M., Preparation and characterization of cellulose nitrate-acetate mixed ester fibers. *Polymer (Guildf)*, 51, 16, 3774–3783, 2010, doi: 10.1016/j.polymer.2010.06.013.
64. Nikolsky, S.N., Zlenko, D.V., Melnikov, V.P., Stovbun, S.V., The fibrils untwisting limits the rate of cellulose nitration process. *Carbohydr. Polym.*, 204, 232–237, Jan. 2019, doi: 10.1016/j.carbpol.2018.10.019.
65. Trache, D., Khimeche, K., Mezroua, A., Benziane, M., Physicochemical properties of microcrystalline nitrocellulose from Alfa grass fibres and its thermal stability. *J. Therm. Anal. Calorim.*, 124, 3, 1485–1496, Jun. 2016, doi: 10.1007/s10973-016-5293-1.
66. Chi, J., *et al.*, Erratum: A bio-inspired photonic nitrocellulose array for ultrasensitive assays of single nucleic acids. *Analyst*, 143, 4559–4565, 2018, DOI: 10.1039/C8AN00939B, Feb. 21, 2020, Royal Society of Chemistry. doi: 10.1039/c9an90125f.
67. Luo, Y., *et al.*, Novel biosensor based on electrospun nanofiber and magnetic nanoparticles for the detection of E. coli O157:H7. *IEEE Trans. Nanotechnol.*, 11, 4, 676–681, 2012, doi: 10.1109/TNANO.2011.2174801.
68. Santamaría, B., *et al.*, Development towards compact nitrocellulose-based interferometric biochips for dry eye mmp9 label-free *in-situ* diagnosis. *Sensors*, 17, 5, May 2017, doi: 10.3390/S17051158.
69. Nartker, S., Hassan, M., Stogsdill, M., Electrospun cellulose nitrate and polycaprolactone blended nanofibers. *Mater. Res. Express*, 2, 3, Mar. 2015, doi: 10.1088/2053-1591/2/3/035401.
70. Li, A., *et al.*, Use of nitrocellulose membranes as a scaffold in cell culture. *Cytotechnology*, 65, 1, 71–81, Jan. 2013, doi: 10.1007/s10616-012-9458-x.
71. Sun, S., *et al.*, Microstructural effects on permeability of Nitrocellulose membranes for biomedical applications. *J. Membr. Sci.*, 595, Feb. 2020, doi: 10.1016/j.memsci.2019.117502.
72. Yew, C.H.T., Azari, P., Choi, J.R., Li, F., Pingguan-Murphy, B., Electrospin-coating of nitrocellulose membrane enhances sensitivity in nucleic acid-based lateral flow assay. *Anal. Chim. Acta*, 1009, 81–88, Jun. 2018, doi: 10.1016/J.ACA.2018.01.016.

73. Altunina, L.K., Tikhonova, L.D., Yarmukhametova, E.G., Method for Deriving Carboxymethyl Cellulose. *Eurasian Chem. Technol. J.*, 3, 1, 49–53, Mar. 2001, doi: 10.18321/ECTJ416.
74. Mohkami, M. and Talaeipour, M., Investigation of the chemical structure of carboxylated and carboxymethylated fibers from waste paper *via* XRD and FTIR analysis. *Bioresources*, 6, 2, 1988–2003, Apr. 2011, doi: 10.15376/biores.6.2.1988-2003.
75. Mondal, M.I.H., Yeasmin, M.S., Rahman, M.S., Preparation of food grade carboxymethyl cellulose from corn husk agrowaste. *Int. J. Biol. Macromol.*, 79, 144–150, Aug. 2015, doi: 10.1016/j.ijbiomac.2015.04.061.
76. Salama, A., Etri, S., Mohamed, S.A.A., El-Sakhawy, M., Carboxymethyl cellulose prepared from mesquite tree: New source for promising nanocomposite materials. *Carbohydr. Polym.*, 189, 138–144, Jun. 2018, doi: 10.1016/J.CARBPOL.2018.02.016.
77. Alatawi, F.S., Monier, M., Elsayed, N.H., Amino functionalization of carboxymethyl cellulose for efficient immobilization of urease. *Int. J. Biol. Macromol.*, 114, 1018–1025, Jul. 2018, doi: 10.1016/J.IJBIOMAC.2018.03.142.
78. Singh, V., Joshi, S., Malviya, T., Carboxymethyl cellulose-rosin gum hybrid nanoparticles: An efficient drug carrier. *Int. J. Biol. Macromol.*, 112, 390–398, Jun. 2018, doi: 10.1016/J.IJBIOMAC.2018.01.184.
79. Zulkifli, F.H., Akma, N., Rani, M., Shahitha, F., Carboxymethyl cellulose nanofibres impregnated with silver nanoparticles for tissue engineering applications. *Materials Today: Proceedings*, 2019, [Online]. Available: www.sciencedirect.comwww.materialstoday.com/proceedings2214-7853.
80. Capanema, N.S.V., *et al.*, Bioengineered carboxymethyl cellulose-doxorubicin prodrug hydrogels for topical chemotherapy of melanoma skin cancer. *Carbohydr. Polym.*, 195, 401–412, Apr. 2018, doi: 10.1016/J.CARBPOL.2018.04.105.
81. Capanema, N.S.V., Mansur, A.A.P., de Jesus, A.C., Carvalho, S.M., de Oliveira, L.C., Mansur, H.S., Superabsorbent crosslinked carboxymethyl cellulose-PEG hydrogels for potential wound dressing applications. *Int. J. Biol. Macromol.*, 106, 1218–1234, Jan. 2018, doi: 10.1016/j.ijbiomac.2017.08.124.
82. Dong, B. and Hadinoto, K., Carboxymethyl cellulose is a superior polyanion to dextran sulfate in stabilizing and enhancing the solubility of amorphous drug-polyelectrolyte nanoparticle complex. *Int. J. Biol. Macromol.*, 139, 500–508, Oct. 2019, doi: 10.1016/j.ijbiomac.2019.08.023.
83. Wang, F., *et al.*, Preparation and characterization of carboxymethyl cellulose containing quaternized chitosan for potential drug carrier. *Int. J. Biol. Macromol.*, 154, 1392–1399, Jul. 2020, doi: 10.1016/j.ijbiomac.2019.11.019.
84. Yang, H., *et al.*, Stable and biocompatible hydrogel composites based on collagen and dialdehyde carboxymethyl cellulose in a biphasic solvent system. *Carbohydr. Polym.*, 222, 114974, 2019.

85. Oun, A.A. and Rhim, J.W., Characterization of carboxymethyl cellulose-based nanocomposite films reinforced with oxidized nanocellulose isolated using ammonium persulfate method. *Carbohydr. Polym.*, 174, 484–492, Oct. 2017, doi: 10.1016/j.carbpol.2017.06.121.
86. Suppiah, K., Leng, T.P., Husseinsyah, S., Rahman, R., Keat, Y.C., The mechanical properties of carboxymethyl cellulose (Cmc) filled halloysite nanotube (hnt) bio-nanocomposite films: Effect of sodium dodecyl sulfate (sds). *Solid State Phenom.*, 280 SSP, 335–339, 2018, doi: 10.4028/WWW.SCIENTIFIC.NET/SSP.280.335.
87. Oun, A.A. and Rhim, J.W., Preparation of multifunctional carboxymethyl cellulose-based films incorporated with chitin nanocrystal and grapefruit seed extract. *Int. J. Biol. Macromol.*, 152, 1038–1046, Jun. 2020, doi: 10.1016/j.ijbiomac.2019.10.191.
88. Sadeghi, S., Nourmohammadi, J., Ghaee, A., Soleimani, N., Carboxymethyl cellulose-human hair keratin hydrogel with controlled clindamycin release as antibacterial wound dressing. *Int. J. Biol. Macromol.*, 147, 1239–1247, Mar. 2020, doi: 10.1016/j.ijbiomac.2019.09.251.
89. Rakhshaei, R., Namazi, H., Hamishehkar, H., Rahimi, M., Graphene quantum dot cross-linked carboxymethyl cellulose nanocomposite hydrogel for pH-sensitive oral anticancer drug delivery with potential bioimaging properties. *Int. J. Biol. Macromol.*, 150, 1121–1129, May 2020, doi: 10.1016/j.ijbiomac.2019.10.118.
90. Bang, S., Ko, Y.G., Il Kim, W., Cho, D., Park, W.H., Kwon, O.H., Preventing postoperative tissue adhesion using injectable carboxymethyl cellulose-pullulan hydrogels. *Int. J. Biol. Macromol.*, 105, Pt 1, 886–893, Dec. 2017, doi: 10.1016/J.IJBIOMAC.2017.07.103.
91. Cai, X., *et al.*, Transglutaminase-catalyzed preparation of crosslinked carboxymethyl chitosan/carboxymethyl cellulose/collagen composite membrane for postsurgical peritoneal adhesion prevention. *Carbohydr. Polym.*, 201, 201–210, Dec. 2018, doi: 10.1016/J.CARBPOL.2018.08.065.
92. He, X., *et al.*, A porous collagen-carboxymethyl cellulose/hydroxyapatite composite for bone tissue engineering by bi-molecular template method. *Int. J. Biol. Macromol.*, 137, 45–53, Sep. 2019, doi: 10.1016/j.ijbiomac.2019.06.098.
93. Matinfar, M., Mesgar, A.S., Mohammadi, Z., Evaluation of physicochemical, mechanical and biological properties of chitosan/carboxymethyl cellulose reinforced with multiphasic calcium phosphate whisker-like fibers for bone tissue engineering. *Mater. Sci. Eng., C*, 100, 341–353, Jul. 2019, doi: 10.1016/j.msec.2019.03.015.
94. Miles, K.B., Maerz, T., Matthew, H.W.T., Scalable MSC-derived bone tissue modules: *In vitro* assessment of differentiation, matrix deposition, and compressive load bearing. *Acta Biomater.*, 95, 395–407, Sep. 2019, doi: 10.1016/j.actbio.2019.01.014.
95. Tamburaci, S., Kimna, C., Tihminlioglu, F., Bioactive diatomite and POSS silica cage reinforced chitosan/Na-carboxymethyl cellulose polyelectrolyte

scaffolds for hard tissue regeneration. *Mater. Sci. Eng., C*, 100, 196–208, Jul. 2019, doi: 10.1016/j.msec.2019.02.104.

96. Qi, P., *et al.*, Fabrication of calcium phosphate-loaded carboxymethyl cellulose non-woven sheets for bone regeneration. *Carbohydr. Polym.*, 189, 322–330, Jun. 2018, doi: 10.1016/J.CARBPOL.2018.02.050.
97. Hasan, A., Waibhaw, G., Saxena, V., Pandey, L.M., Nano-biocomposite scaffolds of chitosan, carboxymethyl cellulose and silver nanoparticle modified cellulose nanowhiskers for bone tissue engineering applications. *Int. J. Biol. Macromol.*, 111, 923–934, May 2018, doi: 10.1016/J.IJBIOMAC.2018.01.089.
98. Maver, U., *et al.*, Carboxymethyl cellulose/diclofenac bioactive coatings on AISI 316LVM for controlled drug delivery, and improved osteogenic potential. *Carbohydr. Polym.*, 230, Feb. 2020, doi: 10.1016/j.carbpol.2019.115612.
99. Sarkar, C., *et al.*, One pot synthesis of carbon dots decorated carboxymethyl cellulose- hydroxyapatite nanocomposite for drug delivery, tissue engineering and Fe3+ ion sensing. *Carbohydr. Polym.*, 181, 710–718, Feb. 2018, doi: 10.1016/J.CARBPOL.2017.11.091.
100. Li, L., Wang, F., Shao, Z., Liu, J., Zhang, Q., Jiao, W., Chitosan and carboxymethyl cellulose-multilayered magnetic fluorescent systems for reversible protein immobilization. *Carbohydr. Polym.*, 201, 357–366, Dec. 2018, doi: 10.1016/j.carbpol.2018.08.088.
101. Adeleke, O.A., Premium ethylcellulose polymer based architectures at work in drug delivery. *Int. J. Pharm. X*, 1, Dec. 2019, doi: 10.1016/j.ijpx.2019.100023.
102. Brady, J., Drig, T., Lee, P.I., Li, J.X., Polymer properties and characterization, in: *Developing Solid Oral Dosage Forms: Pharmaceutical Theory and Practice*, Second Edition, pp. 181–223, Elsevier Inc., Amsterdam, Netherlands, 2017, doi: 10.1016/B978-0-12-802447-8.00007-8.
103. Dürig, T. and Karan, K., Binders in wet granulation, in: *Handbook of Pharmaceutical Wet Granulation: Theory and Practice in a Quality by Design Paradigm*, pp. 317–349, Jan. 2018, doi: 10.1016/B978-0-12-810460-6.00010-5.
104. Ksenofontov, A.A., Guseva, G.B., Stupikova, S.A., Antina, E.V., Novel Zinc(II) Bis(Dipyrromethenate)-Doped Ethyl Cellulose Sensors for Acetone Vapor Fluorescence Detection. *J. Fluoresc.*, 28, 2, 477–482, Mar. 2018, doi: 10.1007/S10895-018-2220-9.
105. Djerafi, R., Masmoudi, Y., Crampon, C., Meniai, A., Badens, E., Supercritical anti-solvent precipitation of ethyl cellulose. *J. Supercrit. Fluids*, 105, 92–98, Oct. 2015, doi: 10.1016/J.SUPFLU.2015.02.033.
106. Shin, T.H., *et al.*, Formulation and *in vivo* pharmacokinetic evaluation of ethyl cellulose-coated sustained release multiple-unit system of tacrolimus. *Int. J. Biol. Macromol.*, 109, 544–550, Apr. 2018, doi: 10.1016/j.ijbiomac.2017.12.111.
107. Paulo, F. and Santos, L., Inclusion of hydroxytyrosol in ethyl cellulose microparticles: *In vitro* release studies under digestion conditions. *Food Hydrocolloids*, 84, 104–116, Nov. 2018, doi: 10.1016/j.foodhyd.2018.06.009.

108. Yang, Y., Wang, H., Li, H., Ou, Z., Yang, G., 3D printed tablets with internal scaffold structure using ethyl cellulose to achieve sustained ibuprofen release. *Eur. J. Pharm. Sci.*, 115, 11–18, Mar. 2018, doi: 10.1016/J.EJPS.2018.01.005.
109. Godakanda, V.U., *et al.*, Tunable drug release from blend poly(vinyl pyrrolidone)-ethyl cellulose nanofibers. *Int. J. Pharm.*, 562, 172–179, May 2019, doi: 10.1016/j.ijpharm.2019.03.035.
110. Mahant, S., Kumar, S., Nanda, S., Rao, R., Microsponges for dermatological applications: Perspectives and challenges. *Asian J. Pharm. Sci.*, May 01, 2020, Shenyang Pharmaceutical University. doi: 10.1016/j.ajps.2019.05.004.
111. Kothai, S. and Umamaheswari, R., Formulation and evaluation of copper nanoparticles loaded microsponges. *Int. J. Pharm. Sci. Drug Res.*, 141–146, Jul. 2019, doi: 10.25004/IJPSDR.2019.110407.
112. Shokri, J. and Adibki, K., Application of Cellulose and Cellulose Derivatives in Pharmaceutical Industries, in: *Cellulose - Medical, Pharmaceutical and Electronic Applications*, InTech, Rijeka, Croatia, 2013, doi: 10.5772/55178.
113. Ahmed, O.A.A. and Badr-Eldin, S.M., Development of an optimized avanafil-loaded invasomal transdermal film: *Ex vivo* skin permeation and *in vivo* evaluation. *Int. J. Pharm.*, 570, Oct. 2019, doi: 10.1016/j.ijpharm.2019.118657.
114. Owusu, F.W.A., Boakye-Gyasi, M.E., Mante, P.K., Ekuadzi, E., Ofori-Kwakye, K., Woode, E., Formulation and evaluation of sustained release matrix tablets of capparis erythrocarpos roots extract to improve patient compliance in management of arthritis. *Sci. Afr.*, 6, Nov. 2019, doi: 10.1016/j.sciaf.2019.e00172.
115. Matamoros-Veloza, A., Hossain, K.M.Z., Scammell, B.E., Ahmed, I., Hall, R., Kapur, N., Formulating injectable pastes of porous calcium phosphate glass microspheres for bone regeneration applications. *J. Mech. Behav. Biomed. Mater.*, 102, Feb. 2020, doi: 10.1016/j.jmbbm.2019.103489.
116. López de Dicastillo, C., Bustos, F., Guarda, A., Galotto, M.J., Cross-linked methyl cellulose films with murta fruit extract for antioxidant and antimicrobial active food packaging. *Food Hydrocolloids*, 60, 335–344, Oct. 2016, doi: 10.1016/j.foodhyd.2016.03.020.
117. Camacho, K.M., Kumar, S., Menegatti, S., Vogus, D.R., Anselmo, A.C., Mitragotri, S., Synergistic antitumor activity of camptothecin-doxorubicin combinations and their conjugates with hyaluronic acid. *J. Controlled Release*, 210, 198–207, Jun. 2015, doi: 10.1016/J.JCONREL.2015.04.031.
118. Chen, Z., He, N., Chen, M., Zhao, L., Li, X., Tunable conjugation densities of camptothecin on hyaluronic acid for tumor targeting and reduction-triggered release. *Acta Biomater.*, 43, 195–207, Oct. 2016, doi: 10.1016/J.ACTBIO.2016.07.020.
119. Quiñones, J.P., Mardare, C.C., Hassel, A.W., Brüggemann, O., Testosterone- and vitamin-grafted cellulose ethers for sustained release of camptothecin. *Carbohydr. Polym.*, 206, 641–652, Feb. 2019, doi: 10.1016/j.carbpol.2018.11.047.

120. Phanthong, P., Reubroycharoen, P., Hao, X., Xu, G., Abudula, A., Guan, G., Nanocellulose: Extraction and application. *Carbon Resour. Convers.*, 1, 1, 32–43, Apr. 2018, doi: 10.1016/J.CRCON.2018.05.004.
121. Sundaram, J., Pant, J., Goudie, M.J., Mani, S., Handa, H., Antimicrobial and Physicochemical Characterization of Biodegradable, Nitric Oxide-Releasing Nanocellulose-Chitosan Packaging Membranes. *J. Agric. Food. Chem.*, 64, 25, 5260–5266, Jun. 2016, doi: 10.1021/acs.jafc.6b01936.
122. Petersen, N. and Gatenholm, P., Bacterial cellulose-based materials and medical devices: Current state and perspectives. *Appl. Microbiol. Biotechnol.*, Sep. 2011, doi: 10.1007/s00253-011-3432-y.
123. (PDF) Microbial Cellulose: Fermentative Production and Applications. Accessed: Oct. 19, 2024. [Online]. Available: https://www.researchgate.net/publication/279896217_Microbial_Cellulose_Fermentative_Production_and_Applications.
124. Chen, L., Hong, F., Xia Yang, X., Fen Han, S., Biotransformation of wheat straw to bacterial cellulose and its mechanism. *Bioresour. Technol.*, 135, 464–468, 2013, doi: 10.1016/j.biortech.2012.10.029.
125. Hong, F. and Qiu, K., An alternative carbon source from konjac powder for enhancing production of bacterial cellulose in static cultures by a model strain Acetobacter aceti subsp. xylinus ATCC 23770. *Carbohydr. Polym.*, 72, 3, 545–549, May 2008, doi: 10.1016/j.carbpol.2007.09.015.
126. Hong, F., Zhu, Y.X., Yang, G., Yang, X.X., Wheat straw acid hydrolysate as a potential cost-effective feedstock for production of bacterial cellulose. *J. Chem. Technol. Biotechnol.*, 86, 5, 675–680, May 2011, doi: 10.1002/JCTB.2567.
127. Cavka, A., Guo, X., Tang, S.-J., Winestrand, S., Jönsson, L.J., Hong, F., Production of bacterial cellulose and enzyme from waste fiber sludge. *Biotechnol. Biofuels*, 2013, [Online]. Available: http://www.biotechnologyforbiofuels.com/content/6/1/25.
128. Liang, L., Bhagia, S., Li, M., Huang, C., Ragauskas, A.J., Cross-Linked Nanocellulosic Materials and Their Applications. *ChemSusChem*, Jan. 09, 2020, Wiley-VCH Verlag. doi: 10.1002/cssc.201901676.
129. Zhou, Y.J., Luner, P., Caluwe', P., Mechanism of Crosslinking of Papers with Polyfunctional Carboxylic Acids. *J. Appl. Polym. Sci.*, 1995.
130. He, X., *et al.*, Citric Acid as Green Modifier for Tuned Hydrophilicity of Surface Modified Cellulose and Lignin Nanoparticles. *ACS Sustainable Chem. Eng.*, 6, 8, 9966–9978, Aug. 2018, doi: 10.1021/acssuschemeng.8b01202.
131. Herrera, M.A., Mathew, A.P., Oksman, K., Barrier and mechanical properties of plasticized and cross-linked nanocellulose coatings for paper packaging applications. *Cellulose*, 24, 9, 3969–3980, Sep. 2017, doi: 10.1007/s10570-017-1405-8.
132. Coma, V., Sebti, I., Pardon, P., Pichavant, F.H., Deschamps, A., Film properties from crosslinking of cellulosic derivatives with a polyfunctional

carboxylic acid. *Carbohydr. Polym.*, 2003. [Online]. Available: www.elsevier.com/locate/carbpol.
133. Shahbazi, M., Ahmadi, S.J., Seif, A., Rajabzadeh, G., Carboxymethyl cellulose film modification through surface photo-crosslinking and chemical cross-linking for food packaging applications. *Food Hydrocolloids*, 61, 378–389, 2016, doi: 10.1016/j.foodhyd.2016.04.021.
134. Kriechbaum, K. and Bergström, L., Antioxidant and UV-Blocking Leather-Inspired Nanocellulose-Based Films with High Wet Strength. *Biomacromolecules*, 21, 5, 1720–1728, May 2020, doi: 10.1021/acs.biomac.9b01655.
135. Wu, Y., *et al.*, Green and biodegradable composite films with novel antimicrobial performance based on cellulose. *Food Chem.*, 197, 250–256, Apr. 2016, doi: 10.1016/j.foodchem.2015.10.127.
136. Beghetto, V., *et al.*, Development of 4-(4,6-dimethoxy-1,3,5-triazin-2-yl)-4-methyl-morpholinium chloride cross-linked carboxymethyl cellulose films. *Carbohydr. Polym.*, 249, Dec. 2020, doi: 10.1016/j.carbpol.2020.116810.
137. Cellulose Nanocomposites, Copyright, Foreword, pp. i–v, 2006, doi: 10.1021/bk-2006-0938.fw001.
138. Cellulose Packaging - Green Compostables. Accessed: Sep. 02, 2024. [Online]. Available: https://www.greencompostables.com/blog/cellulose-packaging.
139. Tan, H.L., Kai, D., Pasbakhsh, P., Teow, S.Y., Lim, Y.Y., Pushpamalar, J., Electrospun cellulose acetate butyrate/polyethylene glycol (CAB/PEG) composite nanofibers: A potential scaffold for tissue engineering. *Colloids Surf B Biointerfaces,* 188, 110713, 2019 Dec 9, PMID: 31884080, doi: 10.1016/j.colsurfb.2019.110713.
140. Prasad, B.S.G., *et al.*, Formulation and evaluation of transdermal patches of tolmetin sodium. *Eur. J. Biomed. Pharm. Sci.,* 5.6, 800–807, 2018.
141. Zhong, Y., Wang, J., Yuan, Z., Wang, Y., Xi, Z., Li, L., Liu, Z., Guo, X., A mussel-inspired carboxymethyl cellulose hydrogel with enhanced adhesiveness through enzymatic crosslinking. *Colloids Surf B Biointerfaces,* 179, 462–469, 2019 Jul 1, PMID: 31005741. doi: 10.1016/j.colsurfb.2019.03.044.

3

Regeneration and Derivatization of Cellulose

Popat Mohite[1*], Abhijeet Puri[1], Riteshkumar Patil[1], Monikkumar Patel[1], Yash Nanda[1], Vijaykumar Prajapati[1] and Sudarshan Singh[2,3]

[1]*AETs St. John Institute of Pharmacy and Research, Palghar, Maharashtra, India*
[2]*Department of Pharmaceutical Sciences, Faculty of Pharmacy, Chiang Mai University, Chiang Mai, Thailand*
[3]*Office of Research Administration, Chiang Mai University, Chiang Mai, Thailand*

Abstract

Cellulose and its derivatives constitute a vital class of biopolymers with widespread applications across diverse industries, driven by their renewable nature and versatile chemical structure. This book chapter explores the molecular structure, dissolution, modifications, derivatization, and sustainability aspects of cellulose and its derivatives. The chapter elucidates its molecular structure and significance in industries like pharmaceuticals. The dissolution of cellulose presents a critical challenge due to its inherent hydrogen-bonded network, yet recent advancements in solvent systems and processing techniques are paving the way for novel applications. The subsequent section highlights the chemical and physical modifications of cellulose, encompassing etherification, esterification, and the production of nanocellulose and aerogels. These modifications enhance cellulose's functionality and broaden its utility in various advanced materials. The derivatization of cellulose further expands its application spectrum, yielding derivatives like methylcellulose and hydroxyethyl cellulose that find uses in pharmaceutical formulations, food products, and personal care items. Sustainability emerges as a central theme, addressing the eco-friendly sourcing of cellulose from renewable biomass and the environmental implications of its derivatives. Biodegradability underscores the imperative for sustainable cellulose technologies. Finally, this chapter aims to provide a holistic understanding of cellulose and cellulose derivatives, emphasizing their pivotal role in advancing sustainable materials and technologies.

**Corresponding author*: mohitepb@gmail.com

Mohd Shabbir (ed.) Regenerated Cellulose: A Renewable Alternative to Petroleum-Based Plastics, (59–90)

***Keywords*:** Cellulose, biopolymers, hydrogen-bonded network, eco-friendly

3.1 Introduction

Cellulose that has undergone chemical treatment and been reshaped into films, fibers, or other shapes is referred to as regenerated cellulose. Because of its natural origin, biodegradability, and environmental benefits, it provides a sustainable substitute for plastics derived from petroleum. Approximately 40% to 50% of the planet's total biomass reserves are composed of cellulose, making it the most abundant natural polymer on earth. Due to its widespread occurrence in microbes and plants, it is a sustainable and renewable resource [1].

The significance of cellulose in natural ecosystems and its potential as a resource are:

1. Cellulose as the most abundant natural polymer
   - Global production
   - Role in photosynthesis
   - Plant biomass [2]
2. Cellulose in the natural environment
   - Storage in plants
   - Presence in microorganisms
3. Environmental sustainability of cellulose
   - Inexhaustible resource
   - Eco friendly decomposition
4. Potential as a green resource
   - Industrial uses
   - Applications of bacterial cellulose [3]

3.1.1 Structure of Cellulose

Cellulose is a natural, natural polysaccharide consisting of units of β-D-glucose, which forms the chief structural component in cell walls of plants and some microorganisms. Being the most abundant biopolymer on earth, cellulose is crucial to biomass with an amount close to 40% to 50% of plant dry weight and contributing major renewable resources. Regeneration of cellulose is the one chemically or physically transformed and reformed into fibers or films retaining its central molecular core.

- **Source:** Found primarily in plant cell walls, cellulose is extracted from wood, cotton, and other plant-based materials for industrial use.
- **Importance:** This resources can be described as renewable, biodegradable, and transformable to varied useful forms like fibers, films, and composites [4].

Cellulose is linear, unbranched, and has repeating β-D-glucose units held together by β(1→4) glycosidic bonds. Thousands of glucose monomers are contained within every cellulose chains are linear and can form strong hydrogen bonds with adjacent chains. This leads to forming of microfibrils and fibrils, which are directly responsible for the rigidity and strength of cellulose [5].

1. **Simple molecular unit**
   - *Glucose monomer*: The repeating unit in cellulose is β-D-glucose, that is, a six-carbon sugar in its pyranose (ring) form [6].
   - *Glycosidic bonds*: These glucose monomers are joined by β (1→4) glycosidic linkages, where the hydroxyl group of one glucose at the 1st carbon (C1) connects to the hydroxyl group at the 4st carbon (C4) of the next glucose unit. This linear structure places cellulose in contrast to starch, whose linkages are α(1→4).
2. **Hydrogen bonding**
   - *Intermolecular bonds*: The hydroxyl groups (–OH) in cellulose also come together to form strong hydrogen bonds across intermolecular distances with neighboring chains. These help to create microfibrils and contribute to the tensile strength of cellulose.
   - *Crystalline and amorphous regions*: Cellulose fibers consist of regions that both are crystalline and amorphous. The crystalline regions are denser and more resistant to chemical degradation, whereas the amorphous regions are relatively more open to reactions such as dissolution [7].
   - *Microfibrils*: Several cellulose chains are agglomerated by hydrogen bonding to make microfibrils, typically 3 to 20 nm in diameter, that can associate to form larger structures.

- *Fibrils*: Microfibrils aggregate to form fibrils that provide mechanical strength and resistance to hydrolysis. Fibrils are embedded in a matrix of other biopolymers, hemicellulose, and lignin within the cell wall of plant cells [8].

3. **Chemical formula:** $(C_6H_{10}O_5)n$
   - **C_6:** Presence of 6 carbon atoms
   - **H_{10}:** Presence of 10 hydrogen atoms
   - **O_5:** Presence of 5 oxygen atoms
   - **n:** Number of repeating glucose units. This number may run into hundreds or thousands depending on the source.

3.1.2 Regeneration of Cellulose

Regenerated cellulose is produced from the natural cellulose found in material such as wood pulp or cotton linters, in a process known as regenerating the dissolved cellulose back into fibers or films. Chemical solvents dissolve the cellulose for subsequent regeneration through chemical reactions, and the precipitation can take any form that can be used as a textile or a film, using either the viscose or lyocell process [9].

3.1.3 Importance in Various Industries (Paper, Textiles, Pharmaceuticals)

Regenerated cellulose is manufactured from natural sources such as wood pulp or cotton linters, which in the last decades has significantly gained relevance in several industrial sectors due to its eco-friendly nature, biodegradability, and diverse applications (Table 3.1).

Due to its versatility, sustainability, and biodegradability, regenerated cellulose has been widely used in various industries. When the mechanical property of paper products is improved within the paper industry, on the other hand, fibers such as rayon and Lyocell offer environment-friendly alternatives to synthetic fibers for the textile industry [13]. The pharmaceutical industry uses cellulose derivatives and hydrogels for drug delivery, wound care, and excipients in formulations. Thus, its wide-range applicability underlines value as a renewable alternative to petroleum-based plastics [14].

3.1.4 Types of Cellulose Derivatives and Derivatization of Cellulose

Structures and forms of cellulose: The degree of polymerization, which varies depends on the cellulose source, determines many features of cellulose.

Table 3.1 Importance in various industries.

Industry	Applications	Advantages	References
Paper industry	• Cellulose fibers: used in paper production to improve strength and durability. • Cellophane: a regenerated cellulose film used in packaging and wrapping paper.	• Sustainability Biodegradable and sourced from renewable resources. • Enhanced properties Improved Mechanical strength Tear strength Print quality.	[10]
Textile industry	• Rayon (Viscose): used in apparel, upholstery, and household textiles. • Lyocell: an eco-friendly alternative used in high-performance textiles. • Modal: used in luxury fabrics like bed linens, towels, and intimate wear.	• Biodegradability: eco-friendly, unlike synthetic fibers. • Comfort and versatility: breathable, moisture-absorbent, and soft. • Sustainability: lyocell has low environmental impact due to closed-loop production.	[11]
Pharmaceutical industry	• Pharmaceutical Excipients: microcrystalline cellulose and carboxymethyl cellulose used as fillers and binders in tablets. • Hydrogels and drug delivery: used in controlled drug release systems and wound dressings. • Bacterial cellulose: applied in wound care products for its healing properties.	• Biocompatibility: safe for medical use. • Controlled drug release: Enables efficient, controlled drug release. • Sustainability: renewable and biodegradable alternative to synthetic polymers.	[12]

The significant intra and intermolecular hydrogen bonding between individual cellulose chains is caused by the high amount of hydroxyl groups on the glucose rings along the skeleton. This causes many cellulose chains to crystallize into insoluble microfibrils and two structural areas, crystalline and amorphous, which are responsible for cellulose's exceptional strength, stiffness, durability and biocompatibility. The hydrophilicity, chirality and biodegradability of cellulose are attributed to the three hydroxyl groups that are present in each monomeric unit and their outstanding reactivity. When part or all the hydroxyl groups on the repeating units in the cellulose react, these hydroxyl groups also help to modify cellulose chemically by and produce cellulose derivatives. This is made feasible by the abundance of hydroxyl groups, which may be functionalized chemically by cyanoethylation, carboxymethylation, etherification, and hydroxypropylation, among other processes. Because of the modified qualities, the resulting cellulose derivatives could be more beneficial than cellulose itself. Cellulose derivatives can be categorized based on the way they are made and the substituents used, such as ether-methyl-cellulose/carboxy-methyl-cellulose by etherification and ester-cellulose–acetate through esterification. Cellulose undergoes chemical processing to produce monomers that are produced from it. This form may be produced by breaking down cellulose into glucose, which can then be transformed into a variety of high value compounds, including sorbitol, lactic acid (LA), levulenic acid (LevA), and S-hydroxymethylfurfural (5-HMF), which are crucial bio-based polymer platforms.

Types of cellulose derivatives: The creation of diverse drug delivery systems frequently makes use of cellulose and its derivatives, including cellulose esters, ethers, and oxycellulose (Figure 3.1). These fall into five main derivatives: sodium-carboxyl-methyl-cellulose (CMC), hydroxy-propyl-cellulose (HPC), methyl cellulose (MC), ethyl cellulose (EC), and hydroxy-propyl-methyl-cellulose (HPMC) [15].

Synthesis methods and conditions

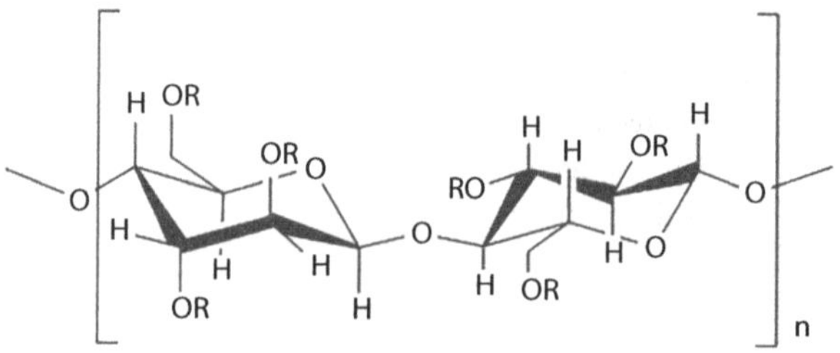

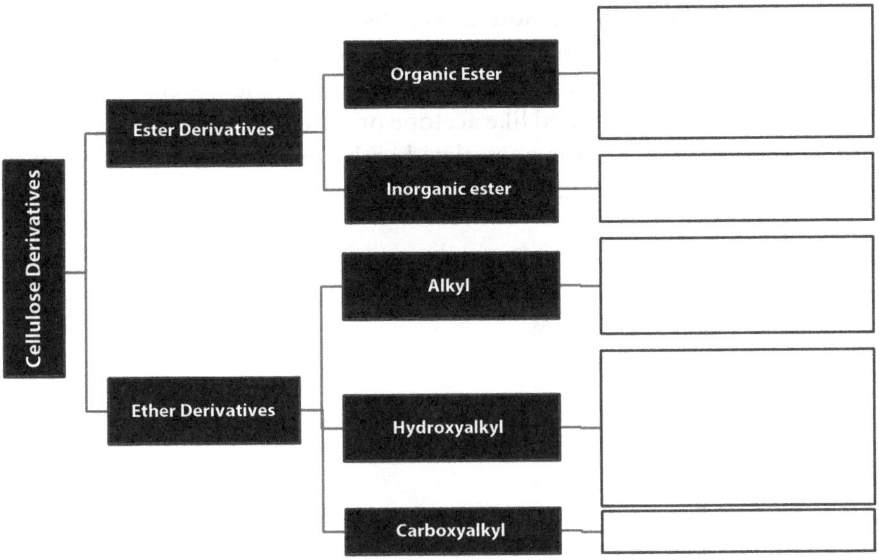

Ethyl cellulose: Most organic solvents require a degree of substitution between 2.2 and 2.6 to achieve solubility. Ethyl cellulose is biodegradable material with outstanding film forming ability, barrier forming qualities, water resistance and no water solubility and toxicity. At a normal temperature ethyl cellulose is very strong and as the temperature rises, this strength rapidly diminishes [16].

The reaction between ethyl chloride and cellulosic alkali yields, ethyl cellulose were 'R' is the cellulosic radical [17]. The chemical structure of ethyl cellulose is based on converting some of the hydroxyl groups on the repeating glucose units into ethyl ether groups. Alkali cellulose and ethyl chloride are reacted for several hours at around 60°C to produce ethyl cellulose [16].

$$R{-}{-}{-}OH + NaOH \longrightarrow R{-}{-}{-}ONa + H_2O$$

$$R{-}{-}{-}ONa + CH_2CH_3Cl \longrightarrow R{-}{-}{-}O{-}{-}{-}CH_2CH_3 + NaCl$$

NaOH, 27 ml of acetone was added dropwise as a solvent. The reaction was conducted at 50°C with the addition of 9 ml of dimethyl sulphide dropwise. After reaction was carried out for 1 hour the, the system was filtered and new reagents were added like acetone or dimethyl sulphide at the same ratios. Then the system was neutralized with 10% acetic acid and washed with acetone three times followed by drying.

$$R\text{---}OH + NaOH \longrightarrow R\text{---}ONa + H_2O$$

$$R\text{---}ONa + CH_3Cl \longrightarrow R\text{---}O\text{---}CH_3 + NaCl$$

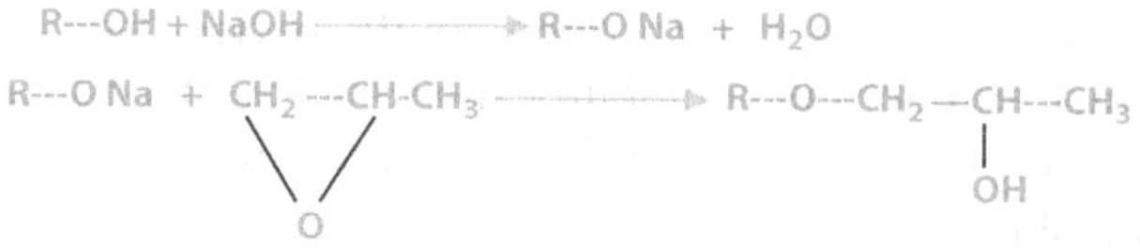

function as an anti-foaming agent is one of the beneficial qualities. Under the regulated medium alkali cellulose reacts with sodium monochloro-acetate or monochloroacetic acid to form carboxyl-methyl-cellulose, where 'R' stands for cellulose radical and R-O-CH_2-COONa stands for sodium-carboxy-methyl-cellulose [17]. Using sugarcane straw cellulose, ethanol, water and NaOH combined in three neck flasks. Mixture agitated at 15±5°C for 1 hour. After alkalinization, ethanol and monochloroacetic acid were added and temperature was then raised to 60°C and continued for several hours. Process ended with filtration and washing with water and methanol to dry at room temperature [19].

$$R\text{---}ONa + Cl\text{---}CH_2\text{---}COONa \longrightarrow R\text{---}O\text{---}CH_2\text{---}COONa + NaCl$$

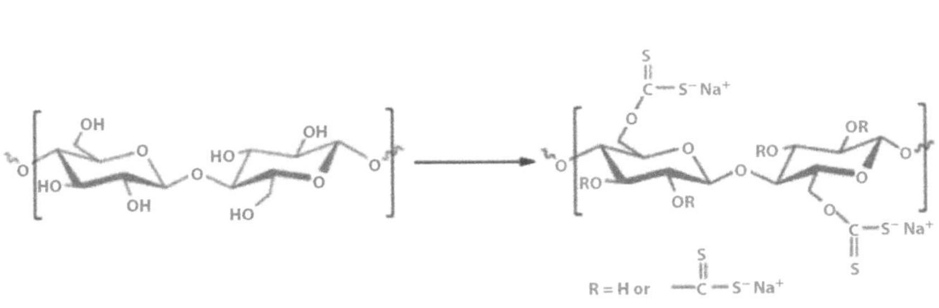

in the process [21]. The substitution of different functional groups at the hydroxyl groups of each glucose ring results in the cellulose derivatives, which becomes less crystalline [18]. Nitrating mixture ($H_2SO_{4\,+}$ HNO_3) is used along with cellulose and glycerol to produce cellulose nitrate, which replaces hydroxyl group with nitrate groups. Lowering nitration degree and adding plasticizers create a versatile plastic material. It is key ingredient in smokeless gun powder. The use of sulphuric acid stabilizes the, nitration mixture [22].

$$HNO_3 + C_6H_{10}O_5 \xrightarrow{H_2SO_4} C_6H_2(NO_2)_3 + 3H_2O$$

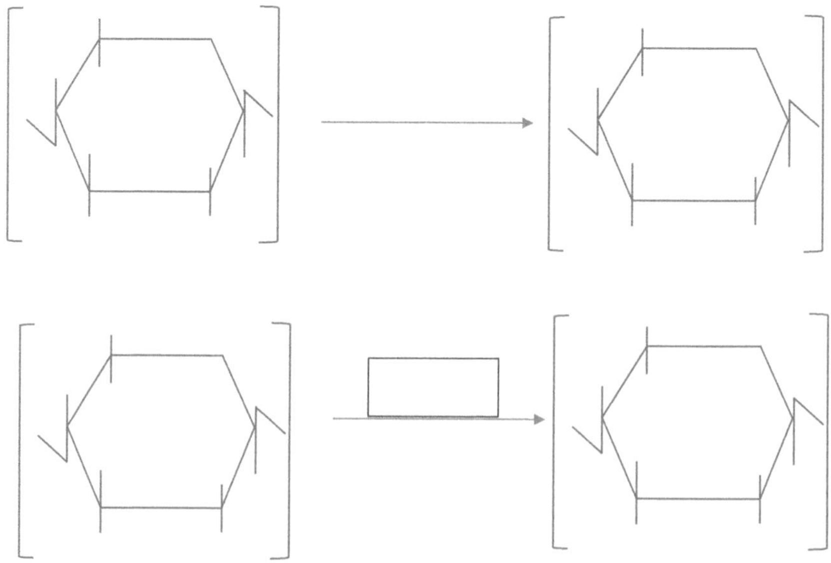

Since cellulose acetate is derived frequently from agricultural byproducts including cotton burrs, sugarcane bagasse hence its reasonably inexpensive. Some of the applications in the biomedical field such as tissue engineering, wound healing and drug delivery system benefit from the fact that current Cellulose acetate preparation techniques do not require additional chemicals or mechanical treatment to separate the remaining cellulose from other components [16].

Cellulose Sulphate:

For activation, 2 gm of dried microcrystalline cellulose was suspended in 10 ml of anhydrous dimethylformamide (DMF) and swirled at room temperature for the whole night. The sulphation agent, a compound of DMF and chlorsulphonic acid, was made by gradually adding 40 ml of chlorsulphonic acid to 360 ml of DMF over the course of 30 to 45 minutes, stirring constantly, and cooling in an ice bath with a temperature maintained between 5°C and 10°C. Sulphation was accomplished by vigorously swirling MCC while adding sulphation agent continuously. The reaction was stopped after 3 hours at thirty degrees Celsius by adding the reaction mixture to 3 litres of saturated ethanolic sodium acetate solution. To acquire raw materials, the precipitate was recovered by centrifugation and then washed with anhydrous alcohol (Figure 3.4) [16, 24].

Functional properties and applications in various fields [16].

1. **Application of cellulose and its derivatives in the pharmaceutical field** [25]
   - **Bioadhesives and Mucoadhesives:** Due to ability to stick to skin and mucus membranes, the medication action is prolonged at site of absorption.
   - **Coating of Drugs:** They are used to coat the drug in solid dosage forms like tablets and pellets, preventing the degradation, regulate drug release and mask bitter taste. Cellulose acetate phthalate is used as they provide acid resistance and release drug in more neutral and alkaline environment.
   - **Osmotic Pressure:** Osmotic delivery devices used to control drug release. Cellulose acetate is used as a semi-permeable membrane is such system.
   - **Gelling Agents:** They are frequently employed as gelling agent in topical formulation or as a building block for more complex drug delivery scheme like liposomes.

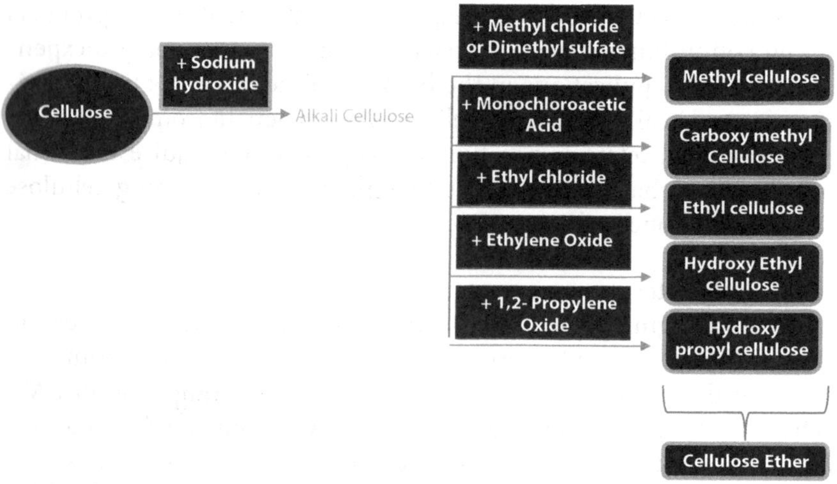

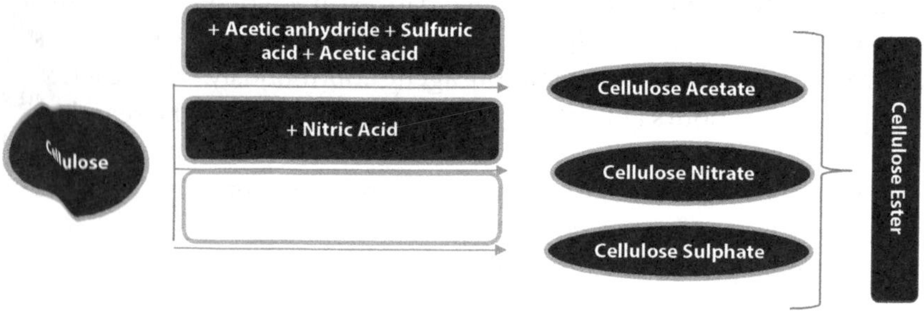

- **Thickeners and stabilizers:** Cellulose ether enhances the product stability and enhances the viscosity of suspensions and emulsions with an aid of consistent dispersion.
- **Binders:** They are used in wet granulation process as binder, making sure that contents come together and form a tablet. Microcrystalline cellulose and HPMC are employed in various strength to improve the mechanical strength of granules.
- **Disintegrating agents**: Tablets containing cellulose derivatives, cross-linked with sodium CMC, designed to break down rapidly in stomach, facilitating faster drug release and absorption.

2. **Application of cellulose in biomedical field**
   Due to the biocompatibility, durability, stability under physiological conditions, bacterial nanocellulose (BNC) has significant application in biomedical field.
   - **Cardiovascular implants:** Since lowers the danger of thrombosis and occlusion, bacterial nanocellulose is advantageous over the conventional materials when used in artificial blood vessels. The anticoagulant qualities of heparin-coated BNC scaffolds make them a suitable material for vascular tissue engineering. BNC may also be used in heart valve replacement procedures.
   - **Bone and connective tissue repair:** The BNC promotes osteoblast and chondroblast development. It has potential for bone tissue regeneration. The development of a BNC and hydroxyapatite membrane has been shown to enhance osteoblast proliferation, alkaline phosphatase activity and bone healing is increased by the HAP crystals in it, which replaces the structure of genuine bone [26].

3.2 Modifications of Cellulose

Chemical modifications:
Numerous chemical changes have been made to cellulose in an effort to maximize its potential in particular fields of interest. The modification ranged from simple molecules to polymers, and they either changed the chemistry or added functional groups to its surface. Over century, aliphatic and aromatic mono- and di-isocyanates have been among the various compounds employed to alter cellulose.

Methods and mechanisms:
Cellulose modification by etherification:
Cellulose etherification modifies cellulose to create ethers used in food, pharmaceuticals and construction. The main types include ionic ethers (Carboxymethyl cellulose), non-ionic ethers (methyl and ethyl cellulose), and silyl ethers (trimethylsilyl cellulose). These ethers are produced through heterogeneous or homogeneous reaction, often using a slurry technique that involves treating swollen cellulose with NaOH solution and an organic solvent for effective modification [27].

Various methods are used for the etherification of α-cellulose like cyanoethylation, carboxymethylation, and hydroxypropylation. Cyanoethylation uses a Michael addition reaction with acrylonitrile in an alkaline medium. Carboxymethylation involves two steps, alkalization and etherification, with optimized for temperature and reactant concentrations. Hydroxypropylation also consists of two steps, adjusting NaOH and propylene oxide levels. These cellulose derivatives are useful for chiral separation in HPLC [40].

Microcrystalline cellulose was dissolved in pyridine and reacted with triphenylchloromethane and 3,5-dimethylphenyl isocyanate to form cellulose-2,3-bis(3,5-dimethylphenylcarbamate) (CBDMPC). This product was further treated with various reagents in THF to form cellulose-2,3-bis(3,5-dimethylphenylcarbamoyl)-6-O-methacrylate (CBDMPC-MA). The derivative was immobilized onto hybrid silica monolith particles, achieving 83% immobilization [28]. Cellulose reacts with triphenylchloromethane (trityl chloride) to selectively modify the primary hydroxyl group at the O-6 position, facilitating research on cellulose derivatives (Figure 3.5). This method has led to the synthesis of compound like 2,3-di-O-methyl-cellulose and 2,3-di-O-ethyl-cellulose [29].

Cellulose modification by esterification: Esterification of cellulose enhances its properties, making it suitable for various applications. This process introduces ester groups, improving solubility, thermal stability, and compatibility. Cellulose esters, created by reacting cellulose with organic acids or anhydrides, are used in textiles, pharmaceutical, coatings, and biodegradable plastics. This underscores the importance of cellulose esterification in developing sustainable, high-performance materials [30].

Employing 1,5-diazabicyclo[4.3.0]non-5-ene acetate as a solvent facilitates efficient and recyclable cellulose acetylation, resulting in low boiling by products. This technique achieves various substitution level (DS) using vinyl carboxylates without the need for additional catalysts, showcasing high DS values and recovery of high purity ionic liquid [31]. A study

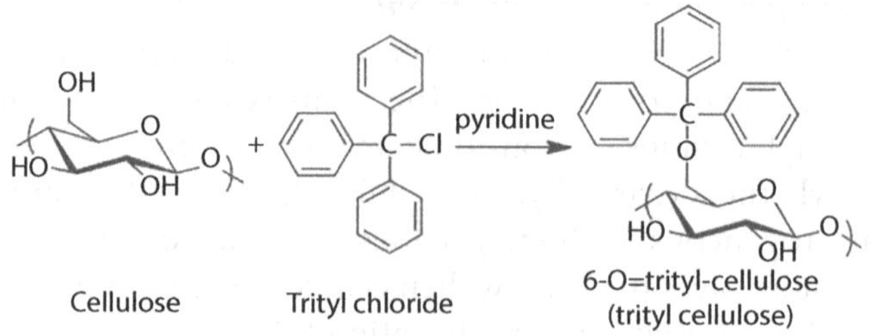

Figure 3.5 Preparation of 6-O-trityl cellulose.

on rayon cellulose esterification with acetic anhydride in lithium chloride-N,N-dimethylacetamide (LiCl-DMAc) solvent, employing p-toluenesulfonyl chloride (p-TsCl) and pyridine as catalysts. Complete substitution occurred at 60°C to 70°C in 10 to 12 hours, with pyridine being more effective. Cellulose degradation was noted at 70°C, while optimal conditions for full acetyl substitution were 50°C to 60°C over 10 hours. p-TsCl may be preferable for higher anhydrides [32]. A study modified cellulose fibers with fatty acids while maintain their structure. It found that longer reaction times and solvent swelling increased the degree of substitution (DS), while longer fatty acid chains decreased DS, leading to greater hydrophobicity and reduced thermal stability Fatty acyl chlorides were produced by reacting fatty acids with thionyl chloride at 115°C. The resulting esterified fibers were filtered, washed and dried [33].

Grafting modification of cellulose: Cellulose, the most abundant natural polymer, has limited application due to poor solubility. Graft modification enhances its functionality, with recent advances in cellulose graft copolymers using controlled radical polymerization enabling well-defined side chains (Figure 3.6). These modifications allow for application in drug delivery and antimicrobial materials. However, challenges in scaling up production and reducing costs persist [34]. Two main grafting approaches are "grafting onto" and "grafting from." Techniques like atom transfer radical polymerization (ATRP) it is first modified to attach initiator groups, with the degree of substitution influencing graft density. Various monomers can then be polymerized from the cellulose backbone under controlled conditions [35].

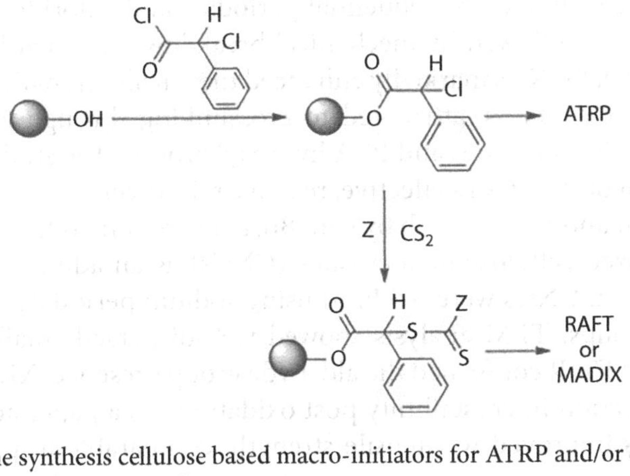

Figure 3.6 The synthesis cellulose based macro-initiators for ATRP and/or chain transfer agent for RAFT and MADIX polymerization.

Reversible addition–fragmentation chain transfer (RAFT) polymerization modifies cellulose by attaching chain transfer agents, influencing grafting. Methods like nitroxide-mediated polymerization (NMP) and single electron transfer living radical polymerization (SET-LRP) provide better control over polymer architecture. Radiation-induced grafting uses ionizing radiation to introduce functionalities by grafting vinyl monomers. Combining radiation with RAFT enhances nanocellulose properties for nanocomposites [36].

The research by Kelly *et al.*, introduces a green method modifies cellulose nanofibrils (CNF) using grafting-through polymerization with various monomers in water. This enhances properties like water absorption and temperature responsiveness while using less polymer. Microscopy revealed changes in morphology and particle size. This technique shows promise for advanced materials and biomedical application [37].

Modification by oxidation and crosslinking:
Oxidative modification adds functional groups like carbonyls and carboxyls to cellulose to cellulose enhancing its reactivity for further processing. Common oxidizing agents include hydrogen peroxide, periodate, and TEMPO. Crosslinking strengthens cellulose by forming covalent bonds between chains, improving durability, thermal stability and chemical resistance. This is achieved using epoxides, citric acid or multifunctional aldehydes. These modifications expand cellulose's use in biomedicine, textiles and environmental engineering for sustainable applications [38].

In a study by Juho Antti, *et al.* investigates the use of bifunctional reactive cellulose nanocrystals (RCNCs), containing carboxyl and aldehyde groups, as reinforcing agents for polyvinyl alcohol (PVA) films. The RCNCs were generated by sequential periodate and chloride oxidation of cellulose fibers, followed by mechanical breakdown. Even at low concentration (0.5%), RCNCs markedly enhanced the tensile strength and young's modulus of PVA films, attributed to crosslinking through acetal bonds between RCNC aldehydes and PVA hydroxyl groups. The study showcases the potential of RCNCs as effective, reactive reinforcement agents for PVA films [39]. In another research by sun, Bo, *et al.* examined the creation and use of oxidized cellulose nanocrystals (CNCs) as an additive to enhance paper strength. CNCs were oxidized using sodium periodate, introducing aldehyde groups. TEM analysis showed well-dispersed smaller oxidized CNCs, while FTIR confirmed the aldehyde group presence. XRD indicated a slight reduction in crystallinity post oxidation. As a paper additive, oxidized CNCs improved wet tensile strength, comparable to polyethylenimine [40].

Carbanilation

Cellulose carbanilation modifies cellulose by attaching carbanilate groups to its structure. This is usually achieved by reacting cellulose with isocyanates or carbonyl chloride alongside a base. The modified cellulose shows enhanced reactivity and better compatibility with various materials [41]. A study by Schlufer *et al.* reveals that bacterial cellulose (BC) with high degree of polymerization can be efficiently dissolved and chemically modified using the ionic liquid 1-N-butyl-3-methylimidazolium chloride. They had achieved homogenous acetylation and carbanilation of cellulose, creating derivatives with high degrees of substitution under mild conditions. They synthesized bacterial cellulose with degree of substitution up to 2.25. This method paves the way for innovative bacterial cellulose structural designs through homogenous phase chemistry enabling the creation of novel cellulose derivatives [42]. In another research examined the oxidative impact of DMSO-based carbanilation mixtures on cellulose, revealing that these mixtures create sulfonium ylide intermediates leading to oxidation and potential degradation of cellulose chain. Through the use of model compounds and trapping agents, researchers identified these oxidizing species and their ability to add carbonyl groups to the cellulose. They devised a color test using a naphthoquinone derivative to detect oxidizing sulfonium species in carbanilation mixtures. This study provide insights into cellulose degradation mechanism during carbanilation and provide practical way to evaluate the oxidative risks of DMSO-based reaction for cellulose modification [43].

Cellulose in ionic liquids (IL) by esterification and acylation:

Ionic liquids are being utilized in innovative cellulose processing technologies to create new materials like fibers, blends, composites, and ion gels, potentially addressing environmental concerns and replacing petroleum-based products [44]. Wu *et al.*, in this study, shows that cellulose may be acetylated uniformly in AIMIMCL without the need for catalysts, yielding cellulose acetates with a broad range of substitution. The reaction is fast and can be controlled by stoichiometric technique, temperature, and reaction time. A significant factor in dissolving of cellulose is high chloride concentration and activity in allyl-3-methylimidazolium chloride (IL). In comparison to other ILs replaced by saturated alkyl, the IL with an alkenyl substituent on nitrogen has a melting point and great thermal stability. The study finds that homogeneous reaction condition can be used to produce one step acetylation of cellulose with a wide degree of substitution without the need for catalysts [45]. In addition to being a convenient a cellulose solvent, 1-ethyl-3-methylimidazolium acetate (EmimOAc) can be used as an activating reagent in a transesterification reaction involving hydroxyl groups in cellulose, where IPA acts as an ester-donating reagent (Figure 3.7).

Kakuchi, Ryohei *et al.*, according to the study, imidazolium-based ionic liquids (Im-ILs) with acetate anions greatly enhance transesterification processes, which has a favorable co-solvent effect on cellulose modifications in imidazolium acetate ionic liquids. Significant improvements in reaction conditions and a favourable co-solvent effect for cellulose modifications in an imidazolium acetate ionic liquid have resulted from the protocol modification of IL-catalyzed transesterification methods [46].

CNCs (Cellulose nanocrystals) were made by Gan *et al.* with a small modification by hydrolysis of native cotton linter in acid. Powdered (Acetylated CNCs) ACN was produced *via* surface acetylation in a nitrogen atmosphere. By solution blending and casting molding, the unaltered and ACNs were used to strengthen by hydrophobic polyester (PHB). The impact of surface acetylation on mechanical improvement of polyester-based composites is the main subject of the investigation [47].

A green synthetic esterification technique was used to generate long-chain esters from microcrystalline cellulose at low doses by J. Lease *et al.* This work focuses at the mechanochemical approach for esterifying microcrystalline cellulose with oleic acid in a p-Toluenesulfonyl chloride and 1-butyl-3-methylimidazolium acetate system, emphasizing the minimal catalyst and solvent needs [48] (Figure 3.8).

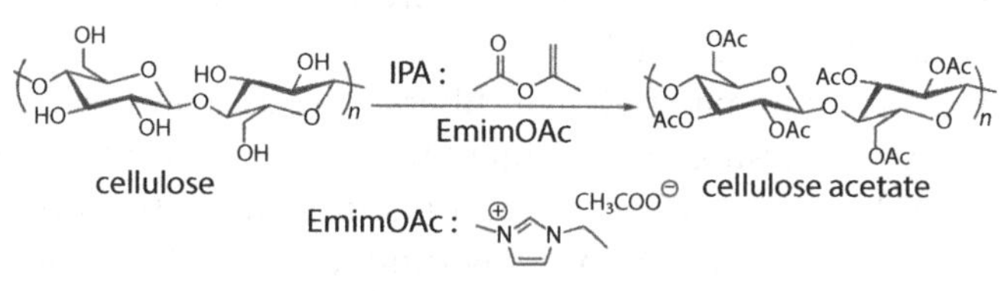

Figure 3.7 Synthesis of cellulose acetate.

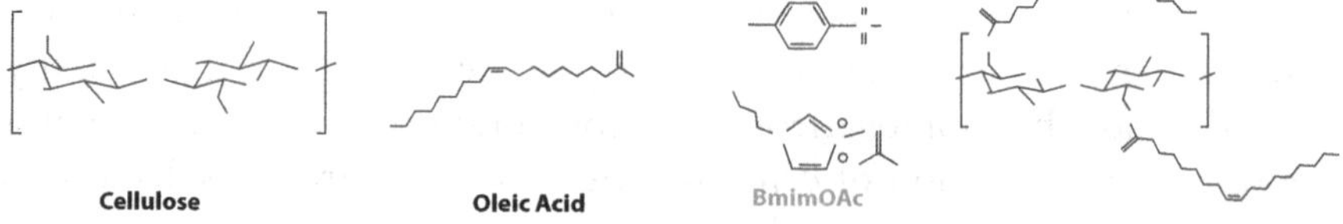

***Modification of cellulose in green solvents*:**
The modification of cellulose with green solvents has become a key area of interest in sustainable material science. These environmentally friendly solvents, including ionic liquids, deep eutectic solvents, and supercritical fluids, offer advantages such as lower toxicity and biodegradability, replacing traditional harmful organic solvents. Techniques like esterification, etherification, and crosslinking using these green solvents improve cellulose's solubility, thermal stability, and mechanical strength [49].

The study also explores use of carboxylic acid-based eutectic solvents (DESs) for producing cellulose nanofibers (CNFs) from hardwood pulp. DESs pretreated cellulose before mechanical processing, yielding CNFs under 100 nm. Higher pretreatment temperatures reduced fiber size but lowered yield. Esterfication during pretreatment improved dispersibility in polylactic acid (PLA) and enhanced composite strength. This eco-friendly approach efficiently produces functionalized CNFs [50].

The use of deep eutectic solvents promising eco-friendly solvents for processing lignocellulosic biomass. Formed by mixing hydrogen bond acceptors and donors, DESs offer advantages over traditional ionic liquids, such as easier preparation and improved biocompatibility. The DESs application in biomass fractionation, lignin extraction, cellulose modification, and converting carbohydrates to platform chemicals like 5-hydroxymethylfurfural (Table 3.2) [51].

3.3 Sustainability Aspects

Sustainable sourcing of cellulose (from wood, agricultural waste)
There are various types of sources available from which the cellulose can be produced, sources include wood, agricultural residues, wood-free biomass and cellulose wastes from factories and municipalities [64]. As the lignocellulose biomass is mostly made up of the 23% to 50% of polysaccharide cellulose and 12% to 29% of hemicellulose and also 13% to 31% of aromatic polymer lignin, cellulose is the prominent and sustainable source of polysaccharides and cellulose is the main building block of plant cell walls [65]. Cellulose from different raw material sources can be extracted and isolated from the other constituents by eliminating the hemicellulose, lignin and other contaminants. The percent amount of cellulose in distinct biomass changes according to the source of that biomass material [66]. The composition of the different sources of cellulose biomass is shown in Table 3.3.

Cellulose from wood sources is obtain from hardwood trees, such as oak, maple, and birch trees and from softwood trees like pine, spruce,

Table 3.2 Applications of modified cellulose.

Modified cellulose	Overview	Application	References
Gold nanoparticles modified cellulose, cellulose cross-linking with epichlorohydrin	Cellulose based materials used for the recovery and detection of heavy metals from wastewater	Metal Adsorption	[52]
AuNPs on cellulose paper for single molecule detection	Utilizes cellulose modified with gold nanoparticle for detecting environmental toxins at low concentrations	Surface-enhanced Raman scattering	[53]
DPPH radical scavenging assays with cellulose-AuNP composites	Cellulose composites that exhibit antioxidant properties, useful in food and health products	Antioxidant Application	[54]
Cyclodextrin glycosyltransferase and alcohol oxidase immobilized on cellulose	Cellulose and its derivatives used to immobilize enzymes for various biochemical application	Enzyme Immobilization	[55]
Enzyme immobilization using AuNPs and cellulose nanocomposites	Cellulose serves as a matrix for enzyme immobilization, allowing for the detection of biological analytes	Biosensors	[56]
Cellulose based hydrogels for drug encapsulation	Modified cellulose used for controlled drug release and targeted delivery in pharmaceutical application	Drug delivery systems	[57]
Cellulose films and coatings in food packaging	Cellulose materials used in biodegradable packaging solutions, offering mechanical and barrier properties	Food packaging	[58]

(*Continued*)

Table 3.2 Applications of modified cellulose. (*Continued*)

Modified cellulose	Overview	Application	References
Cotton, rayon, and lyocell production from cellulose	Cellulose fibers used in the textile industry for sustainable fabric production	Textiles	[59]
Cellulose acetate and other cellulose esters	Chemical modification of cellulose to produce esters for various application, improving solubility and functionality	Esterification	[29]
Cellulose carbanilates for molecular weight distribution analysis	Modification of cellulose using isocyanates to produce carbanilates for various uses	Carbanilation	[60]
Cellulose nanocomposites-reinforced polymer composites for various application	Incorporation of cellulose nanocrystals in polymer to enhance mechanical properties	Nanocomposites	[61]
Cellulose based hydrogels for wound care	Application of cellulose in drug delivery, wound care, and tissue engineering due to its compatibility with biological system	Biomedical uses	[62]
Creation of bioplastics and cellulose based composites	Development of cellulose-derived material as alternatives to petroleum-based products	Sustainable material	[63]

and fir trees, whereas additional plants whose main product is cellulose fiber, such as cotton, jute, hemp, kenaf, flax, these are the wood-free plant sources of cellulose. Waste from agriculture and residues, consisting of rice, wheat, coir from coconut, maize straw, as well as empty fruit bunches from palm trees, are another source of cellulose [66]. There are more other sources including industrial and municipal activities, such as waste from paper production, food and beverage industry by-products, brewer's spent grains as byproducts from brewing, and squeezed sugarcane bagasse [83]. Further ways to obtain cellulose involve certain types of

Table 3.3 Cellulose content in various sources of cellulose.

Biomass type	Source of cellulose	Cellulose (%)	References
Hardwood	Acacia	40–45	[67]
	Eucalyptus	40–45	[68]
	Poplar	35–50	[69]
Softwood	Spruce bark	45.5	[70]
	Douglas fir	40–50	[71]
	Pine	42–50	[72]
Agricultural waste	Corn stalks	35–39.6	[73]
	Rice straw	29.2–34.7	[74]
	Wheat straw	35–39	[75]
	Oat straw	31–35	[75]
	Sorghum straw	32–35	[76]
	Barley straw	36–43	[77]
	Sugarcane bagasse	25–45	[78]
	Corn cobs	33.7–41.2	[79]
	Rice husks	28.7–35.6	[80]
Grasses	Grassess	25–40	[81]
	Switchgrass	35–40	[82]

strains of algae, fungi, and bacteria. Certain bacteria predominantly from genera *Agrobacterium*, *Acetobacter*, and *Sarcina ventriculi* produce cellulose [84]. The difference between plant and bacterial or microbial cellulose is that bacterial or microbial cellulose has a high strength, high purity, and more water-holding capacity. Since the cellulose is naturally produced by bacteria, various new strategies are being studied to enhance the cellulose production from cultures in labs on a larger scale [85]. From algae sources Microalgae like Spirogyra and Chlorella are also good sources of cellulose which has fast growth for rapid production of cellulose.

These days, wood is the primary source of cellulose fibers utilized globally. However, some areas lack access to wood, and wood is being obviously unsustainable, which promotes the use of various non-woody or wood-free sources for a range of purposes, including

Plant hops, hemp, jute, sisal,bamboo, kenaf, cotton.
Algae green, gray, red, and yellow-green algae
Bacteria *Gluconacetobacter xylinus*
Animal: Tunicates, Family of sea animals (Ascidiacea) grasses, flowery plants or aquatic plants, and its byproducts.

Environmental impact of cellulose derivatives

Cellulose derivatives which are derived from the natural polymer cellulose have a variety of applications in various industries because of their biodegradability, renewability, and versatility. Cellulose and its derivatives are having a lesser carbon footprint then the synthetic materials which are produced from fossil fuels therefore cellulose and its derivatives are considered environment friendly. However, the concern is growing for its environmental impact as the demand for cellulose derivatives increases. For environmental protection the cellulose and its derivatives are being produced and used for water treatment, desalination of seawater, purification of air, as well as for textiles, packaging materials, in electrical devices biosensors and biomedical [86]. Mitigating the environmental issues which are associated with the usage of materials made from petroleum and fossil fuels can be replaced by cellulose derivatives. Despite being frequently commended for their low toxicological profile and biodegradability, derivatives of cellulose can nevertheless have considerable detrimental effects on the environment, such as

1. *Degradability of cellulose derivatives*; deacetylation and chain scission are two ways that cellulose derivatives might break down in extremely humid settings, which can result in a loss in structural integrity [87]. This may have an impact on how they handle and are usable, which could result in more waste.
2. *Increased requirement of resources:* The synthesis of cellulose derivatives may require a lot of water, energy, and chemical resources [88]. This may lead to increased ecological footprints and polluting of the environment.
3. *Emission of greenhouse gas*: Breakdown of cellulose derivatives has the potential for releasing greenhouse gases, particularly when significant quantities of the material are accumulated [89]. This can have more significant effects on the ecosystem and leads to global warming.
4. *Problems with management of waste*: The incorrect manner of disposing of cellulose derivatives can contribute to environmental contamination. Despite being biodegradable, their pace of breakdown varies, and in certain situations, they could linger in the environment beyond than anticipated [90].

3.4 Biodegradability and Recycling Considerations

The degradation process may involve multiple mechanisms, including chemical and enzymatic hydrolysis, photodegradation, oxidation, or just biodegradation, based on the environmental conditions [86]. It has been studied how commercial cellulose derivatives degrade in various lab settings as well as in actual natural, agro-industrial, and man-made settings. It is evident that the kinds of modification and substitution levels, along with the actual degradation environment, play a significant role in determining the degradation rate and sensitivity to degradation. One of the design principles that should be considered while creating new materials based on cellulose is the intended end-of-life environment. Considering the recycling of cellulose derivatives, Materials should be developed for "biological recycling" or material recycling in certain predefined environment settings, depending on the intended uses. In one study, Mila Bading *et al.* monitored the environmental biodegradation of commonly used 14 cellulose derivatives as pharmaceutical excipients by employing three different test methods based on the manometric respirometry test (OECD 301F) and closed bottle test (OECD 301D). Every derivatives of cellulose that was examined had different levels of biodegradation, with CODs, i.e., chemical oxygen demand that ranged from 0% to 20.4%. The results obtained from the study concluded that none of the cellulose derivatives fulfilled the requirements to be categorized as "readily biodegradable." They developed a grading classification score (traffic light system) for classifying excipients according to their biodegradation.

This study will help in decision making for selection of cellulose derivatives according to its biodegradability for incorporation into pharmaceutical formulations. It will also help in selection of structures for needful design of derivatives which will have enhanced biodegradability. A number of aspects must be taken into account while recycling cellulose-derived materials in order to maintain both economical and environmentally feasibility. The proper sourcing of raw material like cotton waste is upcycled into carboxymethyl cellulose (CMC) and cellulose acetate (CA) [91]. The impact of recycling processes on environment which consumes energy, water and emission of greenhouse gases it to be considered. There should be use of greener solvents and energy efficient methods for mitigating these impacts [92]. Another aspect to be considered for recycling is the economic viability that is the recycling should be initiated if the cost of recycling processes and the market value of recycled derivatives is feasible [92]. While recycling, it is important to consider the method of recycling the cellulose derivatives. There are various methods of recycling include:

1. *Chemical recycling:* chemical recycling which involve breakdown of cellulose derivative by employing chemicals for example use of enzymatic hydrolysis for conversion of cellulose into cellulose derivative without chemical pretreatment for sustainability [91].
2. *Mechanical recycling*: In this method, the cellulose materials are physically break down into small pieces which are reusable. By using mechanical recycling process the recycled fibers for new products are formed by physically breaking down the cellulose material into small pieces [93].
3. Closed-Loop Systems: This method helps in reducing waste and environmental impact. The dyed cotton textile waste are recycled into colored regenerated cellulose fibers (CRCFs) [92]. The waste materials are reused in this process which helps in reducing the pollution and consumption of resources.
4. Use of eco-friendly solvents: For mitigating the environmental impact of recycling processes, the eco-friendly solvents are utilized. Some commonly used eco-friendly solvents are ionic liquids, deep eutectic solvents, alkaline systems, non-aqueous polar aprotic solvents, mineral acids for recycling of cellulose derivatives [94].

3.5 Conclusion

Cellulose and its derivatives stand at the forefront of sustainable material innovation, offering a unique blend of renewable sourcing, structural versatility, and broad industrial applicability. This chapter has highlighted the fundamental aspects of cellulose, including its intricate molecular structure, the challenges and breakthroughs in its dissolution, and the wide-ranging chemical modifications that enhance its utility. From traditional derivatives, like methylcellulose to cutting-edge materials, such as nanocellulose and aerogels, the adaptability of cellulose continues to unlock new frontiers in pharmaceuticals, food technology, personal care, and beyond. Central to these advancements is the commitment to sustainability—leveraging biodegradable, eco-friendly solutions that align with global environmental goals. As research deepens and technologies evolve, cellulose and its derivatives will remain critical in shaping a more sustainable, circular bioeconomy. This chapter underscores their transformative potential and enduring relevance in modern material science and industry.

AI disclosure statement: The author acknowledges the use of AI-based tools for minor language editing or grammatical correction. However, all intellectual contributions, analyses, and conclusions presented in this chapter are entirely the author's own.

References

1. Heinze, T., Cellulose: Structure and Properties, in: *Cellulose Chemistry and Properties: Fibers, Nanocelluloses and Advanced Materials. Advances in Polymer Science,* vol. 271, Rojas, O. (eds), Springer, Cham, Berlin, Germany, 2015. https://doi.org/10.1007/12_2015_319.
2. Nawrath, C., Poirier, Y., Somerville, C., Plant polymers for biodegradable plastics: Cellulose, starch and polyhydroxyalkanoates. *Mol. Breed.*, 1, 2, 105–22, 1995.
3. Khan, S., Ul-Islam, M.U., Fatima, A., Manan, S., Khattak, W.A., Ullah, M.W., *et al.*, Potential of Food and Agro-Industrial Wastes for Cost-Effective Bacterial Cellulose Production: An Updated Review of Literature. *ES Food Agrofor.*, 13, 905, 2023.
4. McNamara, J.T., Morgan, J.L., Zimmer, J., A molecular description of cellulose biosynthesis. *Annu Rev Biochem.* 84, 895–921, 2015. doi: 10.1146/annurev-biochem-060614-033930.
5. Elango, B., Shirley, C.P., Okram, G.S., Ramesh, T., Seralathan, K.-K., Mathanmohun, M., Structural diversity, functional versatility and applications in industrial, environmental and biomedical sciences of polysaccharides and its derivatives – A review. *Int. J. Biol. Macromol.*, 250, 126193, 2023.
6. Simon, I., Scheraga, H.A., Manley, R.S.J., Structure of cellulose. 1. Low-energy conformations of single chains. *Macromolecules*, 21, 4, 983–90, 1988.
7. Parthasarathi, R., Bellesia, G., Chundawat, S.P.S., Dale, B.E., Langan, P., Gnanakaran, S., Insights into Hydrogen Bonding and Stacking Interactions in Cellulose. *J. Phys. Chem. A*, 115, 49, 14191–202, 2011.
8. Martínez-Sanz, M., Pettolino, F., Flanagan, B., Gidley, M.J., Gilbert, E.P., Structure of cellulose microfibrils in mature cotton fibres. *Carbohydr. Polym.*, 175, 450–63, 2017.
9. Zainul Armir, N.A., Zulkifli, A., Gunaseelan, S., Palanivelu, S.D., Salleh, K.M., Che Othman, M.H., *et al.*, Regenerated Cellulose Products for Agricultural and Their Potential: A Review. *Polymers*, 13, 20, 3586, 2021.
10. Semple, K.E., Zhou, C., Rojas, O.J., Nkeuwa, W.N., Dai, C., Moulded pulp fibers for disposable food packaging: A state-of-the-art review. *Food Packag. Shelf Life*, 33, 100908, 2022.
11. Mohite, P., Shah, S.R., Munde, S., Ade, N., Boban, T., Singh, S., Prajapati, B., Biopolymers in Textile-Based Scaffolding and Wound Healing, in: *Biopolymers in the Textile Industry,* Ahmed, S., Shabbir, M. (eds), Springer, Singapore, Berlin, Germany, 2024. https://doi.org/10.1007/978-981-97-0684-6_7.

12. Rahman, M.S., Hasan, M.S., Nitai, A.S., Nam, S., Karmakar, A.K., Ahsan, M.S., *et al.*, Recent Developments of Carboxymethyl Cellulose. *Polymers* [Internet], 13, 8, 2021.
13. Senthil Kumar, P. and Suganya, S., 1 - Introduction to sustainable fibres and textiles, in: *Sustainable Fibres and Textiles*, S.S. Muthu (Ed.), pp. 1–18, Woodhead Publishing, Sawston, Cambridge, UK, 2017.
14. Sharma, S., Sudhakara, P., Singh, J., Ilyas, R.A., Asyraf, M.R.M., Razman, M.R., Critical Review of Biodegradable and Bioactive Polymer Composites for Bone Tissue Engineering and Drug Delivery Applications. *Polymers* [Internet], 13, 16, 2021.
15. Lukova, P., Katsarov, P., Pilicheva, B., Application of Starch, Cellulose, and Their Derivatives in the Development of Microparticle Drug-Delivery Systems. *Polymers*, 15, 17, 3615, 2023.
16. Seddiqi, H., Oliaei, E., Honarkar, H., Jin, J., Geonzon, L.C., Bacabac, R.G., *et al.*, Cellulose and its derivatives: towards biomedical applications. *Cellulose*, 28, 4, 1893–931, 2021.
17. Choudhury, M.J. and Khan, G.M.A., Utilization of Sawmill By-Product for Making Cellulose and Its Valuable Derivatives, in: *Biomass and Bioenergy: Applications*, K.R. Hakeem, M. Jawaid, U. Rashid (Eds.), pp. 165–86, Springer International Publishing, Cham, 2014.
18. Davi Texeira, R., Anna Karla dos Santos, P., Gessiel Newton, S., Douglas Henrique, P., Plant and Bacterial Cellulose: Production, Chemical Structure, Derivatives and Applications. *Orbital: Electron. J. Chem.*, 11, 5, 2019.
19. Candido, R.G. and Gonçalves, A.R., Synthesis of cellulose acetate and carboxymethylcellulose from sugarcane straw. *Carbohydr. Polym.*, 152, 679–86, 2016.
20. Kennedy, J.F. and Zamir, A., The use of cellulose xanthate for the immobilisation of biological molecules. *Carbohydr. Res.*, 41, 1, 227–33, 1975.
21. Dilshad, E., Waheed, H., Ali, U., Amin, A., Ahmed, I., General Structure and Classification of Bioplastics and Biodegradable Plastics, in: *Bioplastics for Sustainable Development*, M. Kuddus and Roohi (Eds.), pp. 61–82, Springer Singapore, Singapore, 2021.
22. Oprea, M. and Voicu, S.I., Recent advances in composites based on cellulose derivatives for biomedical applications. *Carbohydr. Polym.*, 247, 116683, 2020.
23. Homem, N.C. and Amorim, M.T.P., Synthesis of cellulose acetate using as raw material textile wastes. *Mater. Today: Proc.*, 31, S315–S7, 2020.
24. Puri, A., Mohite, P., Ramole, A., Verma, S., Kamble, M., Ranch, K., Singh, S., Starch Science Advancement: Isolation Techniques, Modification Strategies, and Multifaceted Applications, *Macromol.*, 5, 3, 40, 2025.
25. Shokri, J. and Adibkia, K., Application of Cellulose and Cellulose Derivatives in Pharmaceutical Industries, in: *Cellulose - Medical, Pharmaceutical and Electronic Applications*, T.G.M. Van De Ven and L. Godbout (Eds.), IntechOpen, Rijeka, 2013.

26. Gupta, P.K., Raghunath, S.S., Prasanna, D.V., Priyadharsini, V., Shree, V., C C, *et al.*, An Update on Overview of Cellulose, Its Structure and Applications, in: *Cellulose*, A. Rodríguez Pascual and M.E. Eugenio Martin (Eds.), IntechOpen, Rijeka, 2019.
27. Heinze, T., El Seoud, O.A., Koschella, A., Etherification of Cellulose, in: *Cellulose Derivatives: Synthesis, Structure, and Properties*, pp. 429–77, Springer International Publishing, Cham, 2018.
28. Zhou, Y., Liang, Q., Zhang, Z., Wang, Z., Huang, M., Chiral separations with crosslinked cellulose derivatives attached onto hybrid silica monolith particles *via* the thiol–ene click reaction. *Anal. Methods*, 12, 21, 2727–34, 2020.
29. Fox, S.C., Li, B., Xu, D., Edgar, K.J., Regioselective Esterification and Etherification of Cellulose: A Review. *Biomacromolecules*, 12, 6, 1956–72, 2011.
30. Klemm, D., Heublein, B., Fink, H.-P., Bohn, A., Cellulose: Fascinating Biopolymer and Sustainable Raw Material. *Angew. Chem. Int. Ed.*, 44, 22, 3358–93, 2005.
31. Kakko, T., King, A.W.T., Kilpeläinen, I., Homogenous esterification of cellulose pulp in [DBNH][OAc]. *Cellulose*, 24, 12, 5341–54, 2017.
32. Tosh, B., Saikia, C.N., Dass, N.N., Homogeneous esterification of cellulose in the lithium chloride–N,N-dimethylacetamide solvent system: Effect of temperature and catalyst. *Carbohydr. Res.*, 327, 3, 345–52, 2000.
33. Freire, C.S.R., Silvestre, A.J.D., Neto, C.P., Belgacem, M.N., Gandini, A., Controlled heterogeneous modification of cellulose fibers with fatty acids: Effect of reaction conditions on the extent of esterification and fiber properties. *J. Appl. Polym. Sci.*, 100, 2, 1093–102, 2006.
34. Kang, H., Liu, R., Huang, Y., Graft modification of cellulose: Methods, properties and applications. *Polymer*, 70, A1–A16, 2015.
35. Tang, W., Tsarevsky, N.V., Matyjaszewski, K., Determination of Equilibrium Constants for Atom Transfer Radical Polymerization. *J. Am. Chem. Soc.*, 128, 5, 1598–604, 2006.
36. Barsbay, M. and Güven, O., Surface modification of cellulose *via* conventional and controlled radiation-induced grafting. *Radiat. Phys. Chem.*, 160, 1–8, 2019.
37. Kelly, P.V., Cheng, P., Gardner, D.J., Gramlich, W.M., Aqueous Polymer Modification of Cellulose Nanofibrils by Grafting-Through a Reactive Methacrylate Group. *Macromol. Rapid Commun.*, 42, 3, 2000531, 2021.
38. Lenfeld, P., Brdlík, P., Borůvka, M., Běhálek, L., Habr, J., Effect of Radiation Crosslinking and Surface Modification of Cellulose Fibers on Properties and Characterization of Biopolymer Composites. *Polymers*, 12, 12, 3006, 2020.
39. Sirviö, J.A., Honkaniemi, S., Visanko, M., Liimatainen, H., Composite Films of Poly(vinyl alcohol) and Bifunctional Cross-linking Cellulose Nanocrystals. *ACS Appl. Mater. Interfaces*, 7, 35, 19691–9, 2015.

40. Sun, B., Hou, Q., Liu, Z., Ni, Y., Sodium periodate oxidation of cellulose nanocrystal and its application as a paper wet strength additive. *Cellulose*, 22, 2, 1135–46, 2015.
41. Vera-Loor, A., Walger, E., Marlin, N., Mortha, G., Evaluation of the dissolving ability of cellulosic pulps: Investigation of a novel method using light scattering follow-up during classical cellulose carbanilation. *Holzforschung*, 77, 3, 139–48, 2023.
42. Schlufter, K., Schmauder, H.-P., Dorn, S., Heinze, T., Efficient Homogeneous Chemical Modification of Bacterial Cellulose in the Ionic Liquid 1-N-Butyl-3-methylimidazolium Chloride. *Macromol. Rapid Commun.*, 27, 19, 1670–6, 2006.
43. Henniges, U., Kloser, E., Patel, A., Potthast, A., Kosma, P., Fischer, M., *et al.*, Studies on DMSO-containing carbanilation mixtures: chemistry, oxidations and cellulose integrity. *Cellulose*, 14, 5, 497–511, 2007.
44. Isik, M., Sardon, H., Mecerreyes, D., Ionic Liquids and Cellulose: Dissolution, Chemical Modification and Preparation of New Cellulosic Materials. *Int. J. Mol. Sci.* [Internet], 15, 7, 11922–40, 2014.
45. Wu, J., Zhang, J., Zhang, H., He, J., Ren, Q., Guo, M., Homogeneous Acetylation of Cellulose in a New Ionic Liquid. *Biomacromolecules*, 5, 2, 266–8, 2004.
46. Kakuchi, R., Ito, R., Nomura, S., Abroshan, H., Ninomiya, K., Ikai, T., *et al.*, A mechanistic insight into the organocatalytic properties of imidazolium-based ionic liquids and a positive co-solvent effect on cellulose modification reactions in an ionic liquid. *RSC Adv.*, 7, 16, 9423–30, 2017.
47. Gan, L., Liao, J., Lin, N., Hu, C., Wang, H., Huang, J., Focus on Gradientwise Control of the Surface Acetylation of Cellulose Nanocrystals to Optimize Mechanical Reinforcement for Hydrophobic Polyester-Based Nanocomposites. *ACS Omega*, 2, 8, 4725–36, 2017.
48. Lease, J., Kawano, T., Andou, Y., Esterification of Cellulose with Long Fatty Acid Chain through Mechanochemical Method. *Polymers*, 13, 24, 4397, 2021.
49. Ge, W., Shuai, J., Wang, Y., Zhou, Y., Wang, X., Progress on chemical modification of cellulose in "green" solvents. *Polym. Chem.*, 13, 3, 359–72, 2022.
50. Liu, S., Zhang, Q., Gou, S., Zhang, L., Wang, Z., Esterification of cellulose using carboxylic acid-based deep eutectic solvents to produce high-yield cellulose nanofibers. *Carbohydr. Polym.*, 251, 117018, 2021.
51. Tang, X., Zuo, M., Li, Z., Liu, H., Xiong, C., Zeng, X., *et al.*, Green Processing of Lignocellulosic Biomass and Its Derivatives in Deep Eutectic Solvents. *ChemSusChem*, 10, 13, 2696–706, 2017.
52. Huang, C., Cai, B., Zhang, L., Zhang, C., Pan, H., Preparation of iron-based metal-organic framework @cellulose aerogel by *in situ* growth method and its application to dye adsorption. *J. Solid State Chem.*, 297, 122030, 2021.
53. Zheng, G., Polavarapu, L., Liz-Marzán, L.M., Pastoriza-Santos, I., Pérez-Juste, J., Gold nanoparticle-loaded filter paper: a recyclable dip-catalyst for

real-time reaction monitoring by surface enhanced Raman scattering. *Chem. Commun.*, 51, 22, 4572–5, 2015.
54. Johnston, J.H. and Nilsson, T., Nanogold and nanosilver composites with lignin-containing cellulose fibres. *J. Mater. Sci.*, 47, 3, 1103–12, 2012.
55. Mahmoud, K.A., Male, K.B., Hrapovic, S., Luong, J.H.T., Cellulose Nanocrystal/Gold Nanoparticle Composite as a Matrix for Enzyme Immobilization. *ACS Appl. Mater. Interfaces*, 1, 7, 1383–6, 2009.
56. Schlesinger, M., Giese, M., Blusch, L.K., Hamad, W.Y., MacLachlan, M.J., Chiral nematic cellulose–gold nanoparticle composites from mesoporous photonic cellulose. *Chem. Commun.*, 51, 3, 530–3, 2015.
57. Mohite, P., Puri, A., Pandhare, R., Singh, S., Sonawane, M., Pawar, A., Biomaterials-Based Additive Manufactured Products for Chronic Wound Healing, in: *Biomaterial-based Additive Manufacturing in Tissue Engineering and Regeneration*, pp. 253–278, Springer, Cham, Nature Switzerland, 2025.
58. Amoroso, L., De France, K.J., Milz, C.I., Siqueira, G., Zimmermann, T., Nyström, G., Sustainable Cellulose Nanofiber Films from Carrot Pomace as Sprayable Coatings for Food Packaging Applications. *ACS Sustainable Chem. Eng.*, 10, 1, 342–52, 2022.
59. Puri, A., Mohite, P., Singh, S., Chaudhari, Y., Chaudhari, M., Ansari, Y., Gholap, A., Biomaterials-Based Additive Manufactured Products for Diabetic Wound Healing, in: *Biomaterial-based Additive Manufacturing in Tissue Engineering and Regeneration*, pp. 279–312, Springer, Cham, Nature Switzerland, 2025.
60. Barthel, S. and Heinze, T., Acylation and carbanilation of cellulose in ionic liquids. *Green Chem.*, 8, 3, 301–6, 2006.
61. Zhao, Y., Sun, B., Wang, T., Yang, L., Xu, X., Chen, C., *et al.*, Synthesis of cellulose–silica nanocomposites by *in situ* biomineralization during fermentation. *Cellulose*, 27, 2, 703–12, 2020.
62. Ullah, R., Azam, A., Aziz, T., Farhan, Rehman, H.U., Qiao, S., *et al.*, Peacock Feathers Extract Use as Template for Synthesis of Ag and Au Nanoparticles and Their Biological Applications. *Waste Biomass Valorization*, 13, 1, 659–66, 2022.
63. Aziz, T., Ullah, A., Ali, A., Shabeer, M., Shah, M.N., Haq, F., *et al.*, Manufactures of bio-degradable and bio-based polymers for bio-materials in the pharmaceutical field. *J. Appl. Polym. Sci.*, 139, 29, e52624, 2022.
64. Abolore, R.S., Jaiswal, S., Jaiswal, A.K., Green and sustainable pretreatment methods for cellulose extraction from lignocellulosic biomass and its applications: A review. *Carbohydr. Polym. Technol. Appl.*, 7, 100396, 2024.
65. Sharma, H.K., Xu, C., Qin, W., Biological Pretreatment of Lignocellulosic Biomass for Biofuels and Bioproducts: An Overview. *Waste Biomass Valorization*, 10, 2, 235–51, 2019.
66. Pennells, J., Cruickshank, A., Chaléat, C., Godwin, I.D., Martin, D.J., Sorghum as a novel biomass for the sustainable production of cellulose nanofibers. *Ind. Crops Prod.*, 171, 113917, 2021.

67. Yáñez, R., Romaní, A., Garrote, G., Alonso, J.L., Parajó, J.C., Experimental evaluation of alkaline treatment as a method for enhancing the enzymatic digestibility of autohydrolysed Acacia dealbata. *J. Chem. Technol. Biotechnol.*, 84, 7, 1070–7, 2009.
68. Pappas, C., Tarantilis, P.A., Daliani, I., Mavromoustakos, T., Polissiou, M., Comparison of classical and ultrasound-assisted isolation procedures of cellulose from kenaf (Hibiscus cannabinus L.) and eucalyptus (Eucalyptus rodustrus Sm.). *Ultrason. Sonochem.*, 9, 1, 19–23, 2002.
69. Datta, D., Prajapati, B., Jethva, H., Agrawal, K., Singh, S., Prajapati, B.G., Value-added nanocellulose valorized from fruit peel waste for potential dermal wound healing and tissue regenerative applications, *Regen. Eng. Transl. Med.*, 11, 1, 88–111. 2025.
70. Le Normand, M., Moriana, R., Ek, M., Isolation and characterization of cellulose nanocrystals from spruce bark in a biorefinery perspective. *Carbohydr. Polym.*, 111, 979–87, 2014.
71. Zhu, R. and Yadama, V., Isolation and Characterization of Cellulose Micro/Nanofibrils from Douglas Fir. *J. Polym. Environ.*, 26, 3, 1012–23, 2018.
72. Cavali, M., Soccol, C.R., Tavares, D., Zevallos Torres, L.A., Oliveira de Andrade Tanobe, V., Zandoná Filho, A., *et al.*, Effect of sequential acid-alkaline treatment on physical and chemical characteristics of lignin and cellulose from pine (Pinus spp.) residual sawdust. *Bioresour. Technol.*, 316, 123884, 2020.
73. Ferdous, T., Quaiyyum, M.A., Salam, A., Jahan, M.S., Pulping of bagasse (Saccrarum officinarum), kash (Saccharum spontaneum) and corn stalks (Zea mays). *Curr. Res. Green Sustainable Chem.*, 3, 100017, 2020.
74. Razali, N.A.M., Mohd Sohaimi, R., Othman, R.N.I.R., Abdullah, N., Demon, S.Z.N., Jasmani, L., *et al.*, Comparative Study on Extraction of Cellulose Fiber from Rice Straw Waste from Chemo-Mechanical and Pulping Method. *Polymers*, 14, 3, 387, 2022.
75. Li, H., Zhang, H., Xiong, L., Chen, X., Wang, C., Huang, C., *et al.*, Isolation of Cellulose from Wheat Straw and Its Utilization for the Preparation of Carboxymethyl Cellulose. *Fibers Polym.*, 20, 5, 975–81, 2019.
76. Andrade Alves, J.A., Lisboa dos Santos, M.D., Morais, C.C., Ramirez Ascheri, J.L., Signini, R., dos Santos, D.M., *et al.*, Sorghum straw: Pulping and bleaching process optimization and synthesis of cellulose acetate. *Int. J. Biol. Macromol.*, 135, 877–86, 2019.
77. Kim, T.H., Jeon, Y.J., Oh, K.K., Kim, T.H., Production of furfural and cellulose from barley straw using acidified zinc chloride. *Korean J. Chem. Eng.*, 30, 6, 1339–46, 2013.
78. Asgher, M., Ahmad, Z., Iqbal, H.M.N., Alkali and enzymatic delignification of sugarcane bagasse to expose cellulose polymers for saccharification and bio-ethanol production. *Ind. Crops Prod.*, 44, 488–95, 2013.
79. Shogren, R.L., Peterson, S.C., Evans, K.O., Kenar, J.A., Preparation and characterization of cellulose gels from corn cobs. *Carbohydr. Polym.*, 86, 3, 1351–7, 2011.

80. d Oliveira, J.P., Bruni, G.P., Lima, K.O., Halal, S.L.M.E., d Rosa, G.S., Dias, A.R.G., *et al.*, Cellulose fibers extracted from rice and oat husks and their application in hydrogel. *Food Chem.*, 221, 153–60, 2017.
81. Reddy, K.O., Maheswari, C.U., Dhlamini, M.S., Mothudi, B.M., Kommula, V.P., Zhang, J., *et al.*, Extraction and characterization of cellulose single fibers from native african napier grass. *Carbohydr. Polym.*, 188, 85–91, 2018.
82. Reddy, N. and Yang, Y., Natural cellulose fibers from switchgrass with tensile properties similar to cotton and linen. *Biotechnol. Bioeng.*, 97, 5, 1021–7, 2007.
83. García, A., Gandini, A., Labidi, J., Belgacem, N., Bras, J., Industrial and crop wastes: A new source for nanocellulose biorefinery. *Ind. Crops Prod.*, 93, 26–38, 2016.
84. Fernandes, IAA., Pedro, A.C., Ribeiro, V.R., Bortolini, D.G., Ozaki, M.S.C., Maciel, G.M., *et al.*, Bacterial cellulose: From production optimization to new applications. *Int. J. Biol. Macromol.*, 164, 2598–611, 2020.
85. Abeer, M.M., Mohd Amin, M.C.I., Martin, C., A review of bacterial cellulose-based drug delivery systems: their biochemistry, current approaches and future prospects. *J. Pharm. Pharmacol.*, 66, 8, 1047–61, 2014.
86. Erdal, N.B. and Hakkarainen, M., Degradation of Cellulose Derivatives in Laboratory, Man-Made, and Natural Environments. *Biomacromolecules*, 23, 7, 2713–29, 2022.
87. Romão, S., Bettencourt, A., Ribeiro, I.A.C., Novel Features of Cellulose-Based Films as Sustainable Alternatives for Food Packaging. *Polymers*, 14, 22, 4968, 2022.
88. Biranje, S.S., Kaushik, S., Marewad, D., Yadav, A., Vankundre, V., Panse, M., *et al.*, Applications of nanocellulose and its derivatives in developing sustainable textiles. *Cellulose*, 31, 9, 5343–79, 2024.
89. Machado, B., Costa, S.M., Costa, I., Fangueiro, R., Ferreira, D.P., The potential of algae as a source of cellulose and its derivatives for biomedical applications. *Cellulose*, 31, 6, 3353–76, 2024.
90. Sjahro, N., Yunus, R., Abdullah, L.C., Rashid, S.A., Asis, A.J., Akhlisah, Z.N., Recent advances in the application of cellulose derivatives for removal of contaminants from aquatic environments. *Cellulose*, 28, 12, 7521–57, 2021.
91. Oshikata, M.S.K., Blas, N.S., dL Silva, B., Fukamizu, D.I., da Silva, D.R.B., Gauto, L.P., *et al.*, Cotton waste upcycling: biofuel and cellulose derivatives production. *Cellulose*, 31, 11, 6693–704, 2024.
92. Liu, H., Fan, W., Miao, Y., Dou, H., Shi, Y., Wang, S., *et al.*, Closed-loop recycling of colored regenerated cellulose fibers from the dyed cotton textile waste. *Cellulose*, 30, 4, 2597–610, 2023.
93. Yadav, R. and Kamble, Z., Textile waste-based cellulose composites: A review. *J. Mater. Sci.*, 59, 17, 7147–68, 2024.
94. Acharya, S., Liyanage, S., Parajuli, P., Rumi, S.S., Shamshina, J.L., Abidi, N., Utilization of Cellulose to Its Full Potential: A Review on Cellulose Dissolution, Regeneration, and Applications. *Polymers* [Internet], 13, 24, 2021.

4

Economic Perspective of Cellulose Production and Regeneration

Heena Gupta

Department of Applied Science, G.L. Bajaj Institute of Technology and Management, Greater Noida, India

Abstract

With the increase in global population and development of global economy, the production of waste material has become a major concern all over the globe. This huge amount of nondegradable waste is responsible for damaging the environment which ultimately is a serious threat to human health and destruction of biodiversity. This global trend is driving the search for the alternatives that can maintain the balance between modernization and environmental sustainability. Hence, the sustainable development is need of the era. Regenerated cellulose (RC) is one of the most abundant renewable polymers that has the potential to substitute the pollution causing petroleum-based synthetic polymers. Nowadays, RC fibers are being used abundantly in the rapid growing textile industries as they have a variety of applications such as softness, comfort, breathability, moisture absorption, versatility, drapability, absorbency, dyeability, etc. The biodegradable aspects of RC materials make them suitable option for packaging, clothing, agriculture etc which in turn beneficial from the environmental and economic point of view. The raw materials used for the production of RC fibers (RCF) are the natural ones where no toxic chemicals are used. This production process of RCF is way more environmentally friendly, cost-effective, can also reduce waste, pollution, and resource utilization. These moderated fibers are infinitely recyclable, following the logic of circular sustainability and are of superior quality with potential to be modified according to the requirement. Recycling or upcycling of RC fibers contributes to environmental protection as well as effective reuse can also reduce the production cost. The awareness regarding personal hygiene, biodegradability and sustainability are the key factors which leads to increase the demand of RCF

Email: heena.gupta1290@gmail.com

Mohd Shabbir (ed.) Regenerated Cellulose: A Renewable Alternative to Petroleum-Based Plastics, (91–118) © 2026 Scrivener Publishing LLC

in evolving economies for the production of innovative technologies, upgradation in life styles, rapid modernization and others.

Keywords: Regenerated cellulose, eco-friendly, cost effective, natural raw material

4.1 Introduction

Regenerated cellulose fiber (RCF), obtained from cellulose regeneration, is a kind of man-made fiber that were manufactured from raw material of natural origin, more precisely of cellulose, which is available in abundance. Cellulose is biodegradable polymer and also known as center of carbon cycle (Figure 4.1). Wood, paper, and cotton all are common examples of cellulose material. It can be molded easily to create versatile materials. Industries, such as fashion, health care, packaging, agriculture, etc., are using these cellulose-based RCF materials [1–5]. Blending of cellulose for various applications is one of the phenomenal characteristics. It can be done through a series of chemical reactions. This procedure involves the formation of specific fiber type, such as viscose, lyocell, or modal. The variety of RCF depends on the chemical solvent which we use during manufacturing process. So, RCF can be considered as part natural and part artificial.

There are lots of examples of regenerated fibers around us. They are very soft, comfortable, and breathable fabric, such as lyocell; smooth and moisture wicking fiber (prepared from wood pulp through various chemical processes), viscose & rayon (processed from natural polymers). RCF like

Figure 4.1 Cellulose fibers.

acetate, and triacetate obtained from cell walls of short cotton fibers called liners. All varieties of RCF are quite similar to cotton fibers.

RCF is that category of materials which is obtained *via* processing of natural cellulose to convert into cellulose-based materials, followed by method of regeneration, either in form of fiber (e.g., rayon) or a sheet (e.g., cellophane). This regeneration process of cellulose has various advantages [6], such as environment-friendly; versatility; durability; high tensile strength; not expensive (cheap); nontoxic; biocompatible; perfect water holding capacity for high mechanical performance; suitable option for medical dressing, paper restoration, and blood vessel regeneration, which make them appropriate for biomedical, energy, food packaging purposes, etc.

There are few names of RCF: viscose rayon, cuprammonium rayon, lyocell, acetate, triacetate, alginic, modal, polynosic, etc. Yarn and fibers manufactured from regeneration process are considered to be one of the lowest carbon footprints releasing material in comparison to other synthetic fabrics. The manufacturing of such RCF actually requires very less energy and it is completely biodegradable [7, 8]. Acetate and triacetate are considered as cheap silk as they both have original silk like appearance; lustrous and elegant drape. The potential of absorption makes it suitable option for dye purposes.

These are semi-synthetic fabric (indeed a plant-based), commonly known as fabric of future. It is a perfect alternative for cotton or silk. These fabrics are extremely soft to touch (Figure 4.2). These are manufactured

Figure 4.2 Bamboo lyocell.

from wood pulp obtained from eucalyptus, oak, bamboo and birch trees, highly in demand because of 100% biodegradable and compostable properties (that is if it is not blended with other synthetic fibers).

A British company called "courtaulds fibers" researched [9] about cellulose, lyocell, and manufactured a new fabric known as Tencel. Both fabrics are almost identical, and can be used interchangeably. The only difference lies in the solvent which is used in the manufacturing processes.

4.1.1 Economic and Environmental Aspect of RCF

These cellulose-derived fabrics are actually one of the cheaper alternatives to silk. Natural silk production is sustainable but quite time consuming. It is also quite difficult to produce silk in bulk which drives "artificial silk" revolution that finally ended in the manufacturing of lyocell, Tencel, or rayon. The cost of blending of cellulosic fibers with different solvent is actually a relatively cheaper procedure than the extraction of pure silk or cotton form the raw material. On the other hand, the versatility of such RCF makes them even more economically beneficial [10–12]. There are numerous areas where we can use them (Figure 4.3).

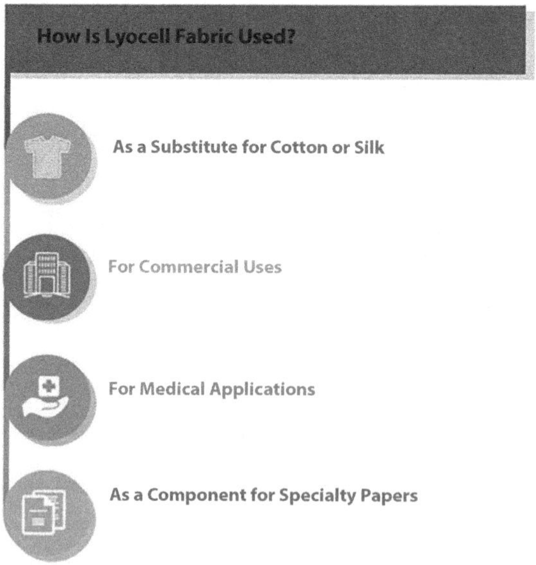

Because of being so strong and resistant, excellent temperature regulation, moisture absorption qualities (perfect for clothing in humid environment), breathability, antibacterial effect, soft-to-the-touch sensation (time and money saver), lightness (does not wrinkle), great elasticity, highly durable, versatile fiber (can be shaped into a variety of designs without compromising quality), lyocell is being used for a variety of industries and products. It is so common to find it in bedding and towels, conveyor belts, specialty paper, activewear, great choice for all kinds of sports and exercising, perfect for both summer and winter clothing due to its thermal regulation properties, suitable for travelling purposes (being light and soft), can be found in medical dressing (wipes, swabs, disposable gowns etc), etc. Lyocell fibers has so much potential that there is still more to explore its wide range of applications [13–15]. While purchasing products made of lyocell material, one should prioritize all its properties rather than the price. On the other hand, one can say it is a long-term investment.

In comparison to other cellulose textiles, RCF are way more eco-friendly materials. These are also called sustainable material, not just because of natural origin (wood pulp) but also because of their manufacturing process. In fact, constituent solvent can be recycled up to 99.5% which means very little amount of chemicals waste is left. This entire process of manufacturing is also known as "closed loop" process. Thanks to this sustainable manufacturing process, lyocell has been awarded with multiple awards, such as the "European Award for the Environment" given by the European Union.

4.2 Need for Regenerated Cellulose Fibers

4.2.1 Global Consumption of Plastics and Pollution

The biggest concern of today's era is the production, management and disposal of huge amount of plastic waste which is non-biodegradable. This petroleum-based plastic waste is affecting and destroying the entire ecosystem from the tallest mountains to the deepest ocean trenches. The plastic-based pollution affects the living beings including human adversely. For this huge production of plastic materials, we, the human is responsible. To fulfil our daily needs in a low cost, plastic-based materials are being produced, utilized and still the consumption of these materials has continued to increase. In the last 70 years, manufacturing of synthetic fibers or petroleum-based plastic materials has been drastically increase and became almost doubled since the beginning of the century. In 1950, the global production was just two million tonnes per year which has now became over 400 million tonnes in 2022 (Figure 4.4). The demand of such materials has raised by six-fold in the last five decades which is again the driving force for the production of this huge amount. The production has been explosively increased from 2 to 400 million tonnes in just a few decades and it is estimated that it will be four times in comparison to 2019 to almost 1480 million tons by 2050. This huge amount is almost three times the total volume of the world's population today [16–19].

All over the planet, Asia is the largest plastics producer in which 32% contribution is done alone by China. In 2022, the fact demonstrates that China is manufacturing 6 and 12 million metric tons of plastic material every month and still the data is increasing [20]. The second largest contribution to the global plastic waste production is by North America with 17% of its share. But fortunately, due to pandemic situation, waste production was a bit lesser in comparison to other years which was actually really good for our planet.

On the records, other than COVID-19–related lockdown measures, we are producing such a huge amount of non-biodegradable waste material which is actually of single use purposes, used for few moments and then thrown away but these will be sustain for the several hundred years on the earth. Several facts [21–28] says that almost 40% of total plastic products are being discarded within just one month. Due to low cost, versatile, and sterile qualities of plastic-based materials, RCF products can be consumed in numerous applications related with constructional activities, domestic

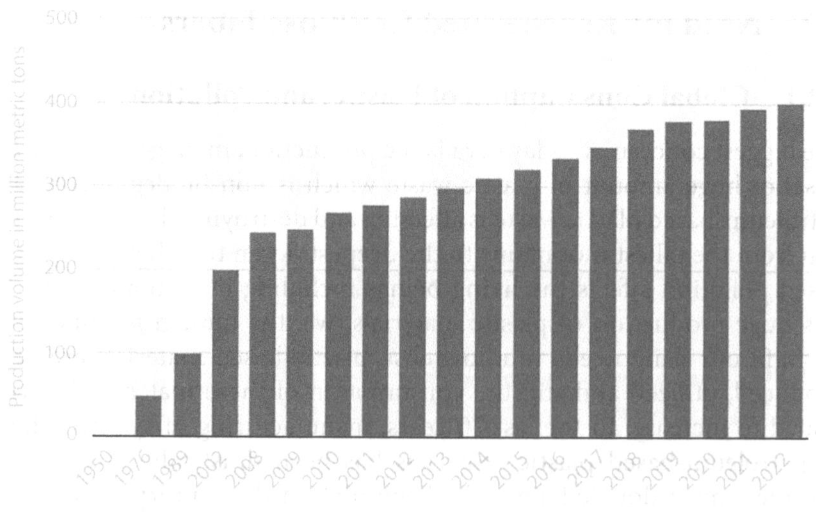

appliances, surgical instruments, packaging materials for food items, etc [29–32]. The manufacturing keeps on increasing day by day ignoring the harmful effects on the earth. However, when it is not being properly managed—not recycled, incinerated, or dumped in oceans or ground—it becomes harmful for the environment, behaves as pollutant affecting wildlife and ecosystems. In this manner, we are destructing the entire planet by putting burden of 300 million tons every year (Figure 4.5). All these facts and figures [33–35] are given by industries themselves, but they did not provide the real picture which can show the contribution of synthetic fibers whose real production is a lot higher.

Unfortunately, we all know the consequences of this plastic pollution but still we are using it. The major problem with the use of these materials is their disposal. It is not easy to manage such amount of million tonnes with the standard protocol. Quite simply, humans have become so much addicted to this nearly indestructible material. The petroleum-based plastic materials are used in a wide variety of products and materials. Several reports say [36, 37] that million tons of discarded plastic waste is being dumped into our oceans and other water resources every year. Plastic never degrades, instead it is break down into smaller pieces (also known as microplastics) until it cannot be seen with the bare eye. Figure 4.6 below shows the type of plastic pollution floating around in ocean.

It is estimated that only 0.5% of the plastic materials can be seen floating over the surface of the waters of oceans or other water resources. The rest of the waste material gets accumulated deeper in the seabed. It is not only because of the breakdown of the existing plastic material or microplastics

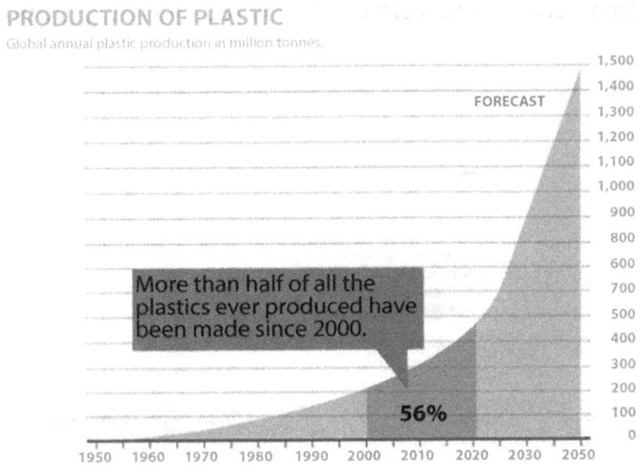

but also from the wearing, tearing and washing of synthetic clothing, consumption and rinsing off toiletries & cosmetics products, wearing of rubber tyres and small plastic granules released by the plastics industry during manufacturing. These days a new term is given to the plastic pollution in the ocean is PLASTIC SOUP. There is a lot of damage that we have already done to our ocean but the harsh reality is microplastics are destroying our crops in the field also with much heavily impact. Nearly half or about 46% of plastic waste [38–40] is dumped into oceans or landfilled, 22% is being decomposed, 17% goes for combustion and remaining 15% is sent for recycling, out of which only 9% is actually recycled after losses (Figure 4.7).

The microplastics were invented for their versatile qualities with lots of applications in cosmetic, packaging, toiletries etc. These microplastics flow from our bathrooms into sewers or deposited on the ground or penetrate

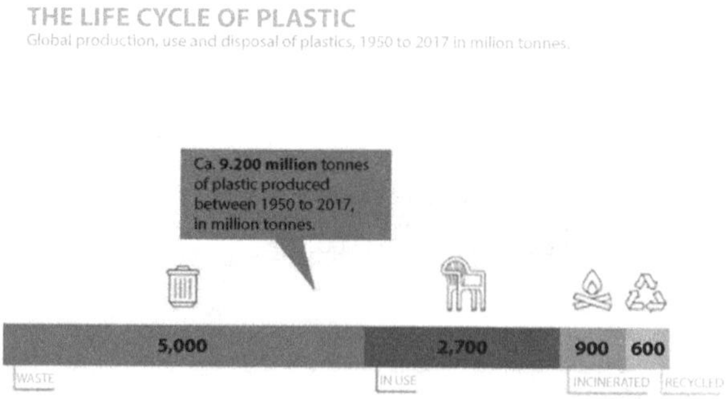

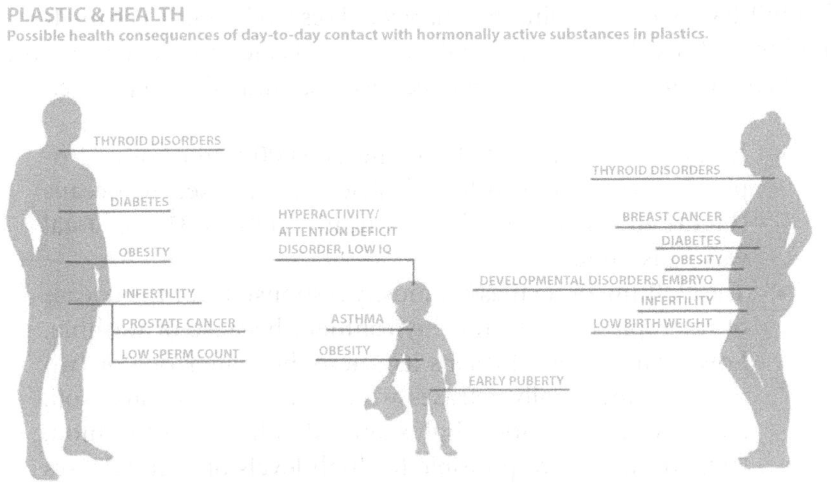

into soil to contaminate underground water. This contamination of sewar water is responsible for the formation of polluted sludge. This contaminated sewer sludge after treatment can be use as fertiliser. Recent studies [41–43] analyse that microplastics are contaminating our vegetation *via* up taking of nutrients, water from contaminated soil. Research shows various harmful impacts of microplastics and the even smaller version in terms of nanoplastics on human health (Figure 4.8) and environment. In our daily life, we all are consuming microplastics and nanoplastics through the intake of contaminated food and drinking water, *via* the inhalation and skin pores. A few surveys [44] among expecting women in the United States of America says that their body consists of average of 56 different chemical substances flow in their blood streams.

Plastic has made up of constituents that can disturb the hormonal balance within the body and also known as "hazardous hormone disruptors." Along with this, the combustion of plastic waste material in the open-air leads to poses several health risks as well as ecological imbalance.

4.2.2 Plastic Pollution from Textile Industry

The consumption and disposal strategies of synthetic fibers in the textile industry is one of the major causes of plastic pollution which ultimately results in microplastics and greenhouse gas emissions. Over the years, textile industries are growing and synthetic waste materials keeps on increasing. Therefore, the textile industries need to adopt waste disposal strategy from a landfill-based to a recycling-based. Several research [45, 46] estimates that landfilling of textile waste materials or microplastics got deposited on the world's ocean floor causing harm to ecosystems, animals and people.

- Every year, between 200,000 and 500,000 tonnes of microplastics from the textile industries are released to oceans (Figure 4.9) globally which is nearly 16% to 35% of total waste disposal.
- The clothing or textiles are mostly responsible for the release of microplastics into the environment because of washing. Low-quality fabric loosens its microfibers after one or two washing and finally releases which further can cause soil, air, or water pollution. High standard of living, fast fashion, latest trends are responsible for high levels of such releases because of consumption only for a short time and discard them off very quickly.
- To prevent the release of synthetic fibers into atmosphere, textile industries should implement strict disposal strategies for it. Sustainable and strict monitoring at every process of manufacturing, processing, disposal can help to reduce or control microplastic emissions.

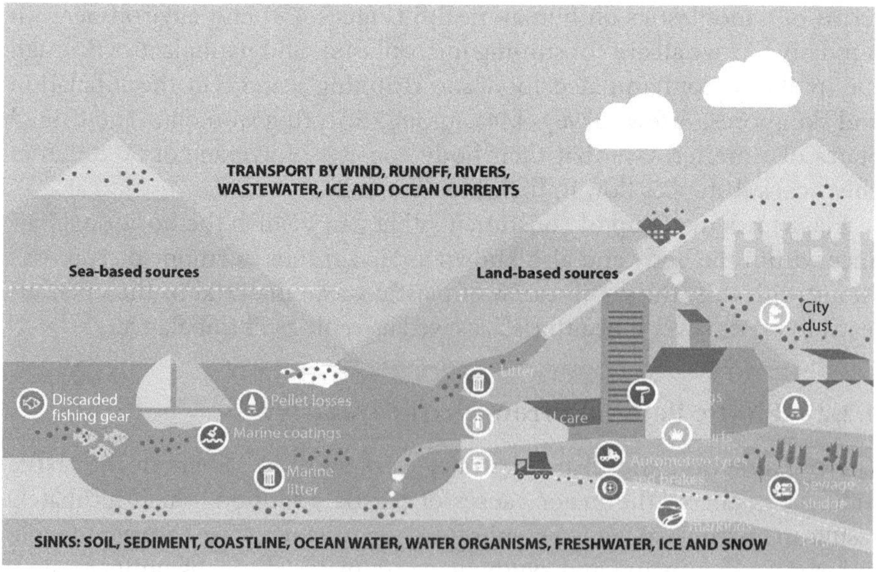

A report submitted by the EEA's European Topic Centre on Circular Economy and Resource Use (ETC/CE) [47] on the release of microplastics into the environment from textiles highlighted few facts:

- Major contribution to global plastic waste materials in last five years are 40% from packaging materials, 12% from consumer items and 11% from clothing and textiles.

We always fail at the disposal strategies. Recycling percentage is very low in comparison to landfilling. Nearly 15% is collected for recycling but unfortunately 40% of that is again dumped into water or ground. Only 9% of plastic waste get the opportunity of recycling. The remaining 19% is incinerated and the rest of them again goes for landfilling, burning or dumped in terrestrial or aquatic ecosystems.

4.2.3 Contribution of Textile Industry to Environmental Damage

Nowadays, we are consuming textile in almost everywhere. Textile does not only mean for clothing or garments purposes, it includes variety of areas whether it is home décor or furnishing, telecommunication or other wires, transportational means or medical appliances/pharmaceuticals, etc. [48–50]. When consumption of textiles is everywhere, then definitely there is a lot of waste too which is further responsible for the microplastic pollution. In the cotton textile industry, processing of raw fiber like cutting, spinning, sizing, resizing, scrubbing, mercerizing, decolorizing, dyeing, printing, and finishing for the production of required textile materials is responsible for the formation of 20% to 49% waste in the form of discarded weaving material, faulty knit fabric, fly yarn and sample waste etc. Textile made up of fibers whether it is natural or synthetic, shed microfibers. Along with these microfibers, the domestic articles and the discarded clothes or fabric are also responsible for such a huge amount of waste. It is predicted [51–53] that textile industry is alone responsible for global production of 0.2 and 0.5 million tonnes of synthetic waste or microplastics every year.

Microfibers are not only the contributor of solid waste production but also responsible for the damage of water quality. The waste water discharged from the textile industry has basic (alkaline) properties which is not suitable for drinking or any other activity without processing. The alkaline water damages the soil natural composition, affects the quality of vegetation and enters into the food chain. Synthetic fibers have petroleum-based origin. The smaller version of these synthetic fibers is also known as Microfibers or nanofibers which has potential threat to human health. Decomposition of these petroleum or plastic-based microfibers take thousands of years as these are non-biodegradable. These microfibers have potential to contaminate the soil, air and water etc. Penetration of contamination into the environment disturbs the food web and hence disrupts the ecological balance. With the rising demand of society for synthetic fibers, the production of related plastic waste is also increasing day by day. It is estimated [54, 55] that every year about 350 million metric tons of discarded plastic is produced globally which is almost similar to weight of over 10 million fully loaded garbage trucks. But unfortunately, only 2% of this huge volume is being recycled.

Generally, microplastics from clothing or textiles industries have a fiber shape which is often known as microfibres. Textiles made of fibres of natural as well as plastic-based origin release microfibres. Moreover, print, dye, button, coatings, glitter, etc. of clothing or garments or any other textile material are also the cause of plastic pollution. Studies, survey, reports, etc. [56, 57] have analyzed that about 0.2 and 0.5 million tonnes of microplastics are discharged into the oceans each year, which actually originated during the washing, manufacturing, processing, disposal, etc. The disposal of synthetic waste from textile industry into the water sources is one of the prime reasons for damaging the aquatic life. Any activity related with textile industry leads to ultimately release of microfibers or microplastics in the atmosphere. These microfibers further are responsible for air, soil and water pollution. This shedding of microfibers is majorly caused by washing of textile materials whether it is done by washing machines or some other traditional methods. With the washing machine effluent, these micro particles enter into sewage system and finally ended up in rivers or oceans which costs the life in marine environment. It is estimated that long wash cycles upsurge wear and tear in elevated temperatures leads to degrades the quality of fabric and damages its structure which ultimately leads to release of high levels of microfibers. The use of certain chemicals such as washing powders and detergents enhance this shedding. On the other hand, the use of fabric softener reduces the release of microfiber during washing, probably by reducing friction and fiber deterioration.

4.3 Microplastics from Textiles: Environmental and Health Impacts

4.3.1 Environmental and Health Impacts

With the industrial evolution, we have grown concerns about the environmental and health issues related with microplastic pollution. In our daily routine, we all are exposing to these microplastics which have undoubtedly negative impact on living beings, including flora or plankton, fauna in both land and aquatic ecosystem. These contamination from air, water and soil are ingested by the living beings that results into several diseases or disorders. Microplastics have become integral part of life these days. It has been reported [58–60] in a wide variety of human foods, beverages, seafood, drinking water, beer, salt, sugar, etc. It has severely infected our agricultural industry. The long-term effects of contaminated crops and other livelihoods, are still largely uncertain. Generally, microplastics exposure is supposed to induce inflammatory reactions and toxicity in the body which further disturb the functioning of vital organs [61].

Microfibers, mainly originating from synthetic textiles, seems to have a higher potential than other fibers to contaminate or enter the food chain, as their size and shape are appropriate to be readily used by small or aquatic organisms. These are more susceptible to entangled in large clots inside the gut that restricts/hampers the usual activity of the organ or can cause blockages [62]. Usually, microplastics can be emitted into the atmosphere at any point in the textile industry from manufacturing to consumption and care, through to disposal. One of the studies reveal [63] by the World Health Organization (WHO) in 2019 has analyzed the presence of microplastics in drinking water (tap or bottled drinking water (WHO, 2019). Another published work [64] shows that approximately 84% of 57 samples of soft drinks, cold tea and energy drinks has contamination of microplastics which influence our health badly. It was also analyzed that nearly 65% microfibers are released during drying and wearing of clothes, again leads to pollute aerial environments [65]. These airborne microplastics get deposited on roads, fields or any surface that contaminates the soil. This soil deterioration is also done by landfilling and dumping of textile waste which in fact damage the quality of vegetable, crops, fruits, etc. Everywhere this contamination has taken its place to deteriorate the environment.

4.3.2 Strategies to Prevent the Emission of Microplastics from Textiles

Although we have gathered so much of information *via* research & study regarding emission and effects of synthetic textiles on the environment and human health, but there are still a lot more to explore. The impact of damage by microfibers depends on its composition, shape, size etc and it can be different for different species. In fact, their impacts and effects are mostly unknown which makes it very difficult to predict the severity of the issue and the measures needed. So, more knowledge is required to solve this impact & effects mystery related with the deterioration of eco-system. To fill this knowledge gap, we need continuous support from the entire society and technologies for the advance research. This will be done only when all the interdisciplinary areas such as technical, behavioural and regulatory measures work together. We all need to consider those potential solutions which include precautions rules and policies to minimize or reduce the release of microfibers from textiles industry. Advanced strategies should be made for sustainable textiles. There are few pathways [66] to reduce microfiber release from plastic and textiles: (1) sustainable design and production; (2) strict guidelines to control microplastic release during consumption; and (3) improved disposal and end-of-life processing and (4) replace synthetic or petroleum base raw materials with natural one, i.e., RCFs. In recent years, RCF is one of the steps forward for the sustainable development in textile industry.

4.4 RCF in Textile Industry

Nowadays, we have become more concerned about the environmental and climatic change. With the ever-increasing demands of the consumer from the nature, environmental is being exploited and deteriorated day by day. Synthetic fibers or microfibers those have petroleum-based origin are majorly responsible for the environmental damage. So, for the sake of sustainable growth of the society, number of apparel companies has started production of novel green material, RCFs out of textile waste, a suitable and better alternative for microfibers. Landfilling or dumping of RCFs (as textile waste) do not pollute the environment. However, we have still lack of knowledge about all the benefits of RCFs. Scientist are keeps on working to explore every possible use of RCF in the textile industry. RCF have potential to fulfil the consumer's need effectively and affect the environment positively which can certainly drive the attention of consumers towards RCF-made apparel. In recent years, consumers have now understood about the adverse impact of synthetic fibers of textile industry on the environment. Fast fashion trends, high standard of living, growing population, improved incomes; all these factors increase the demand of apparel products which in turn leads textile industry to increase the production that ultimately stimulate large volume of textile waste. According to the Ellen McArthur Foundation, more than 100 billion worth of apparel is discarded in the world annually. It was analyzed that only 8% to 12% of discarded waste is being recycled. Generally, we prefer cotton materials, i.e., cellulose for clothing purposes because of obvious reasons being soft, comfortable, durable, breathable, hypoallergenic, biodegradable, odour-free, cling-free, pilling-free, etc. [8, 67]. Despite of all such qualities, conventional cotton is not environment friendly. Actually, cotton is a very pesticide-intensive crop. Lots of harmful chemicals, insecticides, and herbicides are being used for the cultivation of cotton crops. These chemicals are toxic in nature and have adverse effect on living beings and nature. Almost 1 kg of toxic pesticides is being utilized worldwide for every one hectare of cotton cropland. Every year, approximately, US$ 3.3 billion spent on such chemicals which are being sprayed over cotton crops worldwide. For manufacturing of a single pair of clothes, minimum 10 tons of water is required. Cotton cultivation needs a huge amount of water which of course hampers the fulfilment of other water requirements. One can believe it or not, the Aral Sea is almost gone and this is declared as worst environmental disasters in history by United Nations [68]. It is also responsible for the release of 220 million tons of CO_2 every year. By taking into consideration of all these facts, we need an alternative for this commercial cotton. Regenerated cotton fibers is one of the suitable options in this regard. It has numerous advantages for ecosystem and its components, including:

- Negligible consumption of synthetic chemicals
- 91% reduced water consumption
- 62% lesser energy demand
- 46% lower CO_2 emissions
- 26% lesser soil erosion

We have another suitable alternative for cotton which is environment friendly and does not produce any toxic material, low water consumption, pesticides, insecticides free etc i.e. LYOCELL. These cellulose-based fibers can be use in textile industry in place of other synthetic fibers but it has one drawback which limits its use commercially, i.e., its cost. It is way too expensive for customers. One promising solution of all the problems related with textile industry is RCFs generated using cotton waste and linters. The production and consumption of RCF is increasing day by day. Globally, textile industries are using these fibers which helps to reduce their contribution towards plastic pollution. RCF are pure cellulose-based, manufactured from wood pulp *via* different procedures and have better reinforcement potential. RCF is majorly utilized by textile sector, but it can also be consumed in composite materials.

In comparison to natural fibers, the RCF have reproducible properties. One of the Research done by Data Bridge (Figure 4.10), it was concluded that the RCF market was worth of USD 8,200 million in 2021 and is expected to raise USD 12584.43 million by 2029, registering a CAGR of 5.50 % during the forecast period of 2022 to 2029 [69].

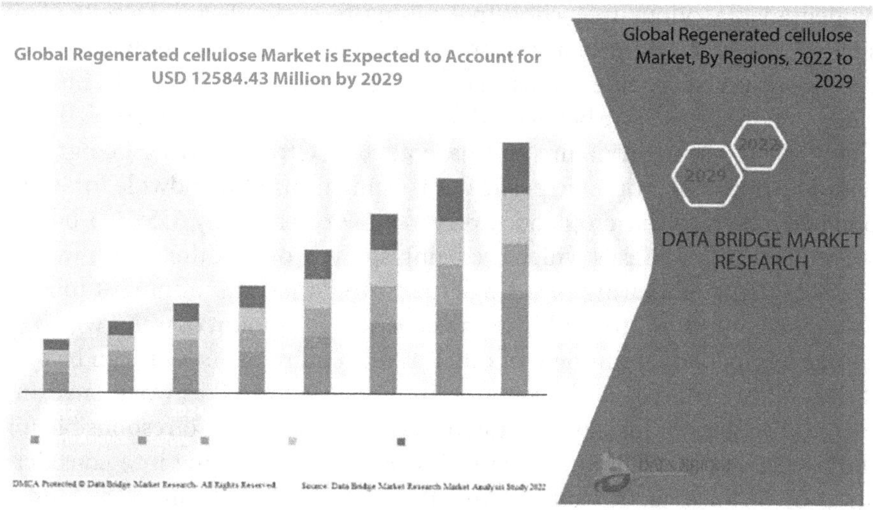

4.4.1 Growth Drivers and Market

Increasing demand for textiles and apparel across the globe is the driving force of RCF. Also, the technological advancement and concept of sustainable environment has encouraged the industries for the manufacturing of biodegradable, comfortable, breathable fibers which can be utilized by customer for clothing purposes and others. With the fastest growing fashion industry, need or demand for RCF materials has increased which ultimately is responsible for its large production by the industries. These days, regenerated cellulose-based materials market has grown so much [69]. Its manufacturing process releases fewer greenhouse gases and other hazardous or toxic materials (considered as pollutants), which helps in regulating the earth's climate and contributes toward the market's growth.

One of another report [70–72] depicts the involvement of RCF in market, not only in the textile industries but also in other industries such as automotive, agriculture, packaging etc. It says that the market has grown strongly in recent years (Figure 4.11) and still it is developing with annual growth rate of (CAGR) 9.7% [69]. This tremendous growth in the market is due to increase demand for sustainable products or fibers, migration from synthetic or petroleum-based to natural fibers, fashion and textile industry trend, market acceptance, enhancing needs in various applications.

The growth trend in upcoming years can be attributed to concern for sustainable development, rigorous environmental regulation, consumer

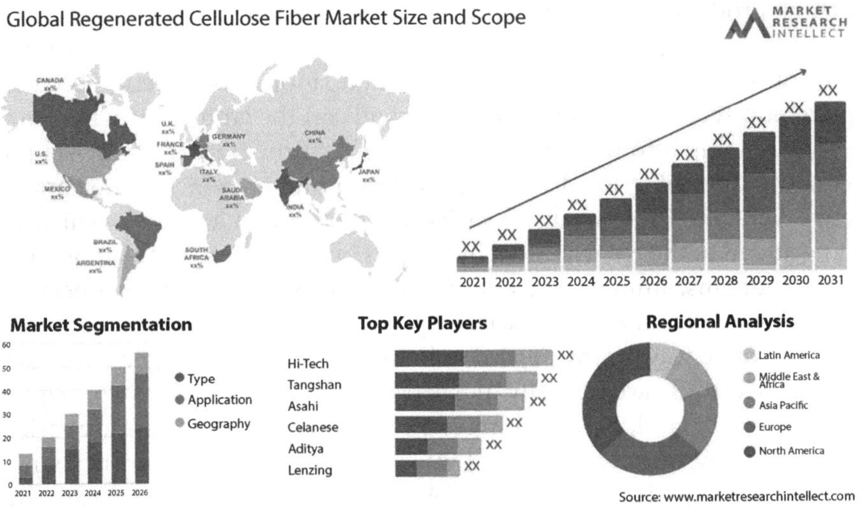

Figure 4.11 Regenerated cellulose market size.

preferences for green products, circular economic initiatives, growing textile industry trends. Various kinds of strategies [73, 74], such as waste reduction initiatives, product innovation collaborations, innovative applications, consumer awareness, etc., are being adopted these days for successful accomplishment of sustainability goals. Nowadays, regenerated cellulose market is being emerging in number of countries such as Australia, China, France, Brazil, India, Germany, Russia, UK, USA, Canada, Italy, Spain belong to different regions of globe like Asia-pacific, Western Europe, Eastern Europe, North America, South America, Middle East, Africa.

4.4.2 Environmental and Economic Perspectives of Cellulose Regeneration

In recent years, one of the major reasons for tremendous growth of regenerated cellulose fibers market is the raw material that we use for its processing, cellulose. It is one of the most abundant biodegradable natural polymers on Earth. Apparently, it has significant environment importance. Due to the most abundance, biodegradability and biocompatible polymer, cellulose is being a promising material for the manufacturing of green products for variety of applications. In old times, consumption of cellulose materials was limited to energy resource (fuel), construction (e.g., wood), clothing purpose (e.g., cotton), etc. [75, 76] but now a days cellulose and its derived products are being used for new emerging opportunities such as production and preparation of biofuels and biopolymeric products (films, aerogels, sponges, etc.), which were previously manufactured from oil-based synthetic polymers. Accumulation of synthetic polymers (petroleum-based nonbiodegradable origin) on the land in terms of landfill or dumping into the water bodies leads to formation of plastic soup is actually responsible for damaging the environment, which is not practically reversible. There are also a variety of chemicals or additives, which are being added to the environment by various human activities that pose hazards to humans, animals, and entire ecosystem. So, we are more concerned about the environment and searching for more bio-based alternatives in every possible way. In this regard, cellulose fits into the category over biopolymers as these are virtually inexhaustible, low in cost, does not require specific type of land, use very limited amount of water etc. There are variety of methods by which agricultural waste can be used as cellulosic sources in numerous applications [75, 76]. Among various plant-based carbohydrates, cellulose is the most resistant polymer to degradation. In this way, these are way more economical preferrable over other substance. These beneficial properties of cellulose drive the manufacturing of cellulose-based materials or cellulose regeneration or production of regenerated cellulose fibers.

Human and their surrounding environment are integrated, therefore, the well-being of one is dependent on the well-being of other. The imperative that humanity faces can be formulated as follows: "without increased incomes and economic development environment protection will fail; without environment protection economic development will be undermined". The threat to the human race due to environment degradation is as great as one from nuclear holocaust. But industrialization/development can-not be stopped. So, we have to go for suitable alternative i.e. Sustainable development (Figure 4.12). Or can say, we have to search for the options of development that helps to maintain higher standard of living (however interpreted) for human beings without the expense of environmental integrity.

Regeneration of Cellulosic fibre is one of the practical strategies in this direction which is actually an environment friendly and cost-effective phenomenon. One of the major raw materials for the production of RCF is cellulose, obtained from natural sources. Due to nature-based origin, cellulose fibers are easily decomposed or can say completely bio-degradable. The trees (bamboo, oak, pine, beech etc.) from which the raw material is obtained are replenishable. With the passage of time, by using rainwater and soil nutrients, trees can be regrown. So, the regeneration and disposal of cellulose fibers, both are completely sustainable. Without causing any

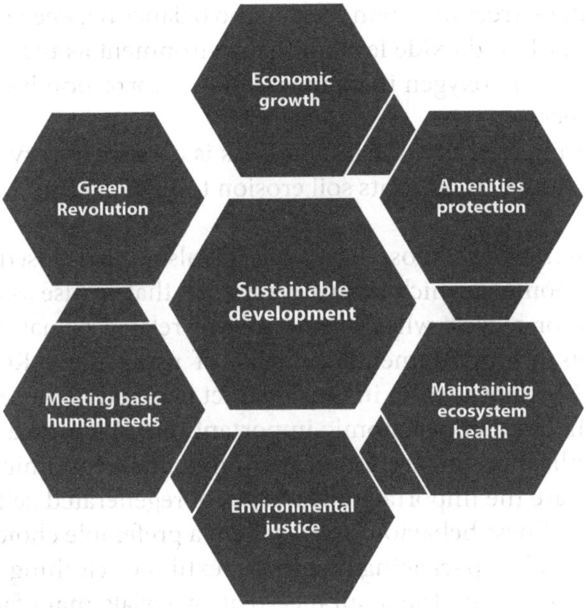

harm to the environment, manufacturing of RCF is entirely eco-friendly and economically beneficially [77].

Solvents and chemicals (such as caustic soda, carbon disulphide [CS_2], sulfuric acid, sodium sulphate, and zinc sulphate) which are being used during the production can be easily recycled. Many by-products released during the process can also be reused. The only problem or concern related with the generation of RCF is the consumption of fossil fuels. If this problem resolves then this process is fully sustainable.

There are many reasons to mostly use bamboo trees for cellulose production and regeneration;

- No additional chemicals are needed for plantation of bamboo trees.
- Only rain water is sufficient for growth of such trees, no extra water or irrigation is not required.
- One time plantation of bamboo trees is enough for very long time.
- Harvesting of bamboo trees only take 3-5 years for the results. These are fast growing trees.

- Bamboo trees maintain a very good balance between Oxygen and carbon dioxide level in the environment as they release 35 % more oxygen in comparison to a corresponding stand of trees.
- Holding capacity of bamboo trees is extremely very well so one can say it prevents soil erosion too.

The production of cellulose-based fibers is also called closed loop manufacturing phenomena which means everything that we use as raw material can replenish or recycle whether it is solvent related or natural cellulose. There are lots of environmental benefits of using these RCF in textile industries [77]. They are also in huge market demand because of their low cost which makes them economic important too. Widespread textile and industrial applications having lots of environmental, economic benefits and sustainability are the important advantages of regenerated cellulose fibers. This eco-friendliness behaviour makes them a preferable choice in numerous applications like packaging materials, textiles or clothing, and disposable products [75, 76]. The assurance that materials manufactured from RCF can be disposed of without spoiling the ecosystem drives a surge in requirement for this environmental favorable substance. Moreover, textile industry actually plays a major role for the growth of regenerated cellulose market. These days we are more concerned about sustainable development, there is a rising demand towards sustainable and biodegradable textiles. Regenerated cellulose fabrics, such as lyocell and viscose, fit perfectly into these trends. Their eco-friendly characteristics, durability, soft texture, and versatility make them appropriate materials for clothing brands looking to satisfy the consumer need and demand for comfortable, environment friendly, trendy, latest and stylish garments. This fashion-driven approach of society and market apparently responsible for rising of the regenerated cellulose market [69].

The extraction of cotton fibers from the discarded materials is also an economical and environment beneficial phenomena. Recycling cost of waste clothing material is comparatively low but unfortunately cotton waste ends up in landfills or incinerators. Processing of discarded cotton waste to produce RCF is eco-friendly method. No harmful, toxic chemicals are used in this manufacturing. This transformation is possible with a series of environment friendly solvents. The resultant fibers are of high quality and can be moulded into the required frame of demand. This is actually a very promising approach of removing discarded cotton waste that will further reduce the contamination of environment and resource consumption. These days industries are developing marketing strategies or policies to promote RCF made apparel, a sustainable approach. Cellulose regeneration can also be done from the discarded paper and board which can reduce the additional pressure on agriculture land for extensive requirement of cellulose for various applications [75, 76]. In this way, cellulose regeneration and processing is participating in circular economy with the concern of sustainable approach. A rough approximation has been made to define the benefits of regenerated cellulose-based materials over traditional synthetic fibers or plastic production from fossil-based oil. One of the most important or significant economic benefits of cellulose regeneration that attracts all industries or companies active in the cellulose production market, is their biodegradability and recyclable through the regeneration process which helps in reducing the pollution on a planetary level. RCF as food packaging materials are also being widely used for their excellent environmental performance. Regenerated cellulose casings are also preferred coating material for cookware because of their high tensile strength other than sausage casings [78]. RCF can tolerate and withstand at high temperatures of 190° and has brilliant resisting properties, has a lower oxygen transmittance (OTR), which can efficiently prevent oxygen from oxidizing food [79]. It can be easily decomposed by the microbes in the soil without causing any damage to the ecosystem, making them green, sustainable, and environment friendly. Currently, a lot of researchers and studies are going on to develop a variety of methods of recycling waste cotton material, including physical recycling or chemical recycling that can be further process for the cellulose regeneration and production. Nowadays, this has become an important way in textile industries as well as other related areas for the green development of the society.

4.5 Conclusion

There has never been a more urgent need to restore damaged ecosystems than now. In the trend of fast fashion, we need strategies for development of society that should have characteristics of intergenerational, i.e., bequeathing the same or improved resource endowment to the future that has been inherited), inter species and inter group equity. There has to be a balance between need/demand of society, improvement of people's living standards and consumption of natural resources.

Cellulose production and regeneration are significant innovation to explore a cleaner and greener textile treatment method to maintain the carrying capacity of the environment and sustainable development. Processing of cellulose-based waste materials is done by mixing variety of eco-friendly

Figure 4.13 Eco-friendly environment.

solvents such as ionic liquids aimed to contribute to the realization of carbon neutrality. Some studies have concluded that microplastic fiber pollution not only has irreversible impact on ecosystem (plantation, microbes in soil, water bodies; ocean and rivers, air composition), but also hampers the activity of vital organs of the individuals such as malfunctioning of digestive tract of organisms, reduces the fertility rate and heart, lungs or gastrointestinal tract related problems. To reduce the pressure on environmental sources for the fulfilling the human need, disposal or recycling strategies related with textile industries have been made. Utilization of discarded cotton waste materials to produce regenerated cellulose fibers has a great economic value and also protect the environment from huge amount of pollution caused by decomposition of waste materials. It aims to promote harmony to achieve a more sustainable and equitable world (Figure 4.13).

AI disclosure statement: The author confirms that no artificial intelligence (AI)-generated content or AI-assisted writing tools were used in the preparation of this chapter (including summaries and the abstract). All text, analysis, and conclusions are the result of the author's original work.

References

1. Yang, L., Meng, J., Xue, T., Wang, Y., Shi, G., Gao, X., Zhi, C., Application of 3D printing cellulose fabrics based on cotton fibers in the textile and fashion industry. *Addit. Manuf.*, 81, 104000, 2024.

2. Kassie, B.B., Mamay Daget, T., Tassew, D.F., Synthesis, functionalization, and commercial application of cellulose-based nanomaterials. *Int. J. Biol. Macromol.*, 278, 134990, 2024.
3. Riseh, R.S., Vazvani, M.G., Hassanisaadi, M., Thakur, V.K., Agricultural wastes: A practical and potential source for the isolation and preparation of cellulose and application in agriculture and different industries. *Ind. Crops Prod.*, 208, 117904, 2024.
4. Huang, H., Zheng, C., Zhang, Z., Huang, C., Adding lignin to bagasse cellulose-based materials increases the wet strength and reduces micro-nano scale wet low-temperature defects. *Ind. Crops Prod.*, 222, 119750, 2024.
5. Hassanisaadi, M., Riseh, R.S., Rabiei, A., Varma, R.S., J.F., Nano/micro-cellulose-based materials as remarkable sorbents for the remediation of agricultural resources from chemical pollutants. *Int. J. Biol. Macromol.*, 246, 125763, Kennedy2023.
6. Santamala, H., Livingston, R., Sixta, H., Hummel, M., Skrifvars, M., Saarela, O., Advantages of regenerated cellulose fibres as compared to flax fibres in the processability and mechanical performance of thermoset composites. *Manufacturing*, 84, 377–385, 2016.
7. Garces, I., Ngo, T.D., Ayranci, C., Boluk, Y., Regenerated cellulose in additive manufacturing. *Cellulose*, 31, 1773–1785, 2024.
8. Khan, A., Mohsin, M., Qutab, H., Abdullah, U., Awais, M., Iftikhar, K., Regenerated Cellulosic Fibres Blended with Cotton to Enhance the Performance and Comfort Properties of Single Jersey Spandex Knitted Fabrics: A Sustainable Approach. *Fiber. Polym.*, 25, 2337–2345, 2024.
9. Geoffrey, O., The rise and fall of great companies: Courtaulds and the reshaping of the man-made fibres industry, in: *Pasold studies in textile history*, Oxford University Press, Oxford, UK, ISBN 9780199592890, 2010.
10. Rosson, L. and Byrne, N., Bicomponent regenerated cellulose fibres: Retaining the colour from waste cotton textiles. *Cellulose*, 29, 4255–4267, 2022.
11. Xie, Y., Cai, P., Cao, X., Pan, Y., High-strength, water-resistant polyvinyl alcohol films reinforced with regenerated cellulose fibers for food packaging. *Ind. Crops Prod.*, 207, 117768, 2024.
12. Lu, D., Liu, X., Shao, J., Wang, Z., Li, Z., Ren, S., Lan, X., Liu, W., Preparation of composite food packaging paper through surface self-assembly of regenerated cellulose/chitosan. *Ind. Crops Prod.*, 218, 118959, 2024.
13. Ozturk, M.K., Comparative Analysis of Different Lyocell Types: Structural, Physical, Mechanical, and Moisture-Related Comfort Properties in Single Jersey Fabrics. 24, 4429–4439, 2023.
14. Zhang, H., Yang, G., Edgar, K.J., Zhong, G., Liu, P., Li, M., Shao, H., Preparation and properties of dual-wavelength excitable fluorescent Lyocell fibers and their applications in papermaking. *Carbohydr. Polym.*, 261, 117861, 2021.
15. Chen, Y., Cui, S., Jin, H., Yang, C., Wang, B., Mao, Z., Feng, X., Zhang, Y., Sui, X., Xu, H., Fabrication of phase change microcapsules with controllable

size *via* regenerated nanochitin stabilized pickering and their applications for lyocell fiber. *Colloids Surf., A*, 655, 130308, 2022.
16. Alves, B., Global waste generation - statistics & facts, Waste Management, Energy & Environment, 2024.
17. Priya, A.K., Muruganandam, M., Imran, M., Gill, R., Reddy, M.R.V., Md Shkir, Sayed, M.A., AlAbdulaal, T.H., Algarni, H., Md Arif, Jha, N.K., Sehgal, S.S., A study on managing plastic waste to tackle the worldwide plastic contamination and environmental remediation. *Chemosphere*, 341, 139979, 2023.
18. Singh, N. and Ogunseitan, O.A., Disentangling the worldwide web of e-waste and climate change co-benefits. *Circ. Econ.*, 1, 100011, 2022.
19. Azad, A.K., Haq, U.N., Md. K. Akter, M., Uddin, Md.A., *Recycling Practices of Pre-Consumer Waste Generated from Textile Industry*, pp. 301–324, Springer Nature, Switzerland, 2024.
20. Slotta, D., Amount of disposed garbage in China in selected years from 1990 to 2022. Energy & Environment, Waste Management, Statista, 2024.
21. Jung, J.M., Kim, C.J., Chung, C.S., Kim, T., Choi, K.Y., Heavy metal characterization of land-based waste dumped at three ocean dumping sites in the Republic of Korea. *Mar. Pollut. Bull.*, 193, 115205, 2023.
22. Chen, D., Mu, G., Wang, X., Zhang, Y., Brief discussion on the new approach of selection and designation of ocean dumping areas in China. *Mar. Policy*, 160, 105949, 2024.
23. Priya, K.L., Iqbal, S., Archana, A.R., Gopika, B., Mina, M., Haddout, S., Madhu, A.M., Implications of solid waste dumps on the microplastic abundance in groundwater in Kollam, India. *J. Environ. Manage.*, 348, 119224, 2023.
24. Eo, S., Hong, S.H., Song, Y.K., Han, G.M., Seo, S., Park, Y.G., Shim, W.J., Underwater hidden microplastic hotspots: Historical ocean dumping sites. *Water Res.*, 216, 118254, 2022.
25. Charrière, E. and Langston, N., Dumping military waste into Lake Superior: The historic legacies of secrecy, censorship, and uncertainty. *Water Hist.*, 15, 173–200, 2023.
26. Mousavi, S.H., Kavianpour, M.R., Alcaraz, J.L.G., The impacts of dumping sites on the marine environment: A system dynamics approach. *Appl. Water Sci.*, 13, 109, 2023.
27. Ryan, P.G., Illegal dumping from ships is responsible for most drink bottle litter even far from shipping lanes. *Mar. Pollut. Bull.*, 197, 115751, 2023.
28. Kakade, A., Mi, J., Long, R., Microplastics in the World Oceans and Strategies for Their Control. *Rev. Environ. Contam. Toxicol.*, 262, 20, 2024.
29. Liu, Y., Chen, R., Li, F., Sun, L., Guo, Z., Jiang, Z., Ren, Y., Asymmetric ionic bond shielding encountering with carboxylate capturing metal ions for enhancing the flame retardant durability of regenerated cellulose fibers. *Int. J. Biol. Macromol.*, 273, 133158, 2024.
30. Saedi, S., Kim, J.T., Lee, E.H., Kumar, A., Shin, G.H., Fully transparent and flexible antibacterial packaging films based on regenerated cellulose extracted from ginger pulp. *Ind. Crops Prod.*, 197, 116554, 2023.

31. Kataoka, L.F.M.S., Leão, R.M., Gontijo, A.B., Falla, M.P.H., Luz, S.M., Regenerated cellulose films from jute fibers applied in conductive nanocomposites. *Mater. Today Commun.*, 33, 104645, 2022.
32. Scheer, A., Fischer, J., Bakhshi, A., Bauer, W., Fischer, S., Spirk, S., Fusion of cellulose microspheres with pulp fibers: Creating an unconventional type of paper. *Carbohydr. Polym.*, 338, 122207, 2024.
33. Falco, Di Pace, E., Cocca, M., Avella, M., The contribution of washing processes of synthetic clothes to microplastic pollution, The Nature. *Sci. Rep.*, 9, 6633, 2019.
34. MacAulay, S., Masud, N., Jones, J.D., Ward, B.D., Cable, J., The impacts of synthetic and cellulose-based fibres and their associated dyes on fish hosts and parasite health. *Environ. Sci. Pollut. Res.*, 30, 121558–121568, 2023.
35. Priyadarshini, S., Jagatee, S., Das, A.P., Renewable Energy Generation and Value Addition from Environmental Microfiber Pollution Through Advanced Greener Solution, in: *Synthetic Fabrics and Microfiber Pollution–An Assessment of Their Global Impact*, pp. 137–157, 2024.
36. Tamta, M. and Kamboj, A., Role of New-Generation Textile Fibres in Reducing the Environmental Impact of Textiles, in: *Climate Action Through Eco-Friendly Textiles*, pp. 41–53, 2024.
37. Ojukwu, V.E., Akaeme, F.C., Ighalo, J.O., Microplastic Pollution in the Changing Climate, in: *Microplastics in African and Asian Environments*, pp. 219–232, 2024.
38. Calmet, D.P. and Bewers, J.M., Radioactive waste and ocean dumping: The role of the IAEA. *Mar. Policy*, 15, 413–430, 1991.
39. Yamahara, S., Viyakarn, V., Chavanich, S., Bureekul, S., Isobe, A., Nakata, H., Open dumping site as a point source of microplastics and plastic additives: A case study in Thailand. *Sci. Total Environ.*, 948, 174827, 2024.
40. Dung Le, D., Dao Le, T.A., Pham, T.G., Vietnam's regulations on sea dumping: Outstanding issues and suggestions for improvement. *Reg. Stud. Mar. Sci.*, 68, 103253, 2023.
41. Corradini, F., Melillán, E., Schoebitz, M., Carrasco, R., Jiménez, M., Assessing on-site plastics fragmentation: Linking macroplastics litter to microplastics pollution in vegetable fields. *Sci. Total Environ.*, 955, 177168, 2024.
42. Kushwaha, M., Shankar, S., Goel, D., Singh, S., Rahul, J., Rachna, K., Singh, J., Microplastics pollution in the marine environment: A review of sources, impacts and mitigation. *Mar. Pollut. Bull.*, 209, 117109, 2024.
43. Othman, R., Mohd Hatta, F.A., Mohd Latif, N.H., Sulaiman, W.S.H.W., Ramya, R., Ali, Q.A.M., Johari, N.N., Bakar, A.E.A., Algal-based Degradation of Microplastic: Prospects for Combating Microplastic Pollution, in: *Microplastic Pollution*, pp. 491–526, 2024.
44. Li, Y., Tao, L., Wang, Q., Wang, F., Li, G., Song, M., Potential Health Impact of Microplastics: A Review of Environmental Distribution, Human Exposure, and Toxic Effects. *Environ. Health*, 1, 249–257, 2023.

45. Yi, Y., Kong, L., Wang, X., Li, Y., Cheng, J., Han, J., Chen, H., Zhang, N., Distribution and characteristics of microplastics in sediment at representative dredged material ocean dumping sites, China. *Mar. Pollut. Bull.*, 193, 115201, 2023.
46. Clark, J.H., Textile waste – an opportunity as well as a threat. *Green Carbon*, 1, 146–149, 2023.
47. ETC-EC - European Topic Centre on Circular Economy and Resource Use 2022-2026 University of Ferrara, 2022.
48. Hossain, K.R., Akter, S., Nanjeba, M., Mahmud, Md.A., Properties and Performance of Biopolymers in Textile Applications, in: *Biopolymers in the Textile Industry*, pp. 41–86, 2024.
49. Nayak, R., Cleveland, D., Tran, G., Joseph, F., Potential of bacterial cellulose for sustainable fashion and textile applications: A review. *J. Mater. Sci.*, 59, 6685–6710, 2024.
50. Karuppuchamy, A., Ramya, K., Siva, R., Novel banana core stem fiber from agricultural biomass for lightweight textile applications. *Ind. Crops Prod.*, 209, 117985, 2024.
51. Perebynos, A., Sepúlveda, D., Ribeiro, C., Study on textile waste generation in the undifferentiated municipal solid waste stream in Guimarães, Portugal. *Waste Disposal Sustainable Energy*, 5, 189–203, 2023.
52. Yen, T.T., Hwang, S.S., Bunekar, N., Environmental Hazards on Textile Waste, in: *Nano-engineered Materials for Textile Waste Remediation*, pp. 153–161, 2023.
53. Sulochani, R.M.N., Jayasinghe, R.A., Priyadarshana, G., Nilmini, A.H.L.R., Ashokcline, M., Dharmaratne, P.D., Waste-based composites using post-industrial textile waste and packaging waste from the textile manufacturing industry for non-structural applications. *Sustainable Chem. Environ.*, 8, 100163, 2024.
54. Robinson, B.H., E-waste: An assessment of global production and environmental impacts. *Sci. Total Environ.*, 408, 183–191, 2009.
55. Suzuki, G., Uchida, N., Tanaka, K., Higashi, O., Takahashi, Y., Kuramochi, H., Yamaguchi, N., Osako, M., Global discharge of microplastics from mechanical recycling of plastic waste. *Environ. Pollut.*, 348, 123855, 2024.
56. Rhodes, C.J., Plastic pollution and potential solutions. *Sci. Prog.*, 101, 207–260, 2018.
57. Brink, P.T., Schweitzer, J.P., Watkins, E., Howe, M., Plastics Marine Litter and the Circular Economy, Institute for European Environmental Policy, 2016.
58. Rahman, Md.A., Haque, Md.M., Tareq, S.M., Abundance and characteristics of microplastics in the landfill leachate of Amin Bazar, Dhaka: A potential risk to aquatic environments. *Phys. Chem. Earth Parts A/B/C/*, 134, 103573, 2024.
59. Kentin, B. and Battaglia, G., Policies and Perspectives on Regulating Microplastic Fibre Pollution, in: *Polluting Textiles*, hapter, ISBN 9781003165385.

60. Phuong, N.N., Pham, Q.T., Ngo, T.X.T., Nguyen, T.M.D., Doan, T.O., Nguyen, X.C., Duong, T.N., Le, T.P.Q., Da Le, N., Phuong, N.A., Gasperi, J., Vergnoux, A.Z., Poirier, L., Vu, T.N., Duong, T.T., Occurrence of microplastics in bivalves from the northern coast of Viet Nam. *Reg. Stud. Mar. Sci.*, 78, 103731, 2024.
61. Koelmans, A.A., Nor, N.H.M., Hermsen, E., Kooi, M., Mintenig, S.M., De France, J., Microplastics in freshwaters and drinking water: Critical review and assessment of data quality. *Water Res.*, 155, 410–422, 2019.
62. Jemec, A., Horvat, P., Kunej, U., Bele, M., Kržan, A., Magna, D., Uptake and effects of microplastic textile fibers on freshwater crustacean. *Environ. Pollut.*, 219, 201–209, 2016.
63. Foglia, A., Pittura, L., Vivani, V., Sgroi, M., DeSimoni, L., Eusebi, A.L., Gorbi, S., Regoli, F., Fatone, F., Microplastics in urban water cycles: Looking for a more scientific approach for sampling and characterization in wastewater and drinking water treatment plants. *Sci. Total Environ.*, 952, 175919, 2024.
64. Shruti, V.C., P.-Guevara, F., Martínez, I.E., Muniasamy, G.K., First study of its kind on the microplastic contamination of soft drinks, cold tea and energy drinks - Future research and environmental considerations. *Sci. Total Environ.*, 726, 138580, 2020.
65. Stanton, T., Stanes, E., Gwinnett, C., Lei, X., Cureg, M.C., Ramos, M., Sallach, J.B., Harrison, E., Osborne, A., Sanders, C.H., Baynes, E., Law, A., Johnson, M., Ryves, D.B., Sheridan, K.J., Blackburn, R.S., McKay, D., Shedding off-the-grid: The role of garment manufacturing and textile care in global microfibre pollution. *J. Cleaner Prod.*, 428, 139391, 2023.
66. Rathinamoorthy, R. and Balasaraswathi, S.R., Microfiber Pollution Prevention—Mitigation Strategies and Challenges, in: *Microfiber Pollution*, pp. 205–243, 2022.
67. Zeng, C., Wang, H., Bu, F., Cui, Q., Li, J., Liang, Z., Zhao, L., Yi, C., A regenerated cellulose fiber with high mechanical properties for temperature-adaptive thermal management. *Int. J. Biol. Macromol.*, 274, 133550, 2024.
68. Witnessing an environmental catastrophe: Reflections from the dried-up Aral sea 2024, United Nations, Convention to Combat Desertification.
69. Global Regenerated Cellulose Market, Industry Trends and Forecast to 2029, Data Bridge market research, 2022.
70. Imani, M., Carreras, I.M.V., Mišić, K.D., Kostić, M., Barceló, E., García, M.A.C., Gane, P., Investigating waste mineral-filled cellulose sourcing in circular economy for regeneration into composite: Matching existing market volumes of oil-based plastics for packaging. *Cleaner Circ. Bioecon.*, 8, 100089, 2024.
71. Huang, Z., Zhang, Y., Xing, T., He, A., Luo, Y., Wang, M., Qiao, S., Tong, A., Shi, Z., Liao, X., Pan, H., Liang, Z., Chen, F., Xu, W., Advances in regenerated cellulosic aerogel from waste cotton textile for emerging multidimensional applications. *Int. J. Biol. Macromol.*, 270, 132462, 2024.

72. Victoria, A., Hine, P.J., Ward, K., Ries, M.E., Design of experiments investigation into the production of all cellulose composites using regenerated cellulosic textiles. *Composites, Part A*, 187, 108510, 2024.
73. Cheng, W., Li, Q., Wu, Q., Ye, F., Jiang, Y., Digital capability and green innovation: The perspective of green supply chain collaboration and top management's environmental awareness. *Heliyon*, 10, 32290, 2024.
74. Etim, E., Leveraging public awareness and behavioural change for entrepreneurial waste management. *Heliyon*, 10, 40063, 2024.
75. George, S., Narayanan, D.P., Anju, T.R., Cellulose as a Material for Future Technological Applications, in: *Handbook of Biomass*, pp. 513–533, 2024.
76. Ghaffar, A., Salman, M., Md. Yameen, Iqbal, S.Z., Altaf, S., Munir, B., Sustainable Biomedical Applications of Cellulose, in: *Regenerated cellulose and composites*, pp. 347–379, 2023.
77. Amjad, A.I., and Kumar, R., Effect of Eco-friendly Bamboo Rayon, Lyocell, and Seacell Fibres on the Properties of Blended Knitted Melange Fabrics. *Fibers Polym.*, 24, 1845–1858, 2023.
78. Gabiatti Jr., C., Neves, I.C.O., Lim, L.T., Bohrer, B.M., Rodrigues, R.C., Prentice, C., Characterization of dietary fiber from residual cellulose sausage casings using a combination of enzymatic treatment and high-speed homogenization. *Food Hydrocolloids*, 100, 105398, 2020.
79. Verde, A., Míguez, J.M., Gallardo, M., Melatonin stimulates postharvest ripening of apples by up-regulating gene expression of ethylene synthesis enzymes. *Postharvest Biol. Technol.*, 195, 112133, 2023.

Part II
ENVIRONMENTAL REMEDIATION

5

Regenerated Cellulose Materials' Sensing Applications for Metal and Inorganic Ions

Anita Devi

Department of Applied Sciences, Advanced Centre of Research & Innovation (ACRI), Chandigarh Engineering College, Chandigarh Group of Colleges-Jhanjeri, Mohali, Punjab, India

Abstract

Water pollution by toxic heavy metals, such as lead, mercury, copper, and arsenic, along with inorganic ions, like iodine, chloride, fluoride, and phosphate, poses critical environmental and health challenges. This chapter explores cellulose-based materials for the detection and removal of both heavy metals and inorganic ions. Cellulose is a renewable and biodegradable polysaccharide which serves as an outstanding scaffold for developing advanced sensors and adsorbents due to its structure tunability and biocompatibility. The discussion centers on cellulose-based nanomaterials, hydrogels, and composites incorporated with metal-organic frameworks (MOFs), which shows enhanced adsorption and detection abilities. This chapter highlights on fluorescent probes, colorimetric sensors, and multifunctional cellulose-based materials for the precise and selective recognition of pollutants, including metal and inorganic ions. These systems meet critical demands in environmental monitoring, industrial applications, and water purification, providing sustainable and efficient solutions for improving water quality.

***Keywords*:** Cellulose, nanomaterials, hydrogels, carbon-dots, metal-organic frameworks, heavy metal detection, inorganic ions detection

Email: anitadevi0025@gmail.com

Mohd Shabbir (ed.) Regenerated Cellulose: A Renewable Alternative to Petroleum-Based Plastics, (121–146) © 2026 Scrivener Publishing LLC

5.1 Introduction

Water pollution from heavy metal such as copper (Cu), lead (Pb), cadmium (Cd), mercury (Hg), chromium (Cr), and arsenic (As) has emerged as a significant concern due to their high toxicity, non-degradability, and ability to bioaccumulate in the food chain [1]. Even minimal concentrations of these pollutants can pose significant threats to environment and human health [2]. The rapid industrialization and economic development have intensified the problem, leading to widespread contamination of freshwater resources. Addressing heavy metal contamination in water systems is essential for preventing further soil pollution, ensuring the availability of clean water.

Similarly, the rapid, sensitive, and selective detection of inorganic ions like fluoride (F^-), chloride (Cl^-), iodine (I^-), phosphate (PO_4^{3-}), hypochlorite (ClO^-), and thiocyanate (SCN^-) is vital for industrial processes, and environmental monitoring [3]. In particular, phosphate is commonly found in food additives, fertilizers, detergents, and wastewater, contributes significantly to water eutrophication [4]. Continuous monitoring of chlorine exposure is essential due to its carcinogenic byproducts like trihalomethanes (THMs) [5]. Furthermore, iodine is vital for thyroid health and is influenced by environmental and dietary factors, with exposure primarily occurring through marine plants, animal products, and disinfectants. This highlights the urgent need for sensors capable of detecting and quantifying both heavy metals and inorganic ions in water.

There are several traditional analytical techniques available for ions detection, such as inductively coupled plasma mass spectrometry, atomic absorption spectroscopy, Neutron activation analysis, and flame atomic absorption spectroscopy [6–8]. Although these methods are sensitive and yield accurate results, they experience issues like lengthy sample processing times, difficult procedure, and high costs, which restrict their widespread application. In contrast, chemo-sensors offers significant advantages by converting ion concentrations into signals that can be readily interpreted accessible instruments or even by individuals without specialized training, particularly when the signal involves a color change [9, 10]. Notably, polymer-based chemo-sensors have garnered increasing interest due to their superior performance compared to small organic molecules. A key advantage of these sensors is their ability to amplify signals, as conjugated polymer chains contain multiple recognition elements that enhance selectivity and binding efficiency for specific analytes.

Moreover, adsorption techniques have gained popularity for their efficiency in removing pollutants that are difficult to treat by conventional methods [11]. Adsorption not only produces high-quality water but also offers a cost-effective purification approach.

5.2 Cellulose: A Sensing Material

Cellulose, a naturally abundant and biodegradable polysaccharide, is widely used in hydrogel formation due to its renewable nature [12]. It is composed of β-glucosidic bonds linking D-glucose units and is the primary component of plant fibers like cotton and linen. Due to its excellent physical and chemical properties of biodegradability, biocompatibility, hydrophilicity, thermal stability, optical stability, and ease of modification, cellulose substances make ideal scaffolds for metal-ion sensing [13, 14]. This chapter provides an overview of cellulose and its derivatives, and their uses in sensing technologies (Figure 5.1).

5.2.1 Types of Cellulose

Cellulose was first isolated from plant matter by French chemist Anselme Payen in 1839. It is classified into several types: Type I, which is extracted from natural sources, and Type II, produced through the regeneration of cellulose fibers, with Type II being more stable than Type I. Although natural cellulose is difficult to dissolve due to its strong intermolecular hydrogen bonding, new solvent systems such as sulfuric acid, lithium chloride, ionic liquids, and others have been developed for effective cellulose dissolution [15].

In recent years, there has been growing interest in developing cellulose-based smart materials for multifunctional applications across various sectors.

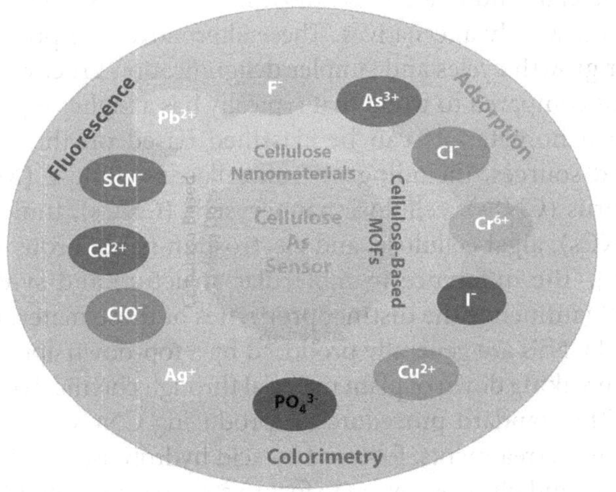

These materials, including fluorescent sensors, nanofibers, hydrogels, films, membranes, nanocrystals, and carbon dots (CDs), offer several advantages such as being renewable, eco-friendly, cost-effective, biodegradable, and easy to fabricate into devices [16]. Importantly, they enable instrument-free applications, making them highly relevant in regions where access to advanced technology is limited.

To improve the physical and chemical properties of cellulose, chromophores containing nitrogen, oxygen, and sulfur units are integrated into its chain structure through various modification techniques, such as oxidation, esterification, etherification, cross-linking, and graft copolymerization [17]. This modification enhances the sensor's performance and adaptability for specific detection needs.

5.3 Preparation of Cellulose-Based Materials

5.3.1 Cellulose Nanomaterials

Cellulose nanomaterials are generally characterized as cellulose structures comprising both crystalline and amorphous regions, with diameters measuring less than 100 nm and lengths extending to several microns. A diverse array of cellulose sources has been utilized to produce these nanomaterials, including various types of wood, seeds, bast fibers, leaf fibers, fruit fibers, algae, fungi, and bacteria [18]. Although wood pulp remains the predominant source for cellulose in most industries and research institutions engaged in the commercial production of cellulose nanomaterials, there is a growing interest in non-woody alternatives. These alternatives are preferred due to their shorter growth cycles and simpler delignification processes, which are advantageous compared to wood that typically has a higher lignin content.

Cellulose nanomaterials can be classified based on their production methods and sources, including microfibrillated cellulose (MFC), cellulose nanofibrils (CNFs), cellulose nanocrystals (CNCs), tunicate-derived CNCs (t-CNCs), algal cellulose, and electrospun nanofibrous membranes (ENMs) [19]. The difference in molecular structure and synthetic techniques greatly influence the distinct properties of these materials.

CNCs and CNFs are generally produced by a top-down approach, which involves the breaking down of plant material through chemical or mechanical treatments. The standard procedure for producing CNCs starts with alkali and bleaching pretreatments, followed by acid hydrolysis, washing, centrifugation, dialysis, and ultrasonication to form a stable suspension [20]. This suspension can undergo lyophilization i.e., freeze-drying or spray-drying based on specific requirements. The factors such as reaction conditions and cellulose source play an important role in determining the properties of CNCs, including crystallinity, aspect ratio, dimensional dispersity, and morphology.

On the other hand, CNFs can be synthesized through various methods, including chemical techniques like TEMPO oxidation and mechanical methods such as high-pressure grinding, ball milling, ultrasonication, and homogenization, often using a combination of these approaches [21]. Pretreatment of wood cellulose fibers can help to reduce energy consumption during mechanical fibrillation, improving the overall degree of fibrillation and efficiency of nanofiber production [22].

The acid hydrolysis is a key technique for fabricating nanocellulose using acids like HCl or H_2SO_4, which produce electrostatically stable CNCs by reacting with the hydroxyl groups of cellulose to form sulfate half-esters [23].

Further, to enhance the performance of CNCs its structural modifications are essential. The incorporation of non-polar or hydrophobic groups increases compatibility with composite materials, while adding positive or negative charges enhances solvent dispersion [24]. These enhancements are generally achieved by chemical modification methods, such as surface treatments and grafting, with functionalization methods including carboxymethylation for ionic surfaces and acetylation, silylation, and etherification for hydrophobic surfaces [25].

Electrospun Nanofibrous Membranes (ENMs) represent an important class of cellulose nanomaterials developed by dissolving cellulose or its derivatives in a solvent *via* a electrospinning techniques [26]. In this process, a high voltage is applied to a viscous solution, resulting in the formation of ultrafine fibers as the solvent evaporates. The resulting non-woven membranes possess high porosity and a large surface area, making them ideal for various applications, especially in optical chemical sensing. This electrospinning process offers control over fiber diameter, alignment, and morphology, allowing tailoring to meet specific application requirements. ENMs can be functionalized by integrating chromophores or other sensing agents into the polymer matrix, which enhances their ability to detect analytes. These membranes display analyte-dependent optical changes, including changes in absorbance and fluorescence in order to improve the stability, reversibility, and sensitivity compared to conventional solution-based systems.

5.3.2 Cellulose-Based Carbon Dots

Carbon dots (CDs) have attracted significant attention due to their advantageous properties, such as cost-effective, excellent biocompatibility, high water solubility, and chemical stability. They are typically synthesized through various methods, such as solvothermal treatment, microwave irradiation, ultrasonic treatment, laser ablation, and electrochemical carbonization. These methods utilize different carbon-based precursors ranging from molecular (like polyaniline and citrates), biomass (like lemon peel and bamboo leaves), as well as waste materials (like carbon paper and bike soot) [27].

To enhance the multifunctional characteristics of CDs, two approaches can be employed. The first involves doping CDs with metallic and non-metallic heteroatoms. The second approach is to select precursor materials that inherently contain carbon, nitrogen, and sulfur units [28]. For example, by treating cellulose with nitric acid, researchers can introduce oxygen and nitrogen units into cellulose molecules while also converting cellulose crystals into nanoparticles, which increases both surface accessibility and the quantum yield of CDs. Liu *et al.* [29] developed a chemosensor based on nitrogen-doped carbon dots (N-CDs) for detecting Fe^{3+} ions. The oxidation process utilizing $HNO_3/H_3PO_4/NaNO_2$ significantly improved the yield and quantum yield of N-CDs derived from softwood dissolving pulp, resulting in spherical N-CDs with a graphite-like structure and abundant functional groups.

5.3.3 Cellulose-Based Hydrogels

Hydrogels are three-dimensional polymer networks capable of absorbing and retaining significant amounts of water. They are used in various biomedical and environmental applications due to their low toxicity, biocompatibility, and high-water content. Hydrogels can be classified based on composition, crosslinking mechanisms, and swelling behavior [30]. By optimizing the synthesis conditions, it is possible to achieve hydrogels with controlled swelling properties, increased porosity, and a high specific surface area.

Hydrogels can be synthesized using crosslinking methods, which can be categorized into physical and chemical types (Figure 5.2).

a. **Physical Crosslinking:** Physically crosslinked hydrogels depend on polymer chain entanglement, hydrogen bonding, and ionic interactions. These reversible hydrogels dissolve at high temperatures and reform under specific conditions. They are easy to prepare, biocompatible, and environmentally friendly due to the absence of crosslinking agents. For example, alginate hydrogels form through ionic interactions, while agarose gels use hydrogen bonding for their structure [31].

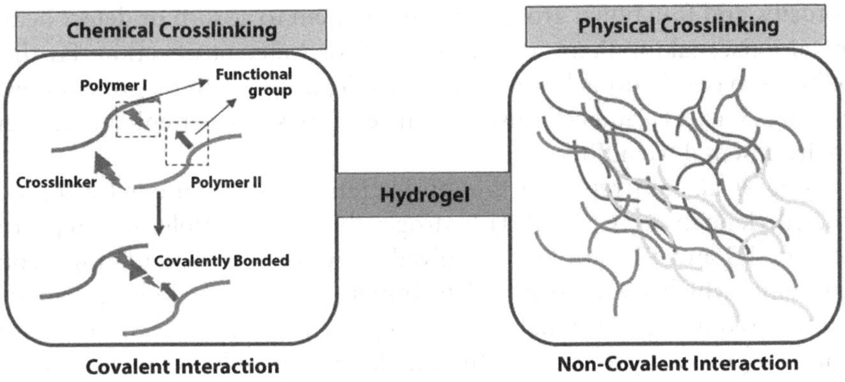

Figure 5.2 Illustration of physical and chemical crosslinking methods.

b. **Chemical Crosslinking:** Chemical crosslinking forms covalent bonds in stable three-dimensional hydrogels, making them resistant to heat and preventing disintegration. It includes methods like copolymerization, graft polymerization, interpenetrating networks, radiation polymerization, and enzymatic polymerization. These hydrogels are structurally robust with good thermal and chemical stability. For example, acrylic acid and acrylamide can be used to form hydrogels with a crosslinking agent. Semi-interpenetrating hydrogels can be made by grafting and crosslinking chitosan, as demonstrated in a temperature-responsive hydrogel using the Michael addition reaction [32]. Chemical crosslinking improves hydrogel stability over physical crosslinking methods.

Hu *et al.* [33a] developed a bilayer hydrogel by combining physical crosslinking of sodium alginate and carboxymethyl cellulose with a chemically crosslinked outer layer, enhancing strength and preventing swelling. Zhao *et al.* [33b] prepared a modified cellulose hydrogel for heavy metal ion adsorption, achieving high capacities for copper (157.51 mg/g), lead (393.28 mg/g), and cadmium (289.97 mg/g).

Cellulose-based hydrogels can be prepared from cellulose ethers like methyl cellulose, hydroxypropyl methyl cellulose, and carboxymethyl cellulose (CMC), which are inexpensive, water-soluble, and environmentally friendly. Chemical modifications, such as incorporating polyethyleneimine (PEI), enhance their ability to remove heavy metals and improve their mechanical properties. These hydrogels possess large surface areas, high porosity, and functional groups that allow them to adsorb or detect heavy metal ions, making them ideal candidate for water purification. Further, cellulose-based hydrogels are biodegradable, as they can be broken down by fungi, bacteria, and cellulose-specific enzymes, further enhancing their environmental friendliness.

Song *et al.* [34] reported the fabrication of carboxymethyl cellulose/polyethyleneimine (CMC/PEI) hydrogel through a simple one-step process. CMC and PEI are first dissolved in water to create homogeneous solutions, which are then mixed to initiate a mild crosslinking reaction without requiring additional crosslinking agents. This leads to the formation of a porous hydrogel structure, making it suitable for environmental applications. The hydrogel showing strong efficiency in removing hexavalent chromium [Cr(VI)] from aqueous solutions under mildly acidic conditions. Its removal process involves adsorption of Cr(VI), reducing it to trivalent chromium [Cr(III)], and subsequently embedding Cr(III) within the hydrogel matrix. This simple method demonstrates the potential of hydrogel for effective heavy metal removal.

Durpekova *et al.* [35a] synthesized a degradable and reproducible cellulose hydrogel using citric acid as the crosslinking agent that exhibits significant swelling potential, helping to maintain soil quality and protecting aquatic life. In another study, Das *et al.* [35b] developed biodegradable hydrogels composed of carboxymethylcellulose sodium salt (CMCNa) and hydroxyethyl cellulose (HEC) employing citric acid (CA) as a crosslinker. Chemical analysis *via* Fourier transform infrared spectroscopy (FT-IR) and the incorporation of cellulose nanocrystals (CNCs) showed that hydrogels with 2% CA exhibited around 600% swelling. However, the dispersion of CNCs negatively affected tensile strength, and optimizing both CA concentration and CNCs distribution could enhance controlled water absorption for agricultural use.

5.3.4 Cellulose-Based Metal Organic Frameworks

Metal-Organic Frameworks (MOFs), also known as porous coordination polymers (PCPs) or porous coordination networks (PCNs) are hybrid porous materials consisting of inorganic metal clusters or nodes connected by organic linkers, forming crystalline structures with varying dimensions (0D, 1D, 2D, and 3D) [36]. These materials have a higher surface area and large pore volume compared to other porous materials like zeolites and covalent organic frameworks (COFs) [37]. MOFs find applications across broad range of fields including adsorption, photocatalysis, fuel cells, environmental applications, and drug delivery.

Combining MOFs with polymers enhances their properties and resolving challenge of low stability. MOFs can be conjugated with different polymers, such as COFs, coordination polymers, and biopolymers like cellulose. This integration enhances MOFs' performance, resulting in materials with improved optical properties, corrosion resistance, conductivity, and mechanical strength. The use of semiconductor COFs further enhances the optoelectronic properties of MOFs. Cellulose-MOFs, hybrid materials can be synthesized using either *in situ* or *ex situ* methods [38] (Figure 5.3).

In the *in situ* approach, MOF precursors are added to cellulose materials, with MOFs growing within cellulose fibers *via* processes like solvothermal or hydrothermal methods. This method enhances the interaction between cellulose and metal ions, improving mechanical properties and controlling MOF crystal morphology. However, controlling the MOF loading is challenging, often resulting in low product yields, which can be increased through layer-by-layer techniques.

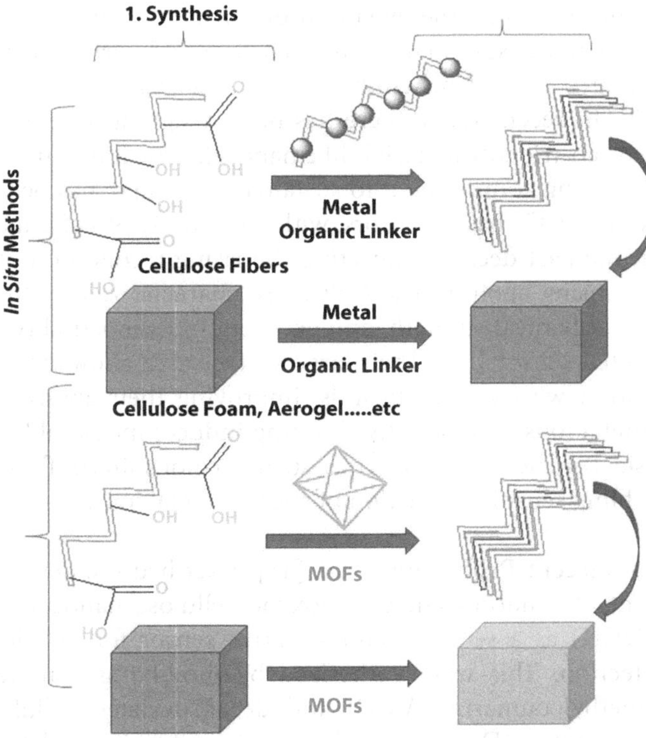

The *ex situ* method involves synthesizing MOFs separately and loading them onto cellulose fibers, offering higher MOF loading and increased accessibility of the MOFs to reactants. However, it requires multiple steps, and MOFs often aggregate, limiting uniform distribution.

The materials combine the high surface area of MOFs with the flexibility and porous nature of cellulose, making them suitable for applications like adsorption, catalysis, and drug encapsulation.

5.3.5 Applications as Metal Ion Sensors

Taking advantage of the presence of many hydroxyl groups in cellulose chains, various functional groups and fluorophores can be incorporated into the cellulose backbone, resulting in a wide variety of cellulose-derived products.

5.3.6 Applications of Cellulose Nanocrystals (CNCs)

The exploration of cellulose-based sensors began with Southern's pioneering work in 1975, which utilized cellulose nitrate filters for specific DNA fragment detection. The evolution of sensing technology since the introduction of biosensors in 1956 has led to a wide array of detection mechanisms, such as electrochemical, optical, piezoelectric, and thermal methods [39]. In recent years, there has been a significant emphasis on incorporating eco-friendly materials like nanocellulose into sensor design, reflecting a growing commitment to sustainability. There has been extensive research on CNCs-based sensors, with around 71.68% of publications published in the past decade, indicating the rising recognition of CNCs' potential in sensing applications. CNCs are characterized by their high aspect ratio, mechanical strength, and excellent biocompatibility, making them well-suited sensor [40]. Their large surface area allows for effective functionalization with various ligands, improving their ability to selectively bind metal ions. This selective binding induces measurable changes in their physical or chemical properties, such as color shifts or fluorescence changes, enabling sensitive detection of specific metal ions.

a. **Fluorescent Probe:** Sun *et al.* [41] described the development of a fluorescent dicarboxylic cellulose nanocrystals (FDCN) as a selective and sensitive sensor for Fe^{3+} ions detection. This was synthesized by modifying 7-amino-4-methylcoumarin (AMC) and decarboxylated cellulose nanocrystals (DCN) through periodate-chlorite oxidation.

The selectivity of sensor was confirmed by examining its behavior towards various metal ions, with Fe^{3+} showing the most distinct effect. As the concentration of Fe^{3+} increases, there is gradual decrease in fluorescence intensity demonstrating a strong linear correlation between the fluorescence intensity ratio (F/F_0) and Fe^{3+} concentration within the range of 1–500 μM ($R^2 = 0.9873$). The probable sensing mechanism involves the formation of a complex between FDCN and Fe^{3+}, resulting to fluorescence quenching. Due to high sensitivity, hydrophilicity, and selectivity, FDCN is suitable for trace-level Fe^{3+} detection in drinking water and shows potential biological applications.

b. **Colorimetric Detection:** Hanif *et al.* [42] reported TA@CNCs, hybrid materials prepared from tannic acid (TA) and cellulose nanocrystals (CNCs) extracted from plants, for the colorimetric detection of Fe^{3+} ions. These materials showed distinct color changes in the presence of Fe^{3+}, making them visually detectable and good for practical applications. The TA@CNCs demonstrated excellent selectivity in a binary metal ion system, showing significant absorbance for Fe^{3+} compared to other metal ions such as Zn^{2+}, Ni^{2+}, Cd^{2+}, and Cu^{2+}, with detection limits of 0.52 ppb in serum and 0.54 ppb in milk. The colorimetric response correlated strongly with Fe^{3+} concentration in both serum and milk, with a linear relationship ($R^2 = 0.993$) observed. This efficient detection method is well-suited for biological and field applications, providing rapid and reliable results.

c. **Adsorption:** CNCs have shown effective performance in the adsorption of metal ions, as demonstrated by several studies. For example, Mohammed *et al.* [43] reported a composite system of bovine serum albumin-protected gold nanoclusters (Au@BSA NCs) embedded in cellulose nanocrystal-alginate hydrogel beads designed to detect and remove toxic heavy metal ions like mercury in water. Practical applications, including batch adsorption and a dip-pen device for *in situ* monitoring, were demonstrated for effective water purification. The high surface area and functionalization capabilities of CNCs make them suitable for adsorbing various metal ions, providing both detection and remediation solutions in environmental applications.

5.3.7 Applications of Electrospun Nanofibrous Membranes (ENMs)

a. **Fluorescent Probe:** Electrospun nanofibers created from fluorophore-doped polymers, such as 9-chloromethylanthracene-doped cellulose acetate, enhance sensing performance through fluorescence quenching. These membranes have been applied to detect Cu^{2+} ions using materials like poly(methylmethacrylate) (PMMA) and ethyl cellulose (EC) as supports for fluorescent probes, achieving sensitivity levels up to 20 times higher than those of thin films, with detection capabilities in the picomolar range [44]. Zhang *et al.* [45] developed chemosensors utilizing ruthenium dye N719 immobilized on titania-coated cellulose nanofibers, demonstrating high selectivity and sensitivity for Hg^{2+} detection with a limit of 10 ppb. Further, Xu *et al.* [46] created solid-state sensors made from rhodamine-modified cellulose filter paper that produce rapid fluorescence responses, with distinct color changes indicating the presence of Hg^{2+}.

b. **Colorimetric Detection:** Nanofibers doped with rhodamine dyes exhibit significant color changes in the presence of Cu^{2+}, showcasing strong selectivity and reusability for heavy metal detection. Similarly, curcumin-loaded cellulose acetate (CC-CA) nanofibers demonstrate high selectivity for Pb^{2+}, with a distinct visual color shift from yellow to orange, making them ideal for low-cost, disposable sensors [47]. The selectivity of CC-CA membranes for Pb^{2+} is clearly seen from UV/Vis spectra changes, where Pb^{2+} shows significantly higher absorption compared to other ions, and this is further confirm by a bar chart quantifying the membrane's response to various heavy metals. Further, fluorescent silver nanoclusters (AgNCs), prepared through UV irradiation of modified poly(acrylic acid) (MPAA) with Ag^{+} ions, provide a sensitive and effective platform for Hg^{2+} sensing. This process involves collecting water samples, incubating them for Hg^{2+} sensing, and using paper-based devices to capture the data, which is then transmitted *via* cloud computing for analysis. These advancements show the versatility and practicality of engineered nanomaterials (ENMs) for real-time, low-cost heavy metal sensing applications [27].

5.3.8 Applications of Cellulose Nanofibrils (CNFs)

Combining CNFs with other materials like metal-organic frameworks (MOFs), can significantly improve their sensing properties.

a. **Adsorption:** The study by Abdelhamid *et al.* highlights the performance of CelloZIFPaper, synthesized *via in situ* and *ex situ* methods, for heavy metal adsorption and sensing applications. This substance showed remarkable adsorption capacities for various metal ions (Cd^{2+}, Cu^{2+}, Fe^{3+}, Pb^{2+}, and Co^{2+}), with the *in situ* variant outperforming the *ex situ* counterpart and TOCNF [28]. Pb^{2+} adsorption showed a saturation capacity of approximately 800 mg/g, demonstrating high binding efficiency. Recyclability tests revealed that Pb^{2+} maintained better adsorption efficiency over three cycles compared to Cu^{2+} and Co^{2+}, although efficiency declined with each cycle. Diffuse reflectance spectra (DRS) provided spectral evidence of metal ion uptake, with significant shifts observed after adsorption of Pb^{2+}, Cu^{2+}, and Co^{2+}. Moreover, visual inspection of CelloZIFPaper revealed distinct color changes following Co^{2+} and Cu^{2+} adsorption, confirming qualitative metal ion uptake. These results highlight the potential of CelloZIFPaper for heavy metal removal from water and as an electrochemical sensor for real-time detection.

In another study by Anirudhan *et al.* [48], a hybrid material was prepared by modifying nanocellulose (NC) with nanobentonite (NB) to create a highly effective Co^{2+} metal adsorbent. The method involved mixing NC and NB in deionized water under stirring conditions, followed by grafting polymer chains of itaconic acid (IA) and methacrylic acid (MAA) onto the composite surface to form P(IA/MAA)-g-NC/NB. This modification enhanced the carboxyl functional groups, enabling efficient adsorption of Co^{2+} ions at a pH of 6. The adsorption process was explained by a two-stage countercurrent batch system, where in the first phase, the adsorbent interacts with a solution of initial concentration (C_0), resulting in an intermediate concentration (C_1) and adsorbed quantity (q_1). In the second phase, the left-out solution contacts fresh adsorbent, further reducing the concentration (C_2) and increasing adsorption (q_2). The adsorption followed a Sips isotherm model under optimum conditions of 30°C, pH 6, an equilibrium time of 2h, and an adsorbent dose of 2 g/L. Especially, even at a low adsorbent dose of 0.25 g/L, this material successfully removed up to 90% of Co^{2+} ions, demonstrating its high efficiency.

b. **Fluorescence Probe:** The luminescence behavior of the 20% Eu-MOF@TOCNF film was studied under excitation at 254 nm in the presence of various metal ions, including Co^{2+}, K^{+}, Ca^{2+}, Na^{+}, Zn^{2+}, Mn^{2+}, Mg^{2+}, Ni^{2+}, and Cu^{2+}. The emission spectra revealed a significant quenching of the fluorescence intensity at 614 nm only in the presence of Cu^{2+}, demonstrating high selectivity for this ion. This selectivity was further confirmed by a bar graph comparing the emission intensity at 614 nm after soaking the film in 10^{-3} M solutions of different metal nitrates for 1 h, where Cu^{2+} showed the most pronounced intensity reduction. Additionally, the luminescence spectra of the film at varying Cu^{2+} concentrations (10–100 μM) exhibited a concentration-dependent quenching effect at 614 nm, establishing the sensitivity of the film towards Cu^{2+}. A linear relationship was observed between the fluorescent intensity ratio ($I_0/I-1$, where I_0 is the initial intensity and I is the quenched intensity) at 614 nm and Cu^{2+} concentration, with a high coefficient of determination ($R^2 = 0.9915$), confirming the film's capability for sensitive and accurate detection of Cu^{2+} in aqueous systems [49].

5.3.9 Applications of Cellulose-Based Carbon Dots

Carbon dots exhibit exceptional potential in sensing applications, particularly for heavy metal detection. Cellulose plays a crucial role in enhancing the performance of these sensors.

a. **Fluorescence Probe:** Yang *et al.* [50a] advanced the application of CDs by developing acidophilic S-doped CDs using cellulose fibers and sulfuric acid as a carbonization agent. These S-CDs showed strong fluorescence and a high quantum yield of up to 32%, maintaining excellent selectivity and sensitivity for detecting Fe^{3+} ions even in strong acidic conditions (pH 0). The experimental results indicated that the aggregation of S-CDs during detection was responsible for fluorescence quenching, providing promising potential for detecting metal cations in low pH environments. The incorporation of heteroatoms into CDs has been shown to enhance their optical and physicochemical properties. For instance, Cheng *et al.* [50b] fabricated nitrogen and sulfur co-doped CDs (N/S-CDs) from willow catkin cellulose bio-waste using a one-step combustion method. The N/S-CDs exhibited high quantum yield (13.3%) and demonstrated excellent selectivity and sensitivity for detecting Fe^{3+} ions, with a detection limit of approximately 0.03 μM. The observed electron transfer mechanism between the N/S-CDs and Fe^{3+} ions confirmed their potential for intracellular detection of metal ions.

Recent advancements also highlight the benefits of agricultural waste as a precursor. Magagula *et al.* [50c] explores the eco-friendly synthesis of nitrogen-doped carbon quantum dots (N-CQDs) using cellulose nanocrystals (CNCs) derived from agricultural waste through a one-step microwave-assisted method. The resulting N-CQDs, averaging 5 ± 2 nm in size, exhibit strong excitation-dependent fluorescence with an emission maximum at 420 nm and pH-dependent stability. The N-CQDs exhibited high sensitivity and selectivity for Fe^{3+} detection in aqueous solutions, achieving a detection limit of 75 nM through fluorescence quenching. This rapid and sustainable approach offers significant potential for environmental monitoring and aligns with green nanotechnology principles.

5.3.10 Applications of Cellulose-Based Hydrogels

a. **Fluorescent Probe:** He *et al.* [51] developed a self-healing fluorescent hydrogel derived from oxidized carboxymethyl cellulose, specifically engineered to detect and adsorb silver ions (Ag^+). The hydrogel exhibited a low detection limit of 3.798 μM for Ag^+, demonstrating its potential for sensitive environmental monitoring. Further, it showed a good adsorption capacity of 407 mg/g, effectively capturing Ag^+ ions, with the adsorption process fitting with the Langmuir isotherm model and following pseudo-second-order kinetics. Remarkably, the hydrogel can self-repair without external intervention and is regenerative up to 7 times without any decrease in adsorption performance. This impressive hydrogel, combining fluorescent detection, high adsorption efficiency, and self-healing properties, offers an excellent solution for wastewater treatment.

b. **Colorimetric Sensing:** Ren *et al.* [52] reported a nanocellulose-based hydrogel (NBH) with silver nanoclusters for sensing and removing hexavalent chromium [Cr^{6+}] ions which is a major environmental pollutant. The recyclability tests showed the NBH maintained over 85% efficiency after four cycles. Moreover, the NBH was seen to be about 6 times cheaper than commercial activated carbon, offering a cost-effective, multifunctional solution for Cr^{6+} removal.
c. **Adsorption:** A CMC/PEI hydrogel was synthesized by carboxymethyl cellulose (CMC) as the matrix, polyethyleneimine (PEI) as the functional component, and epichlorohydrin as the cross-linking agent [34]. This double-network hydrogel was prepared in a one-step method for efficient Cr^{6+} removal. The hydrogel promoted Cr^{6+} reduction by producing hydrogen ions, with Cr^{3+} immobilized on its surface, significantly reducing Cr^{6+} toxicity. Additionally, the hydrogel could be easily separated for practical use.

5.3.11 Applications of Cellulose Based MOFs

Metals play a critical role in human health and nutrition, but imbalances in metal homeostasis can lead to severe health issues, including cancer. Metal-organic frameworks (MOFs) combined with cellulose, have been explored as adsorbents for various toxic metals in water and the environment. The following discusses the adsorption of different metal ions using cellulose based MOFs.

a. **Arsenic (As) Adsorption:** Arsenic (As), a Group 1 carcinogen, is commonly found in water due to natural leaching and industrial activities like mining. Toxic species such as As^{3+} and As^{5+} can be effectively removed using cellulose-based MOFs. For example, MIL-100(Fe) immobilized onto bacterial cellulose (BC) nanofibers exhibited an adsorption capacity of 4.81 mg/g for As^{3+}. The combination of ZnO nanorods with ZIF-8 on cotton fabric also demonstrated an improved arsenic uptake efficiency, achieving 70% compared to 38% using ZnO alone [53].
b. **Lead (Pb) Adsorption:** Lead (Pb^{2+}) contamination in water is a significant concern due to its toxicity, especially in children. Magnetic cellulose nanocrystals (MCNC) combined

with Zn-based MOFs (Zn-BTC) showed an impressive adsorption capacity of 558.66 mg/g for Pb^{2+}, far outperforming MCNC alone [54]. Another cellulose-based MOF composites such as UiO-66-NH_2@cellulose aerogels offered high selectivity and adsorption efficiency for Pb^{2+} removal.

c. **Chromium (Cr) Adsorption:** Hexavalent chromium (Cr^{6+}) is another highly toxic metal that poses a risk to human health. Cellulose-based MOFs, such as ZIF-8@CNF@cellulose foams and ZIF-8@CA composites, demonstrated effective Cr^{6+} removal [55]. ZIF-8@CA achieved removal efficiencies of up to 99.74% at lower Cr^{6+} concentrations. In addition, UiO-66–based composites, such as UiO-66@CNC aerogels, were able to remove Cr^{6+} with adsorption efficiencies reaching 85% depending on the UiO-66 loading.
d. **Copper (Cu) Adsorption:** Cellulose-Based MOFs also show excellent performance for copper (Cu^{2+}) adsorption. UiO-66-NH_2@CA composites exhibited an adsorption capacity of 39.33 mg/g for Cu^{2+}, while UiO-66 aerogels modified with ethylenediaminetetraacetic acid (EDTA) displayed enhanced adsorption behavior for multiple metal ions, including copper [56]. These materials are highly promising for copper removal from contaminated water sources.
e. **Cadmium (Cd) Adsorption:** Cadmium (Cd^{2+}) is another toxic metal that can accumulate in the body over time. ZIF-8@BC foams and Cellulose-ZIF-L composites have been shown to effectively adsorb Cd^{2+} [57]. The porous structure and functional groups of the cellulose-MOF composites provide a large surface area and high affinity for cadmium ions.

5.4 Applications as Inorganic Ion Sensors

5.4.1 Applications of Carbon Dots (CDs) and Cellulose Nanocrystals (CNCs)

Das *et al.* developed fluorescein-based probes combined with carbon dots (CDs) and cellulose nanocrystals (CNCs) for efficient fluorometric detection of fluoride ions (F^-) in water [58]. The limit of detection (LOD) for the CD-based probe was calculated as 35 mM (664.93 mg/L), while the CNC-based probe achieved a much lower LOD of 8 mM (approximately 151.98

mg/L), which is significantly lower than the 1.5-mg/L guideline established by the World Health Organization. Moreover, Li *et al.* synthesized a cellulose membrane-based sensor for the detection and removal of F^- ions [59]. This sensor was prepared by using pyrene-based boronic acid and CDs and further modified with amino groups, demonstrated remarkable adsorption capabilities with 90.2% of fluoride removal efficiency.

5.4.2 Applications of Cellulose-Based Fluorometric Sensor

Li *et al.* developed a fluorogenic cellulose material serving as both a sensor and adsorbent for F^- ion removal from water [60]. The material was able to detect fluoride at concentrations as low as 10 mg/L and reached maximum adsorption within 10 min. The adsorption process adhered to the Langmuir isotherm model and pseudo-second-order kinetics. This method highlights the potential of simple and durable materials for integrated F^- detection and removal in water purification systems. As a broad range of fluorophores available for sensor design, this approach could advance research in anion sensing and removal techniques.

5.4.3 Applications of Cellulose-Based Colorimetric Sensor

A $CsPbBr_3$ PQD/cellulose composite was synthesized by Park *et al.* as a colorimetric sensor for real-time detection of iodine and chlorine ions in aqueous medium [61]. The composite exhibits high photoluminescence efficiency and excellent stability against environmental moisture. The material enables sensitive detection for iodine and chlorine ions, with detection limits of 2.56 and 4.11 mM, respectively, leveraging a fast anion exchange reaction and the porous structure of cellulose fibers. The main advantages of this method include a low-cost one-pot synthesis, high sensitivity, rapid response, naked-eye detection, flexibility for practical applications, and compatibility with water-based systems. The fast color change capability of sensor makes it suitable for real-time and on-site detection. This versatile composite holds significance as a platform for biosensors in food safety, medical diagnostics, environmental monitoring, and point-of-care health systems.

5.4.4 Applications of Cellulose-Based Hydrogel

Hai *et al.* reported a fluorescent lanthanide optical hydrogel (Tb^{3+}-carboxymethyl cellulose, CMC complex) having reversible responsiveness to ClO^- and SCN^- ions, demonstrating potential applications in fingerprint recognition, encryption, and decryption [62]. The fluorescence of fingerprint images could be switched on and off by the successive addition of ClO^- and SCN^-, respectively. When ClO^- was added to fingerprint glass slides, the fluorescence was significantly quenched, but it was restored upon adding SCN^-. These results suggest the hydrogel's potential for secure printing applications. Emission spectra of various compounds showed that the Tb^{3+}-CMC hydrogel exhibited strong fluorescence compared to other derivatives, including the Tb^{3+}-complex-functionalized CMC hydrogel and the 2,6-dimethylpyridin-4-amine-functionalized CMC-binding aptamer. Successive addition of ClO^- to Tb^{3+}-CMC hydrogel resulted in fluorescence quenching, while addition of SCN^- to the Tb^{3+}-CMC hydrogel-ClO^- complex restored fluorescence, showing reversible switching. This fluorescence reversibility was sustained over five cycles of ClO^- and SCN^- additions, indicating the stability of the material. Further, a Chinese character drawn on filter paper demonstrated luminescence switching under UV light, where SCN^- activated fluorescence and ClO^- quenched it. This reversible fluorescence control demonstrates the use of hydrogel in secure and dynamic applications.

Zhang *et al.* prepared a fluorescent cellulose nanofibril-based hydrogel (FCNH) for detection and adsorption of iodine ions (I^-) from wastewater [63]. Composed of cellulose nanofibrils (CNs), silver nanoclusters (AgNCs), and MIL-125-NH_2, it achieved a high adsorption capacity of 373.7 mg/g and retained 88% efficiency after six adsorption-desorption cycles. FCNH also functioned as a fluorescent sensor, detecting I^- with linear ranges of 5 to 200 mg/L and 0.2 to 1.0 μg/L, and a detection limit of 0.11 μg/L. Its performance stemmed from a 3D porous structure. Cost-effective compared to activated carbon, FCNH shows promise as an affordable, efficient, and reusable solution for iodine ion removal.

5.4.5 Applications of Cellulose Based Adsorption Sensor

Shang *et al.* developed a quaternary ammonium anion exchanger derived from cellulose-based sugarcane bagasse (QA-SB) for phosphate adsorption and subsequent use as a slow-release phosphate fertilizer in water or soil [64]. Phosphate adsorption by QA-SB increased with temperature and was most effective at a pH range of 4.33 to 5.88. In fixed-bed column experiments, phosphate capture ranged from 18.9 to 21.4 mg/g in the presence of chloride or nitrate (100 mg/L) but dropped to 9.1 mg/g when sulfate (100 mg/L) was present due to stronger competitive effects. The loaded phosphate could be efficiently desorbed, with the regenerated QA-SB retaining 94.5% adsorption capacity after four cycles. Slow-release studies showed

43.6% of the adsorbed phosphate (11.6 mg/g) was released in water over 12 days, while in soil, equilibrium was reached within 4 days, releasing 27.8% (7.4 mg/g). This highlights QA-SB's potential for phosphate recovery and sustainable fertilizer applications.

AI disclosure statement: The author acknowledges the use of AI-based tools for minor language editing or grammatical correction. However, all intellectual contributions, analyses, and conclusions presented in this chapter are entirely the author's own.

References

1. Liu, J., Su, D., Yao, J., Huang, Y., Shao, Z., Chen, X., Soy protein-based polyethylenimine hydrogel and its high selectivity for copper ion removal in wastewater treatment. *J. Mater. Chem. A*, 5, 8, 4163, 2017.
2. Afkhami, A., Soltani-Felehgari, F., Madrakian, T., Ghaedi, H., Rezaeivala, M., Fabrication and application of a new modified electrochemical sensor using nano-silica and a newly synthesized Schiff base for simultaneous determination of Cd^{2+}, Cu^{2+} and Hg^{2+} ions in water and some foodstuff samples. *Anal. Chim. Acta*, 771, 21, 2013.
3. He, L.W., Dong, B.L., Liu, Y., Lin, W.Y., Fluorescent chemosensors manipulated by dual/triple interplaying sensing mechanisms. *Chem. Soc. Rev.*, 45, 6449, 2016.
4. Warwick, C., Guerreiro, A., Soares, A., Sensing and analysis of soluble phosphates in environmental samples: A review. *Biosens. Bioelectron.*, 41, 1, 2013.
5. Shi, P., Jia, S., Zhang, X.X., Zhang, T., Cheng, S., Li, A., Metagenomic insights into chlorination effects on microbial antibiotic resistance in drinking water. *Water Res.*, 47, 111, 2013.
6. Su, S., Chen, B., He, M., Hu, B., Graphene oxide–silica composite coating hollow fiber solid phase microextraction online coupled with inductively coupled plasma mass spectrometry for the determination of trace heavy metals in environmental water samples. *Talanta*, 123, 1, 2014.
7. Kazi, T.G., Jalbani, N., Arain, M.B., Jamali, M.K., Afridi, H., II, Sarfraz, R.A., Shah, A.Q., Toxic metals distribution in different components of Pakistani and imported cigarettes by electrothermal atomic absorption spectrometer. *J. Hazard. Mater.*, 163, 1, 302, 2009.
8. Gong, T., Liu, J., Liu, X., Liu, J., Xiang, J., Wu, Y., A sensitive and selective sensing platform based on CdTe QDs in the presence of L-cysteine for detection of silver, mercury and copper ions in water and various drinks. *Food Chem.*, 213, 306, 2016.
9. Kumar, V., Kim, H., Pandey, B., James, T.D., Yoon, J., Anslyn, E.V., Recent advances in fluorescent and colorimetric chemosensors for the detection of

chemical warfare agents: A legacy of the 21st century. *Chem. Soc. Rev.*, *52*, 2, 663, 2023.
10. Uniyal, A., Srivastava, G., Pal, A., Taya, S., Muduli, A., Recent advances in optical biosensors for sensing applications: A review. *Plasmonics*, 18, 2, 735, 2023.
11. Petit, C., Present and future of MOF research in the field of adsorption and molecular separation. *Curr. Opin. Chem. Eng.*, 20, 132, 2018.
12. Valencia, L. and Abdelhamid, H.N., Nanocellulose leaf-like zeolitic imidazolate framework (ZIF-L) foams for selective capture of carbon dioxide. *Carbohydr. Polym.*, 213, 338, 2019.
13. (a) Rajesh, S., Shobana, K.H., Anitharaj, S., Mohan, D.R., Preparation, morphology, performance, and hydrophilicity studies of poly (amide-imide) incorporated cellulose acetate ultrafiltration membranes. *Ind. Eng. Chem. Res.*, 50, 9, 5550, 2011. (b) Peng, N., Wang, Y., Ye, Q., Liang, L., An, Y., Li, Q., Chang, C., Biocompatible cellulose-based superabsorbent hydrogels with antimicrobial activity. *Carbohydr. Polym.*, 137, 59, 2016.
14. Lindh, E.L., Bergenstråhle-Wohlert, M., Terenzi, C., Salmén, L., Furó, I., Non-exchanging hydroxyl groups on the surface of cellulose fibrils: The role of interaction with water. *Carbohydr. Res.*, 434, 136, 2016.
15. Jeevanandam, J., Rodrigues, J., Pan, S., Danquah, M.K., Chapter 8- Cellulose-based bionanocomposites: synthesis, properties, and applications, Sharma, B., Thomas, S., Bajpai, P.K., Ghosal, K., Shekhar, S., (Eds.), In Micro and Nano Technologies, *Advances in Bionanocomposites*, pp. 191–210, Elsevier 2024.
16. Ummartyotin, S. and Manuspiya, H., A critical review on cellulose: From fundamental to an approach on sensor technology. *Renew. Sustain. Energy Rev.*, 41, 402, 2015.
17. Chen, X., Liu, Y., Yang, Q.Q., Wu, Y.C., From natural cellulose to functional nanocomposites for environmental applications, *Fundamentals of natural fibres and textiles*, pp. 111–151, Woodhead Publishing, 2021.
18. Sarwer, A., Hamed, S.M., Osman, A.I., Jamil, F., Al-Muhtaseb, A.A.H., Alhajeri, N.S., Rooney, D.W., Algal biomass valorization for biofuel production and carbon sequestration: A review. *Environ. Chem. Lett.*, 20, 5, 2797, 2022.
19. Kassie, B.B., Daget, T.M., Tassew, D.F., Synthesis, functionalization, and commercial application of cellulose-based nanomaterials: A review. *Int. J. Biol. Macromol.*, 278, 134990, 2024.
20. Mishra, R.K., Ha, S.K., Verma, K., Tiwari, S.K., Recent progress in selected bio-nanomaterials and their engineering applications: An overview. *J. Sci.-Adv. Mater. Dev.*, 3, 3, 263, 2018.
21. Rana, A.K., Frollini, E., Thakur, V.K., Cellulose nanocrystals: Pretreatments, preparation strategies, and surface functionalization. *Int. J. Biol. Macromol.*, 182, 1554, 2021.

22. Xie, H., Du, H., Yang, X., Si, C., Recent strategies in preparation of cellulose nanocrystals and cellulose nanofibrils derived from raw cellulose materials. *Int. J. Polym. Sci.*, 2018, 1, 7923068, 2018.
23. Kurniawan, T.W., Sulistyarti, H., Rumhayati, B., Sabarudin, A., Cellulose nanocrystals (CNCs) and cellulose nanofibers (CNFs) as adsorbents of heavy metal ions. *J. Chem.*, 2023, 1, 5037027, 2023.
24. Zhang, Y., Zhang, Y., Xu, W., Wu, H., Shao, Y., Han, X., Zhou, M., Gu, P., Li, Z., Preparation methods of cellulose nanocrystal and its application in treatment of environmental pollution: A mini-review. *Colloid Interface Sci. Commun.*, 53, 100707, 2023.
25. Khoo, R.Z., Chow, W.S., Ismail, H., Sugarcane bagasse fiber and its cellulose nanocrystals for polymer reinforcement and heavy metal adsorbent: A review. *Cellulose*, 25, 4303, 2018.
26. Rahmani, S., Khoubi-Arani, Z., Mohammadzadeh-Komuleh, S., Maroufkhani, M., *Electrospinning of cellulose nanofibers for advanced applications*, pp. 263–296, Springer International Publishing, Cham, 2022.
27. (a) Liu, X., Zong, C., Lu, L., Fluorescent silver nanoclusters for user-friendly detection of Cu^{2+} on a paper platform. *Analyst*, 137, 10, 2406, 2014. (b) Chen, G.H., Chen, W.Y., Yen, Y.C., Wang, C.W., Chang, H.T., Chen, C.F., Detection of mercury (II) ions using colorimetric gold nanoparticles on paper-based analytical devices. *Anal. Chem.*, 86, 14, 6843, 2014.
28. Abdelhamid, H.N., Georgouvelas, D., Edlund, U., Mathew, A.P., CelloZIFPaper: Cellulose-ZIF hybrid paper for heavy metal removal and electrochemical sensing. *Chem. Eng. J.*, *446*, 136614, 2022.
29. Liu, Z., Chen, M., Guo, Y., Zhou, J., Shi, Q., Sun, R., Oxidized nanocellulose facilitates preparing photoluminescent nitrogen-doped fluorescent carbon dots for Fe^{3+} ions detection and bioimaging. *Chem. Eng. J.*, 384, 123260, 2019.
30. Wu, C., Li, J., Zhang, Y.Q., Li, X., Wang, S.Y., Li, D.Q., Cellulose dissolution, modification, and the derived hydrogel: A review. *ChemSusChem*, 16, 21, e202300518, 2023.
31. Zhao, C., Liu, G., Tan, Q., Gao, M., Chen, G., Huang, X., Li, L., Wang, J., Zhang, Y., Xu, D., Polysaccharide-based biopolymer hydrogels for heavy metal detection and adsorption. *J. Adv. Res.*, 44, 53, 2023.
32. Ryu, J.H., Lee, Y., Kong, W.H., Kim, T.G., Park, T.G., Lee, H., Catechol-functionalized chitosan/pluronic hydrogels for tissue adhesives and hemostatic materials. *Biomacromolecules*, 12, 7, 2653, 2011.
33. (a) Hu, Y., Hu, S., Zhang, S., Dong, S., Hu, J., Kang, L., Yang, X., A double-layer hydrogel based on alginate-carboxymethyl cellulose and synthetic polymer as sustained drug delivery system. *Sci. Rep.*, 11, 1, 9142, 2021. (b) Zhao, B., Jiang, H., Lin, Z., Xu, S., Xie, J., Zhang, A., Preparation of acrylamide/acrylic acid cellulose hydrogels for the adsorption of heavy metal ions. *Carbohydr. Polym.*, 224, 115022, 2019.

34. Song, L., Liu, F., Zhu, C., Li, A., Facile one-step fabrication of carboxymethyl cellulose based hydrogel for highly efficient removal of Cr (VI) under mild acidic condition. *Chem. Eng. J.*, 369, 641, 2019.
35. (a) Durpekova, S., Di Martino, A., Dusankova, M., Drohsler, P., Sedlarik, V., Biopolymer hydrogel based on acid whey and cellulose derivatives for enhancement water retention capacity of soil and slow release of fertilizers. *Polymers*, 13, 3274, 2021. (b) Das, D., Prakash, P., Rout, P.K., Bhaladhare, S., Synthesis and characterization of superabsorbent cellulose-based hydrogel for agriculture application. *Starch-Stärke*, 73, 1-2, 1900284, 2021.
36. (a) Stock, N. and Biswas, S., Synthesis of metal-organic frameworks (MOFs): routes to various MOF topologies, morphologies, and composites. *Chem. Rev.*, 112, 2, 933, 2014. (b) Lin, Z.J., Lü, J., Hong, M., Cao, R., Metal–organic frameworks based on flexible ligands (FL-MOFs): structures and applications. *Chem. Soc. Rev.*, 43, 16, 5867, 2014.
37. Ren, X., Liao, G., Li, Z., Qiao, H., Zhang, Y., Yu, X., Wang, B., Tan, H., Shi, L., Qi, X., Zhang, H., Two-dimensional MOF and COF nanosheets for next-generation optoelectronic applications. *Coord. Chem. Rev.*, 435, 213781, 2021.
38. Abdelhamid, H.N. and Mathew, A.P., Cellulose–metal organic frameworks (CelloMOFs) hybrid materials and their multifaceted Applications: A review. *Coord. Chem. Rev.*, 451, 214263, 2022.
39. Southern, E.M., Detection of specific sequences among DNA fragments separated by gel electrophoresis. *J. Mol. Biol.*, 98, 3, 503, 1975.
40. Singh, S., Bhardwaj, S., Tiwari, P., Dev, K., Ghosh, K., Maji, P.K., Recent advances in cellulose nanocrystals-based sensors: a review. *Mater. Adv.*, 5, 2622, 2024.
41. Sun, X., Li, J., He, Q., Xue, Y., Bai, Y., Yang, Y., Wang, X., Wang, S., Li, R., Ferric ion detection mechanism of a dicarboxylic cellulose nanocrystal and a 7-amino-4-methylcoumarin based fluorescent chemosensor. *RSC Adv.*, 12, 26, 16798, 2022.
42. Hanif, Z., Khan, Z.A., Tariq, M.Z., Choi, D., Park, S.J., Coatable tannic acid-deposited cellulose nanocrystals for Fe (III) sensing and its application to a facile, scalable and portable sensing platform. *Dyes Pigm.*, *196*, 109732, 2021.
43. Mohammed, N., Baidya, A., Murugesan, V., Kumar, A.A., Ganayee, M.A., Mohanty, J.S., Chiu Tam, K., Pradeep, T., Diffusion-controlled simultaneous sensing and scavenging of heavy metal ions in water using atomically precise cluster–cellulose nanocrystal composites. *ACS Sustainable Chem. Eng.*, 4, 11, 6167, 2016.
44. Ongun, M.Z., Ertekin, K., Gocmenturk, M., Ergun, Y., Suslu, A., Copper ion sensing with fluorescent electrospun nanofibers. *Spectrochim. Acta, Part A*, *90*, 177, 2012.
45. Zhang, X. and Huang, J., Functional surface modification of natural cellulose substances for colorimetric detection and adsorption of Hg^{2+} in aqueous media. *Chem. Commun.*, 46, 33, 6042, 2010.

46. Xu, L.Q., Neoh, K.G., Kang, E.T., Fu, G.D., Rhodamine derivative-modified filter papers for colorimetric and fluorescent detection of Hg^{2+} in aqueous media. *J. Mater. Chem. A*, 1, 7, 2526, 2013.
47. Raj, S. and Shankaran, D.R., Curcumin based biocompatible nanofibers for lead ion detection. *Sens. Actuators, B*, 226, 318, 2016.
48. Anirudhan, T.S., Deepa, J.R., Christa, J., Nanocellulose/nanobentonite composite anchored with multi-carboxyl functional groups as an adsorbent for the effective removal of Cobalt (II) from nuclear industry wastewater samples. *J. Colloid Interface Sci.*, 467, 307, 2016.
49. Wang, H., Pei, Y., Qian, X., An, X., Eu-metal organic framework@ TEMPO-oxidized cellulose nanofibrils photoluminescence film for detecting copper ions. *Carbohydr. Polym.*, 236, 116030, 2020.
50. (a) Yang, G., Wan, X., Su, Y., Zeng, X., Tang, J., Acidophilic S-doped carbon quantum dots derived from cellulose fibers and their fluorescence sensing performance for metal ions in an extremely strong acid environment. *J. Mater. Chem. A*, 4, 33, 12841, 2016. (b) Cheng, C., Xing, M., Wu, Q., A universal facile synthesis of nitrogen and sulfur co-doped carbon dots from cellulose-based biowaste for fluorescent detection of Fe^{3+} ions and intracellular bioimaging. *Mater. Sci. Eng., C*, 99, 611, 2025. (c) Magagula, L.P., Masemola, C.M., Motaung, T.E., Moloto, N., Linganiso-Dziike, E.C., Synthesis of Cellulose-Based Fluorescent Carbon Dots for the Detection of Fe(III) in Aqueous Solutions. *Processes*, 13, 257, 2025.
51. He, X., Jia, H., Sun, N., Hou, M., Tan, Z., Lu, X., Fluorescent hydrogels based on oxidized carboxymethyl cellulose with excellent adsorption and sensing abilities for Ag^{+}. *Int. J. Biol. Macromol.*, 213, 955, 2022.
52. Ren, T., Peng, J., Yuan, H., Liu, Z., Li, Q., Ma, Q., Li, X., Guo, X., Wu, Y., Nanocellulose-based hydrogel incorporating silver nanoclusters for sensitive detection and efficient removal of hexavalent chromium. *Eur. Polym. J.*, 175, 111343, 2022.
53. Schelling, M., Kim, M., Otal, E., Aguirre, M., Hinestroza, J.P., Synthesis of a zinc–imidazole metal–organic framework (ZIF-8) using ZnO rods grown on cotton fabrics as precursors: arsenate absorption studies. *Cellulose*, 27, 6399, 2020.
54. Wang, N., Ouyang, X.K., Yang, L.Y., Omer, A.M., Fabrication of a magnetic cellulose nanocrystal/metal–organic framework composite for removal of Pb (II) from water. *ACS Sustain. Chem. Eng.*, 5, 11, 10447, 2017.
55. Ma, S., Zhang, M., Nie, J., Tan, J., Song, S., Luo, Y., Lightweight and porous cellulose-based foams with high loadings of zeolitic imidazolate frameworks-8 for adsorption applications. *Carbohydr. Polym.*, 208, 328, 2019.
56. Lei, C., Gao, J., Ren, W., Xie, Y., Abdalkarim, S.Y.H., Wang, S., Ni, Q., Yao, J., Fabrication of metal-organic frameworks@ cellulose aerogels composite materials for removal of heavy metal ions in water. *Carbohydr. Polym.*, 205, 35, 2019.

57. (a) Yang, W., Wang, J., Yang, Q., Pei, H., Hu, N., Suo, Y., Li, Z., Zhang, D., Wang, J., Facile fabrication of robust MOF membranes on cloth *via* a CMC macromolecule bridge for highly efficient Pb (II) removal. *Chem. Eng. J., 339,* 230, 2018. (b) Ma, X., Lou, Y., Chen, X.B., Shi, Z., Xu, Y., Multifunctional flexible composite aerogels constructed through *in-situ* growth of metal-organic framework nanoparticles on bacterial cellulose. *Chem. Eng. J.*, 356, 227, 2019.
58. Das, R. and Mukhopadhyay, B., Carbohydrates in fluoride sensing: Use of cyclodextrin and CNC-based chemical probes. *ChemistrySelect,* 2, 16, 4499, 2017.
59. Li, M., Li, X., Jiang, M., Liu, X., Chen, Z., Wang, S., James, T.D., Wang, L., Xiao, H., Engineering a ratiometric fluorescent sensor membrane containing carbon dots for efficient fluoride detection and removal. *Chem. Eng. J.*, 399, 125741, 2020.
60. Li, M., Liu, Z., Wang, H.C., Sedgwick, A.C., Gardiner, J.E., Bull, S.D., Xiao, H.N., James, T.D., Dual-function cellulose composites for fluorescence detection and removal of fluoride. *Dyes Pigm.*, 149, 669, 2018.
61. Park, B., Kang, S.M., Lee, G.W., Kwak, C.H., Rethinasabapathy, M., Huh, Y.S., Fabrication of $CsPbBr_3$ perovskite quantum dots/cellulose-based colorimetric sensor: dual-responsive on-site detection of chloride and iodide ions. *Ind. Eng. Chem. Res.*, 59, 2, 793, 2019.
62. Hai, J., Li, T., Su, J., Liu, W., Ju, Y., Wang, B., Hou, Y., Anions reversibly responsive luminescent Tb (III) nanocellulose complex hydrogels for latent fingerprint detection and encryption. *Angew. Chem. Int. Ed.*, 57, 23, 6786, 2018.
63. Zhang, X., Qi, X., Ouyang, J., Zuo, Y., Ma, Q., Tan, H., Guo, X., Wu, Y., Fluorescent cellulose nanofibrils-based hydrogel incorporating MIL-125-NH_2 for effective adsorption and detection of iodide ion. *J. Hazard. Mater.*, 474, 134758, 2024.
64. Shang, Y., Guo, K., Jiang, P., Xu, X., Gao, B., Adsorption of phosphate by the cellulose-based biomaterial and its sustained release of laden phosphate in aqueous solution and soil. *Int. J. Biol. Macromol.*, 109, 524, 2018.

6

Reinventing Modern Concrete with Nanocellulose

Mahapara Abbass[1], Shalom Akhai[1], Amandeep Singh Wadhwa[2] and Rajni Verma[3]*

[1]Maharishi Markandeshwar Engineering College, Maharishi Markandeshwar (Deemed to be University), Mullana, Ambala, Haryana, India
[2]University Institute of Engineering and Technology (UIET), Panjab University, Chandigarh, India
[3]Department of Applied Sciences, School of Engineering and Technology (SET), CGC University, Mohali, Punjab, India

Abstract

The worsening problems of climate change, pollution, and unsustainable practices call for efficient green innovation solutions to address the growing environmental concerns. Among the promising substances that seek to address the various environmental challenges are the biodegradable biomass-derived nanomaterials, called as spherical nanocellulose (SNC). This chapter aims to provide an overview of the latest trends in applying SNC to sustainable solutions. In this presentation, the potential, uses, and future trajectories have formed the significant emphasis. SNC-based materials have shown tremendous potential in many environmental applications including purification of water, improvement of air quality, remediation of soil, carbon sequestration, and biomedical applications. They may remove toxins, heavy metals, and pollutants from water sources effectively, promote interior and exterior quality of air, increase fertility of soil, and improve retention capacity of water. SNC-based materials can catch and store CO_2, which helps in reducing climate change. Most of the modification processes that enhance the environmental benefits of SNC, such as chemical functionalization, composite formation, and porous structure design, are applied. Nevertheless, issues such as scalability, cost-effectiveness, standardization, regulation, and life cycle assessment will remain. The research directions for the future are the novel environmental applications, integration of SNC with other green materials, development of new

**Corresponding author*: rajni.verma.edu@gmail.com

Mohd Shabbir (ed.) Regenerated Cellulose: A Renewable Alternative to Petroleum-Based Plastics, (147–180)

characterization tools and models to further elucidate SNC-based environmental systems. Despite these promising results, more work is needed to be done in terms of scalability, cost effectiveness, and life cycle assessments. Spherical nanocellulose has arisen as a versatile and green material to answer most of the environmental challenges. Its unique structure and properties make it the right material for solutions that are themselves sustainable. The current research and development work are directed at overcoming the current barriers toward facilitating widely applicable SNC-based environmental solutions.

***Keywords*:** Spherical nanocellulose, environmental sustainability, green technology, biodegradable materials, environmental remediation, sustainable solutions

6.1 Introduction

Global environmental crisis is mainly caused by climate change and pollution, among other unsustainable practices. Global climate change, as mainly driven by fossil fuel emissions, has increased the degrees of greenhouse gases to such an extent that they degrade habitats as well as lead to losing biodiversity [1]. It is amplified by industrial waste, runoff from agriculture, and debris of plastics, since each one of these leads to air, water, or soil pollution, hence offering devastating effects on human health and ecosystems [2]. Unsustainable production and consumption processes also kill natural resources and cause an ecological imbalance [3]. These environmental issues call for green innovation solutions. Green technologies are based on the efficiency of energy use, with a focus on reducing resource consumption and waste generation [4]. Innovation in material science and engineering will provide alternative solutions that address current environmental issues [5]. The shift towards a circular economy, based on recycling, reusing materials, and relying less on non-renewable resources, offers economic opportunities in the creation of new markets and jobs based on sustainable operation [6]. SNC or spherical nanocellulose is a novel biodegradable material which may be utilized in many applications that can include water purification, improving air quality, remediating soils, and carbon sequestration from natural cellulose [7]. This decreases greenhouse gas emissions through carbon capture, resource efficiency, and material substitution in industrial processes. It can also be applied to sustainable agriculture and forestry through the better quality of soil through reforestation and the recovery of the ecosystem besides preventing soil erosion [8]. Green technology investment enables stakeholders to balance habitat loss and degradation by protecting several species and their ecosystems.

The development of greener technology and processes will allow the mitigation of pollution resulting from industrial operations [9]. Waste and emissions can be reduced through the use of biodegradable materials and advanced filtering systems, while technologies that monitor real-time emissions allow companies to better comply with environmental requirements [10]. Green innovation will be the key to achieving the set goals in the Paris Agreement, most particularly in limiting global warming to below 2°C. Some of these include electric car technologies and smart grid technologies, where society could transition away from fossil fuel use [11]. Eco-friendly farming practices with green technology would potentially enhance carbon absorption within soils, furthering the accomplishment of climate goals. Thus, green innovation would prove pivotal for confronting environmental issues and promoting development that is sustainable [12].

6.2 Synthesis Methods for Spherical Nanocellulose

Spherical nanocellulose refers to the process of making through chemical as well as enzymatic and also mechanical procedures. Each differing in efficiency and characteristics relates to the properties of nanomaterials obtained after the process [13]. Table 6.1 systematically gives an array of synthesis approaches for SNC, procedures, and core properties.

6.2.1 Chemical Methods

Acid Hydrolysis: Acid hydrolysis is a typical chemical process to get SNC. This includes the treatment of cellulose with strong acids, that is, either sulphuric or hydrochloric acid, to break the cellulose fibers into tiny nanocrystals. An illustrative study proved the synthesis of spherical cellulose nanocrystals from micro-cellulose using sulphuric acid hydrolysis, which manifested improved crystallinity and apt particle sizes for various applications [14].

Alkaline Treatment: More efficient methodology is the pretreatment using alkaline solutions, commonly NaOH, followed by an acid hydrolysis for the extraction of nanocellulose. Following such pretreatment, cellulosic materials often get freed from hemicellulose and lignin, increasing the number of purified nanocelluloses along with their characteristics. Many studies have been reported to get spherical nanocellulose particles with ranges between 60 and 570 nm by using an NaOH/DMSO solution, where they exhibited a nearly linear relationship as a function of treatment period [15].

Table 6.1 Synthesis methods for SNC.

Method	Process	Key characteristics	Example study/ findings
Chemical Methods	**Acid Hydrolysis:** Treating cellulose with strong acids (e.g., sulfuric or hydrochloric acid).	- Produces nanocrystals with improved crystallinity. - Suitable particle sizes for multiple uses.	Sulfuric acid hydrolysis of micro-cellulose led to spherical cellulose nanocrystals.
	Alkaline Treatment: Pretreatment with alkaline solutions (e.g., NaOH) followed by acid hydrolysis.	- Removes lignin and hemicellulose. - Produces high-purity nanocellulose. - Particle sizes: 60–570 nm.	Use of NaOH with DMSO demonstrated a near-linear relationship between treatment time and size.
Enzymatic Methods	Utilizes specific enzymes (e.g., cellulases) to selectively degrade cellulose under mild conditions.	- Environmentally friendly. - High purity and functional properties. - Operates under mild conditions.	Enzymatic treatment of pulp fibers successfully produced spherical cellulose nanocrystals.
Mechanical Methods	Processes such as high-pressure homogenization or grinding to physically reduce cellulose fibers.	- Involves shear forces to break down cellulose. - Can be combined with chemical pretreatment.	Mechanical grinding with chemical pretreatment produced SNC with desirable characteristics.

6.2.2 Enzymatic Methods

The process specifically involves enzymes like cellulases, breaking the cellulose down only into nanocrystalline forms. This process is more environmentally friendly since it occurs under milder conditions and requires fewer aggressive chemicals used in chemical treatments. Pure and more functional spherical cellulose nanocrystals have resulted from some cases where pulp fibers have been demonstrated upon enzymatic treatment [16].

6.2.3 Mechanical Methods

SNC can be generated through mechanical methods, namely high-pressure homogenization or grinding. These methodologies break down cellulose into nanoscale dimensions because of shear pressures. Recently, spherical nanocellulose with desirable properties for many uses has been reported to be produced based on mechanical grinding combined with chemical pretreatment [17].

6.3 Functionalization and Surface Modification of SNC

Modification of surface and functional group of SNC is of extreme importance to develop properties for specific applications. Different approaches can be used to modify the functional groups on the surface of SNC for performance tuning for target applications [18].

6.3.1 Thiol Functionalization

Thiol-functionalization is attaching thiol groups (-SH) onto the surface of SNC. It can potentially enhance the ability of SNC to adsorb heavy metals and organic pollutants. Thus, the interaction between SNC and contaminants in water becomes more favorable, hence the potential use of this material in the purification of water [19].

6.3.2 Carboxylation

Introducing carboxylic acid groups (–COOH) to the surface of SNC increases its hydrophilicity and the interaction of SNC with water molecules, which can participate in hydrogen bonding. Therefore, modification could improve dispersion stability in aqueous solutions of SNC, thus finding practical application in drug delivery or in biomedical fields [20].

6.3.3 Amine Modification

These usually imply attachment of amine groups (–NH2) in SNC, thus enhancing its anion affinity such as the resultant pollutants or drugs.

As the functionalities enhance its interactivity to adsorption effectiveness or serve as drug-delivering vehicle to facilitate the target species-biomolecule interaction [21]. These are the synthesis techniques of spherical nanocellulose, which include chemical, enzymatic, and mechanical processes. These represent different synthesis routes through which this material may be synthesized with tailored properties [22]. Functionalization techniques, such as thiol-functionalization, carboxylation, and amine modification, improve the versatility of SNC for application in environmental remediation, drug delivery, and biomedicine, among others [23]. As the methodology is advanced by research, it will significantly contribute to the solution of sustainable development problems across sectors.

6.4 Environmental Applications of SNC

SNCs are one class of highly advanced materials designed along nanotechnology principles with sustainability in mind. They are increasingly used in most applications for environmental reasons, largely owing to their origin from renewable sources or engineered to have little ecological impact [24]. Super nanocomposites possess several outstanding properties, such as high surface area, biocompatibility, mechanical strength, and reactivity, making them the best candidates to tackle some urgent environmental problems [24]. This paper is going to present the environmental applications of SNCs in five major areas, including water treatment, air quality improvement, soil remediation, carbon capture, and biomedical uses. Table 6.2 discusses the environmental applications of smart nanocomposites; their role in environmental remediation and biomedical applications; new innovative applications toward sustainability and health care; the multi-function approach toward challenges in water, air, soil, and biomedicine; and it provides a general overview.

SNCs revolutionize materials for environmental uses. They possess a very large surface area, selective reactivity, and environment-friendly sustainability, making them apt to solve the problems worldwide [25]. Superparamagnetic nanoparticles are transforming water purification, enhancing air quality, and facilitating soil remediation. They are essential for carbon sequestration to mitigate climate change and provide unique biological treatments for wound healing, medication delivery, and tissue engineering [27]. The inclusion of SNCs in environmental and biological applications supports sustainable development goals and contributes to a cleaner, healthier future [28].

Table 6.2 Environmental applications of smart nanocomposites: mechanisms, examples, and benefits.

Environmental application	Description	Mechanisms	Examples of SNCs	Benefits
Water Purification	Removing heavy metals, organic pollutants, and toxins from contaminated water.	• High surface area for adsorption • Catalytic degradation of pollutants • Selective binding of contaminants	• Metal oxide nanocomposites (ZnO, Fe_2O_3) • Graphene-based nanocomposites (GO, rGO) • Biopolymeric SNCs (chitosan, cellulose) • Nanoscale Zero-Valent Iron (nZVI)	• Cost-effective and scalable • Biodegradable and eco-friendly • Effective removal of emerging contaminants
Air Quality Improvement	Enhancing indoor and outdoor air quality by removing pollutants, gases, and particulates.	• Adsorption of VOCs, NOx, SOx, and PM • Photocatalytic degradation under light • Enhanced filtration efficiency	• TiO_2 and ZnO nanocomposites • Activated carbon-based SNCs • Graphene-based filters • Antimicrobial nanocomposites (Ag, Cu)	• Durable and efficient • Minimal energy requirements • Scalable for various applications

(Continued)

Table 6.2 Environmental applications of smart nanocomposites: mechanisms, examples, and benefits. (*Continued*)

Environmental application	Description	Mechanisms	Examples of SNCs	Benefits
Soil Remediation	Restoring soil health and fertility while removing contaminants and improving water retention.	• Immobilization of heavy metals and pollutants • Nutrient delivery to soil • Enhanced water retention using nanocomposites	• Biochar-graphene hybrids • Biopolymeric SNCs (chitosan-based composites) • Nanoclays • Nanoscale Zero-Valent Iron (nZVI)	• Sustainable solutions for soil restoration • Increased agricultural productivity • Efficient water management in arid regions
Carbon Sequestration	Mitigating climate change by capturing and storing carbon dioxide.	• Adsorption and capture of CO_2 • Chemical conversion of CO_2 into fuels/chemicals • Biomimetic carbon fixation approaches	• Metal-Organic Frameworks (MOFs) • Functionalized graphene composites • Nanocatalysts for CO_2 conversion	• High efficiency and scalability • CO_2 valorization for energy production • Supports carbon-neutral processes
Biomedical Applications	Utilizing SNCs for biomedical applications to improve health and healing.	• Antimicrobial action for wound healing • Controlled and targeted drug delivery • Scaffolds for tissue regeneration	• Silver nanoparticles (Ag) • Hydrogel nanocomposites (chitosan, cellulose) • Mesoporous silica nanocarriers • Hydroxyapatite-based SNCs	• Biocompatible and reduced toxicity • Accelerated healing and regeneration • Efficient drug delivery with reduced side effects

6.4.1 Water Purification: Removal of Toxins, Heavy Metals, and Pollutants

Water pollution is a global issue, due to industrialization, urbanization, and agricultural activities. Heavy metals, organic contaminants, and poisons deteriorate water quality and pose health threats [29]. Nanomaterials, including composites, have emerged as potential options for effective water filtration due to the enhanced surface area and adsorption properties [30, 31]. These materials show catalytic activity such as photocatalytic titanium dioxide composites which break down organic contaminants through acceleration of oxidation process upon sun exposure [32]. SNCs with chemical moieties improve efficiency in removal processes. Some examples of the use of SNC for water purification are: metal oxide nanocomposites, graphene-based nanocomposites, biopolymeric nanocomposites, and nanoscale zero-valent iron [33]. Water filter SNCs advantages would include economical, scalable solution suitable for large-scale applications, being biocompatible and environmentally benign with respect to traditional treatment schemes [34]. SNCs has bright promise in delivering clean water since it is efficient in getting rid of impurities from water with low environmental cost [35].

6.4.2 Air Quality Improvement: Promotion of Interior and Exterior Air Quality

Sustainable nanomaterials (SNCs) are a vital part of improving air quality, especially by removing harmful pollutants, gases, and particles. SNCs possess impressive adsorption properties of VOCs, NOx, SOx, and particulate matter. It may also degrade air pollutants by ultraviolet or visible light irradiation, allowing cleaner air [36]. Superparamagnetic nanoparticles can be incorporated into air filters to enhance nanoparticle retention and removal efficiency of allergens and pollutants [37]. Nanocomposite coatings may improve indoor air quality with activated carbon, graphene, or metal oxides used in HVAC systems, masks, and air purifiers at home [38]. SNCs selectively adsorb and neutralize VOCs emitted from household products, paints, and cleaning solutions. It exhibits antibacterial activity by inhibiting the growth of microorganisms [39]. Photocatalytic coatings on building facades and pavements may improve outdoor air quality by degrading airborne pollutants such as NOx and Sox [40]. Smart nanocomposites may improve the effectiveness of outdoor filtration systems and urban pollution abatement mechanisms. Using SNCs may result in significant strides toward preserving breathable air and reducing health risks due to air pollution [41].

6.4.3 Soil Remediation: Increase of Soil Fertility and Water Retention Capacity

Soil degradation, caused by pollution, erosion, and nutrient loss, threatens food security and ecological integrity. Soil remediation using nanocomposites (SNCs) involves the chelation and immobilization of heavy metals, pesticides, and organic pollutants, thereby reducing their bioavailability [42]. Soil nutrient carriers (SNCs) act as conduits for essential nutrients, increasing soil fertility. They increase water retention by reducing the need for irrigation. The applications of SNCs are the removal of heavy metals, nutrient control, and improvement in water retention [43]. Biopolymeric nanocomposites provide a progressive release of nutrients, thereby contributing to soil fertility and agricultural yield enhancement [45]. Soil nutrient complexes enhance soil porosity and water retention especially in arid regions [46]. SNCs offer sustainable and nontoxic means for the rehabilitation of soil health, higher agricultural output, and improved water management in arid regions. SNCs significantly contribute to soil remediation and sustainable agriculture in rehabilitation [47].

6.4.4 Carbon Sequestration: Capture and Storage of CO_2

One of the most significant mitigation strategies to be implemented for carbon sequestration in greenhouse gas emissions control is control of climate change. Synthesis nanoparticles are highly efficient and sustainable in carbon dioxide collection and storage [47]. Solid nanocomposites consisting of MOFs and graphene-based nanomaterials are proven to have an enhanced surface area and porosity to effectively absorb CO_2. It can also catalyze the process that will help in the conversion of CO_2 into useful products [48]. Biomimetic approaches resemble natural carbon capture mechanisms for CO_2 usage [49]. SNCs may offer the capability for air direct harvesting, carbon storage, and catalytic conversion, providing for sustainable sources of energy. Some examples of SNCs are porous MOFs which possess highly efficient capabilities in CO2 adsorption, as well as functionalized graphene composites [50]. The benefits of SNCs are the efficiency in their large-scale application, since they could be compatible with renewable energy systems for carbon-neutral operation and potentially support the circular economy through CO_2 valorization [51].

6.4.5 Biomedical Applications: Wound Healing, Drug Delivery, and Tissue Engineering

SNCs have renovated biomedical applications through their unique characteristics, such as biocompatibility, regulated drug release, and regeneration potential. They are used in wound healing, pharmaceutical delivery, and tissue engineering [52]. Antimicrobial nanocomposites, including silver nanoparticles, zinc oxide, and copper oxide, facilitate wound cleaning and healing [53]. Hydrogel-based SNC hydrogels provide a humid environment, expedite healing, and minimize scarring [54]. SNCs promote cellular proliferation and tissue regeneration at wound locations. They serve as vehicles for pharmaceuticals, delivering medicinal substances to specific locations gradually [55]. Functionalized SNCs provide targeted medication delivery to particular tissues, minimising adverse effects. SNCs are used in tissue engineering as scaffolds for regeneration and osseous repair [56]. The benefits of SNCs involve reduced toxicity, enhanced therapeutic efficacy, improved wound healing, and sustainable, biodegradable materials suitable for prolonged use [57].

6.5 Modification Processes for Enhanced Environmental Benefits

Modifications to SNC can significantly enhance this material for its applicability in diverse environmental scenarios. Such methods include chemical functionalization, composite formation, and design of its porous structure [58]. All these enhance SNC maximally towards achieving the desired environmental benefits as a multi-faceted solution to ecological challenges [59]. Table 6.3 presents different modification processes to enhance SNC properties include surface and chemical/physical modifications for environmental applications, functionalization to enhance performance, composite formation, and porous structure design [60].

Spherical nanocellulose (SNC) undergoes modification by chemical functionalization, composite synthesis, and the design of porous structures, hence augmenting its environmental advantages [61]. These tactics enhance adsorption and filtering efficacy, broadening their applicability in sustainable solutions such as water purification, air quality enhancement, and soil remediation [62]. Additional research is anticipated to augment SNC's potential.

Table 6.3 Design considerations for enhancing more environment-friendly applications of SNCs.

Modification process	Description	Benefits	Examples
Chemical Functionalization	Introduction of functional groups onto SNC surfaces to enhance adsorption, catalysis, and selectivity.	• Enhanced pollutant adsorption • Improved catalytic activity • Selective binding of target molecules	• Functionalized graphene oxide (GO) • Amino-silane functionalized nanomaterials • Carboxylated carbon nanotubes (CNTs)
Composite Formation	Combination of SNCs with other materials to achieve enhanced and synergistic properties.	• Enhanced mechanical strength • Improved chemical stability • Multi-functionality for diverse applications	• Metal-oxide nanocomposites (ZnO-TiO_2) • Polymer-nanocomposite membranes • Hybrid bio-nanocomposites
Porous Structure Design	Development of porous structures in SNCs to increase surface area and enhance adsorption and filtration.	• High surface area for adsorption • Improved pollutant filtration • Enhanced catalytic performance	• Mesoporous silica nanomaterials • Metal-Organic Frameworks (MOFs) • Porous carbon-based SNCs

6.5.1 Chemical Functionalization

Chemical functionalization is a method in which some functional groups are introduced onto the surface of nanocellulose, known as SNC. This could potentially improve the properties of this material for certain applications and, therefore, increase the capacity of the material to absorb contaminants and react. Amine-functionalized SNC has increased the efficiency of the adsorption of heavy metals such as lead and cadmium [63]. Chemical functionalization may also increase the catalytic activity of SNC in the chemical processes involved either in the degradation of harmful substances or in the treatment of waste chemicals to result in useful chemicals [64]. This includes advantageous chemical processes especially involved in wastewater treatment in situations with organic contaminant removal. Efficient SNC remediation agents function with the help of chemical functionalities [65].

6.5.2 Composite Formation

The process of composite formation is the method of incorporating other materials with nanoparticles or SNC to form hybrid systems that can enhance their characteristics and capabilities. This procedure yields composites that can enhance their mechanical strength and endurance, thereby making it useful for building or packaging [66]. Other added materials make it easy to develop composites with capabilities in targeted applications, which includes adsorption and filtration. It displays synergetic effects since there is the interplay between SNC and the other constituents that elevate the performance beyond what would be expected as if each material was put to use individually; thereby, composites are worth for multiple functions in a given environmental application. Such opens wider applicability of SNC into water treatment, air purifications, and soil cleanings [67].

6.5.3 Porous Structure Design

Porous structures' design involves creating porous topologies inside solid nanocrystals (SNC) in order to increase their efficiency in adsorption and filtration processes. It is designed for an increased surface area with easier paths for fluid flow, which supports greater interaction between pollutants and the substance [68]. Porously structured SNC improves adsorption capacity by promoting superior access to binding sites for contaminant adsorption, like the uptake of heavy metals and organic compounds. They have superior filtering efficiencies by selective elimination of some contaminants while permitting free passage of clean fluids [69]. Porous networking enhances the linked network to improve the process of mass transfer through accelerating faster interactions between the contaminant and the adsorbent material, leading to high rates of removal [70]. SNC functions as a useful tool to clean up the environment [71].

6.6 Comparative Analysis with Other Nanomaterials

Spherical Nanocellulose or SNC is a biodegradable nanomaterial and offers clear-cut advantages over other traditionally used nanomaterials, such as carbon nanotubes (CNTs), graphene, and metal oxides [72]. This comparative review discusses the properties, environmental footprint, and performance of SNC in comparison with other materials, thereby underlining its benefits in terms of biodegradability, sustainability, and cost-effectiveness [73]. Table 6.4 gives a comparison of physical and chemical properties of

Table 6.4 Properties of spherical nanocellulose (SNC) vs. other prominent nanomaterials.

Property	Spherical nanocellulose (SNC)	Carbon nanotubes (CNTs)	Graphene	Metal oxides
Biodegradability	Highly biodegradable	Not biodegradable	Limited biodegradability	Generally non-biodegradable
Mechanical Strength	Moderate	Extremely high	Very high	Variable (depends on type)
Electrical Conductivity	Low	High (can be metallic or semiconducting)	Very high	Variable (depends on type)
Thermal Conductivity	Moderate	High	Extremely high	Variable (depends on type)
Surface Area	High	Very high	Extremely high	Moderate to high
Cost	Generally low	High	High	Variable (depends on type)

SNC with others nanomaterials CNTs, graphene, and metal oxides and pointed out their potential application to environmental issues and their respective physical and chemical properties.

Environmental Impact

- **Spherical Nanocellulose (SNC):** SNC are obtained from renewable biomass sources and, it is considered as an eco-friendly approach. As it biodegradable so it is found that it does not create any long-term pollution, and no waste remains in the environment. Also, its utilization in agriculture enhances the soil quality and usage for carbon sequestration [74].
- **Carbon Nanotubes (CNTs):** Though CNTs have desirable mechanical and electrical characteristics, their environmental impact is a matter of concern. As they are not biodegradable, so they tend to remain in the environment for a long time and pose toxicity issues to wildlife and humans. Their production processes are also found to have concerns of harmful chemicals [75].
- **Graphene:** Similar to CNTs, graphene also has excellent properties however it is found to have environmental concerns in that it is not highly biodegradable. Moreover, graphene production may also produce toxic byproducts which may cause harm to ecosystems [76].
- **Metal Oxides:** The environmental impact of metal oxides materials varies largely. Some metal oxides are toxic and can tend to accumulate in the environment, while others are found to be useful applications, such as photocatalysis. However, they are generally not biodegradable [77].

Performance in Applications

- **SNC:** SNC are found as a successful material for wide range of environmental applications, like water purification, soil remediation, and carbon capture. Its high surface area enables it to effectively adsorb pollutants while being environmentally benign [78].

- **CNTs:** Owing to their desirable strength and electrical conductivity, CNT are popular in electronics and material science usage, although it has several toxicities limiting their usage directly in human application or through environmental exposure [79].
- **Graphene:** Graphene has an exceptional conductivity thereby making it suitable for electronic applications and energy storage devices. But as it has a high production cost, which limits its widespread use [80].
- **Metal Oxides:** These are one of the most universal materials found in catalysis and the semiconductor family. The features within them are sometimes very varied from one metal oxide to another [81].

Advantages of SNC:

- **Biodegradability:** The nature of biodegradability in SNC reduces environmental impact in comparison to others that are non-biodegradable, such as CNTs, graphene, etc. [82].
- **Sustainability:** Being derived from renewable biomass sources aligns SNC with sustainable development goals, promoting the use of eco-friendly materials.
- **Cost-effectiveness:** Generally, the production costs of SNC are lower than those of CNTs and graphene, thus making it more affordable for different applications [85].
- **Versatility:** SNC can be functionally modified and composed, depending on the requirements through chemical functionalization to boost its properties for particular usages in environmental remediation and sustainable practices [86].

In conclusion, with regards to carbon nanotubes, graphene, and metal oxides, one could argue that SNC has some benefits for particular applications—it is biodegradable and sustainable, and not very expensive. Such properties place SNC in the line of bright alternatives to overcome modern challenges of the environment in a creative way. Table 6.5 provides a brief comparison of nanomaterials.

Table 6.5 Comparison of nanomaterials in brief.

Category	Spherical nanocellulose (SNC)	Carbon nanotubes (CNTs)	Graphene	Metal oxides
Environmental Impact	Biodegradable, renewable	Non-biodegradable, potential toxicity	Limited biodegradability, potential toxicity	Varies; some toxic, some beneficial
Performance	Water purification, soil remediation, carbon capture	High strength, electrical conductivity	High electrical and thermal conductivity	Catalysis, semiconductors, photocatalysis
Advantages	Biodegradable, renewable, cost-effective, versatile	High strength, electrical conductivity, potential for energy storage	High electrical conductivity, high surface area, versatile	Wide range of applications, performance enhancement
Disadvantages	Lower mechanical strength compared to CNTs and graphene	Potential toxicity, high production costs	High production costs	Varying environmental impact depending on specific metal oxide

6.7 Case Studies of Spherical Nanocellulose (SNC) Applications

Spherical Nanocellulose (SNC) has demonstrated significant potential in various environmental applications, including water purification, soil remediation, air quality improvement, and biomedical uses. Below are real-world examples and case studies showcasing the successful implementation of SNC in these areas. Table 6.6 presents case studies of spherical nanocellulose applications in environmental and biomedical fields, compares SNC-based materials for environmental remediation, highlights key properties for sustainable applications, and compares performance metrics in water, soil, and air purification.

6.7.1 Water Purification

Case Study: Removal of Heavy Metals
A study highlighted the use of spherical nanocellulose as an effective adsorbent for the removal of heavy metals from wastewater. The research demonstrated that SNC modified with thiol groups exhibited rapid and selective adsorption capabilities for mercuric ions. The SNC-based adsorbent has been found to have high removal efficiencies, notably decreasing mercury concentrations in the contaminated water samples. Such an application highlights the possible utilization of SNC in industrial wastewater treatment as an efficient and sustainable heavy metal remediation approach [86].

Case Study: Nanocellulose Membranes for Filtration
Past studies indicated that cellulose nanomaterials, including SNC, are successful in the development of advanced filtration membranes for water treatment. Such membranes utilize the high surface area and functionalizability of nanocellulose to effectively remove contaminants from water [86]. According to the study, membranes made from nanocellulose could achieve high rejection rates for various pollutants while maintaining excellent permeability, making them suitable for sustainable water treatment technologies.

6.7.2 Soil Remediation

Case Study: Enhancing Soil Fertility
SNC have been used in agricultural practice to improve soil quality and fertility. In a field study, SNC was added to soil amendments, and this resulted in improved water retention and nutrient availability for crops [87].

Table 6.6 Case studies of SNC applications in environmental and biomedical fields.

Application area	Case study	Description	Key findings
Water Purification	Removal of Heavy Metals	SNC modified with thiol groups have been used as an effective adsorbent for the removal of heavy metals, particularly mercuric ions, from wastewater.	High removal efficiencies; selective adsorption for mercury, showcasing SNC's potential for industrial wastewater treatment.
	Nanocellulose Membranes for Filtration	Advanced filtration membranes developed using SNC showcased exceptional pollutant rejection rates while maintaining high permeability for water treatment applications.	Effective contaminant removal; high permeability, making SNC membranes suitable for sustainable water filtration.
Soil Remediation	Enhancing Soil Fertility	SNC were incorporated into soil for its modifications being utilized for agricultural practices, leading to improved water retention, nutrient availability, and reduced dependence on synthetic fertilizers.	Enhanced crop growth and soil quality; promoted sustainable agriculture practices with biodegradable SNC-based materials.
Air Quality Improvement	Air Filtration Applications	SNC are also successfully integrated into air filtration systems to enhance the capture of airborne pollutants, including volatile organic compounds (VOCs) and particulate matter, particularly for indoor environments.	Improved pollutant adsorption; highlighted the biodegradable and sustainable nature of SNC for air purification.
Biomedical Uses	Drug Delivery Systems	SNC have been utilized as a drug delivery vehicle due to its biocompatibility and functionalizability, enabling controlled encapsulation and release of therapeutic agents.	Enhanced treatment efficacy with controlled drug release, minimizing side effects; demonstrated versatility of SNC.

The use of SNC promoted plant growth while minimizing the application of synthetic fertilizers, thus proving to be a biodegradable alternative that supports sustainable agriculture [88].

6.7.3 Air Quality Improvement

Case Study: Air Filtration Applications
SNC have been investigated as a filler in air filter media because of its capacity to capture airborne pollutants [89]. Studies have shown that addition of SNC to the filter media improves the adsorption of VOCs and particulate matter. The application is highly relevant to indoor environments where air quality is crucial for health and well-being. The biodegradable nature of SNC makes it an attractive, sustainable material for air purification technologies.

6.7.4 Biomedical Uses

Case Study: Drug Delivery Systems
In biomedical applications, SNC have been studied for potential applications as a drug delivery carrier. It is an interesting candidate for targeted drug delivery systems due to its biocompatibility and possibility of being functionalized [90]. Research shows that SNC can encapsulate therapeutic agents in its core and release these therapeutic agents in a controlled manner to improve treatment efficiency while reducing side effects. This application is thus further evidence of the potential for SNC beyond applications for environmental remediation.

Thus these successful case studies on spherical nanocellulose in water purification, soil remediation, improvement in air quality, and biomedical applications present the multi-faceted nature of the sustainable material. While ongoing research studies seek to further develop its properties, SNC is apt to significantly contribute to environmental solutions as a promoter of sustainability in numerous fields.

6.8 Challenges and Future Directions

Even though spherical nanocellulose, SNC, have all the potential to be one of the most effective solutions to environmental problems, there are still a number of challenges that need to be addresses before it can be widely applied in actual practice. The integration of AI and ML can enhance process optimization and predictive modeling for SNC applications, enabling better resource management and performance analysis [91–94]. Additionally,

Taguchi methods can be employed to systematically optimize production parameters, improving yield and quality while minimizing [95–98]. This part discusses SNC's scalability, cost-effectiveness, standardization, regulation, life cycle evaluation, and innovative applications, and the significant problems and future research initiatives associated with these aspects of the technology.

6.8.1 Scalability: Large-Scale Production of SNC-Based Materials

The key challenge in SNC production is scalability. Currently, most production methods for SNC are limited to lab-scale operations because they demand a lot of energy and take much time [99]. This calls for the development of efficient and cost-effective manufacturing processes that can produce large amounts of SNC without compromise in quality. Research on optimized extraction methods and use of abundant biomass sources can make SNC production scalable enough for industrial needs.

6.8.2 Cost-Effectiveness: Reduction of Production Costs for Widespread Adoption

The cost of production of SNC-based materials is the main limitation for its wide usage. The raw material cost, processing technique, and energy consumption costs are very high and reduce the competitiveness of SNC in the market. Innovations in ways of reducing these costs—such as using waste biomass as feedstock or improving the efficiency of processing—would be important to make SNC a viable alternative material. Economically feasible large-scale methods of synthesis will also determine the eventual acceptance and use across different applications [100].

6.8.3 Standardization: Development of Standards for SNC-Based Materials

The development of standards for SNC-based materials is very importance to ensure quality control and reproducibility in both research and industrial applications. Current challenges in the evaluation of SNC properties and performance are mainly due to the lack of standardized characterization methods among different studies. Clear guidelines for measuring parameters, such as particle size, surface area, and functionalization, will better compare studies and enhance the credibility of SNC products in the marketplace [101].

6.8.4 Regulation: Establishment of Regulations for the Use of SNC-Based Materials

There is a for determining regulations for the safe use of SNC-based materials in a protected manner and in line with environmental compliances. Development of regulatory frameworks that determine health risks in usage entails production and application of the new material. This comes with the need for more collaboration between researchers, those in the industry, and regulatory bodies in creating thorough guidelines that ensure safety concern while encouraging innovation.

6.8.5 Life Cycle Assessment: Evaluation of the Environmental Impact of SNC-Based Materials

A detailed life cycle assessment is required to analyze the environmental implications of SNC-based materials from raw material extraction all through to end-of-life disposal [102]. LCA can thus highlight areas where improvement is needed with regard to resource efficiency and environmental performance. With a full understanding of the ecological footprint of SNC products, decisions can be made for improvements in sustainability with minimum impact on ecosystems.

6.9 Future Research Directions

Future investigations must be focussed on following key areas if the application of SNC is to be advanced further for wider suitability.

- **Novel Environmental Applications:** New applications of SNC must explored in bioremediation, carbon capture technologies, and sustainable packaging, all of which can expand the impact of SNC towards environmental sustainability.
- **Integration with Other Green Materials:** Investigate must be carried out on the synergistic effects of merging SNC with other biodegradable or sustainable materials can create improved properties and functionalities.
- **Development of New Characterization Tools and Models:** The development of new characterization tools and models for SNC would aid in better understand its behavior in various applications and lead to better quality control measures.

Thus, SNC applications can be expanded while tackling these above-mentioned impediments and pursuing novel research avenues due to the potential to become a cornerstone material in green technologies aimed at dealing with major environmental concerns [103, 104].

6.10 Conclusion

Spherical nanocellulose have a good potential in addressing environmental concerns such as water purification, air quality improvement, remediation of soils, and carbon capture due to a high surface area, good biocompatibility, its mechanical strength, and ability to react. Its versatile and sustainable properties have made progress in terms of reducing pollution, increasing ecosystem health, and global sustainability. However, other factors, such as scalability, cost-effectiveness, standardization, regulation, and life cycle evaluation, must be approached with focused research and development in order to ensure full exploitation of the capabilities of SNC. Innovations in production methods, integration with other green materials, and advanced characterization tools will further enhance its capabilities and broaden its applicability in industrial and environmental solutions. In conclusion, SNC is a highly advanced, eco-friendly material with huge potential to revolutionize sustainable technologies. Overcoming the challenges it faces and investing in further research will make SNC a cornerstone in the global transition toward greener, cleaner, and more sustainable solutions for a healthier planet.

Summary

- Climate change, pollution, and unsustainable activities are major factors responsible for the global environmental disaster. Green innovation solutions aim at minimizing the utilization of resources and generation of wastes. Spherical nanocellulose is biodegradable; it comes from natural cellulose; in water purification, improvement of air quality, soil remediation, and carbon sequestration. The green innovation emits less greenhouse gas emission while improving resource use efficiency, advocating sustainable agriculture, and forestry.
- Investment in green technology allows stakeholders to mitigate habitat loss and degradation by protecting species and

their ecosystems. The advancement of sustainable technology and methodologies will help reduce pollution related to industrial operations. Biodegradable materials and improved filtration systems may reduce waste and emissions, while technologies that allow for real-time emission monitoring enhance corporate compliance with environmental standards.

- Green innovation is the only way to achieve the goals set forth in the Paris Agreement, particularly in limiting global warming to under 2°C. Electric vehicle technology and smart grid systems are examples that make a transition from fossil fuels possible. Environmentally sustainable agricultural practices using green technology could increase carbon sequestration in soils, which is furthering climate goals.
- Spherical nanocellulose are multifunctional material that have tuneable characteristics. These characteristics can be modified by different processes like thiol functionalization, carboxylation, and amine modification. Adjustments increase the adsorption efficacy of SNC toward heavy metals and organic pollutants, thus making it a promising material for applications in water purification.
- The use of SNCs have been found to be substantial for the development of nanomaterials for air-quality improvement, soil remediation, carbon sequestration, and biomedical applications. SNCs show excellent adsorption capacity for VOCs, NOx, SOx, and particulate matter. They can be added to the air filters to enhance removal efficiency.
- The biomedical applications of SNCs include wound healing, drug delivery, and tissue engineering. Modifications procedures for enhanced environmental benefits of SNCs including chemical functionalization, composite preparation and the formation of porous architectures. Further research is likely to improve the prospects of SNC.
- SNC are biodegradable nanomaterial having exceptional advantages over the many other common nanomaterials such as carbon nanotubes, graphene, and metal oxides. The origin of SNC is renewable biomass, making it an environmentally friendly alternative. Biodegradable, it does not only lead to the accumulation of persistent pollutants or waste but also helps in enhancing soil quality and facilitating carbon sequestration in agricultural contexts. Efficacy

of SNC has been established in several environmental applications, such as water purification, soil remediation, and carbon capture.

- However, SNC cannot be used widely because of their potential toxicity and high cost of production. The SNC advantages include biodegradability, sustainability, cost-effectiveness, and flexibility. Chemical functionalization and composite development may be considered in order to enhance the properties for specific applications in environmental remediation and sustainable practices.
- There are huge applications of SNC in water purification, soil remediation, air quality improvement, and biological use. The case studies demonstrate that SNC has been developed as an efficient adsorbent for heavy metals to be extracted from wastewater and transformed into sophisticated filtration membranes for water treatment. Soil amendments have also used this for improving the quality and fertility of soils. In biomedical applications, SNC has been used as a drug delivery vehicle encapsulating medicinal chemicals that release them in a controlled manner to enhance the effectiveness of treatment and reduce adverse effects.
- Despite these benefits, a few challenges that need to be overcome include scalability, cost-efficiency, standardization, regulation, life cycle assessment, and innovative applications. By solving these problems and pursuing new research directions, SNC might become the basis of green technology to mitigate some of the most serious environmental issues.

AI disclosure statement: The author confirms that no artificial intelligence (AI)-generated content or AI-assisted writing tools were used in the preparation of this chapter (including summaries and the abstract). All text, analysis, and conclusions are the result of the author's original work.

Bibliography

1. Jige, S.B., Impact of Development on Climate Change, in: *Multidisciplinary Approaches to Sustainable Human Development*, pp. 206–219, IGI Global, Bogotá, Colombia, 2023.
2. Rathod, Sandip V., ed., *Eco-restoration of Polluted Environment: A Biological Perspective*, CRC Press, Boca Raton, FL, 2024.

3. Xiong, Y., Guo, H., Nor, D.D.M.M., Song, A., Dai, L., Mineral resources depletion, environmental degradation, and exploitation of natural resources: COVID-19 aftereffects. *Resour. Policy*, *85*, 103907, 2023.
4. Li, C., Ahmad, S.F., Ayassrah, A.Y.B.A., Irshad, M., Telba, A.A., Awwad, E.M., Majid, M.I., Green production and green technology for sustainability: The mediating role of waste reduction and energy use. *Heliyon*, *9*, 12, 2023.
5. Gidiagba, J.O., Daraojimba, C., Ofonagoro, K.A., Eyo-Udo, N.L., Egbokhaebho, B.A., Ogunjobi, O.A., Banso, A.A., Economic impacts and innovations in materials science: a holistic exploration of nanotechnology and advanced materials. *Eng. Sci. Technol. J.*, *4*, 3, 84–100, 2023.
6. Moreno-Mondejar, L., Triguero, A., Cuerva, M.C., Exploring the association between circular economy strategies and green jobs in European companies. *J. Environ. Manage.*, *297*, 113437, 2021.
7. Channab, B.E., El Idrissi, A., Essamlali, Y., Zahouily, M., Nanocellulose: Structure, modification, biodegradation and applications in agriculture as slow/controlled release fertilizer, superabsorbent, and crop protection: A review. *J. Environ. Manage.*, *352*, 119928, 2024.
8. Das, R., Lindstrom, T., Sharma, P.R., Chi, K., Hsiao, B.S., Nanocellulose for sustainable water purification. *Chem. Rev.*, *122*, 9, 8936–9031, 2022.
9. Zhang, L., Xu, M., Chen, H., Li, Y., Chen, S., Globalization, green economy and environmental challenges: state of the art review for practical implications. *Front. Environ. Sci.*, *10*, 870271, 2022.
10. Onifade, M., Zvarivadza, T., Adebisi, J.A., Said, K.O., Dayo-Olupona, O., Lawal, A.I., Khandelwal, M., Advancing toward sustainability: The emergence of green mining technologies and practices. *Green Smart Min. Eng.*, *1*, 2, 157–174, 2024.
11. Teske, S., *Achieving the Paris climate agreement goals: Global and regional 100% renewable energy scenarios with non-energy GHG pathways for+ 1.5 C and+ 2 C*, p. 491, Springer Nature, Elsevier, Amsterdam, Netherlands, 2019.
12. Shang, M. and Xie, J., Agricultural sustainable development: Soil, water resources, biodiversity, climate change, and technological innovation. *Adv. Resour. Res.*, *4*, 2, 181–204, 2024.
13. Kaur, P., Sharma, N., Munagala, M., Rajkhowa, R., Aallardyce, B., Shastri, Y., Agrawal, R., Nanocellulose: Resources, physio-chemical properties, current uses and future applications. *Front. Nanotechnol.*, *3*, 747329, 2021.
14. Gopakumar, D., Lyczko, N., Maria, H.J., Nzihou, A., Thomas, S., Extraction and Characterization of Nanomaterials from Agrowaste, in: *Handbook on Characterization of Biomass, Biowaste and Related By-products*, pp. 841–897, 2020.
15. Xie, H., Du, H., Yang, X., Si, C., Recent strategies in preparation of cellulose nanocrystals and cellulose nanofibrils derived from raw cellulose materials. *Int. J. Polym. Sci.*, *2018*, 1, 7923068, 2018.

16. Xu, J., Liao, Q., Ren, H., Yuan, B., Zhang, H., A new facile and efficient strategy for high-yield preparation of spherical cellulose nanocrystals *via* enzymolysis-mechanical synergy. *Cellulose, 30*, 14, 8909–8927, 2023.
17. Ng, S.W., Chong, W.T., Soo, Y.T., Tang, T.K., Ab Karim, N.A., Phuah, E.T., Lee, Y.Y., Pickering emulsion stabilized by palm-pressed fiber cellulose nanocrystal extracted by acid hydrolysis-assisted high pressure homogenization. *PLoS One, 17*, 8, e0271512, 2022.
18. Thakur, V., Guleria, A., Kumar, S., Sharma, S., Singh, K., Recent advances in nanocellulose processing, functionalization and applications: A review. *Mater. Adv., 2*, 6, 1872–1895, 2021.
19. Dang-Bao, T., Nguyen, T.M.C., Hoang, G.H., Lam, H.H., Phan, H.P., Tran, T.K.A., Thiol-surface-engineered cellulose nanocrystals in favor of copper ion uptake. *Polymers, 15*, 11, 2562, 2023.
20. Mujtaba, M., Negi, A., King, A.W., Zare, M., Kuncova-Kallio, J., Surface modifications of nanocellulose for drug delivery applications; a critical review. *Curr. Opin. Biomed. Eng., 28*, 100475, 2023.
21. Kareem, F., Chandrawati, R., Ahmed, M.U., Tailoring nanocellulose: A comprehensive exploration of functionalisation with small, macro, and inorganic molecules for enhanced sensing and dual-mode biosensing. *Measurement*, 115499, 2024.
22. Yadav, V., Singh, S., Garg, M.P., and Akhai, S., Effect of Natural Sapindus mukorossi Treatment Process on Bio-waste Bagasse Fibers for Biocomposite Fabrication and Application Purposes, Industrial Crops and Products, 219, 119066, 2024.
23. Rahmani, S., Khoubi-Arani, Z., Mohammadzadeh-Komuleh, S., Maroufkhani, M., Electrospinning of cellulose nanofibers for advanced applications, in: *Handbook of Nanocelluloses: Classification, Properties, Fabrication, and Emerging Applications*, pp. 263–296, Springer International Publishing, Cham, 2022.
24. Akhai, S., Biogas Plants for Sustainable Municipal Waste Management: A Brief Review, in: *Handbook of Research on Safe Disposal Methods of Municipal Solid Wastes for a Sustainable Environment*, IGI Global, Hershey, PA, USA, pp. 162–179, 2023.
25. Zhang, J., Hassane Hamadou, A., Chen, C., Xu, B., Encapsulation of phenolic compounds within food-grade carriers and delivery systems by pH-driven method: A systematic review. *Crit. Rev. Food Sci. Nutr., 63*, 19, 4153–4174, 2023.
26. Srivastava, S., Advancements in Nanocellulose Derived from Plant Waste: Strategies for Sustainable Innovations and Applications. *J. Solid Waste Technol. Manage., 50*, 3, 543–563, 2024.
27. Harish, V., Tewari, D., Gaur, M., Yadav, A.B., Swaroop, S., Bechelany, M., Barhoum, A., Review on nanoparticles and nanostructured materials: Bioimaging, biosensing, drug delivery, tissue engineering, antimicrobial, and agro-food applications. *Nanomaterials, 12*, 3, 457, 2022.

28. Reshmy, R., Philip, E., Madhavan, A., Arun, K.B., Binod, P., Pugazhendhi, A., Sindhu, R., Promising eco-friendly biomaterials for future biomedicine: Cleaner production and applications of Nanocellulose. *Environ. Technol. Innovation*, *24*, 101855, 2021.
29. Sharma, K., Rajan, S., Nayak, S.K., Water pollution: Primary sources and associated human health hazards with special emphasis on rural areas, in: *Water Resources Management for Rural Development*, pp. 3–14, Elsevier, Amsterdam, Netherlands, 2024.
30. Ibrahim, Q., Creedon, L., Gharbia, S., A literature review of modelling and experimental studies of water treatment by adsorption processes on nanomaterials. *Membranes*, *12*, 4, 360, 2022.
31. Zhao, W., Chen, I.W., Huang, F., Toward large-scale water treatment using nanomaterials. *Nano Today*, *27*, 11–27, 2019.
32. Tanos, F., Razzouk, A., Lesage, G., Cretin, M., Bechelany, M., A Comprehensive Review on Modification of Titanium Dioxide-Based Catalysts in Advanced Oxidation Processes for Water Treatment. *ChemSusChem*, *17*, 6, e202301139, 2024.
33. Thamarai, P., Kamalesh, R., Saravanan, A., Swaminaathan, P., Deivayanai, V.C., Emerging trends and promising prospects in nanotechnology for improved remediation of wastewater contaminants: Present and future outlooks. *Environ. Nanotechnol. Monit. Manage.*, 100913, 2024.
34. Abbass, M., and Akhai, S., Exploring the Sustainability Potential of Geopolymer Concrete with Coal Bottom Ash and Basalt Fibres, Multiscale and Multidisciplinary Modeling, Experiments and Design, 8, 3, 199, 2025.
35. Abbass, M., and Akhai, S., Eco-friendly Geopolymer Bricks Synthesized from Alkali-activated Sugarcane Bagasse Ash and Rice Husk Ash, Industrial Crops and Products, 233, 121410, 2025.
36. Kasbaji, M., Mennani, M., Oubenali, M., Benhamou, A.A., Boussetta, A., Ablouh, E.H., Moubarik, A., Bio-based functionalized adsorptive polymers for sustainable water decontamination: A systematic review of challenges and real-world implementation. *Environ. Pollut.*, 122349, 2023.
37. Elella, M.H.A., Castro-Riquelme, C.L., López-Maldonado, E.A., Mohammed, D.M., Abady, M.M., Abu-Thabit, N.Y., Samy, M., Application of waste-synthesized nanoparticles in capture of air pollutants, in: *Waste-Derived Nanoparticles*, pp. 251–280, Elsevier, Amsterdam, Netherlands, 2024.
38. Mallakpour, S., Azadi, E., Hussain, C.M., Fabrication of air filters with advanced filtration performance for removal of viral aerosols and control the spread of COVID-19. *Adv. Colloid Interface Sci.*, *303*, 102653, 2022.
39. Bijanu, A., Rajak, G., Paulose, R., Challenges of Nanofillers in Environment, Legislation, Health, and Safety, in: *Handbook of Nanofillers*, pp. 1–25, Springer Nature Singapore, Singapore, 2024.
40. Lee, T.X., *Applications and future perspectives of photocatalytic coatings for air purification and self cleaning* (Doctoral dissertation, UTAR), 2022.

41. Chakraborty, D., Sen, K., Mukhopadhyay, K., Dalal, K., Application of Nanotechnology in the Mitigation of Air Pollution Exposure for Better Public Health, in: *Emerging Sustainable Nanomaterials for Biomedical Applications*, pp. 269–299, Springer Nature Switzerland, Cham, 2024.
42. Wahab, A., Muhammad, M., Ullah, S., Abdi, G., Shah, G.M., Zaman, W., Ayaz, A., Agriculture and environmental management through nanotechnology: Eco-friendly nanomaterial synthesis for soil-plant systems, food safety, and sustainability. *Sci. Total Environ.*, 171862, 2024.
43. Guha, T., Gopal, G., Kundu, R., Mukherjee, A., Nanocomposites for delivering agrochemicals: a comprehensive review. *J. Agric. Food. Chem.*, *68*, 12, 3691–3702, 2020.
44. Kaur Takkar, M. and Gumber, K., Agro-nanocarriers: An emerging trend for better plant nutrition. *J. Plant Nutr.*, *47*, 2, 332–350, 2024.
45. do Nascimento, I.V., Fregolente, L.G., de Araújo Pereira, A.P., do Nascimento, C.D.V., Mota, J.C.A., Ferreira, O.P., Costa, M.C.G., Biochar as a carbonaceous material to enhance soil quality in drylands ecosystems: A review. *Environ. Res.*, *233*, 116489, 2023.
46. Lewicka, K., Szymanek, I., Rogacz, D., Wrzalik, M., Łagiewka, J., Nowik-Zając, A., Rychter, P., Current Trends of Polymer Materials' Application in Agriculture. *Sustainability (2071-1050)*, *16*, 19, 2024.
47. Tiwari, T., Kaur, G.A., Singh, P.K., Balayan, S., Mishra, A., Tiwari, A., Emerging bio-capture strategies for greenhouse gas reduction: Navigating challenges towards carbon neutrality. *Sci. Total Environ.*, 172433, 2024.
48. Zhang, X., Zhang, S., Tang, Y., Huang, X., Pang, H., Recent advances and challenges of metal–organic framework/graphene-based composites. *Composites, Part B*, *230*, 109532, 2022.
49. Sillu, D. and Achal, V., Carbon dioxide sequestration with carbonic anhydrase nanobiocatalysts: a review. *Environ. Chem. Lett.*, 1–27, 2024.
50. Hou, J., Wang, H., Zhang, H., Zirconium metal–organic framework materials for efficient ion adsorption and sieving. *Ind. Eng. Chem. Res.*, *59*, 29, 12907–12923, 2020.
51. Cao, T.N.D., Mukhtar, H., Yu, C.P., Bui, X.T., Pan, S.Y., Agricultural waste-derived biochar in microbial fuel cells towards a carbon-negative circular economy. *Renewable Sustainable Energy Rev.*, *170*, 112965, 2022.
52. Gopinath, V., Kamath, S.M., Priyadarshini, S., Chik, Z., Alarfaj, A.A., Hirad, A.H., Multifunctional applications of natural polysaccharide starch and cellulose: An update on recent advances. *Biomed. Pharmacother.*, *146*, 112492, 2022.
53. Pachaiappan, R., Rajendran, S., Show, P.L., Manavalan, K., Naushad, M., Metal/metal oxide nanocomposites for bactericidal effect: A review. *Chemosphere*, *272*, 128607, 2021.
54. Ahovan, Z.A., Esmaeili, Z., Eftekhari, B.S., Khosravimelal, S., Alehosseini, M., Orive, G., Gholipourmalekabadi, M., Antibacterial smart hydrogels: New hope for infectious wound management. *Mater. Today Bio*, *17*, 100499, 2022.

55. Mohamed, R.R., Hydrogel-Nanocomposite in Sustained Drug Release: Synthesis and Properties, in: *Novel Bio-nanocomposites for Biomedical Applications*, pp. 227–260, Springer International Publishing, Cham, 2024.
56. Pereira, C., Lazar, S.V., Wang, A., Bioengineering approaches for nerve graft revascularization: Current concepts and future directions. *WIREs Mech. Dis.*, *15*, 4, e1609, 2023.
57. Laubach, J., Joseph, M., Brenza, T., Gadhamshetty, V., Sani, R.K., Exopolysaccharide and biopolymer-derived films as tools for transdermal drug delivery. *J. Controlled Release*, *329*, 971–987, 2021.
58. Lee, J., Lee, T.H., Jang, H.W., Park, H.S., Chemical modification of ordered/disordered carbon nanostructures for metal hosts and electrocatalysts of lithium-air batteries. *InfoMat*, *4*, 1, e12268, 2022.
59. Srinivas, A., Chandran, K.R., Nehra, S., Kumar, D., Future Aspects in the Field of Bio-nanomaterials Toward Environmental Assessment, in: *Bio-Nanomaterials in Environmental Remediation: Industrial Applications*, pp. 295–317, 2025.
60. Kumar, Y., Paswan, K.K., Nayan, K., Pandurangappa, G., Dwivedi, D., Sangwai, J.S., Introduction to Functional Materials: Synthesis, Properties, Environmental Sustainability, and General Applications, in: *Functional Materials for the Oil and Gas Industry*, pp. 1–22, CRC Press, American Chemical Society (ACS), Washington, DC, USA, 2023.
61. Szafraniec, M., Grabias-Blicharz, E., Barnat-Hunek, D., Landis, E.N., A critical review on modification methods of cement composites with nanocellulose and reaction conditions during nanocellulose production. *Materials*, *15*, 21, 7706, 2022.
62. Gobi, M., Kumar, A., Singh, J., Singh, S., Ramamurthy, P.C., Nanocellulose-Based Adsorption for the Removal of Heavy Metal from Wastewater—A Review. *Water Conserv. Sci. Eng.*, *9*, 1, 24, 2024.
63. Al-Gethami, W., Qamar, M.A., Shariq, M., Alaghaz, A.N.M., Farhan, A., Areshi, A.A., Alnasir, M.H., Emerging environmentally friendly bio-based nanocomposites for the efficient removal of dyes and micropollutants from wastewater by adsorption: a comprehensive review. *RSC Adv.*, *14*, 4, 2804–2834, 2024.
64. Gong, X., Xie, J., Pan, X., Luo, X., S, N co-doped carbon material functionalized catalytic membrane for efficient peroxymonosulfate activation and continuous refractory pollutants flow-treatment. *Chem. Eng. Sci.*, *282*, 119353, 2023.
65. Saleh, T.A., Materials, nanomaterials, nanocomposites, and methods used for the treatment and removal of hazardous pollutants from wastewater: Treatment technologies for water recycling and sustainability. *Nano-Struct. Nano-Objects*, *39*, 101231, 2024.
66. Mayakannan, S., Raj, J.B., Raja, V.L., Nagaraj, M., Effectiveness of silicon nanoparticles on the mechanical, wear, and physical characteristics of PALF/sisal fiber–based polymer hybrid nanocomposites. *Biomass Convers. Biorefin.*, *13*, 14, 13291–13305, 2023.

67. Ihsanullah, I., Bilal, M., Jamal, A., Recent developments in the removal of dyes from water by starch-based adsorbents. *Chem. Rec.*, *22*, 7, e202100312, 2022.
68. Le, T., Chen, X., Dong, H., Tarpeh, W., Perea-Cachero, A., Coronas, J., Esfahani, M.R., An evolving insight into metal organic framework-functionalized membranes for water and wastewater treatment and resource recovery. *Ind. Eng. Chem. Res.*, *60*, 19, 6869–6907, 2021.
69. Tripathy, J., Mishra, A., Pandey, M., Thakur, R.R., Chand, S., Rout, P.R., Shahid, M.K., Advances in Nanoparticles and Nanocomposites for Water and Wastewater Treatment: A Review. *Water*, *16*, 11, 1481, 2024.
70. Kumar, N., Gusain, R., Pandey, S., Ray, S.S., Hydrogel nanocomposite adsorbents and photocatalysts for sustainable water purification. *Adv. Mater. Interfaces*, *10*, 2, 2201375, 2023.
71. Bhavya, G., Belorkar, S.A., Mythili, R., Geetha, N., Shetty, H.S., Udikeri, S.S., Jogaiah, S., Remediation of emerging environmental pollutants: a review based on advances in the uses of eco-friendly biofabricated nanomaterials. *Chemosphere*, *275*, 129975, 2021.
72. Gulati, S., Amar, A., Chhabra, L., Katiyar, R., Sahu, T., Varma, R.S., Greener nanobiopolymers and nanoencapsulation: environmental implications and future prospects. *RSC Sustainability*, *2*, 10, 2805–2832, 2024.
73. Alshammari, O., Basem, A., Hameed, A.I., Agarwal, D., Shawabkeh, A., Kenjrawy, H.A., Jerbi, H., Advanced eco-friendly power and cooling cogeneration-thermal energy storage utilizing phase change materials and chemisorption in renewable-based configurations. *Therm. Sci. Eng. Prog.*, *54*, 102813, 2024.
74. Akhai, S., Bansal, S.A., and Singh, S., A Critical Review of Thermal Insulators from Natural Materials for Energy Saving in Buildings, *Journal of Critical Reviews*, 7, 19, 278–283, 2020.
75. Hitam, C.N.C. and Jalil, A.A., Recent advances on nanocellulose biomaterials for environmental health photoremediation: An overview. *Environ. Res.*, *204*, 111964, 2022.
76. Obayomi, K.S., Lau, S.Y., Mayowa, I.E., Danquah, M.K., Zhang, J., Chiong, T., Rahman, M.M., Recent advances in graphene-derived materials for biomedical waste treatment. *J. Water Process Eng.*, *51*, 103440, 2023.
77. Gautam, S., Agrawal, H., Thakur, M., Akbari, A., Sharda, H., Kaur, R., Amini, M., Metal oxides and metal organic frameworks for the photocatalytic degradation: A review. *J. Environ. Chem. Eng.*, *8*, 3, 103726, 2020.
78. Taneja, T., Bhatia, B.S., and Akhai, S., Assessing Groundwater Quality in Rampura-Bathinda for Physicochemical Parameters and Heavy Metal Concentrations Using Electronic Equipment, in: *2024 1st International Conference on Sustainability and Technological Advancements in Engineering Domain* (SUSTAINED), IEEE, Faridabad, India, pp. 791–796, 2024.
79. Kim, M., Goerzen, D., Jena, P.V., Zeng, E., Pasquali, M., Meidl, R.A., Heller, D.A., Human and environmental safety of carbon nanotubes across their life cycle. *Nat. Rev. Mater.*, *9*, 1, 63–81, 2024.

80. Olabi, A.G., Abdelkareem, M.A., Wilberforce, T., Sayed, E.T., Application of graphene in energy storage device–A review. *Renewable Sustainable Energy Rev., 135,* 110026, 2021.
81. Farooq, U., Ahmad, T., Naaz, F., Islam, S.U., Review on metals and metal oxides in sustainable energy production: progress and perspectives. *Energy Fuels, 37,* 3, 1577–1632, 2023.
82. Gamage, A., Thiviya, P., Mani, S., Ponnusamy, P.G., Manamperi, A., Evon, P., Madhujith, T., Environmental properties and applications of biodegradable starch-based nanocomposites. *Polymers, 14,* 21, 4578, 2022.
83. Sarangi, P.K., Mishra, S., Mohanty, P., Singh, P.K., Srivastava, R.K., Pattnaik, R., Sahoo, U.K., Food and fruit waste valorisation for pectin recovery: Recent process technologies and future prospects. *Int. J. Biol. Macromol., 235,* 123929, 2023.
84. Karim, N., Sarker, F., Afroj, S., Zhang, M., Potluri, P., Novoselov, K.S., Sustainable and multifunctional composites of graphene-based natural jute fibers. *Adv. Sustainable Syst., 5,* 3, 2000228, 2021.
85. Amani, A.M., Tayebi, L., Vafa, E., Abbasi, M., Vaez, A., Kamyab, H., Bazargan-Lari, R., On the horizon of greener pathways to travel into a greener future portal: Green MXenes, environment-friendly synthesis, and their innovative applications. *J. Cleaner Prod.*, 140606, 2024.
86. Si, R., Pu, J., Luo, H., Wu, C., Duan, G., Nanocellulose-based adsorbents for heavy metal ion. *Polymers, 14,* 24, 5479, 2022.
87. Khan, M.A., Basir, A., Shah, S.T., Rehman, M.U., Hassan, M.U., Zheng, H., Mancinelli, R., Sustainable soil management in alkaline soils: The role of biochar and organic nitrogen in enhancing soil fertility. *Land, 13,* 11, 1869, 2024.
88. Manibusan, S.N.C., *Development and Application of Ferroelectric Film-Based Bioaerosol Samplers* (Doctoral dissertation, Rutgers The State University of New Jersey, School of Graduate Studies)., M. R., Islam, M. T., & Sarker, A. (2024). of Sustainable Agriculture. *Soil Bacteria: Biofertilization and Soil Health*, 3, 2022.
89. Fadil, F., Affandi, N.D.N., Misnon, M.I., Bonnia, N.N., Harun, A.M., Alam, M.K., Review on electrospun nanofiber-applied products. *Polymers, 13,* 13, 2087, 2021.
90. Verma, D. and Sharma, S.K., Recent advances in guar gum based drug delivery systems and their administrative routes. *Int. J. Biol. Macromol., 181,* 653–671, 2021.
91. Motia, K., Kumar, R., Akhai, S., AI and Smart Manufacturing: Building Industry 4.0, in: *Modern Management Science Practices in the Age of AI*, pp. 1–28, IGI Global, Elsevier, Amsterdam, Netherlands, 2024.
92. Akhai, S. and Abbass, M., Toward Resilient Futures: The Role of AI-Driven Strategies in Climate Adaptation, in: *Nexus of AI, Climatology, and Urbanism for Smart Cities*, pp. 325–340, IGI Global Scientific Publishing, Frontiers Media SA, Lausanne, Switzerland, 2025.

93. Abbass, M., Abbas, U., Jafri, R., Arif, S.M., Akhai, S., AI and Machine Learning Applications in Sustainable Smart Cities, in: *Sustainable Smart Cities and the Future of Urban Development*, pp. 1–32, IGI Global Scientific Publishing, Woodhead Publishing (Elsevier), Cambridge, UK, 2025.
94. Akhai, S., Towards Trustworthy and Reliable AI: The Next Frontier, in: *Explainable Artificial Intelligence (XAI) in Healthcare*, CRC Press, Elsevier, Amsterdam, Netherlands, pp. 89–99, 2024.
95. Wadhwa, S.A., Mahapara, A., Akhai, S., Kumar, D., Kumar, P., Integrating Taguchi Optimization for Multi-Criteria Decision Making in Engineering Applications, in: *Recent Theories and Applications for Multi-Criteria Decision-Making*, vol. 1, no. 1, pp. 125–150, IGI Global, Hershey, PA, USA. 2024.
96. Kumar, H., Wadhwa, A.S., Akhai, S., Kaushik, A., Parametric analysis, modeling and optimization of the process parameters in electric discharge machining of aluminium metal matrix composite. *Eng. Res. Express*, *6*, 2, 025542, 2024.
97. Kumar, H., Wadhwa, A.S., Akhai, S., Kaushik, A., Parametric optimization of the machining performance of Al-SiCp composite using combination of response surface methodology and desirability function. *Eng. Res. Express*, *6*, 2, 025505, 2024.
98. Akhai, S., A Review on Optimizations in μ-EDM Machining of the Biomedical Material Ti6Al4V Using the Taguchi Method: Recent Advances Since 2020, in: *Latest Trends in Engineering and Technology*, pp. 395–402, 2024.
99. Bhatia, R.K., Sakhuja, D., Mundhe, S., Walia, A., Renewable energy products through bioremediation of wastewater. *Sustainability*, *12*, 18, 7501, 2020.
100. Lee, X.J., Hiew, B.Y.Z., Lai, K.C., Lee, L.Y., Gan, S., Thangalazhy-Gopakumar, S., Rigby, S., Review on graphene and its derivatives: Synthesis methods and potential industrial implementation. *J. Taiwan Inst. Chem. Eng.*, *98*, 163–180, 2019.
101. Perović, K., Morović, S., Jukić, A., Košutić, K., Alternative to Conventional Solutions in the Development of Membranes and Hydrogen Evolution Electrocatalysts for Application in Proton Exchange Membrane Water Electrolysis: A Review. *Materials*, *16*, 18, 6319, 2023.
102. dos Santos, G.Z.B., Caldas, L.R., de Almeida Melo Filho, J., Monteiro, N.B.R., Rafael, S.I., da Silva, N.M., Circular alternatives in the construction industry: An environmental performance assessment of sisal fiber-reinforced composites. *J. Build. Eng.*, *54*, 104603, 2022.
103. Akhai, S., Navigating the Potential Applications and Challenges of Intelligent and Sustainable Manufacturing for a Greener Future. *Evergreen*, *10*, 4, 2237–2243, 2023.
104. Khang, A. and Akhai, S., Green Intelligent and Sustainable Manufacturing: Key Advancements, Benefits, Challenges, and Applications for Transforming Industry, in: *Machine Vision and Industrial Robotics in Manufacturing*, pp. 405–417, 2024.

7

Regenerated Cellulose for Catalytic Removal of Pollutants from Wastewater

Azam Ali[1]*, Jiri Militky[1], Alyaa Abdulhussein Alsaedi[2], Maria Murtaza[3], Narmeen Zahra[4] and Sidra Ghafoor[5]

[1]*Department of Material Sciences, Technical University of Liberec, Czech Republic, Liberec, Europe*
[2]*Department of Microbiology, Collage of Veterinary Medicine, University of Al-Qadisiyah, Diwaniyah, Iraq*
[3]*Department of Physics, International Islamic University, Islamabad, Pakistan*
[4]*Department of Chemistry, International Islamic University, Islamabad, Pakistan*
[5]*Department of Chemistry, Riphah International University, Faisalabad, Pakistan*

Abstract

Nowadays, various types of industries can be found in nearly every region globally. These industries provide us with different products and services and generate huge waste (either in wastewater effluent or as solid waste). It is necessary to find effective remedial solutions for problems related to water pollution, as it gives rise to many health and environmental issues. The current work focuses on the catalytic removal of pollutants from wastewater through the use of regenerated cellulose. Due to the unique properties of cellulose, it is considered a potential substrate for the treatment of sewage. Secondly, it is the most abundant polymer found in nature. Here, in this work, cellulose has been taken to eliminate dyes heavy metals, and organic compounds from wastewater. Moreover, different strategies have been adopted to assess the catalytic performance of regenerated cellulose including incorporation of active sites, incorporation of metal nanoparticles, modifications in surface functionalization, etc. Significant pollutant removal mechanisms are also mentioned to assess the suitability of each mechanism for specific pollutant treatment. Hence, it summarizes the applications of cellulose-based material for remedial purposes. Furthermore, this work also highlights the future challenges and perspectives for high-potential wastewater treatment. The application of regenerated cellulose as a versatile and ecologically friendly

**Corresponding author*: azam.ali@tul.cz

Mohd Shabbir (ed.) Regenerated Cellulose: A Renewable Alternative to Petroleum-Based Plastics, (181–198)

material for catalytic wastewater treatment holds great importance for advancing toward more sustainable and effective environmental remediation technologies.

Keywords: Nanocellulose-based composites, wastewater treatment, biotechnology, bacterial pollutants, sustainable nanotechnology

7.1 Introduction

The release of harmful substances due to increased human activities has a significant impact on the environment. Water, being highly susceptible to pollution, has become a key focus in addressing critical environmental issues. Various sources, such as dyes, heavy metals, organic compounds, and pathogens, contribute to the deterioration of water quality. The importance of clean water has led to an emphasis on water treatment technologies. By employing physical, chemical, and biological methods, pollutants can be reduced, thereby minimizing health and environmental repercussions. This includes discussing various wastewater treatment technologies, with a focus on pre-treatment and post-treatment processes. Techniques like reverse osmosis, filtration, sedimentation, degasification, precipitation, flocculation, and adsorption play a crucial role in removing pollutants from water. Membrane separation and adsorption methods are more efficient in combating water pollution compared to traditional approaches. This review highlights various methods that efficiently remove pollutants from wastewater [1]. The heavy metals concentration in the soil is significantly greater than in areas without effluents. The sludge from sewage treatment plants (STPs) has both positive and negative effects on agriculture. It contains high levels of toxic heavy metals and pesticides but also has beneficial nutrients like nitrogen (N), phosphorus (P), and potassium (K) that act as fertilizers. The studied sludge had levels of cadmium, chromium, and nickel above the acceptable limits for agricultural and land use [2]. Treatment methods have made considerable developments in removing emerging contaminants (ECs) from wastewater [3]. While individual methods may not be entirely effective, hybrid systems, combining biological, chemical, and physical approaches, have shown greater efficiency [4]. Methods such as membrane separation and adsorption are very effective in controlling water pollution when compared to traditional techniques. These new approaches have demonstrated significant potential for eliminating pollutants from wastewater, emphasizing their importance in water treatment. In other research conducted by Le Pivert *et al.* where ZnO nanoparticles were manufactured on a zinc substrate and applied to a roof surface for the removal of organic pollutants from rainwater [5]. The researchers used a hydrothermal technique to create ZnO nanoparticles on the Zn roof in a rather short period of time. The existence of Zn(II) ions and a natural oxide layer on the Zn surface provided active sites for the development of ZnO nanoparticles. The efficiency of photocatalysis was evaluated using methylene blue under UV light and Acid Red 14 under solar light. The use of renewable agricultural products and biomass feedstock in wastewater treatment reduces carbon emissions as well as potentially lowers treatment costs. Both natural and modified cellulosic materials are utilized to remove organic and inorganic contaminants from wastewater for example, cellulose-based adsorbents are used to extract heavy metals, while cellulose membranes are utilized for water purification [6].

7.1.1 Wastewater Pollutants

Wastewater pollutants are contaminants that enter into water systems from various sources, including industrial, domestic, urban activities, and agricultural. These pollutants pose risks to human health, disrupt ecosystems, and degrade water quality. Understanding the various types of pollutants is critical for successful wastewater treatment and environmental protection.

It is concerning that water pollution has become a major global issue, necessitating a continual review of water resource regulations to combat it. The impact of water pollution is severe, with deaths and diseases occurring worldwide, leading to approximately 14,000 deaths daily [7]. Wastewater acquires different types of pollutants to include in water. These contaminants can be organic pollutants like fats, oils, insecticides detergents, and inorganic pollutants like salts, heavy metals, and acids. Additionally, excessive nutrients, such as nitrogen and phosphorus, as well as pathogens (including viruses, bacteria, and parasites). The remaining contaminants may include industrial chemical waste and solvents. It is concerning that pollutants present on the surface can penetrate and pollute groundwater sources when they seep underground water bodies [8, 9]. When fecal-contaminated water with pathogens infiltrates the ground, the contamination of water by pathogens can make it unsafe for drinking. Groundwater that is polluted with pathogens may contain protozoa, viruses, bacteria, and sometimes even helminth eggs. Drinking such contaminated water can lead to illnesses, like diarrhea and cholera [10].

The nitrogen-rich fertilizers can lead to dissolved oxygen deficiency in coastal areas, rivers, and lakes causing harm to marine life. Since 2006, regulations in the United States and Northwest Europe have aimed to standardize nitrogen fertilizer usage to avoid groundwater pollution [11, 12]. Selenium, a heavy element found naturally in soil, accumulates through irrigation practices, potentially leading to groundwater pollution. This accumulating selenium can cause groundwater pollution. make its way to water bodies, providing a high poisoning danger for animals and humans. Effective management of fertilizers, pesticides, and heavy metals is essential for safeguarding water quality and protecting ecosystems [13].

Small particles in the air, like carbon dioxide from burning fossil fuels, can mix with water molecules in the atmosphere to create sulfuric acid. Similarly, sulfur dioxide from sources like volcanoes and industries can react with water to create sulfuric acid. Burning coal and petroleum products can also generate sulfur dioxide. Nitrogen dioxide can mix with water to produce nitric acid. Moreover, airborne particles are significant contributors to water pollution as they can be transported to water sources through rainfall [7].

Harmful chemicals from factories and industrial waste are significant contributors to water pollution. These chemicals are often by-products of manufacturing processes and can have detrimental effects on water bodies [14]. Hazardous chemical waste can exist in gases, liquids, and solids and their characteristics, like corrosiveness, ignitability, toxicity, and reactivity, make them hazardous to the environment [15]. The problem of industrial waste and chemical contamination dates back to the Industrial Revolution when mass production started [14]. It is crucial to recognize that industrial waste chemicals require particular treatment facilities for efficient disposal, as they cannot be effectively processed by traditional sewage treatment facilities. Proper management of industrial waste is crucial to prevent further pollution of water bodies and protect the environment [16–18].

7.2 Treatment Approaches for Wastewater

Wastewater is treated utilizing physical, chemical, and biological processes, with options for primary, secondary, and tertiary treatment. Primary treatments involve screening and sedimentation of larger particles, while intermediate treatments include aerobic and anaerobic lagoons, and tertiary treatments include chemical approaches. Adsorption, due to its low cost and simplicity, is the most successful technology for removing toxins from wastewater and water [18]. In modern research, eco-friendly approaches are being focused on treating the pollutant water. Such as, the leather and tanning industries are concerned about water pollution because the vast amounts of water necessary for animal skin treatment produce dangerous chemicals and salts. About 85% of this water is dumped as effluent, posing health risks to humans and animals. Conventional wastewater treatment procedures are insufficient for total pollution removal, whereas biological treatment approaches are slow and limited for non-biodegradable contaminants. Nanotechnology is another eco-friendly and fast-emerging technology that uses nanoscale materials in

Table 7.1 Wastewater treatment processes and possible disadvantages [18].

Treatment	Disadvantage	Process details
Ion exchange	Expensive process, improper removal of ions.	Metallic ions in aqueous solutions are replaced by ions influenced by electrostatic forces which accumulate on exchange sites.
Adsorption	Chemical regeneration was required, as was an adsorbent for suspended materials, and the disposal of exhausted air was a difficulty.	The process involves the adherence of molecules or atomic coating on the surface of the adsorbent substrate.
Chemical precipitation	Limitation due to the generation of slug, expensive to deal with unwanted slug.	Normally, coagulants such as iron salt, limes, alum, and organic polymers are added to produce the precipitation of metal ions.
Reverse osmosis	Energy-consuming and costly, due to the use of pumps (pressure generation), the requirement to repair membrane films.	Metal ions are separated by a semipermeable membrane at a pressure that exceeds the internal osmotic pressure generated by the dissolved particles.
Electro dialysis	Membrane fouling and increased consumption of energy lead to high operational costs.	Metal ions are separated by using special membranes that only allow certain ions to pass through. When there is a difference in electrical charge between the two electrodes it leads to the separation of negatively charged ions (anions). And positively charged ions (cations).

(*Continued*)

Table 7.1 Wastewater treatment processes and possible disadvantages [18]. *(Continued)*

Treatment	Disadvantage	Process details
Coagulation/ flocculation	Increase in chemical usage leads to high operational cost and the generation of more slug volume.	Coagulant is added in water to promote the adhesion of colloidal particles, which attract suspended materials.
Flotation	Needs high operating costs for repairing and maintenance, extra capital expenses.	Bubble attachment is utilized to isolated solids or dispersed liquid phases.

different industries such as pharmaceutics, agriculture, medicine, smart textiles, and photocatalytic degradation, with biological approaches being preferred due to biocompatibility and simplicity of handling [19]. Table 7.1 is giving the concise overview about the different techniques adopted for wastewater treatment, their process details and possible disadvantages.

7.3 Cellulose and Their Limitations in Treating Water Pollutants

Cellulose fiber is the most abundant naturally occurring biopolymer on the earth and it has a good ability to serve as a highly efficient, cheap, and environment-friendly solution for the treatment of wastewater. Commonly, cellulose is found in plants and many aquatic species. The structure of cellulose (a carbohydrate polymer) is made up of D-glucose-β-1,4-linked anhydro units with the accumulation of three (OH) groups attached to each unit [20]. Its properties are shaped by hydrogen bonding systems, hydrophobic regions, and both intramolecular and intermolecular hydrogen bonds. Cellulose II, a denser and less reactive allomorph, occurs in two polymorphs: cellulose Iβ and cellulose Iα. Figure 7.1 shows the hierarchical structure of cellulose found in plants and trees, which ranges from the meter to the nanoscale scale.

The dimensions of cellulose are interconnected to its morphology; MCC (Microcrystalline Cellulose) is rod-like whereas NCC (Nanocrystalline cellulose) and NFC (Nanofibrillar Cellulose) mark needle/whisker-like and fibril-shaped characteristics. The final morphology is directly affected by the extraction method, tailored based on the structure and characteristics

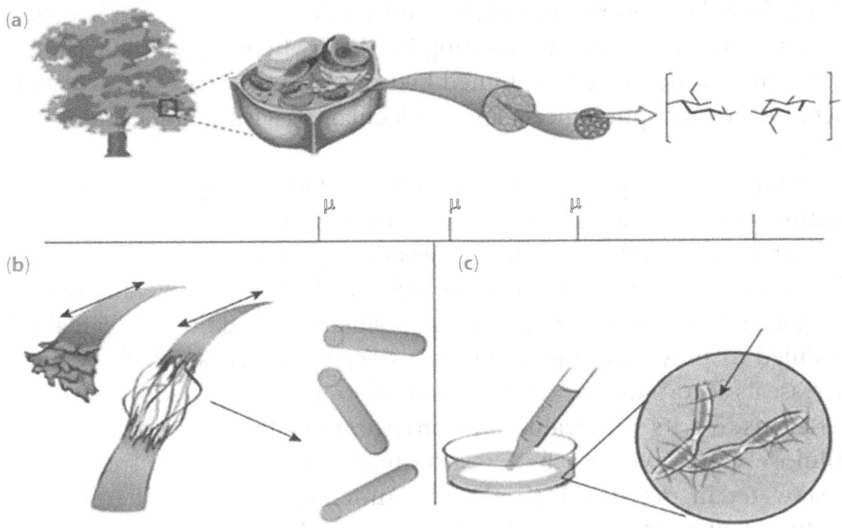
(a)
μ
μ
μ
(b)
(c)

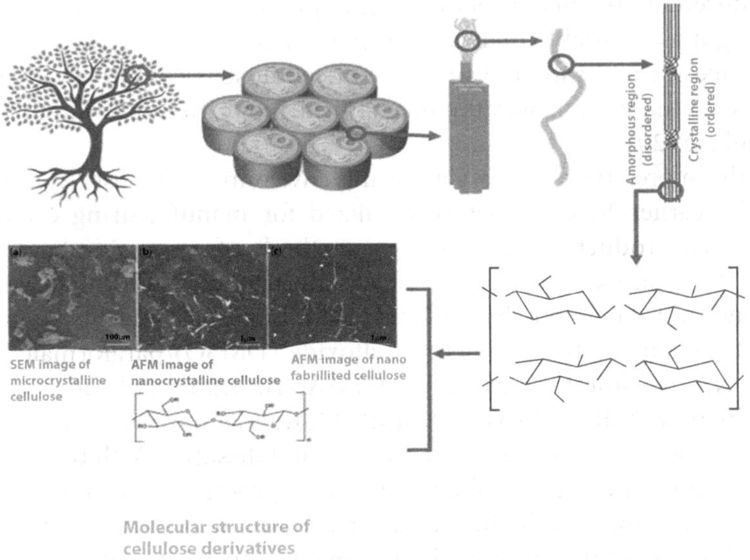
Amorphous region
(disordered)
Crystalline region
(ordered)
SEM image of
microcrystalline
cellulose
AFM image of
nanocrystalline cellulose
AFM image of nano
fabrillited cellulose
Molecular structure of
cellulose derivatives

of cellulose. For controlling cellulose dimensions, mineral acid hydrolysis is the traditional technique resulting in MCC(Microcrystalline Cellulose) being microscale caused by diluted hydrolysis [22]. Figure 7.2 depicts the hierarchical, morphological, and molecular structures of cellulose and its derivatives.

Throughout the past, the use of cellulose has rapidly increased for the treatment of wastewater because of its various properties such as abundance, biodegradability, and its potential to absorb a specific range of contaminants. Natural cellulose can behave as a bio-sorbent which effectively removes dyes, heavy metals, and many other organic pollutants. In addition, it can also be modified or processed into nanocellulose or membranes, which eventually increase its adsorption capacity, mechanical strength, and chemical resistance.

Because of its renewable and environment-friendly characteristics, cellulose can serve as a better sustainable choice for the treatment of wastewater in many industrial and municipal activities. Therefore, the treatment of wastewater with pure cellulose fibers has some disadvantages such as limited adsorption, degradation of cellulose, and weak chemical bonding which are eventually overcome by modifying the structure of cellulose. Thus, it was necessary to modify the structure of cellulose and convert it into regenerated cellulose (which includes breaking the structure of cellulose and transforming it into a useful and stable form of fiber. Regenerated cellulose, a flexible material is generally made up of different chemical processes being processed on natural cellulose, allowing it as a suitable material for many fields. Through the regeneration process, the crystalline structure of cellulose is completely changed which eventually improves the mechanical properties and functionality of cellulose [20].

Furthermore, there were significant environmental costs connected with the earlier Viscose methods utilized for manufacturing cellulose-regenerated products. Therefore, new methods of processing have been explored for an extended period in an effort to avoid costly processing pathways and dangerous byproducts [24]. Numerous solvents, comprising [24], alkali solutions dimethyl sulfoxide (DMSO)/paraformaldehyde (PF), N, N-dimethylacetamide (DMAc)/lithium chloride (LiCl), N-methylmorpholine-N-oxide (NMMO [20], and ionic liquids [ILs]), were recognized as causing cellulose to broaden and dissolve. With the increase in population, the demand for textile-based products is also increasing, eventually leading to a higher need for textile raw materials. Regardless of avoiding the challenging method of manufacturing traditional viscose fibers, the direct dissolution method of regenerated cellulose production has shown many unique features [25].

7.4 Regenerated Cellulose to Treat Industrial Wastewater

Regenerated cellulose is known to be a promising material for wastewater treatment in many industries because of its unique properties such as effective adsorption properties and sustainability. Recently, research has shown the ability of regenerated cellulose to remove dyes, heavy metals as well as pollutants from contaminated water [20].

The application of regenerated cellulose materials for dye adsorption from wastewater has been extensively researched due to their cost-effectiveness, reusability, and energy efficiency in remediation processes Regenerated cellulose materials also show great potential in absorbent applications. Their exceptional physical characteristics, such as high porosity and surface area [26], low crystallinity, simple functionalization and composite creation, low toxicity, and abundance, make them appropriate absorbents for dyes, heavy metal ions, oils, and CO2. For example, Dassanayake *et al.* used aero-celluloses in combination with activated carbon for CO2 adsorption [27].

Similarly, Wang *et al.* investigated the use of porous and hydrophobic silanized cellulose, synthesized by reacting microcrystalline cellulose (MCC) with hexadecyltrimethoxysilane (HDTMS), for the separation of oil and water. They attained a separation effectiveness of 99.93% for a vegetable oil-water mixture, with only a minor decrease to 99.77% after ten recycling cycles [28]. Zhan *et al.* used a polyethyleneimine-modified crosslinked cellulose-sodium alginate composite to adsorb Zn(II), Cu(II), and Pb(II) in aqueous solutions. The greatest adsorption capacities obtained for Zn(II), Cu(II), and Pb(II) ions were 110.2, 177.1, and 234.2 mg/g, respectively, due to the presence of chelating carboxylate and amino groups, as well as the absorbents' high porosity. Sun *et al.* investigated cellulose-chitosan composites made by co-dissolving cellulose and chitosan in [C4mim]Cl IL and found adsorption capabilities of 0.303, 0.417, 0.225, 0.251, and 0.127 mmol/g for Zn(II), Cu(II), Ni(II), Cr(VI), and Pb(II), respectively [29]. Ren *et al.* used a solution mixing-regeneration and freeze-drying procedure to create regenerated cellulose-graphene oxide composite aerogels that had excellent methylene blue adsorption efficiency [30]. Takada and Takada and Kadokawa [31], observed that during cellulose dissolution in the ionic liquid, the crystalline structure of the cellulose was broken, and surplus water surrounding the cellulose was absorbed into the ionic solution. This resulted in the aggregation of cellulose crystalline components, which could act as crosslinking locations for the development of an ion gel Figure 7.3.

In another study, Sun *et al.* [33], ionic liquid was employed to produce composite biosorbents from cellulose and chitin for heavy metal adsorption. However, this method has several disadvantages. Researchers considered employing ionic liquid for dissolving cellulose due to its excessive cost. Furthermore, the researchers were inexperienced with the procedure of recovering and recycling the ionic liquid, and part of it may have been lost in the process [34]. Cellulose is also known to be effective in the form of nanocrystals. It has numerous applications in the industry, which continue to grow. Yu *et al.* [35] used hydrolysis to generate cellulose crystals from cotton fiber. These cellulose nanocrystals were chemically engineered to remove Pb^{2+} and Cd^{2+} from aqueous solutions. The modification used succinic anhydride to form sodium bicarbonate ($NaHCO_3$) to form NaSCNCs, a sodium-based nano adsorbent and SCNCs (succinic anhydride cellulose nanocrystals). A model water containing Ag^+ was created, and removal percentages were calculated for all three types of adsorbents. CNCs had the best removal effectiveness (64% at neutral pH), followed by CHNCs (37%), and CNFs (27%). In another study, Zhou *et al.* [36] created magnetic chitosan hydrogel beads for Pb(II) adsorption from aqueous solutions. These beads were made of carboxylated cellulose nanofibrils, and the researchers found that the inclusion of carboxylate groups increased the adsorbent's adsorption ability. Additionally, nanoporous cellulose has been used for color treatment in wastewater [37]. Pillai *et al.* [38] employed xanthate nano banana cellulose (XNBC) to absorb Cd(II) from aqueous solutions. The adsorbent's sulfur-containing groups caused a significant absorption of heavy metals. The usage of cellulose derivatives chitosan and other materials is widespread for removing heavy metals from wastewater.

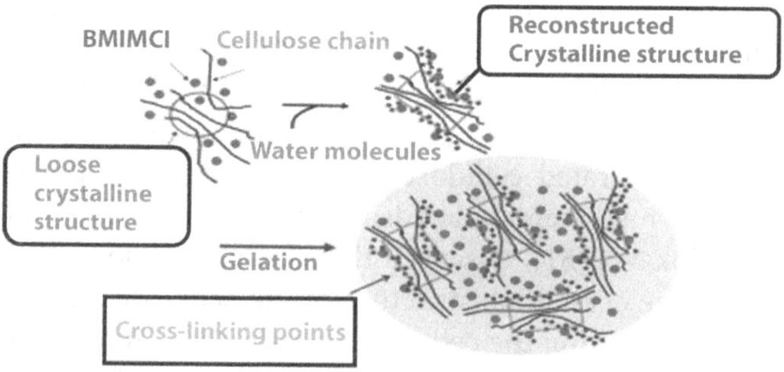

The wastewater contained Ag^{+}, Cu^{2+}, and Fe^{3+} ions. Both the adsorption capacity and speed were increased, leading to up to 99% removal of Cu^{2+} and Fe^{3+}. This process proved to be effective for solutions containing single metals as well as mixtures of metal ions [39].

One way to utilize cellulose for heavy metal removal is through a derivative called carboxymethyl cellulose (CMC), which forms complexes to remove heavy metals from wastewater. An experiment with CMC to extract Cr(III), Ni(II), and Cu(II) from synthetic wastewater solutions. Singha *et al.* [40], devised another excellent approach for producing heavy metal adsorbents, this time using graft copolymers derived from cellulose biopolymers. Using this derived adsorbent, heavy metals, such as Zn^{2+}, Pb^{2+}, Cd^{2+}, and Cu^{2+}, were successfully removed from wastewater [41]. Nanocrystalline cellulose has also been found to be effective and is being incorporated in many industries as well. To prepare cellulose nanocrystals from cotton, Yu *et al.* [35] employed a hydrolysis process in the preparation of nanocrystals. The Pb^{2+} and Cd^{2+} in water were successfully precipitated using the chemically treated nanocrystals. The modification applied succinic anhydride to form sodium bicarbonate ($NaHCO_3$) and SCNCs (succinic anhydride cellulose nanocrystals) to prepare NaSCNCs sodium-based nano adsorbent.

A research work described the model water solution containing Ag^{+} and searched the removal efficiencies of three adsorbent kinds. Among the removal effectiveness of three groups of adsorbents, CNCs recorded the highest removal effectiveness at neutral pH (64%), followed by CHNCs (37%), and CNFs (27%). Zhou *et al.* [36] developed magnetic chitosan hydrogel beads with carboxylated cellulose nanofibrils for the adsorption of Pb (II) from water in a different work. According to the sections of the study, the introduction of carboxylate groups enhanced the adsorption capacity of the adsorbent. Additionally, recent studies have shown that nanoporous cellulose can also be utilized for the removal of dyes from wastewater [37]. In another study, Pillai *et al.* [38] employed xanthate-modified (XNBC) nano banana cellulose to remove Cd (II) from aqueous solutions. The inclusion of sulfur-containing groups in the adsorbent resulted in a high heavy metal absorption rate. Cellulose derivatives, chitosan, and other compounds have been shown to behave as heavy metal absorption agents. The wastewater contained Ag^{+}, Cu^{2+}, and Fe^{3+} ions. Both the adsorption capacity and speed were increased, leading to up to 99% removal of Cu^{2+} and Fe^{3+} ions. This process proved to be effective for solutions containing single metals as well as mixtures of metal ions [39]. From the various applications of cellulose, one of the effective methods includes the use of carboxymethyl cellulose (CMC) to chelate complexes which can then be utilized to remove the heavy metals from the wastewater. In one investigation, CMC was utilized to selectively precipitate Ni (II), Cr (III), and Cu (II) ions from synthetic wastewater solutions. Singha *et al.* [40] also designed another new technique for synthesizing heavy metal adsorbents from cellulose biopolymers through the graft copolymerization process. This showed that adsorbent was effective for the elimination of heavy metal ions such as Zn^{2+}, Pb^{2+}, Cd^{2+}, and Cu^{2+} from wastewater [41] as illustrated in Figure 7.4.

Zhang *et al.* [43] observed that porous graphene oxide or carboxymethyl cellulose is an excellent adsorbent for Ni^{2+} ions. The adsorbent was a solid created using unidirectional freeze-drying. Wei *et al.* [44] employed carboxymethyl cellulose fibers to adsorb Cd(II). The adsorbent was developed as a thin fiber with a porous and rough surface. These surface characteristics facilitated the adsorption process. Khan *et al.* [45] discovered another novel cellulose-based adsorbent. Nanoparticles of zirconium dioxide have been created on a cellulose matrix. This technique was used to remove eight types of metal ions, including, Cd^{2+}, Cr^{3+}, Cu^{2+}, Fe^{3+}, Zn^{2+}, Ni^{2+}, and Zr^{4+}. However, the adsorbent only demonstrated efficiency for Ni^{2+} [46]. Table 7.2 summarizes various regenerated cellulose-based materials, their pollutant removal types and preparation methods for the treatment of industrial wastewater.

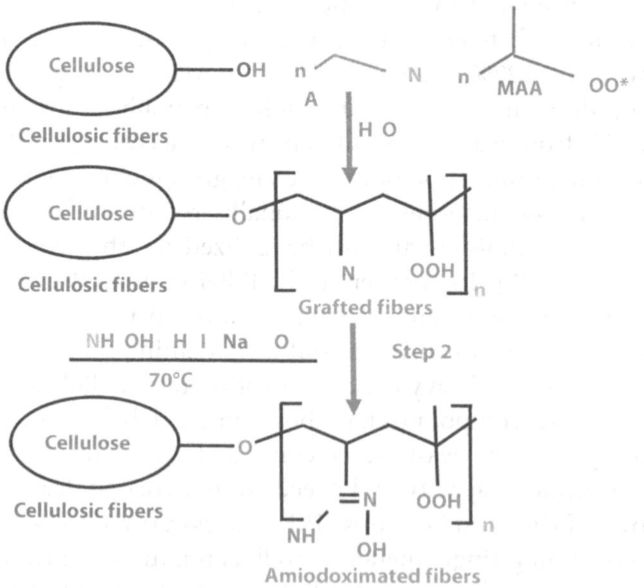

Table 7.2 Different regenerated cellulose and adopted methodologies to treat water pollutants.

Sr. no.	Regenerated cellulose material	Waste removal type	Method	References
1	Aero celluloses combined with activated carbon	CO_2 adsorption	Adsorption	[27]
2	Porous hydrophobic silanized cellulose (reaction between microcrystalline cellulose and hexadecyltrimethoxysilane)	Oil-water separation	Silanization of cellulose or hydrophobic method	[28]
3	Polyethyleneimine-modified crosslinked cellulose	Zn(II), Cu(II), and Pb(II)	Adsorption	[29]
4	Cellulose-chitosan composites	Zn(II), Cu(II), Ni(II), Cr(VI), and Pb(II)	Co-dissolving cellulose and chitosan	[29]
5	Graphene oxide composite aerogels	methylene blue adsorption	Freeze-drying process	[30]
6	Biosorbents from cellulose and chitin	heavy metals	Adsorption	[33]
7	Cotton to produce cellulose crystals	Pb^{2+} and Cd^{2+} from water solutions	Hydrolysis	[35]
8	Magnetic chitosan hydrogel beads	Pb(II) from water solutions	Adsorption	[36]
9	Nanoporous cellulose	dye treatment in wastewater	Adsorption	[37]
10	(XNBC) Xanthate nano banana cellulose	Cd(II) from water solutions	Adsorption	[38]
11	Carboxymethyl cellulose (CMC) to chelate complexes	Cr (III), Ni (II), and Cu (II) ions from synthetic wastewater	Precipitation	[40]
12	Cellulose biopolymers	Pb^{2+}, Zn^{2+}, Cu^{2+}, and Cd^{2+} from Wastewater	Graft copolymerization	[41]
13	Cotton to produce cellulose	adsorb Pb^{2+} and Cd^{2+} from water solutions	Hydrolysis	[35]

(*Continued*)

Table 7.2 Different regenerated cellulose and adopted methodologies to treat water pollutants. (*Continued*)

Sr. no.	Regenerated cellulose material	Waste removal type	Method	References
14	Porous graphene oxide or carboxymethyl cellulose	Ni^{2+} ions	Unidirectional freeze-drying	[43]
15	Zirconium dioxide on cellulose matrix	Heavy metal ions, such as Co^{2+}, Cd^{2+}, Cr^{3+}, Fe^{3+}, Cu^{2+}, Zn^{2+}, Ni^{2+}, and Zr^{4+}	Nanocomposite adsorption	[45]

7.5 Conclusion

There are different types of industries found throughout the globe. These industries not only provide us with different types of products but also generate harmful waste. The untreated water when enters into the nearby water bodies. This wastewater frequently comprises harmful chemicals, organic pollutants, and heavy metals which can pollute drinking water bodies and lead to health problems like gastrointestinal diseases, cancer, and reproductive issues. Regenerated cellulose is used to remove the catalytic removal of pollutants from the wastewater. It is an eco-friendly approach and hence, it is used to treat wastewater. Different types of materials like cotton, cellulose biopolymer, porous graphene oxide or carboxymethyl cellulose, xanthate nano banana cellulose (XNBC) and magnetic chitosan hydrogel beads zirconium dioxide on cellulose matrix, and cellulose-chitosan composites act as regenerated cellulosic material and they remove the Pb^{2+}, Cd^{2+}, Pb(II), Ni^{2+} ions, Heavy metal ions, methylene blue Co^{2+}, Cd^{2+}, Cr^{3+}, Fe^{3+}, Cu^{2+}, Zn^{2+}, Ni^{2+}, and Zr^{4+}, Cd(II) from water solutions and Zn(II), Cu(II), Ni(II), Pb(II), and Cr(VI), ions, respectively. Different types of methods are adopted by regenerated cellulosic structures such as adsorption, hydrolysis, nanocomposite adsorption, graft copolymerization, and precipitation to remove the pollutants from wastewater. This shows great importance in future perspective due to their effective environmental remediation technologies.

AI disclosure statement: The author acknowledges the use of AI-based tools for minor language editing or grammatical correction. However, all intellectual contributions, analyses, and conclusions presented in this chapter are entirely the author's own.

References

1. Chaudhry, F.N. and Malik, M., Factors affecting Water Pollution: A Review. *J. Ecosyst. Ecography*, 2017.
2. Azizullah, A., Khattak, M.N.K., Richter, P., Häder, D.P., Water pollution in Pakistan and its impact on public health - A review. *Environ. Int.*, 37, 2, 479–497, 2011. doi: 10.1016/j.envint.2010.10.007.
3. Muga, H.E. and Mihelcic, J.R., Sustainability of wastewater treatment technologies. *J. Environ. Manage.*, 88, 3, 437–447, 2008. doi: 10.1016/j.jenvman.2007.03.008.
4. Ahmed, S.F., *et al.*, Recent developments in physical, biological, chemical, and hybrid treatment techniques for removing emerging contaminants from wastewater. *J. Hazard. Mater.*, 416, 2021. doi: 10.1016/j.jhazmat.2021.125912.
5. Le Pivert, M., Piebourg, A., Bastide, S., Duc, M., Leprince-Wang, Y., Direct One-Step Seedless Hydrothermal Growth of ZnO Nanostructures on Zinc: Primary Study for Photocatalytic Roof Development for Rainwater Purification. *Catalysts*, 12, 10, 1231, Oct. 2022. doi: 10.3390/catal12101231.
6. Abdel-Fatah, M.A., Nanofiltration systems and applications in wastewater treatment. *Ain Shams Eng. J.*, 9, 4, 3077–3092, 2018.
7. Letchinger, M., Pollution and Water Quality, Neighbourhood water quali-tyassessment, in: *Project oceanography*, 2000.
8. Pink, D.H., Investing in tomorrow's liquid gold. *World J. Anal. Chem.*, 2006.
9. Larry, W., World water day: A billion people worldwide lack safe drinking water, 2006.
10. Lennon, M.A., Whelton, H., O'Mullane, D., Ekstrand, J., *Fluoride in drinking-water*, World Health Organization, Geneva, Switzerland, 2004.
11. Van Grinsven, H.J.M., Ten Berge, H.F.M., Dalgaard, T., Fraters, B., Durand, P., Hart, A., Willems, W.J., *et al.*, Management, regulation and environmental impacts of nitrogen fertilization in northweste. *Biogeosciences*, 9, 12, 5143–5160, 2012.
12. U.S. EPA, An Urgent Call to Action - Report of the State-EPA Nutrient Innovations Task Group, in: *US Environ. Prot. Agency*, p. 35, 2009, [Online]. Available: https://www.epa.gov/sites/production/files/documents/nitgreport.pdf.
13. Ganje, T.J., Selenium, in: *Diagnostic Criteria for Plants and Soils*, Chapman, H.D. (Ed.), pp. 394–404, 1966.
14. Maczulak, A., *Pollution: Treating Environmental Toxins*, p. 120, Infobase Publishing, New York, 2010.
15. Kumar, A., Thakur, A. K., Gaurav, G. K., et al., A critical review on sustainable hazardous waste management strategies: A step towards a circular economy. *Environ. Sci. Pollut. Res.*, 30, 105030–105055, 2023. Springer, Berlin, Germany. https://doi.org/10.1007/s11356-023-29511-8.
16. https://www3.epa.gov/npdes/pubs/ pretreatment_program_intro_2011.pdf.
17. Sabri, N.A., *et al.*, Fate of antibiotics and antibiotic resistance genes during conventional and additional treatment technologies in wastewater treatment plants. *Sci. Total Environ.*, 741, 2020. doi: 10.1016/j.scitotenv.2020.140199.

18. Younas, F., *et al.*, Current and emerging adsorbent technologies for wastewater treatment: Trends, limitations, and environmental implications. *Water (Switzerland)*, 13, 2, 2021. doi: 10.3390/w13020215.
19. Fouda, A., Hassan, S.E.D., Saied, E., Azab, M.S., An eco-friendly approach to textile and tannery wastewater treatment using maghemite nanoparticles (γ-Fe2O3-NPs) fabricated by Penicillium expansum strain (K-w). *J. Environ. Chem. Eng.*, 9, 1, 2021. doi: 10.1016/j.jece.2020.104693.
20. Peng, B., Yao, Z., Wang, X., Crombeen, M., Sweeney, D.G., Tam, K.C., Cellulose-based materials in wastewater treatment of petroleum industry. *Green Energy Environ.*, 5, 1, 37–49, 2020. doi: 10.1016/j.gee.2019.09.003.
21. Miyashiro, D., Hamano, R., Umemura, K., A review of applications using mixed materials of cellulose, nanocellulose and carbon nanotubes. *Nanomaterials*, 10, 2, 2020. doi: 10.3390/nano10020186.
22. Joseph, B., Sagarika, V.K., Sabu, C., Kalarikkal, N., Thomas, S., Cellulose nanocomposites: Fabrication and biomedical applications. *J. Bioresour. Bioprod.*, 5, 4, 223–237, 2020. doi: 10.1016/j.jobab.2020.10.001.
23. He, X., *et al.*, Cellulose and cellulose derivatives: Different colloidal states and food-related applications. *Carbohydr. Polym.*, 255, 2021. doi: 10.1016/j.carbpol.2020.117334.
24. Fink, H.P., Weigel, P., Purz, H.J., Ganster, J., Structure formation of regener_ated cellulose materials from NMMO-solutions. *Prog. Polym. Sci.*, 26, 1473–1524, 2001.
25. ص. ف. دیعس, No Titlرد کاتالیزگری, in: *Codex Aliment. Comm. Ensuring Food Saf. Nutr. Secur. Over 50 Years*, p. 368, 1390, doi:] A.M.A. Pintor, V.J.P. Vilar, C.M.S. Botelho, R.A.R. Boaventura, *Chem. Eng. J.*, 297, 229–255, 2016.
26. Parajuli, P., Acharya, S., Hu, Y., Abidi, N., Cellulose-based Monoliths with Enhanced Surface Area and Porosity. *J. Appl. Polym. Sci.*, 137, 1–12, 2020.
27. Dassanayake, R.S., Gunathilake, C., Jackson, T., Jaroniec, M., Abidi, N., Preparation and Adsorption Properties of Aerocellulose_Derived Activated Carbon Monoliths. *Cellulose*, 23, 1–12, 2016.
28. Wang, X., Xu, S., Tan, Y., Du, J., Wang, J., Synthesis and Characterization of a Porous and Hydrophobic Cellulose-Based Composite for Efficient and Fast Oil-Water Separation. *Carbohydr. Polym.*, 140, 188–194, 2016.
29. Zhan, W., Xu, C., Qian, G., Huang, G., Tang, X., Lin, B., Adsorption of Cu(Ii), Zn(Ii), and Pb(Ii) from Aqueous Single and Binary Metal Solutions by Regenerated Cellulose and Sodium Alginate Chemically Modified with Polyethyleneimine. *RSC Adv.*, 8, 187, 2018.
30. Ren, F., Li, Z., Tan, W.Z., Liu, X.H., Sun, Z.F., Ren, P.G., Yan, D.X., Facile Preparation of 3D Regenerated Cellulose/Graphene Oxide Composite Aerogel with High-Efficiency Adsorption towards Methylene Blue. *J. Colloid Interface Sci.*, 532, 58–67, 2018.
31. Takada, A. and Kadokawa, J.I., Fabrication and characterization of polysaccharide ion gels with ionic liquids and their further conversion into value-added sustainable materials. *Biomolecules*, 5, 244, 2015.

32. Kulkarni, S.J. and Kaware, J.P., Analysis of packed bed adsorption column with low cost adsorbent for cadmium removal. *Int. J. Therm. Environ. Eng.*, 9, 17–24, 2015.
33. Sun, X., Peng, B., Ji, Y., Chen, J., Li, D., Chitosan (chitin)/cellulose composite biosorbents prepared using ionic liquid for heavy metal ions adsorption. *AIChE J.*, 55, 2062–2069, 2009.
34. Rahman, N.S.A., Yhaya, M.F., Azahari, B., *et al.*, Utilisation of natural cellulose fibres in wastewater treatment. *Cellulose,* 25, 4887–4903, 2018. https://doi.org/10.1007/s10570-018-1935-8.
35. Yu, X., Tong, S., Ge, M., Wu, L., Zuo, J., Cao, C., Song, W., Adsorption of heavy metal ions from aqueous solution by carboxylated cellulose nanocrystals. *J. Environ. Sci.*, 25, 5, 933–943, 2013. DOI: 10.1016/s1001- 0742(12)60145-4.
36. Nayak, J.N., Chen, Y., Kim, J., Removal of Impurities from Cellulose Films after Their Regeneration from Cellulose Dissolved in DMAc/LiCl Solvent System. *Ind. Eng. Chem. Res.*, 47, 1702–1706, 2008.
37. Karim, Z., Mathew, A.P., Grahn, M., Mouzon, J., Oksman, K., Nanoporous membranes with cellulose nanocrystals as functional entity in chitosan: Removal of dyes from water. *Carbohydr. Polym.*, 112, 668–676, 2014. DOI: 10.1016/ j.carbpol.2014.06.048.
38. Gao, Q., Shen, X., Lu, X., Regenerated Bacterial Cellulose Fibers Prepared by the NMMO/H2O Process. *Carbohydr. Polym.*, 83, 1253–1256, 2011.
39. Ono, Y., Ishida, T., Soeta, H., Saito, T., Isogai, A., Reliable dn/dc Values of Cellulose, Chitin, and Cellulose Triacetate Dissolved in LiCl/N,N-Dimethylacetamide for Molecular Mass Analysis. *Biomacromolecules,* 17, 192–199, 2016.
40. Varghese, A.G., Paul, S.A., Latha, M.S., Remediation of heavy metals and dyes from wastewater using cellulose-based adsorbents. *Environ. Chem. Lett.*, 17, 2, 867–877, 2019. doi: 10.1007/s10311-018-00843-z.
41. Henniges, U., Kostic, M., Borgards, A., Rosenau, T., Potthastt, A., Dissolution Behavior of Different Celluloses. *Biomacromolecules*, 12, 871–879, 2011.
42. Singha, A.S. and Guleria, A., Chemical modification of cellulosic biopolymer and its use in removal of heavy metal ions from wastewater. *Int. J. Biol. Macromol.*, 67, 409–417, 2014. doi: 10.1016/j.ijbiomac.2014.03.046.
43. Zhang, Y., Liu, Y., Wang, X., Sun, Z., Ma, J., Wu, T., Xing, F., Gao, J., Porous graphene oxide/carboxymethyl cellulose monoliths with high metal ion adsorption. *Carbohydr. Polym.*, 101, 392–400, 2014.
44. Aulin, C., Ahola, S., Josefsson, P., *et al.*, Nanoscale cellulose films with different crystallinities and mesostructures their surface properties and interaction with water. *Langmuir*, 25, 7675–7685, 2009.
45. Gindl, W., Emsenhuber, G., Maier, G., Keckes, J., Cellulose in Never-Dried Gel Oriented by an AC Electric Field. *Biomacromolecules*, 10, 1315–1318, 2009.
46. Jamshaid, A., *et al.*, Cellulose-based Materials for the Removal of Heavy Metals from Wastewater – An Overview. *ChemBioEng Rev.*, 4, 4, 240–256, 2017. doi: 10.1002/cben.201700002.

8

Regenerated Cellulose for Packaging

Muntajena Nanjeba[1], Md. Osman Goni Shihan[1], Rifat Jahangir[1], Joyanto Sutradhar[1] and Md. Arif Mahmud[2]*

[1]*Department of Textile Engineering, Ahsanullah University of Science and Technology, Dhaka, Bangladesh*

[2]*Department of Materials, The University of Manchester, Manchester, United Kingdom*

Abstract

The fabrication of regenerated cellulose-based materials has garnered significant interest due to their versatile applications. This chapter will systematically evaluate the impact of cellulose sources, morphologies, and various fabrication process on the key properties of regenerated cellulose, including thermal stability, mechanical strength, antibacterial, and barrier properties (water vapor, oxygen permeability, and light transmittance) for packaging applications. Fabrication is the most crucial part that affect functional attributes of regenerated cellulose. Proper process selection and used chemicals in the fabrication result in optimum properties required for packaging purposes. Key aspects of crystallization, structure, and orientation of regenerated cellulose are also intertwined to understand their impact on the material properties. In addition, challenges associated with the fabrication process will be critically assessed with a focus on process complexity, recyclability of waste, safety assessment concerning potential migrant from fabricated product into packaged goods, and the inherent low solubility of cellulose. The solution and innovative approaches to overcome these challenges are essential, providing a comprehensive understanding of potential limitations of regenerated cellulose in packaging applications, with an emphasis on environment concerns. In this study, by offering a detailed examination of the factors affecting the properties of regenerated cellulose and addressing the associated challenges, this chapter will serve as an essential resource for researchers to contribute ongoing research on regenerated cellulose. Overall, it aims to enhance the understanding and application of sustainable, high-performance packaging materials derived from regenerated cellulose, contributing to advancements in eco-friendly packaging solutions.

**Corresponding author*: arif.shawon@hotmail.com

Mohd Shabbir (ed.) Regenerated Cellulose: A Renewable Alternative to Petroleum-Based Plastics, (199–242)

Keywords: Biodegradable packaging, sustainable materials, barrier properties, food packaging applications, thermal stability, nanocellulose reinforcement, intelligent packaging, circular economy

8.1 Introduction

Plastic pollution is a global problem, finding its way into deserts, farmlands, mountaintops, the deep ocean, tropical dumps, and even Arctic snow [1]. This problem is not only on the rise but is also expected to persist, even in some of the most optimistic projections for reducing plastic waste. In 2016, global estimates indicated that between 9 and 23 million metric tons of plastic waste was entering rivers, lakes, and oceans annually. That was also about how much, 13 to 25 million metric tons a year, was being emitted into land environments. These figures are a reminder of the great and persistent challenge of tackling plastic pollution [2, 3].

Plastics are often regarded as a long-lived and stable type of material, meaning that these particles can be stable and accumulate as they spread in the environment over time [3, 4]. Research has found that our world is now covered in plastics, from the water we drink to the very air we breathe, to the ground on which we walk [4]. The negative impact of macroplastics, microplastics, and nanoplastics on the environment and organisms are due to a combination of their inherent properties and toxicological nature. These plastics leach toxic additives, chemicals, and persistent pollutants that lasso to the debris, compounding their effects. While plastic pollution is a major environmental problem, many solutions exist to mitigate, manage, and eliminate this global issue [5].

Garbage is what you have when a plastic object reaches the end of its life (EoL) and approaches all the steps leading to a meaningful waste. As a first step, the best we could do with an object that cannot be used anymore is reusing it. If you cannot reuse, there is recycling, which is good but not as good as reuse. However, with sustainability efforts on the rise, alternative means of managing plastic waste have become increasingly important [6]. A potential solution for this problem that we developed a few years ago is the use of biodegradable polymers in replacement of traditional plastics. This emphasizes that renewable materials are a sustainable alternative to petroleum-derived plastics. But because of their persistence in natural ecosystems and difficulties with decomposition, plastic has severe environmental downsides. Cellulose, on the other hand, is a renewable plant polymer and an attractive substitute for petroleum-based plastics. Cellulose-based materials are environment-friendly as they are biodegradable and renewable, which could help reduce plastic waste and mitigate reliance on fossil fuels. These are chemically modified forms of cellulose, which scientists have made into products that retain many of the good aspects of plastic—like, for example, its flexibility, durability, and versatility for use in packaging and elsewhere. Using cellulose-based substitutes can promote a greener production process and provide environmentally conscious solutions for disposal [7].

8.2 Background

8.2.1 Packaging Materials History and Its Context

Packaging is a critical and dynamic sector, with a more than 400 billion Euros in global yearly production. It is a vital component of various sectors like food and beverages, personal care, pharmaceuticals, electronics, and chemicals. When it comes to storing, protecting, and shipping goods, packaging is a must, but it can also make or break the commercial success of consumer and industrial products.

It is an industry rich with diversity and innovation. As the first step of touch between any product and its customer, packaging design effectively decides whether or not the customer would pick up or say no to it over another item, driving their buying decision. For example, chocolate boxes and milk cartons have easily identifiable shapes, and luxury goods, such as jewellery or perfume, often have unique and striking packaging.

Aside from its aesthetics, packaging has to be functional—sturdy enough to protect whatever's inside, while being efficient in size, as well as shape. Good packaging is a combination of artistic design and sophisticated engineering processes, which must satisfy both logistical and marketing goals [8].

8.2.2 Packaging: Key Requirements and Standards

Packing acts as a sheath that prevents the interior of the cell, i.e., electrodes and electrolytes, from exposing to outside conditions and atmosphere. Some of the top requirements from an effective packaging structures perspective are:

- Strong barrier against permeation from both liquid and vaporized electrolytes, as well as from moisture, carbon dioxide, and oxygen.
- High durability and mechanical stability.
- Resistance to electrolytes and chemical inertness.
- Electrochemical stability suited to the specific electrochemical system.

- Thermal resilience.
- Affordable manufacturing costs.
- Ease of workability and processability [9].

8.2.3 Common Types of Packaging Materials

Here are a few common types of packaging materials:

a. Cardboard Boxes: Probably the most recognized form of packaging, very useful for moving and storing. Secure packaging from a supermarket or pharmacy The tricky thing with buying secure packaging off the internet is that you want it to be strong enough to hold what you're carrying, but light-weight and disposable too. They are typically inexpensive, making them an economical option for mover and storage. Cardboard gets damaged and mangled while moving but it is also easy to replace.
b. Glass: The glass is well known for its chemical stability of properties, and because it is an excellent barrier to oxygen. While glass is a great choice for certain packaged goods, it does have some potential drawbacks; unlike many other materials, it can be heavy and fragile, breaking easily.
c. Tin cans: These are generally used to pack food if being pasteurized or sterilized. Tin cans protect food, but there are trace amounts of tin, as the coating that prevents rust can slowly leach into the food.
d. Plastic packaging: Plastic packaging is available in multiple types rigid and laminate. Flexible plastic packages can (i) control moisture or nominally block(3) (ii) oxygen. Materials, such as polypropylene (PP) and polyethylene terephthalate (PET), are top choices due to their lighter weight and durability in contrast to glass. Plastics are also durable and come in a variety of shapes and sizes, but can be more expensive and heavier than cardboard, making them less feasible for moving. For storage and organization, though, plastics are pretty good. Be also aware that reuse of plastic bottles could become a health hazard, particularly if the bottle previously contained something toxic, since substances migrate from plastic to whatever is in the bottle now.

Plastic packaging and coatings are frequently highlighted as major factors in today's environmental challenges. In response, many packaging companies are exploring new methods to lessen their environmental footprint, such as developing biodegradable packaging. Biodegradability means that a product can be broken down into simpler components by natural processes, minimizing its environmental impact. Unlike biodegradable options, petrochemical plastics remain in the environment for an extended period without breaking down [8].

e. Bioplastics: Bioplastics are plastics made from biodegradable materials or plant-based raw materials. Recently, there has been growing interest in using bioplastics due to their environmental benefits. The process involves extracting starch from these plants and then using microorganisms to convert it into lactic acid, which forms long-chain polymers known as polylactic acid (PLA). These plant-based plastics, also called bioplastics, are not only biodegradable but also made from renewable resources, which helps lessen the demand on limited petrochemical supplies. Despite their advantages, bioplastics are significantly more expensive, costing up to four times more than conventional petroleum-based plastics. In 2004, global production of bioplastics was only 250,000 tons. While bioplastics serve similar purposes as traditional plastics, their key difference is that they are designed to break down during specific composting processes [8].

Bioplastics present an effective solution to the issues associated with disposable plastic applications, mainly due to their biodegradability, allowing them to be recycled and reused. Many industries are shifting toward bioplastics as a preferred raw material over traditional plastics, given the numerous benefits they offer. Bioplastic production draws from a broad range of high-quality raw materials, often at lower costs, which enhances competitiveness. Additionally, bioplastics manufacturing enables the use of innovative technologies, opening new business opportunities for niche products. Due to their biodegradable nature, bioplastics are eco-friendly, providing an ideal packaging option for items like organic food products in compostable packaging. This makes bioplastics a highly favorable choice in sustainable packaging solutions [8].

8.2.4 Types of Bioplastics

Bioplastics can be made from a variety of renewable materials, including both edible and nonedible sources. These materials range from starch and

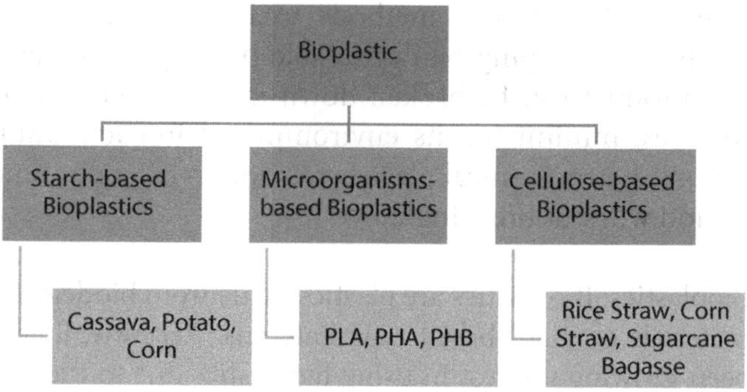

cellulose to proteins and certain bacteria such as PHA, PHB, and PLA. Bioplastics are typically categorized into three main types: starch-based, cellulose-based, and microbe-based. Each type uses different raw materials for production, contributing to the diversity in bioplastic options available [10]. Figure 8.1 shows the classification of bioplastics.

8.2.5 Required Properties of Packaging Materials

Petroleum-based plastic materials have generally been considered in packaging applications for many decades. Conventional plastics rely on nonrenewable resources and are non-biodegradable. The low density, low cost, excellent mechanical strength, and barrier properties make it an appropriate candidate for various industrial packaging applications [11]. However, the massive consumption of such material leads to environmental challenges mainly littering and global warming [12]. Growing public awareness of the ecological issues associated with conventional plastic materials, along with consumer pressure for greater sustainability, has spurred the development of alternative food packaging.

It is challenging to find suitable alternatives to plastic packaging as they must meet certain criteria, primarily in terms of mechanical strength and barrier properties [13]. When evaluating the effectiveness of packaging materials, the properties of interest are addressed as follows.

8.2.5.1 Mechanical Strength

Tensile strength and elongation at break are crucial property for food packaging material to remain integrity during transportation, handling, and storage of food [14]. The most important factors influencing mechanical performance are the molecular weight, the structure of polymer backbone chains such as the presence of linear and/or branched polymer chains, and the ratio between the amorphous region and the crystalline region. In addition, the orientation of polymer chains, like the mono-orientation or biorientation of polymer chains, can enhance mechanical strength [13].

Polymer films are generally used in packaging that withstand the tension of being pulled without stretching and breaking. For example, the addition of plasticizer into regenerated cellulose (RC) film showed improved of tensile strength and elongation at break. The addition of plasticizer relaxes the amorphous region and increases the mobility of the polymer chain [15]. The most studied plasticizer are glycerol and polyethylene glycol which are used to make flexible film [16, 17]. Another effective method to enhance mechanical strength is by incorporating a crosslinking agent into polymer chain, leading to the formation of a three-dimensional structure through hydrogen bond or covalent bond. This approach significantly increases elongation at break while reducing tensile strength [17].

8.2.5.2 Barrier Properties

Barrier properties such as oxygen permeation (OP) and water vapor permeation (WVP) play a significant role, especially in food packaging. Because the deterioration of food is often driven by aerobic food bacteria growth and lipid oxidation involving gases like oxygen, as well as the permeation of water vapor, so effective barrier properties in packaging materials are essential for preserving food quality [18]. The solubility and diffusion of penetrant depend on the nature of the penetrant, molecular architecture (branches and molecular weight) and morphology (crystallinity and molecular orientation) of the polymer, and the selection of processing method to fabricate are crucial factors that determine barrier properties (OP and WVP) [19].

Modification of the polymer chain by Incorporating small chemical groups into the polymer backbone chain can significantly improve the barrier characteristics. For example, introducing a nonpolar bulky group at the low barrier end of the permeability spectrum, or the incorporation of self-interacting group at the high barrier end of the permeability spectrum, can improve the barrier properties [20].

Cohesive energy density is a useful physical factor that assists in understanding and predicting the barrier properties of polymeric materials. The cohesive energy of a substance in its condensed state is referred to the increase in internal energy when all intermolecular forces are removed. For low molecular weight substance, this energy is determined by the heat of evaporation whereas for polymer it can be estimated using the additive group contribution model. As a result, this parameter provides insight into the strength of the interactions between molecules and how the interaction is affected by the introduction of numerous chemical groups to the polymer chain. Another important factor is free volume which refers to the void spaces within polymer matrix. Permeant, such as gas and small molecules, uses these pathways to diffuse through polymer material. This transport properties depends on the number and size of these void spaces [20].

8.2.5.3 Biodegradability

Biodegradation is a natural process that is mainly formed by microorganisms including bacteria, fungi and actinomycetes. These microorganisms utilize organic matter as their food and produce byproduct (carbon-dioxide). This natural process requires the availability of essential elements like carbon, oxygen, nitrogen and water [21].

Water or moisture can influence the degradation of polymer chain. Firstly, water is a crucial element for the growth and expansion of microorganisms and secondly moisture-rich environment increases hydrolysis reaction which increases accessible site for microorganism to attack. Moreover, temperature affects the hydrolysis reaction. As temperature increase both hydrolysis and microbial activity increase. However, if the temperature is excessively high, the microbial activity decreases or comes to a stop [22].

The flexibility of polymer chains significantly influences their decomposition. Flexible polymer chains allow more availability of water for hydrolysis, and the polymer can more easily shape into the active sides of enzymes, both enhancing the biodegradation rate. Polymer flexibility is influenced by the presence of bulky side groups and the types of linkage in the backbone polymer chain. Polymers with more sides group tend to degrade faster than those with fewer side groups. Additionally, the crystallinity and amorphous region within polymer structure affect substantially in degradation. Water cannot easily diffuse through the crystalline region while in the amorphous region water easily penetrate. As a result, amorphous region is more responsive to both hydrolysis and biodegradation [21].

8.2.5.4 Thermal Stability

Thermal stability is another important requirement for packaging materials which is influenced by polymer type, intermolecular bond (H bond), structure (crystallinity and amorphous region) and chemical composition of polymeric substances. Higher thermal properties result from greater H bond interaction between polymeric chains and greater crystalline structure. In addition, chemical content of polymer affects thermal properties significantly. For example, hemicellulose in the cellulosic fiber responsible for lower thermal degradation [23].

According to previous studies, there are several approaches an improvement in thermal stability. Fiber treatment with chemicals (like alkali, acid, and coupling agents) enhance thermal properties of polymeric materials. The changes contribute to the removal of impurities in the polymer composition which are lower molecular weight and can promote ignitability at lower temperature [24].

8.3 Cellulose: A Promising Candidate for Packaging

Cellulose is a complex carbohydrate polymer that is the primary structural component within plant cell walls. It is a linear polymer consisting of glucose subunits linked by β-1,4-glycosidic bonds [25]. The molecular structure can be represented by a formula where "n" indicates the degree of polymerization, which typically ranges from hundreds to thousands of glucose monomers in naturally occurring cellulose. Cellulose is the predominant naturally occurring organic polymer on the planet and is employed in numerous industrial processes, encompassing paper production, textile manufacturing, and biofuel generation [26].

Cellulose is known for being highly crystalline and hydrophilic, which results in a brittle material with low flexibility and tensile strength, making it unsuitable for packaging. However, derivatives of cellulose, such as hydroxypropyl cellulose, hydroxypropyl methylcellulose, carboxymethylcellulose, and methylcellulose, are well suited for use in packaging due to their improved properties [10].

8.3.1 Chemical Composition

The chemical structure features (shown in Figure 8.2):

- **I.** **β-1,4-Glycosidic Bonds**: The individual glucose units are linked together through β-1,4-glycosidic bonds, producing lengthy, straight polysaccharide polymers [26].
- **II.** **Linear Chains**: The molecular chains consist of a repeating disaccharide unit of cellobiose, linked *via* a β-1,4 glycosidic bond [26, 27].

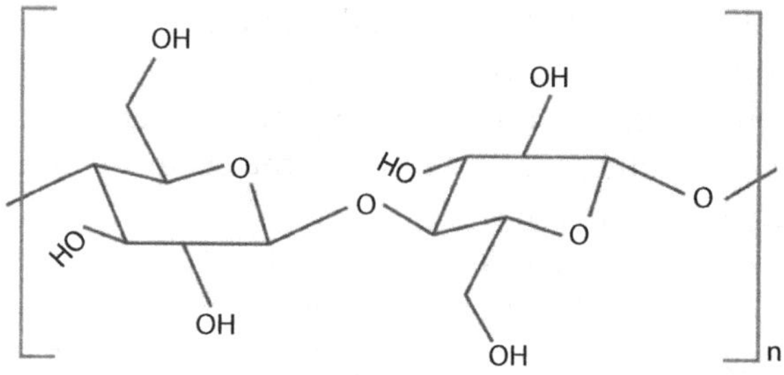

Figure 8.2 The repeat unit of cellulose.

III. **Hydrogen Bonding**: The cellulose fibers form extensive hydrogen bonds with each other, which increases the overall strength and rigidity of the fiber structure [26, 27].

8.3.2 Natural Sources

Cellulose, a naturally occurring macromolecule, is pervasive throughout the plant kingdom and functions as a primary structural component of plant cell walls. Here are some of the most common sources of cellulose in nature:

a. **Wood:** Wood by weight is 40% to 50% porous cellulose, the most abundant cellulose source for paper, textiles, etc. [25].
b. **Cotton:** Cotton fibers are nearly pure cellulose (90%) so it is a good source for textiles and other cellulose derivative production [28].
c. **Hemp:** Hemp has a high cellulose content, and is used for paper production, textiles, and biodegradable plastics [29].
d. **Bamboo:** Bamboo consists of 50% to 70% cellulose, making it an excellent feedstock for the production of textiles, paper, and bio-based composite materials. This plant has a rapid growth rate and is currently used widely in these applications due to their environment-friendly nature [30].
e. **Grasses:** As per Vogel & Jung, switchgrass and sugarcane are abundant in cellulose which makes them good biofuel producers [31].
f. **Flax:** Flax fibers are used to make linen and have 65% to 75% cellulose. They are also strong and durable [32].

The total global production of cellulose by plants is estimated to be around 1.5 trillion tons annually [25]. Cellulose makes up about 33% of all plant matter, which makes it the most significant component of the plant cell wall [33].

8.3.3 Properties

Cellulose exhibits a range of properties that render it a versatile material, facilitating its use across diverse industries including textiles, paper, food, and biofuels. The following outlines some of the key characteristics of this significant biopolymer:

- **Tensile Strength:** Cellulose fibers have high tensile strength, while wood roots have a range of 132 to 201 MPa [34].
- **Elastic Modulus:** The elastic modulus of cellulose is typically high, ranging from 10 to 20 GPa, depending on the source and processing methods [35].
- **Crystallinity:** Cellulose exhibits a semi-crystalline morphology, comprising both highly organized crystalline domains and disordered amorphous regions. The degree of crystallinity affects its mechanical strength, thermal stability, and reactivity. The typical crystallinity index of natural cellulose ranges from 50% to 90% [36].
- **Density**: The density of cellulose is about 1.5 g/cm^3, contributing to its firmness and strength [36].
- **Thermal Stability:** Cellulose is stable up to 200°C to 260°C, beyond which it decomposes, forming volatile compounds like levoglucosan, furfural, and carbonaceous char [37].
- **Water Absorption:** Cellulose's hydroxyl groups make it highly hydrophilic, so it absorbs water easily. This is useful for absorbent materials, but it can also lead to swelling and degradation in moist environments [25].
- **Solubility:** Native cellulose is insoluble in water and they are insoluble in most organic compounds; however, they can be dissolved in a variety of solvents such as ionic liquids or strong alkaline solutions allowing further processing and convenience in utilizing this type of material [38].
- **Transparency:** Fibrous films of cellulose, particularly when prepared similarly to nanocrystalline cellulose, can exhibit remarkably high optical clarity. This property is important for, for example, packaging materials and flexible displays [39].

8.3.4 Uses

Cellulose is a very versatile and multifaceted material and finds numerous applications in various industrial sectors. Here is a few of the main uses of this common material:

Paper and Cardboard Production: Cellulose, obtained from wood pulp, is the main ingredient of paper and cardboard and is utilized in forming different sort of paper products including writing paper, newsprint, and packaging materials [40].
Textile Industry: Cotton, flax (linen) and hemp are cellulose fibers forming the foundation of the textile industry. Cellulose is also used to produce semi-synthetic fibers, such as rayon and acetate, which can be processed into various fibers for use in clothing, upholstery and other fabrics [40].
Pharmaceuticals: In the pharmaceutical industry, cellulose is used as a binder, disintegrant, and filler in tablet formulations. Microcrystalline cellulose is particularly common in this role. Cellulose derivatives play an important role in helping drugs work better with controlled-release systems [41].
Biofuels: Cellulose is crucial in the production of biofuels, particularly cellulosic ethanol. This renewable fuel is created by converting cellulose into sugars, which are then fermented into ethanol. It is considered a more sustainable option compared to fossil fuels [42].
Plastics and Composites: Cellulose derivatives, such as cellulose acetate, are used to produce biodegradable plastics and composites. These materials are utilized in various applications, including packaging, photographic films, and as eco-friendly alternatives to conventional plastics [43].

8.4 Regenerated Cellulose

The concept of molecule regeneration refers to restoring or reforming a molecule to its original state or creating a new molecule through transformation [44]. Cellulose regeneration employs chemical processes to convert native cellulose into reconstituted cellulose materials. This process enables the production of sustainable materials, such as films and fibers, making cellulose suitable for various applications, especially in sustainable packaging and textiles [45].

8.4.1 Differences between Cellulose and Regenerated Cellulose

- Source
    a. Cellulose: Naturally occurring in plants, algae, and some bacteria [25].

   b. Regenerated cellulose: Produced through chemical processes, often from wood pulp or cotton [46].
- Structure
   a. Cellulose: The plant cell wall is fortified by cellulose, a structural polysaccharide characterized by linear chains of β-linked D-glucose subunits. Owing to its highly crystalline and water-insoluble nature, cellulose exhibits a rigid, fibrous architecture that bolsters the overall strength and stability of the plant cell wall [25].
   b. Regenerated cellulose: Regenerated cellulose like rayon and cellophane has less crystalline and more amorphous cellulose chains, making it more flexible and easier to process into various products [47].
- Comparison between cellulose and regenerated cellulose properties
   a. Cellulose: The processing of cellulose is burdened by its insolubility in water and most organic solvents. However, cellulose possesses high tensile strength, facilitating its use in unprocessed form for applications such as paper and textiles, in addition to its employment as a reinforcement component in composite materials [25].
   b. Regenerated cellulose: Regenerated cellulose is more flexible and can be spun into fine fibers or moulded into thin films. However, it may lose some of the strength and crystallinity of natural cellulose but gains enhanced versatility in applications such as textiles, films, and packaging [28].

8.4.2 The Historical Progression of Regenerated Cellulose

In 1846, the foundational work on cellulose began with Christian Friedrich Schönbein's discovery of nitrocellulose, also known as guncotton [48]. After that, British chemists Charles Frederick Cross, Edward John Bevan, and Clayton Beadle developed the viscose process. This was the first significant method for dissolving and reconstituting cellulose into a new form [49]. The initial large-scale commercial manufacture of viscose rayon was initiated in 1905 by the British enterprise Courtaulds Ltd. It was initially marketed as 'artificial silk' because of its similar appearance and texture [50]. Since World War II, the production of synthetic fiber has surged, revolutionizing industries and creating new possibilities. Nylon and polyester have surged in popularity due to their durability and versatility. Meanwhile, rayon remains a top choice for its exceptional absorbency and luxurious softness [51]. In the 20th century, people were focused on sustainability and that is why lyocell was developed as a more environmentally friendly [28, 52].

8.4.3 Properties of Regenerated Cellulose

Cellulose is a linear chain composed of two anhydroglucose rings $(C_6H_{10}O_5)_n$, linked together by an oxygen atom covalently bonded to the C1 of one glucose ring and the C4 of the adjacent ring (1–4 linkage). This bond known as the β 1-4 glucosidic bond gives rise to a flat ribbon-like conformation [53–55]. Generally, RC is made directly from cellulose *via* the dissolution regeneration process. This process is sustainable and environmentally friendly as most of the agents (such as solvents and coagulants agent) can be recycled and reused. By adjusting the regeneration parameters (cellulose extraction, dissolution and coagulation), the RC vary in properties and shapes such as powder, fiber, film, and hydrogels.

Thus biopolymers (such as starch, cellulose and RC) have gained significant attention and have renewable, biocompatibility and environmentally friendliness nature [56]. From an environmental perspective, regenerated cellulose is considered a viable alternative to petroleum-based plastics [57]. Moreover, cellulose offers excellent film forming abilities. However, poor flexibility takes it backseat [58]. Table 8.1 shows the mechanical properties and preparation methods for various regenerated cellulose films.

8.4.3.1 Mechanical Strength

The property of RC film depends on the chemical structure, crystallinity, and morphology of the polymer chain which are reflected through FTIR, XRD, and SEM analysis [16, 60]. In RC-2, adding plasticizer into cellulose solution during dissolution significantly improved mechanical strength which could be explained by XRD analysis [16]. RC-2 film was less crystalline (due to plasticizer) than RC-1 (XRD) resulting in improvement at elongation at break as mobility of polymer chain increased [15]. Moreover, no new peak was formed in FTIR of RC-2 film as the rearrangement of intermolecular bond (hydrogen bond) [16]. This rearrangement of the H-bond may affect in improvement of tensile strength in RC-2 film.

Mechanical properties have been affected (significant improvement of tensile strength of RC-3 film) as chemical treatment affects to remove hemicellulose, lignin and other contaminant which are proved by FTIR spectrum [61] and reduce molecular weight. Moreover, cellulose source has been affected significantly on mechanical strength (RC-3 compared

Table 8.1 Properties of RC film.

Code	Cellulose extraction	Regeneration process (parameter and used chemicals)	Mechanical strength	References
RC-1	Source: wood pulp	Dissolution: chemical: NaOH/ Urea (7/12 wt%) Coagulation: solvent evaporation	Tensile strength (Mpa): 30–35 Elongation at break (%): 6–10	[16]
RC-2	Source: wood pulp	Dissolution: chemical: NaOH/ Urea (7/12 wt%) and plasticizer (polyethylene glycol) Coagulation: solvent evaporation	Tensile strength (Mpa): 40–45 Elongation at break (%): 10–15	[16]
RC-3	Source: MCC (micro crystal cellulose) Chemical treatment: water, methanol and DMAc	Dissolution: chemical: DMAc/ LiCl (8%) Coagulation: Chemical: water	Tensile strength (Mpa): 141.7 Elongation at break (%): 3.04	[59]
RC-4	Source: ginger pulp Chemical treatment: dewaxed, alkalized, bleached, acid hydrolyzed	Dissolution: chemical: DMAc/ LiCl (8%) Coagulation: Chemical: ethanol	Tensile strength (Mpa): 51–53 Elongation at break (%): 4.9–5	[60]
RC-5	Source: ginger pulp Chemical treatment: dewaxed, alkalized, bleached, acid hydrolyzed	Dissolution: chemical: DMAc/ LiCl (8%) Coagulation: Chemical: water	Tensile strength (Mpa): 29–30 Elongation at break (%): 8–10	[60]

with RC-5) as MCC are highly crystalline and contain fewer impurity [59]. As a result, the SEM images (RC-3) showed the homogenous, defect-free, nonporous, and intact film as chemical treatment aids cellulose to dissolve successfully during dissolution process [59, 62].

The selection of a coagulant agent has a marked influence on the microstructure of surface morphology (SEM analysis), with water as a coagulant agent leading to segregated formation on the top surface of RC film which substantially affected mechanical strength (decrement of tensile strength of RC-5 compared with RC-4). When water mixed with cellulose solution during coagulation, the top surface of cellulose solution underwent an immediate phase inversion rate compared with the bottom surface. The formation of RC film on the top layer regulates the diffusion rate of water into lower layer result in a lower phase inversion rate in the lower layer of RC film. Consequently, lower phase inversion leads to a greater dense structure compared with immediate phase inversion [60].

8.4.3.2 Barrier Properties

The barrier properties (WVP) of RC film are considerably lower due to high crystallinity and numerous H-bonds between cellulose chains [63]. High WVP result comes from the high concentration of hydrophilic hydroxyl groups. Hence, many attempts have been devoted to decrease the WVP through surface modifications or layer-by-layer coating [64, 65]. Incorporating inorganic nanofiller, particularly in the RC film may enhance the WVP barrier of the films [63].

The WVP depends on solubility and diffusivity of water molecules in the polymer matrix [66]. Generally, Plasticizer are used in the Rc film fabrication process which decrease the barrier of WVP. As plasticizer is a hydrophilic material which swells, enlarges intermolecular chain space and creates H-bond which facilitates water adsorption [66]. As a result, RC film exhibits poor performance in terms of the barrier of WVP.

The OP can be determined by chain mobility and free volume in the material as well as plasticizer content and water [67]. A high number of H-bonding between cellulose chains can enhance the barrier of OP. However, the breakdown of H-bond enables the solubilization and diffusivity of oxygen and boosts the mobility of oxygen molecule within the materials [66].

UV blocking related to the surface morphology of RC film. Typically, light transmittance % of these film ranges from 70 to 80. This property could be triggered by incorporating nanoparticles. However, the tendency of nanoparticles to agglomerate result in decreased transparency [68].

8.4.3.3 Thermal Stability

Thermal degradation of RC film occurs in two steps. Initial decomposition occurs in the range of 0°C to 100°C and the second step's rapid decomposition occurs 265°C to 395°C. The first degradation is responsible for moisture content and absorbed water in RC film while the second weight loss is for polymer degradation and the decomposition of glycosidic linkage [62, 69].

As previously discussed, the incorporation of plasticizer in the RC film fabrication process allows water in the intermolecular chain space which leads to more weight loss in the initial decomposition during thermal compared to its nonplasticized counterpart under identical thermal conditions [16, 66].

8.4.3.4 Biodegradability

RC is becoming more and more known for its superior water adsorption and biodegradability qualities, which greatly expand its potential as an environmentally friendly packaging material. According to recent studies, RC decomposes more than 80% in just eighteen days, which makes it a desirable substitute for traditional plastics [16]. Its hydrophilic character, which results from the many hydroxyl (–OH) groups found in its chemical structure, is primarily responsible for its rapid breakdown. Although these qualities are advantageous for maintaining a sustainable environment, they can present difficulties in applications where moisture control is crucial, such as food packaging [70].

The chemical structure of RC showing the higher density of (–OH) groups helps in its biodegradability. RC can absorb more moisture due to the hydrogen interactions that these groups form with water molecules. Microorganisms utilize the water absorbed to metabolize cellulose when exposed to environmental conditions; thus, rapidly breaking down cellulose chains [71].

- Water adsorption's effect on biodegradability

The hydrophilic nature of RC greatly affects its biodegradation process. Hydroxyl groups make the materials more porous with a higher susceptibility to microbial attack and also facilitates water adsorption.

Hydrogen bonding: when water molecules and hydroxyl groups create hydrogen bonds, the substance becomes more wet-able, which encourages microbial colonization and activity of enzymes that breakdown cellulose strands.

Moisture sensitivity: water absorption promotes biodegradation, but it can also cause difficulties for applications like food packaging that need to control moisture. Excessive moisture can cause products to deteriorate, grow mold, and lose quality [72].

8.4.4 Ways of Regenerating Cellulose

New technologies are employed to reload cellulose, taking advantage of the natural cellulose to return it into many functional types and fit for an array of applications. Viscose process is one of the well-established processes by which cellulose is treated with sodium hydroxide and carbon disulfide yielding cellulose xanthate, that is regenerated to fibers or films in an acidic coagulation bath. Another approach is the lyocell process, which dissolves cellulose with N-methyl morpholine N-oxide (NMMO) and then regenerates it in water to produce high-quality fibers. Additionally, cellulose can be regenerated into nanoparticles by solubilizing it in sodium hydroxide/urea solutions and then regenerating it in water or other solvents. Other ways include employing ionic liquids to dissolve cellulose, which allows to produce films, membranes, and nonwoven materials. Each of these processes not only improves the qualities of regenerated cellulose, such as strength and biodegradability, but also fits with sustainability goals, making regenerated cellulose a desirable commodity in the packaging sector and elsewhere [73].

8.4.4.1 Viscose Process

Viscose fibers are the most widely utilized synthetic fibers, playing an important role in the textile industry and other applications. Viscose fiber manufacture is a very complex process that needs exact control at several points to produce the appropriate product qualities and attributes. Because different process parameters can be changed at different stages to produce different types of fibers specifically suited for a given application.

The process of making viscose fibers begins with converting wood pulp, which includes cellulose staple fibers, into a spinnable solution. This transformation is carried out by a succession of precisely controlled and conditioned stages, each of which is intended to prepare the cellulose for regeneration into long, continuous filaments. The employment of chemical treatments to breakdown cellulose and regenerate it into a fiber form makes the viscose rayon process especially notable. The process is shown in Figure 8.3.

To initiate the process, cellulose must first dissolve in a strong alkaline solution, such as sodium hydroxide (NaOH) and then it must interact with carbon disulfide (CS_2). The swelling and dissolving of the wood pulp is the first significant phase in the process. This crucial step breaks down the structure of the cellulose and increases its reactivity, readying it for other chemical reactions. This stage is critical for the interaction between the cellulose and the alkaline solution.

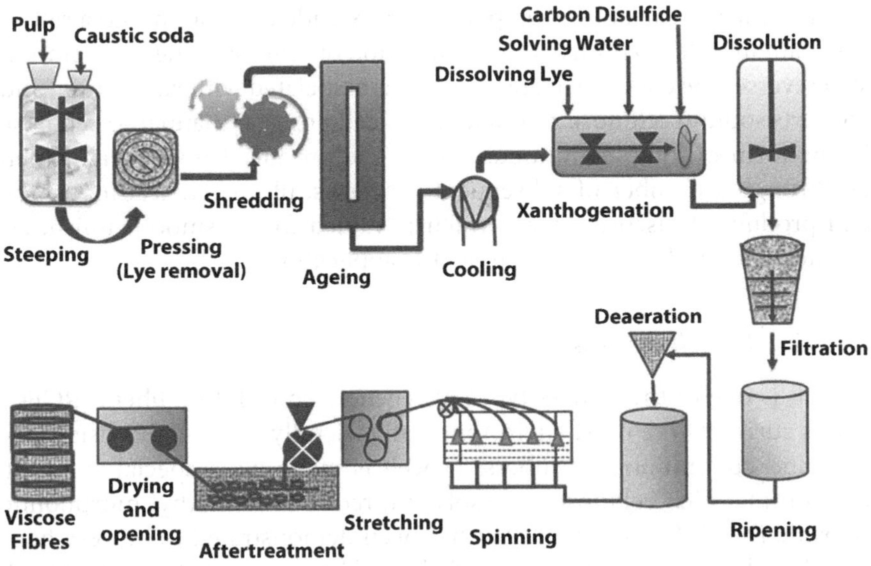

Figure 8.3 Schematic illustration of viscose process. Reproduced with permission from Shen *et al.* [73].

When exposed to a NaOH solution, cellulose swells significantly due to the formation of hydrates in the presence of both NaOH and water. The strong intramolecular and intermolecular hydrogen bonds that bind the cellulose molecules together are effectively broken by this hydration process. The solvent, in this case an alkaline solution, penetrates the amorphous parts of the cellulose, causing them to swell. Further changes in structure result from the alkali ions diffusing into the crystalline areas of the cellulose as the swelling increases.

One of the primary results of this interaction is the creation of an intermediate molecule known as alkali cellulose. This intermediate is important because it experiences mercerization or permanent enlargement. This procedure modifies the physical structure of cellulose, but also enhances its reactivity, making cellulose more active in other chemical processes [74].

Xanthation, a vital chemical reaction in which mercerized cellulose is combined with gaseous carbon disulfide (CS_2), is the second step in the viscose process. Xanthation converts cellulose into sodium cellulose xanthate, which is soluble in dilute alkali solutions. The dissolution of sodium cellulose xanthate into a uniform viscose dope—thick viscous solution, that once extruded through bulbed spinneret produces filament in continuity—is a key step of process.

Once the viscose dope is prepared, it is extruded and spun into a salt or acid bath in order to initiate the regeneration of cellulose. The regeneration step reverses the chemical transformations associated with xanthation, and converts sodium cellulose xanthate (Na+–cellulose xanthate) into cellulose in the form of long, continuous fibers. The produced fibers might then go through a number of subsequent processes, ultimately leading to the end-product of viscose rayon, which is valued for its smooth finish, air permeability, and versatility as to textile applications [73].

8.4.4.2 Lyocell Process

Lyocell is one of the newest types of regenerated cellulose fibers (RCFs), and is currently the most environmentally friendly of the RCFs using viscose and cuprammonium fibers. Lyocell is made *via* a special spinning method in which cellulose is dissolved directly in N-methyl morpholine N-oxide (NMMO), a solvent that has been demonstrated to be very beneficial in this application. The word "lyocell" itself comes from the Greek word "lyein," which means "to dissolve," paired with the word "cell," which is short for cellulose. This suggests that the process depends on the dissolving of cellulose as an essential stage [75]. The process is shown in Figure 8.4.

NMMO is the most effective solvent for this process because of cellulose's better solubility in it, even though several organic solvents, like dimethyl

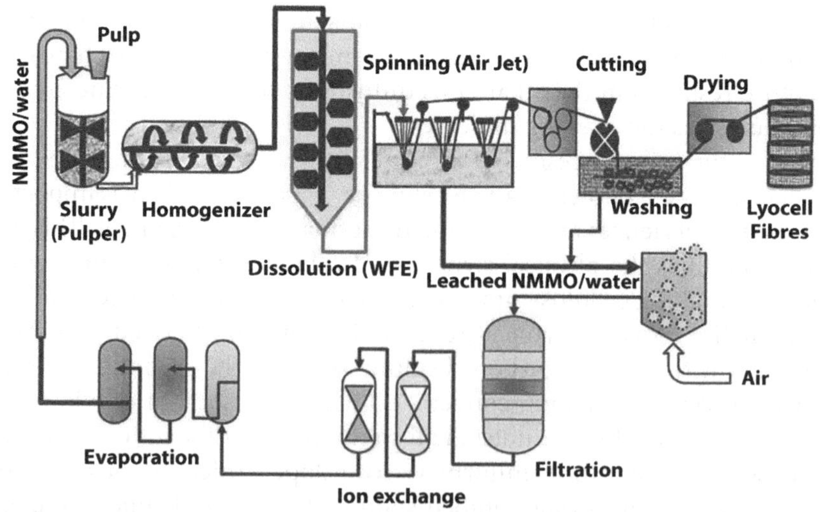

Figure 8.4 Schematic illustration of lyocell process. Reproduced with permission from Shen *et al.* [73].

sulfoxide (DMSO), dimethylformamide (DMF), dimethylacetamide (DMAc), N-nitrosodimethylamine (NDMA), and hexamethylphosphoramide (HMPA), can dissolve cellulose. The procedure of dry-jet wet spinning is used to produce lyocell fibers. Using an air gap, the liquid fibres in this method are submerged in a coagulation bath where they solidify into RCFs. This approach has been in use since the 1980s and represents a significant improvement in the mass manufacturing of regenerated cellulose fibers (RCFs), particularly because it provided a more sustainable alternative to previous processes such as the viscose process, which are less environmentally friendly.

The requirement to obtain a consistent, spinnable cellulose solution prior to spinning is an essential part of the lyocell process. Its chemical structure—more especially, the strong dipolarity of the nitrogen–oxygen (N–O) bond—is what makes NMMO an efficient solvent. Because of its dipolarity, the oxygen group in NMMO can establish hydrogen bonds with cellulose's anhydroglucose units (AGUs). Theoretically, NMMO can simultaneously target each of a cellulose molecule's three hydroxyl (-OH) groups, disrupting the network of hydrogen bonds that connects the polymer chains. Strong interacts with arise between the N–O group of NMMO and the -OH groups of cellulose because of this disruption. Consequently, the crystalline areas of the cellulose become penetrated by NMMO and water, which causes the cellulose to swell and eventually results in the total breakdown of the crystalline structure, forming a homogenous cellulose solution [73, 76].

In the lyocell process, RCFs are produced by dissolving cellulose in NMMO to generate a viscous solution. After passing through a spinneret, this solution is stretched into fine liquid strands that go through an air gap and then enter an aqueous NMMO solution, where the fibers renew. Over 99% of the solvent used in the spinning process is recovered, significantly lowering wastewater discharge, making the minimum environmental effect of the lyocell method one of its major advantages. Furthermore, the process' byproduct, NMMO, is nontoxic, making it safe for the environment [73].

8.4.4.3 Nonderivative Organic Solvent Systems

LiCl/DMAc system

DMAc (N,N-dimethylacetamide) is known for its ability to dissolve several cellulose derivatives, including cellulose acetate (CA) and hydroxypropyl cellulose (HPC). DMAc is not effective enough to dissolve pure cellulose. Cellulose is dissolved by adding a small amount of lithium chloride (LiCl) to the DMAc solvent. In 1980, McCormick found that the LiCl/DMAc system could dissolve cellulose. This invention established a non-degrading solvent system, able to solvate cellulose without cleavage of its molecular structure. This property was especially useful in determining the molecular weight distribution of cellulose, enabling them to ascertain the properties of the polymer with higher accuracy. The economic feasibility of using this LiCl/DMAc solvent system for the production of RCF was studied by Rayonier and coworkers in 1981. They developed a more effective method of cellulose dissolution under a high concentration of cellulose (5wt% in LiCl) in a relatively short period of time (4–6 hours), even at elevated temperatures (85°C), thus the LiCl/DMAc system represented a very useful method for cellulose processing, mainly for RCF synthesis. This process is illustrated in Figure 8.5.

Regeneration of cellulose fibers using the LiCl/DMAc system involves a number of key steps. The first step is activation of cellulose, which is required before it can be broken down. This includes the transformation of the cellulose structure to a more relaxed conformation, enabling the solvent to more readily access the dense crystalline regions. After activation, the solvent is prepared by mixing LiCl and DMAc followed by adding the activated cellulose to form a slurry [77].

Once the cellulose is completely dissolved, the subsequent steps are precipitation, regeneration, and solvent recovery. In the dissolution process, cellulose is swollen in a polar solvent (water, for example), which makes the polymer structure accessible. Then, the expanded cellulose is dissolved into LiCl/DMAc mixture at room temperature. A few studies analyzed the dissolution mechanism of this inorganic/organic mixed system. It includes the interaction of Cl− and Li+ ions with cellulose chains. Chloride anions accumulate along the cellulose polymer, working as a negative charge, and the Li+–DMAc

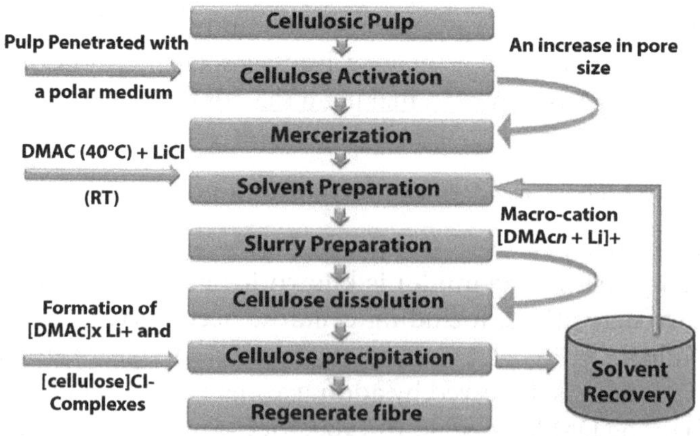

Figure 8.5 Schematic illustration of Licl/DMAc process. Reproduced with permission from Shen *et al.* [73].

complex serves as a counterions. This complex continuously interacts with cellulose hydroxyl (–OH) groups, rupturing the hydrogen bonds, which are crucial to binding polymer chains together and resulting in separation [73].

The salinity causes an osmotic pressure to form, allowing solvent molecules to continually flow into the solution, disrupting the internal associations of the cellulose until it is fully dissolved. Genuine cellulose solutions capable of being manipulated into fibers can also be obtained with this efficient dissolution method.

Hong and his mates carried out a detailed study of RCFs from LiCl/DMAc solutions. They showed that the RCFs produced from this system especially with 7% cellulose had different crystalline morphologies based on the treatment temperature. Annealing the liquid cellulose solution prior to electrospinning was found to enhance the mechanical properties of the resulting fibers. Fibers treated with acetone also resulted in the highest mechanical strength (4.32 g d^{-1}). The improvement in mechanical properties is attributed to the cellular chain reintegration and the formation of more stable and well-organized crystalline fractions of fibers during the annealing process [78].

8.4.4.4 Alkali/Urea Aqueous Solutions

Although the viscose process is great for creating cellulose fibers; it already has an intensive energy use and negative environmental influence. This approach to production is far less sustainable due to both the use of harsh chemical inputs and the vast amounts of processing, all of which increases environmental pollution and energy resource use. However, a more eco-friendly alternative has recently emerged: the alkali/urea aqueous solution system. Therefore, this technology is an environmentally friendly and long-term regeneration method for producing regenerated cellulose products due to its low price, nontoxic and fast dissolution.

Due to its ability to generate both green and sustainable RCFs, the alkali/urea aqueous solution system has attracted attention as an alternative to the viscose process [73].

Kamida *et al.*, as early as 1984, noted that low molecular weight, non-crystal cellulose was able to dissolve in a sodium hydroxide (NaOH) aqueous solution (8–10%) at low temperature. The potential of NaOH as a cellulose solvent was highlighted in this study, as it promotes interaction within the cellulose chains, separates polymeric chains, and breaks hydrogen bonds, allowing cellulose to dissolve in concentrations of alkali solution. This method, however, had two significant disadvantages: it could only solubilize low molecular weight cellulose, and the generated solution was unstable and characterized by rapid gelatinization [79].

Drawing from this information, the researchers worked to optimize their NaOH aqueous solution system. This changed when a group reported that urea, when added to aqueous NaOH, could not only dissolve cellulose of high molecular weight, but also improve the stability of the solution. Developing green, nontoxic, low-cost alkali/urea systems [80] were a significant breakthrough, hence a significant increase cellulose breakdown processes [80].

Two formulations of this system have been discussed: 7% NaOH/12% urea and 4.6% lithium hydroxide (LiOH)/15% urea solutions, both of which have shown excellent catalytic activity. These formulations can cleave cellulose within 3 to 5 minutes at −12°C; and the disintegration mechanism in these systems is fascinating: NaOH or LiOH hydrates at low temperature disrupt intra- and interhydrogen bonds in cellulose chains but create new hydrogen bonds with the –OH groups of the cellulose. Added support for this is present in the form of urea hydrates, which enclose the freshly formed NaOH or LiOH-bonded cellulose. This urea–cellulose complex can then dynamically self-assemble using weaker van der Waals forces to form tubular structures in the admixture, before ultimately dissolving the cellulose.

Two formulations that have demonstrated the highest efficiency for this system include 7% NaOH/12% urea and 4.6% lithium hydroxide (LiOH)/15% urea solutions. These formula have been shown to dissolve cellulose in 3 to 5 minutes at −12°C. The disintegration mechanism in these systems is interesting: At low temperatures, NaOH or LiOH hydrates disturb the intrahydrogen and interhydrogen bonds within cellulose chains and form new hydrogen bonds with the -OH groups of cellulose. Urea hydrates encircle the newly formed NaOH or LiOH-attached cellulose, which provides even more evidence for this connection. The urea–cellulose complex is said to undergo a dynamic self-assembly *via* van der Waals forces into tubular structures within the mixture that functions to disintegrate the cellulose until they are dissolved [73].

8.4.4.5 Cuprammonium Solution

The cuprammonium industrial process, a fundamental breakthrough in producing synthetic fibers, was first developed in the 1890s. This approach originated in the work of M. E. Schweizer (proposed in 1857), but it became famous for producing silky fibers. The cuprammonium method was changed by Louis-Henri Despeissis who spun fibers from the solution in 1890.

This process includes dissolving cellulose in an aqueous cuprammonium hydroxide solution to yield a viscous viscose cellulose solution, which can be spun into fibers. This solution is spun through a spinneret, a device with tiny holes, into a coagulation bath containing weak acid, ethanol and concentrated cresol solution. This process solidifies cellulose creating RCF. Using cuprammonium process was new then and that gave a new option for manmade fibers in the form of silky texture like natural silk [76, 81].

The cellulose–cuprammonium solution was extruded from a defined distance above a mobile coagulation bath during the spinning process. Due to the intrinsic instability of the solution when meeting the coagulation bath, an automatic template of a spiral structure formed at the surface. Helical-shaped (diameter, 100–400 μm; helix, 300–700 μm) RCFs were formed through acid treatment and air-drying.

These helical RCFs exhibits remarkable mechanical properties, including high elongation at break and large elasticity within a range of strain, both of which were influenced by their unique winding structure. These fibers exhibited a tensile strength of around 21 MPa and an elongation at break of 160%, or more. RCFs possess strength as well as flexibility along with a porous structure, which these make them suitable candidates for various other advanced applications, such as nanoparticle scaffolds and spiral templates [82].

Despite the unique nature and versatility of cuprammonium rayon fiber, its manufacture has considerable limitations. Although it has successfully replaced natural silk in many applications and is widely used in the clothing, decorative, and fabric sectors, the manufacturing method is both expensive and environmentally damaging. One of the primary environmental problems is the outflow of wastewater during production, which contains high levels of copper, a material that causes considerable environmental dangers. The procedure also uses numerous more dangerous compounds, which adds to its negative environmental impact [83].

8.5 Comparison with Other Packaging Materials

The RC derives from renewable sources (cotton, wood pulp, and pine pulp) and is biodegradable. At the same time, conventional plastic material (HDPE, polystyrene, polypropylene and PET) comes from nonrenewable resources that are not biodegradable and create environmental issues like global warming and deforestation. Like conventional polymers, RC is not thermoplastics, first degradation occurs in the temperature range 0°C to 100°C. However, the main drawback of RC packaging is poor barrier

Table 8.2 Comparison between RC and conventional plastics.

Attributes	**RC [59]**	**LDPE [84]**	**HDPE [84]**	**Polystyrene [84]**	**Polypropylene [84]**	**PET [84]**
Sources	Renewable	Nonrenewable	Nonrenewable	Nonrenewable	Nonrenewable	Nonrenewable
Barrier grade WVP	Poor	Medium	High	poor	Medium	Medium
Barrier grade OP	Medium	Poor	Poor	Poor	Medium	Medium
Biodegradability	yes	No	No	No	No	No
Cost	High	low	low	low	low	low
Thermal stability, T_m	0–100°C (first degradation)	100°C (T_m)	135°C (T_m)	100°C (T_m)	160°C (T_m)	

properties and high fabrication cost. On the other hand, the low price, adequate barrier, and excellent thermal stability of conventional plastics make it a suitable material for food packaging. Table 8.2 compares regenerated cellulose to common conventional plastics based on important material attributes.

With the boosting of convenience foods in our modern society, packaging materials are responsible for 40% [13] of global plastic production and it is expected to reach USD 411.3 billion by 2025 in the worldwide market [85]. According to the European market, most plastics in packaging are thermoplastics such as HDPE, LDPE, polystyrene, polypropylene and PET. As previously discussed, they come from nonrenewable resources and are not biodegradable resulting in serious environmental impact. Table 8.3 provides data on the market demand for different polymer materials in Europe.

Packaging materials (profoundly plastic bottles and food wrapping), due to their low density, are accountable for most of the environmental litter. The detrimental effect of plastics litter has also been growing on the marine environment. According to the United Nations Group of Experts on the Scientific Aspects of Marine pollution, chemical, and plastics can be identified from beaches to the deepest ocean [12].

Water pollution arises from the disposal of plastics waste discharge on landfills and rivers during manufacturing process. Basic water polluting parameters, such as biological oxygen demand (BOD), chemical oxygen demand (COD), volatile suspended solid (VSS), and total suspended solid (TSS) are affected by the dispose of plastics waste in freshwater. Additionally, the main source of air pollution emission of greenhouse gases (carbon dioxide, methane, and nitrous oxide) during manufacturing which are responsible for global warming through ozone layer depletion [12].

Table 8.3 European polymer demands, 2007 [84].

Material	Packaging (tonnes)	Other market (tonnes)
Low density polyethylene	5,976	3,110
High density polyethylene	3,357	2,609
Polypropylene	3,414	5,789
Polystyrene	1,511	2,218
PET	3,096	147
Other	132	3,084

8.6 Current Market

In the food supply chain, packaging plays a vital role since it keeps food fresh by retaining its color, taste and consistency. It also stops microbes from growing and stops nutrients from being lost throughout storage, distribution and transit. Normally foodstuff is well protected from oxygen and moisture by traditional packing materials like glass and metal. But metal is prone to corrosion and denaturation, whereas glass is also weak and brittle. That is why, petroleum-based plastic has become more popular since it is cheap, durable and simple to process. Nevertheless, these advantages, plastic's continuous usage and inappropriate disposal create serious threats to the environment.

Most of this disposal poses serious environmental issues due to their low capacity for recycling and reuse as well as poor biodegradability. As a result, food packaging composed of natural polymers has become progressively more common as a viable substitute. Of these, cellulose is the most frequent biodegradable and renewable polymer on Earth. Its low density, nontoxicity, affordability, adaptability and superior mechanical qualities are only a few of its many advantages [86].

Among all the cellulosic derivatives regenerated cellulose (RC) is being used as one of the most significant biodegradable packaging materials. These films can successfully replace conventional plastic films in food packaging, according to recent studies. With growing awareness of environmental issues, there is a high demand for materials that degrade organically after use. Regenerated cellulose films are an excellent substitute of plastics, which increase landfill trash and environmental damage, because they may decompose into nontoxic byproducts [87, 88].

Regenerated cellulose sheets as biodegradable packaging materials that retain food quality, reduce environmental impact, are breathtaking. The films are suitable for packaging perishables such as fruits and vegetables, where such packaging requires sufficient permeability to extend shelf life and avoid spoilage. They have also good mechanical and barrier properties.

In addition to these technologies, the RC packaging utilizes ingredients that actively engage with what's inside the food and help maintain the freshness of the food, and preserve its quality. RC films can be incorporated with antioxidants, oxygen scavengers or antimicrobials to develop active packaging systems. (RC)are effective against bacteria in food [89].

Research has demonstrated that germs like Staphylococcus aureus and E. coli can be efficiently inhibited by adding nanoparticles, such as silver

(Ag), zinc oxide (ZnO), and copper oxide, (CuO) to regenerated cellulose films; silver nanoparticles being the most effective.

Research also emphasizes the adoption of safer substitutes for synthetic additives: natural antimicrobials such essential oils. For instance, cellulose films containing oregano oil had a 99.99% inhibition rate against Listeria and *E. coli* [89–91].

Another significant system, that is food packaging with intelligent features are made to keep an eye on the state of the food they package and to report on its safety and freshness. Films made of regenerate cellulose can be used as substrates for sensors that monitor variations in the pH, temperature or gas composition inside the container.

To allow the packaging to change color in reaction to contamination or spoiling, pH-sensitive dyes, can be added to cellulose films that have been reconstituted. This feature improves food safety and lowers food waste by giving customers visual indications about the product's freshness. Furthermore, intelligent films made of regenerated cellulose have been created to track the freshness of food. These films react to pH variations by changing hue, showing the deterioration of foods including shrimp, minced beef and lamb meat [89, 92].

Another advantage of using RC sheets is its mechanical and barrier qualities. Regenerated cellulose films are equivalent to traditional plastic films due to their superior tensile strength, flexibility and transparency. It has been found that regenerated cellulose films' barrier qualities can be modified to satisfy certain packaging needs. For instance, adding nanofillers, like cellulose nanocrystals or nano clay, can improve the regenerated cellulose layers' ability to withstand gasses and moisture. This change is especially crucial for food packaging, as it keeps the product fresh and keeps it from spoiling [93].

Regenerated cellulose is regaining favor as a respectable substitute for synthetic plastics as environmental concerns continue to drive demand for sustainable packaging solutions. Among the creative uses are:

Compostable packaging: Films made from regenerated cellulose decompose naturally in the environment and don't leave any toxic behind. This makes them the perfect option for packaging applications like shopping bags and single-use food packaging when disposal at the end of the product's life is a worry.

Eco-friendly laminates: Multilayered packaging solutions with improved barrier qualities and little environmental impact can be produced by laminating regenerated cellulose with additional biodegradable materials. Applications where moisture and oxygen barriers are essential, like snack packaging, demand for the usage of these laminates.

In conclusion, the demand for sustainable alternatives is the main reason behind the growth of the regenerated cellulose market for packaging. An increasing amount of consumers are switching to renewable and biodegradable packaging options due to the knowledge of the environmental effects of traditional plastics. Stricter legislation regarding the use of plastics, especially single-use plastics, is also pressuring manufacturers to use eco-friendly materials such as regenerated cellulose. Regenerated cellulose is gaining popularity because to its excellent properties like barrier qualities, biodegradability and renewability. Regenerated cellulose is becoming more widely accepted and used in packaging as consumers place a higher priority on product quality, safety and environmental benefits. Manufacturers are allocating resources toward research to enhance its characteristics, rendering it a competitive substitute for traditional plastics [89, 94].

8.7 Challenges and Future Scope

Regenerated cellulose is becoming more and more popular as a packaging material because of its sustainable and eco-friendly qualities. Despite its numerous advantages, there exist substantial obstacles that restrict its extensive integration, specifically due to its mechanical strength, permeability to water vapor, variation in its biodegradability and others.

8.7.1 Mechanical Strength

Cellulose loses its crystalline structure when it regenerates. The regeneration process involves dissolving cellulose in a solvent. As it comes out of the solution, its once-ordered crystal arrangement reforms. Instead, the cellulose becomes more shapeless leading to a less regular pattern of its chains. This uneven alignment of cellulose chains means fewer molecules can interact so fewer hydrogen bonds form in this shapeless state. This decrease in hydrogen bonding makes the cellulose less rigid and more bendable. As a result, the cellulose's mechanical strength lowers a bit [95].

Cellulose has a tough well-organized crystal structure because of the many strong hydrogen bonds that form between the hydroxyl groups on nearby chains. The dense packing of cellulose gives it high tensile strength, which keeps it rigid and resistant to deformation [96]. Hydrogen bonds both within chains and between chains have an impact on its structural integrity. This mix of bonding creates a strong material, which suits applications like packaging that need strength and stability [97].

Normally this extremely organized structure is altered by the regeneration process. Solvents, such as N-methylmorpholine N-oxide (NMMO) or ionic liquids, are commonly used to dissolve cellulose. To reshape cellulose into new forms, like films or fibers, dissolution is required; however, as the cellulose chains break during this process, the hydrogen bonds are destroyed. The chains reassemble once the cellulose has been precipitated out of the solution, either by means of a coagulation bath or alternative techniques [73]. But they don't go back to the highly organized crystalline condition that they were in before. The cellulose chains are more dispersed and spaced farther apart than in native cellulose, forming an amorphous or semi-crystalline structure instead [74].

The material is immediately weakened by this loss of crystallinity. The material gets more flexible and less stiff as the number of hydrogen bonds that may form between chains decreases. It also indicates that regenerated cellulose is less appropriate for uses requiring great mechanical strength since it is more likely to shred or deform in real-life circumstances [98].

Various approaches are explored to enhance the mechanical properties of regenerated cellulose without sacrificing its environmental benefits. Now, these solutions are primarily based on transforming the cellulose chemical structure or implementing reinforcing chemicals to strengthen the regenerated material.

Chemical Cross-linking: Cross-linking agents such as glutaraldehyde and epichlorohydrin are used to form covalent links between cellulose chains. These bonds increase the stiffness of the polymer network, improving the stability and tensile strength of the regenerated cellulose. Additionally, this approach contributes to recover some of the lost crystallinity by forming more ordered structures [99].

Reinforcement with Nanocellulose: Cellulose nanocrystals (CNCs) or cellulose nanofibrils (CNFs) can be incorporated as nanofiller in regenerated cellulose matrices. Nanocellulose has a high aspect ratio and crystalline structure This enables it to form a network of the strong interactions between the nanocellulose fibers and the polymer matrix, which can enhance the mechanical properties of the regenerated cellulose [100].

Blending with Polymers: Regenerated cellulose can also be blended with other biodegradable polymers as polyhydroxybutyrate (PHB) or polylactic acid (PLA) in order to improve mechanical strength. These polymers can give the material greater stiffness without compromising its biodegradability. The advantage of blending, therefore, is that material qualities can be adapted to particular packaging requirements [101].

8.7.2 Permeability to Water Vapor

The ability of regenerated cellulose to let water vapor pass through that is permeability, is a major drawback when using it for packaging material. This feature allows moisture to get inside, which causes issues in packaging for food products. Keeping food dry and in a controlled setting is key to making it last longer on shelves. When moisture gets in, it can harm foods that is susceptible to humidity, perishables, or dry foods to lose their quality. This can lead to food going bad growing mold changing how it feels or losing its taste or nutritional value.

The hydrophilic nature of regenerated cellulose comes from the many hydroxyl groups (–OH) in its chemical structure. These groups form hydrogen bonds with water molecules fast letting water vapor pass through the material. This can be advantageous in situations where breathability is necessary (like in some medical uses), but it's a problem for food packaging that needs to keep moisture out [102].

Several tactics can be used to improve the suitability of regenerated cellulose for food packaging, especially for items that are sensitive to moisture.

8.7.2.1 Surface Coatings

One of the best solutions for this problem is to coat the surface of regenerated cellulose with coatings that are resistant to moisture. These coatings have the ability to create a layer of barrier that stops water vapor from entering the substance. Typical coatings consist of:

- Biodegradable polymers such as polylactic acid (PLA) or polyvinyl alcohol (PVA). When put as thin layers above cellulose films, these polymers offer superior moisture resistance in addition to being biodegradable.
- Wax coatings, which are often used in food packaging, can also be applied on regenerated cellulose to increase water vapor barrier qualities. But, if waxes are not chosen appropriately, they could lower the total biodegradability [103, 104].

8.7.2.2 Blending with Hydrophobic Polymers

Combining regenerated cellulose with hydrophobic polymers can considerably reduce water vapor permeability. By mixing it with materials, such as polyhydroxyalkanoates (PHA) or polypropylene, the composite material can preserve some of cellulose's desirable biodegradable properties while increasing moisture resistance. These mixes produce materials with tailored permeability levels ideal for a variety of packaging applications, notably those requiring higher moisture barriers [103].

8.7.2.3 Chemical Modification

Changing the chemical structure of cellulose can enhance moisture resistance. Acetylation of cellulose, for example, replaces part of its hydrophilic hydroxyl groups with hydrophobic acetyl groups, lowering the material's attraction for water. This method can improve cellulose's resilience to moisture while retaining some biodegradability. Cellulose acetate, a derivative of regenerated cellulose, is an example of a material that has been chemically modified to improve its moisture resistance [104].

8.7.3 Biodegradability Variability

The environment plays a big role in how well regenerated cellulose breaks down, which can make it tricky to use as a green packaging option or sustainable option. In the right settings, like commercial composting facilities regenerated cellulose decomposes well. These places keep a close eye on temperature, moisture and microbe activity. Under these conditions, the material breaks down fast making it a promising option for biodegradable packaging [105].

However, in less-than-ideal places, like landfills dry areas, or regions without good composting systems, the biodegradation process slows down a lot. This change in how fast things break down limits how useful regenerated cellulose can be in areas where the environment doesn't help things decompose. For instance, not having enough oxygen, water or the right tiny organisms in landfills can slow down biodegradation. This means the material stays around longer than expected, which reduces its benefits for the environment.

Scientists are still working hard to improve how regenerated cellulose breaks down in tough conditions. They're trying to make it degrade better outside of controlled environments [106].

The biodegradability of regenerated cellulose presents another difficulty. Although regenerated cellulose is biodegradable, the environment has a significant impact on how quickly and well it breaks down. Regenerated cellulose decomposes relatively quickly in optimal composting conditions, such as commercial composting facilities with regulated temperature, humidity, and microbial activity. However, under less-than-ideal circumstances, such as in landfills, arid areas, or places lacking the necessary infrastructure for composting, the decomposition process can be greatly slowed down. The material is less effective as a truly sustainable solution because of this fluctuation. Regenerated cellulose may not break down as intended in areas where natural decomposition is slow moving, which could reduce its beneficial effects on the environment [107].

8.7.4 Cost-Effectiveness through Improved Production Processes

In addition to functional improvements, regenerated cellulose places a strong emphasis on cost-effectiveness. The cost of regenerated cellulose has long been a significant obstacle to its widespread application in comparison to synthetic polymers such as polyethylene and polypropylene. Other advancements in manufacturing are helping to reduce those costs, though. Manufactures are minimizing the overall cost of the material by optimizing manufacturing processes and increasing the effectiveness of cellulose regeneration. This is occurring by making regenerated cellulose a more appealing alternative to standard polymers, thus advancing the utility of the material in an ever-expanding number of industries [108].

On the flip side, the future market for regenerated cellulose in packaging is projected to grow due to increased regulatory pressure against the use of traditional plastics and increasing customer demand for sustainable materials. However, in order to maximize the benefits of regenerated cellulose in conductive and transparent packaging, several problems need to be solved, although it shows great potential [109].

However, the future market for regenerated cellulose in packaging is growing as more regulators are putting pressure on the use of conventional plastics and consumers are demanding more sustainability in materials. However, there are some issues that need to be resolved, despite the potential of regenerated cellulose in the packaging industry.

8.7.5 Regulatory Support and Initiatives

In order to lessen environmental pollution governments all around the world have been implementing more stringent legislation on plastic packaging. The European Union set directives to cut down on single-use plastics and encourage biodegradable alternatives. These regulatory frameworks create a positive environment for the application of regenerated cellulose in packaging applications. Governments will offer some financial benefits and assistance for corporations investing into environmentally friendly packaging solutions such as regenerated cellulose [86].

8.7.6 Technological Advancements

Regenerated cellulose is a dynamic research and development field. Advances in production techniques, including nanotechnology and ionic liquids, are hallmarks of the promising properties of regenerated cellulose films that make them a competitor to conventional plastics. It has been suggested that, by incorporating nanofillers into regenerated cellulose films, they would display improved mechanical strength and barrier properties, thus rendering them suitable for a wider range of packaging applications. The use of regenerated cellulose as the substrate for sensors have been explored for developing smart packaging that allow for real-time monitoring of food quality and safety [89].

8.7.7 Consumer Trends toward Natural Products

The consumers are increasingly opting for natural and eco-friendly products. This trend is good news for regenerated cellulose: a biodegradable, renewable resource based polymer used in packaging. It may help you to sell more to environmentally conscious consumers if you have packaging that made using regenerated cellulose. For companies looking to differentiate themselves in a crowded space, effectively communicating the sustainability benefits of their packaging can be an important conduit.

8.7.8 Growing Demand for Sustainable Packaging Solutions

Sustainable packaging options are becoming increasingly popular as consumer environmental awareness increases. Regenerated cellulose matches in perfectly with this trend because it is biodegradable and made from renewable resources. The need for eco-friendly materials will probably increase in parallel with the predicted growth of the worldwide packaging sector. The need for sustainable materials in a variety of industries, including packaging, is expected to drive the growth of the global regenerated cellulose market, which is projected to reach approximately USD 42.24 billion by 2032 at a compound annual growth rate (CAGR) of 9.2% from 2023 to 2032, according to a report by Precedence Research [109].

8.8 Conclusion

The importance of regenerated cellulose is highlighted as a viable and sustainable substitute for petroleum-based plastics, which have long been the standard material in a variety of industries, especially packaging. Regenerated cellulose has drawn interest because of its potential for innovation, adaptability and environmental benefits. Due to this, it is an essential component of the continuous drive to implement more environmentally friendly industrial processes and reduce dependency on plastics derived from fossil fuels.

Regenerated cellulose's biologically degradable and self-renewal nature make it an impactful green alternative to plastics. Unlike traditional types of plastics, which persist in the environment for hundreds of years, regenerated cellulose degrades naturally, which leads to a reduced amount of plastic material polluting landfills and the ocean. Derived from ubiquitous natural resources, such as cotton and wood, its renewable roots support a more sustainable production process and correspond with the world-wide drive for sustainable products and reducing the environmental impact. Another significant aspect of the material is that it is versatile and it can be utilized in different fields like textiles and packaging. The composition of regenerated cellulose films makes them suitable for food packaging, as they provide good barriers to gases and Odors in packaging. There is still research being done to enhance the properties of regenerated cellulose, like its moisture resistance and production costs, which presents a massive opportunity for innovation. Such advances could expand the material's possible applications and elevate it beyond that of conventional polymers. However, there remain barriers to widespread acceptance of the material, such as issues with its permeability for water vapor, which can limit its utility in moisture-sensitive applications, such as some food packaging. Biodegradability can also be dependent upon environmental conditions, which can be a disadvantage. Regenerated cellulose might, however, act less aerodynamically in regions where there are no composting facilities or in less-than-ideal conditions, therefore diminishing its value as an eco-friendly product. These challenges notwithstanding, ongoing R&D initiatives are providing prospectives for alternatives with advanced coatings that increase moisture resistance, as well as advances in manufacturing methods that reduce the cost of the regenerated cellulose. There is also an increasing demand for sustainable alternatives to traditional plastics due to government requirements and consumer choices. Regenerated cellulose is uniquely situated to fulfill this desire because of its versatility and sustainability benefits. As it is a sustainable alternative that meets consumer and environmental needs, it is a preferred alternative for companies looking for the best way to improve their environmental footprint.

The likelihood of long-term success for regenerated cellulose as a sustainable material will rely on continued innovation, working together between companies, researchers and government, and major investments in integrated ecosystem like manufacturing processes and composting sites. Regenerated cellulose represents a sustainable material for the future where industrial performance and environmental sustainability can coexist in harmony. Regenerated cellulose can become the backbone material of the new circular economy which favors the production of materials that can be recycled, reused and returned to the environment without destructive damage. Such a paradigm shift could Perhaps, turbocharge the industrial ecosystem toward more sustainable and responsible outlooks.

AI disclosure statement: The author acknowledges the use of AI-based tools for minor language editing or grammatical correction. However, all intellectual contributions, analyses, and conclusions presented in this chapter are entirely the author's own.

References

1. Ostle, C., Thompson, R.C., Broughton, D., Gregory, L., Wootton, M., Johns, D.G., The rise in ocean plastics evidenced from a 60-year time series. *Nat. Commun.*, 10, 1, 1622, 2019.
2. Borrelle, S.B., *et al.*, Predicted growth in plastic waste exceeds efforts to mitigate plastic pollution. *Sci. (1979)*, 369, 6510, 1515–1518, 2020.
3. Lau, W.W.Y., *et al.*, Evaluating scenarios toward zero plastic pollution. *Sci. (1979)*, 369, 6510, 1455–1461, 2020.
4. do Sul, J.A.I. and Costa, M.F., The present and future of microplastic pollution in the marine environment. *Environ. Pollut.*, 185, 352–364, 2014.
5. Iroegbu, A.O.C., Ray, S.S., Mbarane, V., Bordado, J.C., Sardinha, J.P., Plastic pollution: a perspective on matters arising: challenges and opportunities. *ACS Omega*, 6, 30, 19343–19355, 2021.
6. Galati, A. and Scalenghe, R., Plastic end-of-life alternatives, with a focus on the agricultural sector. *Curr. Opin. Chem. Eng.*, 32, 100681, 2021.
7. Wani, W.A., Pathan, S., Bose, S., The Journey of Alternative and Sustainable Substitutes for 'Single-Use' Plastics. *Adv. Sustain. Syst.*, 5, 12, 2100085, 2021.
8. Веремейчик, O.V. and Якимович, IE.B., Packaging. History. Packaging materials. Bionic design, 2014. БНТУ.
9. Herrmann, M., Packaging-materials review, in: *AIP Conference Proceedings*, pp. 121–133, American Institute of Physics, 2014.
10. Kumari, S., Rao, A., Kaur, M., Dhania, G., Petroleum-Based Plastics Versus Bio-Based Plastics: A Review. *Nat. Environ. Pollut. Technol.*, 22, 3, 2023.

11. Fredi, G., Rigotti, D., Bikiaris, D.N., Dorigato, A., Tuning thermo-mechanical properties of poly(lactic acid) films through blending with bioderived poly(alkylene furanoate)s with different alkyl chain length for sustainable packaging. *Polym. (Guildf)*, 218, 123527, 2021. doi: https://doi.org/10.1016/j.polymer.2021.123527.
12. Kutz, M., The environmental impacts of packaging, in: *Environmentally conscious materials and chemicals processing*, pp. 237–278, 2007.
13. Siracusa, V. and Rosa, M.D., 8 - Sustainable Packaging, in: *Sustainable Food Systems from Agriculture to Industry*, C.M. Galanakis (Ed.), pp. 275–307, Academic Press, 2018, doi: https://doi.org/10.1016/B978-0-12-811935-8.00008-1.
14. Rouhi, M., Razavi, S.H., Mousavi, S.M., Optimization of crosslinked poly(vinyl alcohol) nanocomposite films for mechanical properties. *Mater. Sci. Eng. C*, 71, 1052–1063, 2017. doi: https://doi.org/10.1016/j.msec.2016.11.135.
15. Wang, B., C., J., P., H., G., J., K., J., C., Y., Investigation on changes in the miscibility, morphology, rheology and mechanical behavior of melt processed cellulose acetate through adding polyethylene glycol as a plasticizer. *J. Macromol. Sci. Part B*, 894–907, 2016.
16. Zhang, Y., *et al.*, Performance-enhanced regenerated cellulose film by adding grape seed extract. *Int. J. Biol. Macromol.*, 232, 123290, 2023. doi: https://doi.org/10.1016/j.ijbiomac.2023.123290.
17. Chu, Y., Popovich, C., Wang, Y., Heat sealable regenerated cellulose films enabled by zein coating for sustainable food packaging. *Composites, Part C: Open Access*, 12, 100390, 2023. doi: https://doi.org/10.1016/j.jcomc.2023.100390.
18. Yang, Q., Fukuzumi, H., Saito, T., Isogai, A., Zhang, L., Transparent Cellulose Films with High Gas Barrier Properties Fabricated from Aqueous Alkali/Urea Solutions. *Biomacromolecules*, 12, 7, 2766–2771, 2011. doi: 10.1021/bm200766v.
19. Meares, P., The Diffusion of Gases Through Polyvinyl Acetate1. *J. Am. Chem. Soc.*, 76, 13, 3415–3422, 1954. doi: 10.1021/ja01642a015.
20. C., R., G., R., Lagaron, J.M., Structural characteristics defining high barrier properties in polymeric materials. *Mater. Sci. Technol.*, 1–7, 2004.
21. Kale, G., K., T., A., R., R., M., S., S.E., S., S.P., Compostability of bioplastic packaging materials: an overview. *Macromol. Biosci.*, 7, 255–277, 2007.
22. P. I., A.L., H., P.N., Ho, K.-L.G., Effects of Temperature and Relative Humidity on Polylactic Acid Plastic Degradation. *J. Environ. Polym. Degrad.*, 7, 83–92, 1999.
23. Poletto, M., Ornaghi Junior, H.L., Zattera, A.J., Native cellulose: Structure, characterization and thermal properties. *Materials*, 7, 9, 6105–6119, 2014.
24. Krishnasamy, S., *et al.*, Recent advances in thermal properties of hybrid cellulosic fiber reinforced polymer composites. *Int. J. Biol. Macromol.*, 141, 1–13, 2019.

25. Klemm, D., Heublein, B., Fink, H., Bohn, A., Cellulose: Fascinating biopolymer and sustainable raw material. *Angew. Chem. Int. Ed.*, 44, 22, 3358–3393, 2005.
26. Saxena, I.M. and Brown Jr, R.M., Cellulose biosynthesis: current views and evolving concepts. *Ann. Bot.*, 96, 1, 9–21, 2005.
27. Heinze, T., Cellulose: Structure and properties. *Adv. Polym. Sci.*, 271, 1–52, 2015. doi: 10.1007/12_2015_319.
28. Lewin, M., *Handbook of fiber chemistry*, CRC press, 2006.
29. Roulac, J.W., *Hemp horizons: The comeback of the world's most promising plant*, 1997.
30. Scurlock, J.M.O., Dayton, D.C., Hames, B., Bamboo: An overlooked biomass resource? *Biomass Bioenergy*, 19, 4, 229–244, 2000.
31. Vogel, K.P. and Jung, H.-J.G., Genetic modification of herbaceous plants for feed and fuel. *CRC Crit. Rev. Plant Sci.*, 20, 1, 15–49, 2001.
32. Kozłowski, R., *Handbook of natural fibres: Types, properties and factors affecting breeding and cultivation*, Woodhead Pub, 2012.
33. Lynd, L.R., Weimer, P.J., Van Zyl, W.H., Pretorius, I.S., Microbial cellulose utilization: fundamentals and biotechnology. *Microbiol. Mol. Biol. Rev.*, 66, 3, 506–577, 2002.
34. Genet, M., *et al.*, The influence of cellulose content on tensile strength in tree roots. *Plant Soil*, 278, 1–9, 2005.
35. Eichhorn, S.J., *et al.*, Current international research into cellulose nanofibres and nanocomposites. *J. Mater. Sci.*, 45, 1–33, 2010.
36. French, A.D., Idealized powder diffraction patterns for cellulose polymorphs. *Cellulose*, 21, 2, 885–896, 2014.
37. Yang, H., Yan, R., Chen, H., Lee, D.H., Zheng, C., Characteristics of hemicellulose, cellulose and lignin pyrolysis. *Fuel*, 86, 12–13, 1781–1788, 2007.
38. Heinze, T. and Koschella, A., Solvents applied in the field of cellulose chemistry: A mini review. *Polímeros*, 15, 84–90, 2005.
39. Habibi, Y., Lucia, L.A., Rojas, O.J., Cellulose nanocrystals: chemistry, self-assembly, and applications. *Chem. Rev.*, 110, 6, 3479–3500, 2010.
40. Biermann, C.J., *Handbook of pulping and papermaking*, Elsevier, 1996.
41. Rowe, R.C. and Sheskey, P.J., *Marian E Quinn Handbook of Pharmaceutical Excipients*, Pharmaceutical press, 2009.
42. Himmel, M.E., *et al.*, Biomass recalcitrance: Engineering plants and enzymes for biofuels production. *Sci. (1979)*, 315, 5813, 804–807, 2007.
43. De Menezes, A.J., Siqueira, G., Curvelo, A.A.S., Dufresne, A., Extrusion and characterization of functionalized cellulose whiskers reinforced polyethylene nanocomposites. *Polym. (Guildf)*, 50, 19, 4552–4563, 2009.
44. Chaudhari, K.B., TYB Sc.(Biochemistry).
45. Wang, S., Lu, A., Zhang, L., Recent advances in regenerated cellulose materials. *Prog. Polym. Sci.*, 53, 169–206, 2016.
46. Klemm, D., Philpp, B., Heinze, T., Heinze, U., Wagenknecht, W., *Comprehensive cellulose chemistry. Volume 1: Fundamentals and analytical methods*, 1998.

47. Williams, P.A. and Phillips, G.O., Gum arabic, in: *Handbook of hydrocolloids*, pp. 252–273, Elsevier, 2009.
48. Abel, F.A., XIV. Researches on gun-cotton.—on the manufacture and composition of gun-cotton. *Philos. Trans. R Soc. Lond.*, 156, 269–308, 1866.
49. Cross, C.F. and Bevan, E.J., Researches on cellulose, 1922.
50. Varian, B.D., Protection and the British rayon industry during the 1920s. *Bus. Hist.*, 64, 6, 1131–1148, 2022.
51. Hearle, J.W.S. and Morton, W.E., *Physical properties of textile fibres*, Elsevier, 2008.
52. Woodings, C., *Regenerated cellulose fibres*, Elsevier, 2001.
53. Roy, D., Semsarilar, M., Guthrie, J.T., Perrier, S., Cellulose modification by polymer grafting: A review. *Chem. Soc. Rev.*, 38, 7, 2046–2064, 2009. doi: 10.1039/B808639G.
54. Azizi Samir, M.A.S., Alloin, F., Dufresne, A., Review of Recent Research into Cellulosic Whiskers, Their Properties and Their Application in Nanocomposite Field. *Biomacromolecules*, 6, 2, 612–626, 2005. doi: 10.1021/bm0493685.
55. Moon, R.J., Martini, A., Nairn, J., Simonsen, J., Youngblood, J., Cellulose nanomaterials review: structure, properties and nanocomposites. *Chem. Soc. Rev.*, 40, 7, 3941–3994, 2011. doi: 10.1039/C0CS00108B.
56. Mallakpour, S., Sirous, F., Hussain, C.M., A journey to the world of fascinating ZnO nanocomposites made of chitosan, starch, cellulose, and other biopolymers: Progress in recent achievements in eco-friendly food packaging, biomedical, and water remediation technologies. *Int. J. Biol. Macromol.*, 170, 701–716, 2021. doi: https://doi.org/10.1016/j.ijbiomac.2020.12.163.
57. Alizadeh Sani, M., *et al.*, Titanium dioxide nanoparticles as multifunctional surface-active materials for smart/active nanocomposite packaging films. *Adv. Colloid Interface Sci.*, 300, 102593, 2022. doi: https://doi.org/10.1016/j.cis.2021.102593.
58. Xu, S., *et al.*, Preparation and High Performance of Cellulose Acetate Films by Grafting with Imidazole Ionic Liquid. *ACS Omega*, 6, 19, 12500–12506, 2021. doi: 10.1021/acsomega.0c06361.
59. Saedi, S., Shokri, M., Kim, J.T., Shin, G.H., Semi-transparent regenerated cellulose/ZnONP nanocomposite film as a potential antimicrobial food packaging material. *J. Food Eng.*, 307, 110665, 2021. doi: https://doi.org/10.1016/j.jfoodeng.2021.110665.
60. Saedi, S., Kim, J.T., Lee, E.H., Kumar, A., Shin, G.H., Fully transparent and flexible antibacterial packaging films based on regenerated cellulose extracted from ginger pulp. *Ind. Crops Prod.*, 197, 116554, 2023. doi: https://doi.org/10.1016/j.indcrop.2023.116554.
61. Chen, G.-Y., Y., H.-Y., Z., C.-H., Z., Y., Y., J.-M., A universal route for the simultaneous extraction and functionalization of cellulose nanocrystals from industrial and agricultural celluloses. *J. Nanopart. Res.*, 18, 1–14, 2016.

62. Zhao, G., Lyu, X., Lee, J., Cui, X., Chen, W.-N., Biodegradable and transparent cellulose film prepared eco-friendly from durian rind for packaging application. *Food Packag. Shelf Life*, 21, 100345, 2019. doi: https://doi.org/10.1016/j.fpsl.2019.100345.
63. Farmahini-Farahani, M., B., A.H., P., Y., X., H., E., M., C., F., Cellulose/nanoclay composite films with high water vapor resistance and mechanical strength. *cellulose*, 3941–3953, 2015.
64. Li, W., W., S., W., W., Q., C., W., M., Facile preparation of reactive hydrophobic cellulose nanofibril film for reducing water vapor permeability (WVP) in packaging applications. *Cellulose*, 26, 3271–3284, 2019.
65. Forsman, N., Lozhechnikova, A., Khakalo, A., Johansson, L.-S., Vartiainen, J., Österberg, M., Layer-by-layer assembled hydrophobic coatings for cellulose nanofibril films and textiles, made of polylysine and natural wax particles. *Carbohydr. Polym.*, 173, 392–402, 2017. doi: https://doi.org/10.1016/j.carbpol.2017.06.007.
66. Ghanbarzadeh, B., Musavi, M., Oromiehie, A.R., Rezayi, K., Razmi Rad, E., Milani, J., Effect of plasticizing sugars on water vapor permeability, surface energy and microstructure properties of zein films. *LWT Food Sci. Technol.*, 40, 7, 1191–1197, 2007. doi: https://doi.org/10.1016/j.lwt.2006.07.008.
67. Hansen, N.M.L., B., T.O.J., H., M.S., *et al.*, Properties of plasticized composite films prepared from nanofibrillated cellulose and birch wood xylan. *Cellulose*, 9, 2015–2031, 2012.
68. John, A., K., H.-U., K., D.-G., K., J., Preparation of cellulose-ZnO hybrid films by a wet chemical method and their characterization. *Cellulose*, 18, 675–680, 2011.
69. Priyadarshi, R., Sauraj, Kumar, B., Negi, Y.S., Chitosan film incorporated with citric acid and glycerol as an active packaging material for extension of green chilli shelf life. *Carbohydr. Polym.*, 195, 329–338, 2018. doi: https://doi.org/10.1016/j.carbpol.2018.04.089.
70. Radu, E.-R., *et al.*, The soil biodegradability of structured composites based on cellulose cardboard and blends of polylactic acid and polyhydroxybutyrate. *J. Polym. Environ.*, 29, 2310–2320, 2021.
71. Ahsan, W.A., Hussain, A., Lin, C., Nguyen, M.K., Biodegradation of different types of bioplastics through composting—a recent trend in green recycling. *Catalysts*, 13, 2, 294, 2023.
72. Hubbe, M.A., Lavoine, N., Lucia, L.A., Dou, C., Formulating Bioplastic Composites for Biodegradability, Recycling, and Performance: A Review. *Bioresources*, 16, 1, 2021.
73. Shen, H., Sun, T., Zhou, J., Recent progress in regenerated cellulose fibers by wet spinning. *Macromol. Mater. Eng.*, 308, 10, 2300089, 2023.
74. Cai, J., Kimura, S., Wada, M., Kuga, S., Zhang, L., Cellulose aerogels from aqueous alkali hydroxide–urea solution. *ChemSusChem Chem. Sustainability Energy Mater.*, 1, 1-2, 149–154, 2008.

75. Sayyed, A.J., Mohite, L.V., Deshmukh, N.A., Pinjari, D.V., Effect of ultrasound treatment on swelling behavior of cellulose in aqueous N-methylmorpholine-N-oxide solution. *Ultrason. Sonochem.*, 49, 161–168, 2018.
76. Fink, H.-P., Weigel, P., Purz, H.J., Ganster, J., Structure formation of regenerated cellulose materials from NMMO-solutions. *Prog. Polym. Sci.*, 26, 9, 1473–1524, 2001.
77. Bikova, T. and Treimanis, A., Problems of the MMD analysis of cellulose by SEC using DMA/LiCl: A review. *Carbohydr. Polym.*, 48, 1, 23–28, 2002.
78. Medronho, B. and Lindman, B., Competing forces during cellulose dissolution: From solvents to mechanisms. *Curr. Opin. Colloid Interface Sci.*, 19, 1, 32–40, 2014.
79. Kamida, K., Okajima, K., Matsui, T., Kowsaka, K., Study on the solubility of cellulose in aqueous alkali solution by deuteration IR and 13C NMR. *Polym. J.*, 16, 12, 857–866, 1984.
80. Zhou, J. and Zhang, L., Solubility of cellulose in NaOH/urea aqueous solution. *Polym. J.*, 32, 10, 866–870, 2000.
81. Woodings, C.R., The development of advanced cellulosic fibres. *Int. J. Biol. Macromol.*, 17, 6, 305–309, 1995.
82. Jia, B., Yu, L., Fu, F., Li, L., Zhou, J., Zhang, L., Preparation of helical fibers from cellulose–cuprammonium solution based on liquid rope coiling. *RSC Adv.*, 4, 18, 9112–9117, 2014.
83. Sayyed, A.J., Deshmukh, N.A., Pinjari, D.V., A critical review of manufacturing processes used in regenerated cellulosic fibres: viscose, cellulose acetate, cuprammonium, LiCl/DMAc, ionic liquids, and NMMO based lyocell. *Cellulose*, 26, 2913–2940, 2019.
84. Emblem, A., 13 - Plastics properties for packaging materials, in: *Packaging Technology*, A. Emblem and H. Emblem (Eds.), pp. 287–309, Woodhead Publishing, 2012, doi: https://doi.org/10.1533/9780857095701.2.287.
85. Wu, F., Misra, M., Mohanty, A.K., Challenges and new opportunities on barrier performance of biodegradable polymers for sustainable packaging. *Prog. Polym. Sci.*, 117, 101395, 2021. doi: https://doi.org/10.1016/j.progpolymsci.2021.101395.
86. Liu, X., Qin, Z., Ma, Y., Liu, H., Wang, X., Cellulose-Based Films for Food Packaging Applications: Review of Preparation, Properties, and Prospects. *J. Renew. Mater.*, 11, 8, 2023.
87. Mendes, A.C. and Pedersen, G.A., Perspectives on sustainable food packaging:–is bio-based plastics a solution? *Trends Food Sci. Technol.*, 112, 839–846, 2021.
88. Jiang, Z. and Ngai, T., Recent advances in chemically modified cellulose and its derivatives for food packaging applications: A review. *Polym. (Basel)*, 14, 8, 1533, 2022.
89. Huang, K. and Wang, Y., Recent applications of regenerated cellulose films and hydrogels in food packaging. *Curr. Opin. Food Sci.*, 43, 7–17, 2022.

90. Ebrahimi, Y., Peighambardoust, S.J., Peighambardoust, S.H., Karkaj, S.Z., Development of antibacterial carboxymethyl cellulose-based nanobiocomposite films containing various metallic nanoparticles for food packaging applications. *J. Food Sci.*, 84, 9, 2537–2548, 2019.
91. Gómez-García, M., *et al.*, Antimicrobial activity of a selection of organic acids, their salts and essential oils against swine enteropathogenic bacteria. *Porcine Health Manag.*, 5, 1–8, 2019.
92. Ding, L., *et al.*, A naked-eye detection polyvinyl alcohol/cellulose-based pH sensor for intelligent packaging. *Carbohydr. Polym.*, 233, 115859, 2020.
93. Cazón, P., Vázquez, M., Velázquez, G., Regenerated cellulose films with chitosan and polyvinyl alcohol: Effect of the moisture content on the barrier, mechanical and optical properties. *Carbohydr. Polym.*, 236, 116031, 2020.
94. Versino, F., Ortega, F., Monroy, Y., Rivero, S., López, O.V., García, M.A., Sustainable and bio-based food packaging: A review on past and current design innovations. *Foods*, 12, 5, 1057, 2023.
95. Atalla, R.H., *Structures of cellulose*, ACS Publications, 1987.
96. Gardner, K.H. and Blackwell, J., The structure of native cellulose. *Biopolymers Original Res. Biomol.*, 13, 10, 1975–2001, 1974.
97. Atalla, R.H. and Vanderhart, D.L., Native cellulose: a composite of two distinct crystalline forms. *Sci. (1979)*, 223, 4633, 283–285, 1984.
98. Klemm, D., Heinze, T., Philipp, B., Wagenknecht, W., New approaches to advanced polymers by selective cellulose functionalization. *Acta Polym.*, 48, 8, 277–297, 1997.
99. Gao, M., Xie, X., Huang, T., Zhang, N., Wang, Y., Glutaraldehyde-assisted crosslinking in regenerated cellulose films toward high dielectric and mechanical properties. *Cellulose*, 29, 15, 8177–8194, 2022.
100. Chen, Y., Gan, L., Huang, J., Dufresne, A., Reinforcing mechanism of cellulose nanocrystals in nanocomposites, in: *Nanocellulose: from fundamentals to advanced materials*, pp. 201–249, 2019.
101. Sarkar, M., Upadhyay, A., Pandey, D., Sarkar, C., Saha, S., Cellulose-Based Biodegradable Polymers: Synthesis, Properties, and Their Applications, in: *Biodegradable Polymers and Their Emerging Applications*, pp. 89–114, Springer, 2023.
102. Singh, A.K., Itkor, P., Lee, Y.S., State-of-the-art insights and potential applications of cellulose-based hydrogels in food packaging: Advances towards sustainable trends. *Gels*, 9, 6, 433, 2023.
103. Xie, F., Biopolymer-based multilayer films and coatings for food preservation: An update of the recent development. *Curr. Food Sci. Technol. Rep.*, 1, 1, 1–12, 2023.
104. Ferreira, P.S., Ribeiro, S.M., Pontes, R., Nunes, J., Production methods and applications of bioactive polylactic acid: a review. *Environ. Chem. Lett.*, 22, 4, 1831–1859, 2024.

105. Leppänen, I., Vikman, M., Harlin, A., Orelma, H., Enzymatic degradation and pilot-scale composting of cellulose-based films with different chemical structures. *J. Polym. Environ.*, 28, 458–470, 2020.
106. Ülger-Vatansever, B., Onay, T.T., Demirel, B., Evaluation of bioplastics biodegradation under simulated landfill conditions. *Environ. Sci. Pollut. Res.*, 31, 12, 17779–17787, 2024.
107. Brunšek, R., Kopitar, D., Schwarz, I., Marasović, P., Biodegradation properties of cellulose fibers and PLA biopolymer. *Polym. (Basel)*, 15, 17, 3532, 2023.
108. Cui, X., Lee, J.J.L., Chen, W.N., Eco-friendly and biodegradable cellulose hydrogels produced from low cost okara: Towards non-toxic flexible electronics. *Sci. Rep.*, 9, 1, 18166, 2019.
109. Regenerated Cellulose Market Size, Share, and Trends 2024 to 2034, Oct. 2023.

9

Regenerated Cellulose: A Sustainable Solution for Storage and Packaging

Arti A. Bagada*, **Monika B. Sangani and Priya V. Patel**

Department of Pharmaceutical Sciences, Saurashtra University, Rajkot, Gujarat, India

Abstract
Nowadays, nonbiodegradable plastic packaging materials have caused serious disposal and pollution challenges, endangering human health and development. The use of biodegradable packing materials can assist in solving the problem. Cellulose (CEL) is plentiful and sustainable, hence CEL-based materials have been proposed as an alternative to petroleum-based plastic food packaging. Regenerated CEL food packaging is becoming increasingly popular due to its renewable raw ingredients and superior environmental performance. Finally, the biodegradability and safety of CEL-based materials are discussed, along with future potential and difficulties in this busy research field. The present chapter focuses on contemporary methodologies for producing RC-based shapes and their prospective use in food and healthcare product packaging. Biodegradability and toxicity issues are discussed, as well as future potential, challenges, and research paths.

Keywords: Regenerated cellulose, sustainable packaging, cellulose food packaging, environmental impact, biodegradability, healthcare product packaging

9.1 Introduction

Because of the pollution generated by petroleum-based polymers and the loss of non-renewable resources, the 21st century has transitioned toward ecologically friendly materials, renewable resources, and sustainable practices [1]. These days, using non-biodegradable plastic packaging materials

**Corresponding author*: arti910@gmail.com

Mohd Shabbir (ed.) Regenerated Cellulose: A Renewable Alternative to Petroleum-Based Plastics, (243–264)

has led to serious pollution and disposal problems that endanger human growth and health. One way to alleviate the problem is to use packing materials that decompose naturally. Since CEL is plentiful and sustainable, products based on it was proposed as an option to plastic food packaging made of petroleum. Numerous regenerated CEL films and hydrogels have been created for diverse purposes with the advancement of CEL solvents [2].

For pharmaceuticals to be stable, effective, and safe, their packaging is essential. It has primary, secondary, and tertiary packaging to shield the product from damage caused by the environment and during transit. To fulfil consumer needs and provide user-friendly packaging, modern packaging technology must update its methods and incorporate features like child- and tamper-resistant packaging for various dosage forms [3]. Merely 20% of Canadian home garbage is collected for recycling and reuse, with food packaging making up the majority of waste at 33%. The renewable and biodegradable polymer CEL provides an alternative with the benefits of cheap cost, low density, and non-toxicity [2]. Because of its mechanical qualities, sustainability, and biodegradability, CEL is a renewable biopolymer that finds extensive usage in packaging and other applications. It is made from cellulosic agricultural and food waste, cotton, wood, and crops. At the moment, wood and cotton are the basic resources of CEL; however, the restricted availability of wood stems from its protracted development cycle. To produce important elements like CEL, research has been done on the valorization of plant waste, including agricultural and food waste from sources including sugarcane bagasse, durian rind, and vegetable waste [4]. The current chapter highlights the various types of regenerated CEL, their application in food and healthcare field packaging with future prospects.

9.2 Types and Forms of Regenerated CEL

A type of synthetic fiber called regenerated CEL fiber is made from CEL as a raw material, usually from wood or plant fibers. Regenerated CEL fiber, formerly referred to as "artificial silk" in the 1850s when manufacturers were attempting to develop a silk substitute, was the first synthetic fiber utilized in the textile and apparel industries.

In 1924, the generic name Rayon was assigned to regenerated CEL fiber by the U.S. Department of Commerce and other industrial groups. "Ray"—indicated the fiber's brightness, and "-on" indicated its structure, which was comparable to cotton. Although regenerated CEL fiber is chemically significantly different from silk, it yet looks shiny and smooth [5]. Structure and manufacturing steps of CEL fiber are shown in Figure 9.1.

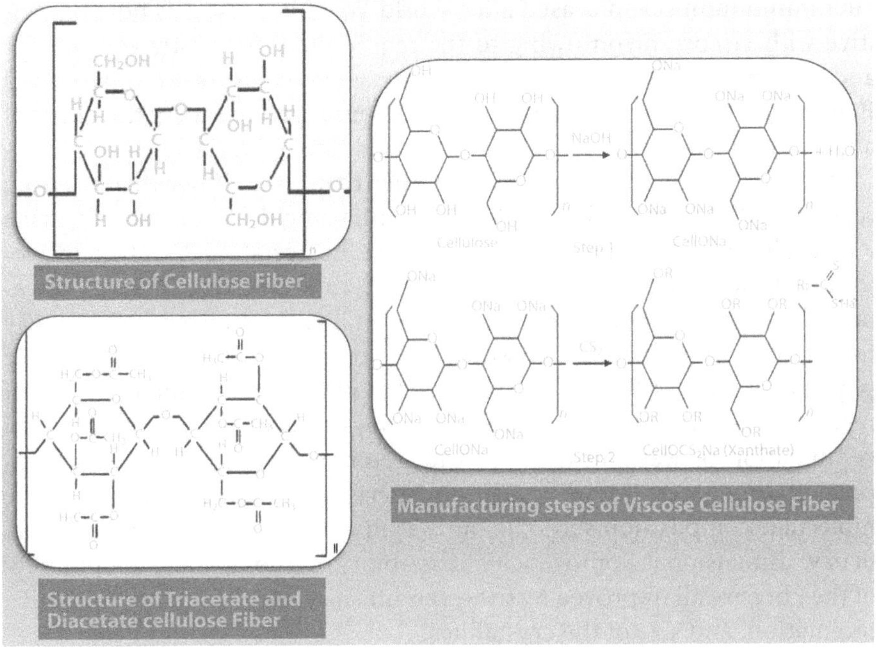

Regenerated CEL fiber has a smooth, glossy look and absorbs a lot of water, exactly like cotton, despite its chemical differences from silk. Because of their softness and high degree of drapability, textiles made from regenerated CEL fiber are frequently used in suits, blouses, jackets, skirts and pants, among other apparel items. However, the development of new synthetic fibers has resulted in a decline in the market share of regenerated CEL fiber. The amount of regenerated cellulosic fiber produced worldwide has been decreasing since it peaked in the early 1980s at 2.96 million metric tons.

Regenerated CEL, also known as CEL-derived fiber, comes in four main varieties: viscose, lyocell, cupro, and acetate. These various regenerated CEL fibers are categorized according to the process used to produce them. Alkali CEL is converted to CEL xanthate by reacting it with carbon disulphide to create viscose rayon fiber. With over 93% of the market for regenerated CEL and CEL-derived fibers, viscose rayon is the most common type of regenerated CEL fiber. CEL is immediately dissolved in the solvent N-methylmorpholine-N-oxide (NMMO) to create lyocell rayon fiber. A new generation of regenerated CEL fiber with better fiber qualities and environmentally friendly processes is called lyocell rayon. By dissolving CEL in a cuprammonium solution and then wet-spinning to regenerate Cupro rayon fiber is made from CEL. Commercial development of cuprammonium rayon ceased after World War I because it is not competitive with viscose rayon owing to the requirement for costly cotton CEL and copper salts. These days, a few businesses (such as Asahi in Japan and Bemberg in Germany) create cuprammonium rayon to satisfy the needs of a niche market for artificial silk and medical filter materials [6].

Acetate is a fiber made from CEL rather than a CEL fiber that has been regenerated. It is created by acetylating CEL using liquid acetic anhydride and a sulphuric acid catalyst. The resulting CEL acetate is dissolved in acetone and then spun using a dry spinning method to create fiber. CEL acetate fiber, which accounts for 2% of the total CEL fiber market, is solely used in cigarette filters [7]. Dissolving the original CEL *via* a chemical or physical process is the first of two crucial phases in the manufacture of regenerated CEL fiber. There are three ways to spin CEL: wet, dry, or a combination of wet and dry-jet. After re-crystallization, the dissolving–precipitating process modifies the polymeric molecular structure, but the molecular composition stays the same as in raw CEL. The strength, absorbency, dimensional homogeneity, drapability, dyeability, and tailorability of the fiber are all improved by these modifications to the CEL crystallinity, orientation, and size of the crystallites.

9.2.1 Viscose Rayon

According to the U.S. Federal Trade Commission (FTC) regarding generic names and definitions for manufactured fibers, rayon is defined as [8] a synthetic fiber composed of regenerated CEL and synthetic fibers composed of regenerated CEL in which substituents have replaced no more than 15% of the hydrogen atoms in the hydroxyl groups. British scientists Charles Cross, Edward Bevan, and Clayton Beadle were the first to patent viscose rayon fiber in 1892. It is a regenerated CEL fiber created using viscose technology [9].

They discovered that CEL xanthate could be produced by reacting raw CEL from wood or cotton with an alkali and carbon disulphide. The resulting CEL xanthate solution may likewise be precipitated in an ammonium sulphate solution, and it can be neutralized with diluted sulphuric acid to return to CEL. Like other textile fiber materials, viscose rayon fiber's tensile strength and breaking elongation are two of its most important mechanical properties. Tensile strength of fibers is commonly expressed in terms of tenacity, which is measured in force per denier [9].

They discovered that CEL xanthate could be produced by reacting raw CEL from wood or cotton with an alkali and carbon disulphide. The resulting CEL xanthate solution may likewise be precipitated in an ammonium sulphate solution, and it can be neutralized with diluted sulphuric acid to return to CEL. Like other textile fiber materials, viscose rayon fiber's tensile strength and breaking elongation are two of its most important mechanical properties. Tensile strength of fibers is commonly expressed in terms of tenacity, which is measured in force per denier. Moisture content (%) and moisture regain (%) are commonly used to describe water absorption capability. The formula for moisture content is the weight of absorbed water divided by the weight of the fiber, determined at a certain temperature and humidity. This fiber weight is the sum of the fiber's dry weight and the weight of the absorbed water. Moisture recovery is defined as the weight of absorbed water divided by the dry weight of the fiber. The moisture regain for viscose rayon fibre is between 12% and 14% [10]. This value is higher than the moisture regains for all natural and synthetic fibers except wool, indicating that viscose rayon fibers possess superior water absorption.

9.2.2 Lyocell

Lyocell (marketed as Tencel in the United States) is another type of regenerated CEL fiber derived from wood pulp. But the method for creating CEL solution is very different from the method for creating viscose rayon fiber. The Federal Trade Commission defines lyocell as CEL fiber that separates from an organic solution without the need for chemical intermediaries or the replacement of hydroxyl groups. Lyocell fiber belongs to the rayon category [9]. Researchers are searching for new methods to make CEL solutions because of environmental concerns. It was discovered that NMMO is a solvent that has the ability to directly dissolve CEL pulp. This invention was first mentioned in a patent that detailed a straightforward procedure for dissolving CEL using the NMMO solvent [11]. For the following 10 years, from 1969 to 1979, American Enka, another American business, researched spinning regenerated CEL fiber using the NMMO CEL solution, but they were unable to commercialize the process. In the end, an R&D team led by Pat White from Courtaulds in the UK successfully created an engineering strategy for CEL solution spinning. In 1982, Courtaulds built the first small pilot plant in Coventry, England, capable of producing up to 100 kg of lyocell fiber each week. In 1984, the product capacity of this pilot line was increased to one tonne per week. In 1988, Grimsby, England, saw the launch of a semi-commercial production line capable of producing 25 tonnes per week. In 1992, Courtauds in Mobile, Alabama, achieved its maximum commercial production capacity by manufacturing.

Austria's Lenzing AG is another significant European firm that produces lyocell fiber. In 1990, Lenzing, which had previously focused on producing viscose rayon fiber, set up a prototype facility to begin producing lyocell fibre [12]. In 1997, Lenzing opened a full-scale production facility at Heiligenkreuz that could produce 12,000 metric tonnes of Lenzing Lyocell®, a lyocell staple fiber, annually. The TENCEL® Group, which had two plants in Grimsby and Mobile, Alabama, was fully acquired by Lenzing in 2004. As of right now, Lenzing is the biggest producer of lyocell fiber in the world, with an annual production capacity of over 130,000 metric tonnes for the rayon market worldwide. Numerous changes were made to improve the lyocell fibers' performance quality, including Applying etherification-modified ultra-low formaldehyde 2D resin to lyocell fabric can enhance the fiber's anti-fibrillation properties by increasing the lateral force of contact between its macromolecular chains [13].

To prepare the fabric for antibacterial activity, nano zinc oxide was added during the second Lyocell fiber-forming process. The fabric's chemical and crystal structures were investigated using Fourier transform infrared spectroscopy and X-ray diffraction. The findings show that although the fiber's crystallinity was reduced, its chemical structure remained mostly unchanged. The thermo gravimetric test results show that the thermal stability of the fabric doped with nano-ZnO remains unchanged [14].

Environmental friendliness, durability, and efficiency are critical factors in the field of flame retardant textile modification. An innovative technique was developed to produce a gallic acid-derived intumescent flame retardant (GADPP) that contained reactive (–P(=O)(O–NH4+)2) and (–P(=O)(OCH2CH3)2) groups to facilitate functional modification of lyocell fabric. The GADPP modification greatly improved the lyocell fabric's flame retardancy and ultraviolet (UV) resistance [15].

Adenosine triphosphate (ATP), chitosan (CS), and polyethyleneimine (PEI) were effectively deposited layer by layer in a bio-based coating to produce excellent fire-resistant and antibacterial lyocell textiles (LBL/Lyocell). After adding 11.5 weight percent, the final textiles achieved the limiting oxygen index (LOI) [16]. Lyocell fabric with a reduced propensity for fibrillation thanks to an effective, low-temperature, and environmentally friendly crosslinking technique. Here, fibrillation resistance is achieved by crosslinking N1, N6-bis(oxiran-2-ylmethyl)hexane-1,6-diamine with lyocell fabric [17]. Low carbon emissions are one tactic for sustainable development; lyocell is a carbon-free, renewable, biodegradable CEL fiber. The flammability of lyocell fabric puts people's lives and property at danger. To address this issue, a bio-based flame retardant called tyramine polyphosphate (APP-LA) was developed and used to lyocell textiles. APP-LA was made using an ecologically friendly, organic solvent-free synthesis method. The damage length in the combustion test was just 6.7 cm, and the APP-LA treated lyocell fabrics' limiting oxygen index rose to 35.8% from 18.5% for the control lyocell [18].

Through continuous thermostabilization catalyzed by sulphuric acid and subsequent carbonization, carbonized fibers with quantifiable mechanical properties were produced from ecologically benign lyocell fibers (LFs). A low mass yield (~ 12 weight percent) was obtained after carbonization as a result of the virgin LF breaking down into tiny molecules that might evaporate at high temperatures during thermal treatment. On the other hand, at low temperatures, distinct chemical processes took place in the LF that was impregnated with sulphuric acid (ILF). The fiber's protons and water served as catalysts for these reactions. In addition to the heat dissociation of glycosidic bonds, these catalyzed processes involved the elimination of breakdown products, the dehydration of decomposed products to 5-hydroxymethylfurfural, and its conversion. A series of chemical studies revealed that linear CEL macromolecular chains were transformed into highly cross-linked network topologies composed of broken products.

The lyocell spinning technique may be used to increase the regenerated CEL fiber's dry and wet tensile strengths. The dry tenacity rating of lyocell fiber is higher than that of viscose and HWM rayon fibers and almost comparable to that of polyester fiber. Furthermore, when wet, lyocell fiber maintains 85% of its dry toughness. With a wet tensile strength greater than cotton's, it is the only CEL fiber ever created. In comparison to viscose rayon fiber, lyocell fiber has a significantly lower elongation value, which is somewhat greater than that of HWM rayon fiber.

The high degree of CEL crystallinity and crystal orientation, which produces microfibrils aligned parallel to the fiber axis, is responsible for the fibrillary structure of lyocell fiber. Because of this, lyocell fiber may readily form a fibrillated surface when subjected to mechanical abrasion [19]. Lyocell fiber has improved qualities in terms of softness, drapability, dimensional stability, dye absorption, and colorfastness, although it still has many physical characteristics with viscose rayon fiber. Lyocell fiber has a moisture regain of about 11%, which is little less than viscose rayon [20]. This is mainly because the lyocell spinning process produces a higher CEL crystallinity.

9.2.3 Cupro

A cellulosic fiber called cupro has been regenerated from wood pulp or cotton linters. CEL is dissolved in a solution containing copper salts and ammonia, and the resulting slurry is then pushed through a spinneret and into a coagulation bath, where it regenerates to create multifilament yarn.

Thus, Cuprammonium Rayon is another name for it. Since Cupro utilizes what would have been a waste product and transforms it into something beautiful and useful with little energy and emissions, it is considered an eco-textile. Additionally, cupro doesn't damage the environment and dissolves rapidly in soil. Another specialty yarn is cuprammonium rayon, which is used to create textured textiles with slubbed, uneven surfaces in lightweight summer dresses and blouses. It is occasionally combined with cotton [21].

9.2.4 CEL Acetate

CEL acetate fiber is another type of CEL-related fiber that is created. Acetate fiber is no longer considered a regenerated CEL fiber since it is made of acetate (CEL ester), not CEL, as the polymer formula. In 1865, Paul Schützenberger tried dissolving cotton in a solution of acetic anhydride and found the method for producing CEL acetate. When this solution was mixed with water, trigacetate, or white flakes, developed. However, this discovery had no direct bearing on the development of acetate fiber spinning until 1904, when George Miles discovered a method to convert triacetate into secondary acetate, or diacetate [22]. The issue of immediately dissolving triacetate for fiber spinning was resolved by this innovation. The British CEL Co. Ltd. in Spondon, England, started producing acetate filament on a commercial scale for the first time in 1921. Acetate fibre manufacturing in the United States was started a year later by Camille Dreyfus, one of the shareholders of the British CEL Co. Ltd. Nowadays, acetate fiber is little used in the clothing and textile industries. Regular acetate, also known as secondary acetate or diacetate, and triacetate are the two varieties of acetate fiber. The chemical formulae of these two fibers. These two kinds of fibres are produced using two different processes. Triacetate synthesis is frequently the initial stage in the creation of acetate fiber. Purified raw CEL (wood pulp or cotton lint) is mixed with liquid acetic acid and acetic anhydride to undergo acetylation. To speed up the acetylating process, sulphuric acid is used as a catalyst. Following the completion of the acetylation process, CEL triacetate is a thick, viscous solution. To make triacetate fiber, the triacetate solution is poured into water, which causes triacetate flakes to precipitate. After washing and drying, the triacetate flakes are dissolved in methylene chloride to generate the spinning dope. Triacetate fiber is produced by forcing the dope through the spinnerets and into a chamber containing the process of extruding the dope *via* the spinnerets into a heated air chamber produces triglycemic fiber.

To precipitate as triacetate flakes and yield ordinary acetate, acetylated CEL (triacetate) needs to be hydrolyzed with water. Acid hydrolysis caused by diluted access acetic acid at 95% concentrations is linked to a 20-hour ripening time. When one-sixth of the acetate groups (CH3COO–) are replaced by hydroxyl groups (–OH) during hydrolysis, triacetate is converted to diacetate. After hydrolysis, diacetate flakes are precipitated by adding more water. To make spinning dope, dried and cleaned diacetate flakes are dissolved in acetone. The same dry spinning method used to make triacetate fiber is also used to make regular acetate fiber.

Similar to viscose fiber, normal acetate and triacetate fibers feature irregularly lobed contours in their cross-section. There are also some longitudinal striations on the surface of regular acetate and triacetate fibers. Ordinary acetate and triacetate fibers can have their lustre altered to create bright, dull, and semi-dull variants. The volume density of both acetate fibers is 1.32 g/cm3, which is lower than that of viscose fibers. The strength of regular acetate fiber is 0.9–1.0 g/denier when wet and 1.2–1.4 g/denier when dry. Triacetate fiber has a wet tenacity of 0.8–1.0 g/denier and a dry tenacity of 1.1–1.4 g/denier. On the other hand, triacetate fiber typically regains between 3.2% and 3.5% of its moisture content, which is comparable to that of nylon [23]. Both regular acetate and triacetate fibers provide a good dimensional stability with some stretch and resistance to shrink. In terms of wrinkle resistance, triacetate performs slightly better.

9.3 Applications of Regenerated CEL for Food Packaging

CEL, one of the most common natural materials in the world, has lately acquired favor as a packaging material that is safe for the environment. CEL is a naturally occurring linear polymer composed of D-glucose units joined by β-(1,4) linkages. D-glucose units in this macromolecular sugar chain range from 500 to 14000 [24, 25]. The many hydroxyl groups in CEL molecule chains promote the formation of hydrogen bonds, allowing the chains to stack one on top of the other to form tightly packed CEL microfibrils [26]. For usage in bioengineering applications, CEL's conformation gives it the required mechanical properties, biocompatibility, and biodegradability. Moreover, CEL has a high heat resistance and might include antioxidant and antibacterial compounds [27]. The widespread development of CEL is hindered by three problems. First, solubility. CEL's complicated network of hydrogen connections and partially crystalline region make it difficult to dissolve in conventional solvents. Solubilization is necessary for the development of CEL into desirable products. The second problem is high hydrophilicity, and the third is the lack of antibacterial action. Every one of these challenges is being addressed. For example, studies on green solvents that break the hydrogen bonds in the CEL structure have been conducted [28]. Polymers have been added to CEL-based films to enhance their bacterial inhibitory capacity and decrease their hydrophilicity [29]. Research is also being done on CEL and bioplastic combinations in composite films [30, 31]. Application of regenerated CEL in packaging shown in Table 9.1.

Because of its exceptional barrier characteristics and ability to tolerate temperatures as high as 190°, regenerated CEL food packaging meets the basic standards for food packaging. Because the regenerated CEL film has a lower oxygen transmittance (OTR) than conventional polymers, it can effectively prevent oxygen-induced food oxidation [32]. It also has many wonderful natural benefits. Because its raw ingredients are renewable biomass, which degrades swiftly in the soil by microorganisms without posing a threat to the environment, they are considered environmentally benign and green [33].

To prevent the drinking contents from oxidizing and decarbonizing, polymers used in the packaging of carbonated beverages must have strong O_2 and CO_2 barriers. Transporting CO_2 in other items is more easy than transporting O_2 or liquid water. Due to their complexity, food items need specific packing methods, and as food is delivered across ever-greater distances between producers and consumers, the demands on the package industry will only rise [34]. Since no one pure polymer can meet all of the mechanical and tribological requirements required for any potential food packaging, complex multilayer films or composites are frequently used. When an application requires exceptionally strong oxygenation barriers throughout a wide temperature range, EVOH, a water control material, can be made by combining two layers of a moderately hydrophobic polymer, such polyethene [34].

Marilena Carbone *et al.* [35] reviewed a study in which the researcher discovered that edible pullulan films coated with nanoparticles (AgNPs, ZnO NPs) and essential oils (EOs) such as oregano or rosemary oil shown antibacterial activity in the preservation of deli meat in Turkey. AgNPs and oregano oil edible films were found to be much more effective than ZnO nanoparticles and rosemary oil safe-to-eat films. The preservation and prolongation of orange juice's long-term storage were investigated using Ag and ZnO nanoparticles in a decreased polyethylene (LDPE) polymer matrix. When combined with heat treatment at pasteurized temperatures, this active nanocomposite showed remarkable antibacterial capabilities [36].

Table 9.1 Applications of regenerated CEL for packaging.

Type of regenerated CEL	Characteristics	Example of product	Use in packaging	References
Edible pullulan films	Incorporation of essential oils (EOs) like rosemary or oregano and Ag/ZnO nanoparticles	Deli meat	Antibacterial activity	[35]
Reduced polyethylene (LDPE) polymer matrix	Ag and ZnO nanoparticles	Orange juice	Antibacterial properties	[36]
CEL hydrogel	Formation of 12-hydroxyoctadecanoic acid (HOA) crystals	Biodegradable packaging	Excellent hydrophobicity and biodegradability	[37]
NanoCEL-reinforced chitosan nanocomposites film	Inhibitory impact against gram-positive and gram-negative bacteria	Ground beef	Reduces lactic acid bacteria population	[38]
NanoCEL in the agar matrix	Higher tensile strength and modulus	Nanobiocomposite films	Enhanced mechanical properties	[39]
Tannin CEL films (TCFs)	Strong mechanical properties, UV-blocking, and antioxidant qualities	Active food packaging	Provides protection and extends shelf life	[40]

It is crucial to contrast CEL-based materials with synthetic ones in light of the volume of research on CEL-based materials. The most common material is plastic, particularly when it comes to packaging films [41]. Over nine billion metric tonnes of plastic products were manufactured globally by 2017. It is estimated that at least 76% of plastic waste is landfilled or released into the environment annually since only 10% to 24% of plastic objects are recycled [42]. To help with this problem, researchers are searching for bio-based remedies. CEL is one of these materials. The fact that CEL breaks down spontaneously is one of its advantages. Because CEL films are biodegradable and compostable, they may decompose into their constituent parts under the right conditions. However, because plastics may persist in the environment for hundreds of years and are difficult to biodegrade, they contribute to pollution. CEL films are created from renewable resources, whereas plastics are often made from natural gas or petroleum. This gives the CEL films a competitive advantage in terms of biodegradability and sustainability. Despite these benefits, the CEL films have poorer barrier, flexibility, transparency, and economic properties. CEL-based films frequently have superior barrier properties against gases, moisture, and other environmental factors when compared to plastic films. Additionally, compared to CEL films, the polymers offer greater transparency. Compared to conventional polymers, CEL films can frequently be more expensive. In conclusion, because CEL films are more sustainable and biodegradable than conventional plastics, they are a more environmentally friendly choice. However, a number of factors, such as consumer preferences, legal issues, and specific packaging specifications, may have an impact on their acceptability. Furthermore, they might not necessarily have the same properties as plastics [43–45].

9.3.1 Examples of CEL Packaging for Food and Beverage

Cellophane which is made from regenerated CEL, cellophane is a translucent, flexible film. It is frequently used as a wrapping for candies, nuts, and baked products. Cellophane comes in a wide variety of forms, including bags, cartons, and more. CEL-based films, are made from CEL that has been regenerated, which is a type of CEL that comes from wood pulp. These films' strong barrier properties, flexibility, and transparency make them ideal for food packing. CEL-based wrappers, are frequently used to package foods including baked goods, sandwiches, and snacks. You can see what's inside since they are clear and translucent. They are durable and robust, and they may be heat-sealed to maintain the freshness of food. CEL-based wrappers are thicker and more opaque than CEL-based films.

CEL-based pouches are typically larger than wrappers and contain a resealable closure. For this reason, they are an excellent choice for encasing dry products such as nuts, granola, and trail mix. CEL-based moulded pulp packaging is made by moulding CEL fibers into the proper form. Fruits, vegetables, and eggs are frequently packaged in these types of containers.

9.3.2 Advantages of CEL Food Packaging

9.3.2.1 Environmental Benefits

- Biodegradable and compostable: Food packaging made of CEL is both biodegradable and compostable as it comes from natural plant sources. This implies that it will decompose organically over time, leaving behind no trace of potentially harmful microplastics that might contaminate the environment.
- Renewable: The resource CEL is a renewable one. It may be cultivated and collected without reducing the quantity of other resources available. The production of CEL has less effect on greenhouse gas emissions and environmental degradation compared to petroleum-based polymers.
- Minimize waste: Products made of CEL packing can break down organically. Therefore, the amount of plastic trash released into the environment may be decreased by employing CEL food packaging goods rather than plastic packaging.
- Reduce pollution: Unlike plastic, CEL, which is obtained organically and is non-toxic, does not produce microplastics or other hazardous materials that are detrimental for the environment or people's health.

9.3.2.2 CEL Food Packaging's Advantages for Businesses

- Moisture resistant: CEL food packaging efficiently protects food commodities from moisture loss, increasing shelf life and keeping freshness. This is especially helpful for items like fruits, vegetables, and baked products that are sensitive to moisture.
- Oil resistant: Because CEL food packaging is so oil resistant, it may be used to package fatty or oily foods without worrying about leaks or spoiling. This is especially advantageous for goods like fried meals, nuts, and seeds.
- Versatile: CEL may be used in a wide range of packaging applications since it can be formed into several shapes, such

as films, sheets, molds, and composite materials. Because of its flexibility, innovative packaging solutions that meet the needs of certain items may be created.
- Encourages sustainability: By utilizing CEL food packaging, companies can demonstrate their dedication to environmental responsibility and sustainability. They may draw in eco-conscious clients and enhance their business image [46].

A family of naturally occurring carbohydrate polymers known as CEL is found in an almost limitless supply of raw materials, including plants, agricultural waste, the shells of marine animals, and microbes (Figure 9.2a) [47, 48]. Depending on the origin of the plants, the CEL fibers in the plant cell wall have varying ratios of CEL embedded as microfibrils in a matrix of hemiCEL, lignin, pectin, ash, and other extractives [47, 49, 50]. A type of linear, stereoregular, semicrystalline polysaccharides, CEL is made up of D-anhydroglucose units (AGU) that are β-1,4-linked and have three hydroxyl groups (–OHs) per unit [51, 52]. CEL-based materials exhibit superior film-forming properties, a lower coefficient of thermal expansion, and enhanced mechanical strength and thermal stability at high temperatures when juxtaposed with traditional petroleum-based polymers like

Figure 9.2 (a) Sources of CEL.

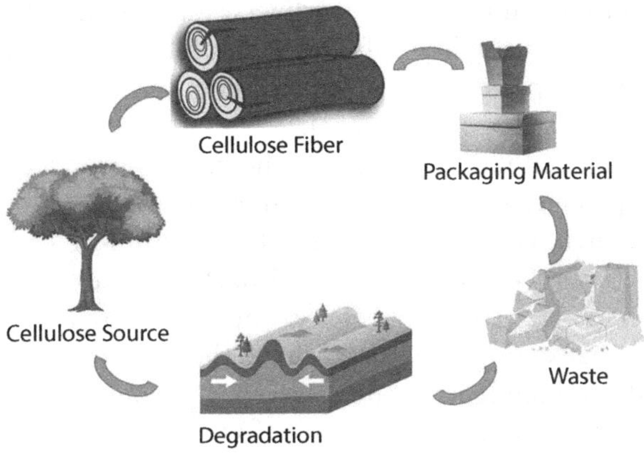

Figure 9.2 (b) Life cycle of CEL.

polyethylene (PE), polypropylene (PP), and polystyrene (PS) [53–55]. This material is very promising for use in many practical fields, such as packaging, straws, foams, flexible electronics, and tissue engineering materials, for the realization of sustainable product solutions and achieving a low carbon footprint because of its intrinsic renewability and biodegradability (Figure 9.2b) [56].

The total biodegradability of CEL and its plentiful natural origin are the benefits of CEL films. It is imperative to assess and utilize chemical modification's capacity to lessen CEL's hydrophilic properties and guarantee that CEL-based packaging performs on par with petroleum-based alternatives in terms of mechanical and barrier qualities, all the while preserving the films' biodegradability. It has proven possible to attain mechanical and barrier properties that are on par with—if not better than—corresponding petroleum-based polymers like LDPE using several methods. Techniques for creating packaging made of active CEL, such as antibacterial and antioxidant packaging, have also been developed. It has been shown that combining cellulosic material with additives like plasticizers and active agents enhances its mechanical properties and extends its capacity to store food without greatly reducing its biodegradability [56].

9.4 Applications of Regenerated CEL in Healthcare Product Packaging

Packaging has an impact on producers, stockists, retailers, and customers by being essential for product identification and distinction.

Patients can grasp the intended use of diverse packaging systems since different delivery routes employ different methods. Additionally, packaging shields medical components from the elements and maintains their security until the final user opens them [3].

He *et al.* employed CEL hydrogel as the base to produce hydrophobic CEL film composites. They used hair-inspired crystalline development in spongy CEL matrix cavities to promote the creation of 12-hydroxyoctadecanoic acid (HOA) crystals on the hydrogel's surface, giving the hydrogel its highly water-resistant properties. The films were a good option for biodegradable packaging materials since they demonstrated excellent hydrophobicity and biodegradability, as well as greater tensile strength and elongations at break than the CEL matrix [37].

From lignoCEL biomass, nanoCEL—a kind of CEL nanocrystal or CEL nanofibril—can be isolated and utilized as nano reinforcement in polymer nanocomposites. Understanding and manipulating matter at 1100 nm dimensions is what nanotechnology entails, opening up new application possibilities. Cellulosic nanoparticles, cellulosic nanocrystals, and bacterial nanoCEL are examples of nanomaterials with superior mechanical qualities, abundance, and capacity for reinforcement. They also have low densities, are biodegradable, and comprises an substantial amount of hydroxyl groups. These characteristics allow CEL nanoparticles to be modified to suit different polymeric matrix materials, which makes them perfect candidates for reinforcing agents in polymeric matrix materials [57].

The antibacterial properties of a nanoCEL-reinforced chitosan nanocomposites film were investigated by Dehnad *et al.* Their investigation demonstrated that nanocomposites films have an inhibitory impact against both gram-positive as well as gram-bacteria like Staphylococcus aureus and Salmonella enteritidis. When compared to a nylon-packaged sample stored for six days at 25C, the lactic acid bacteria population in the ground beef was reduced by up to 3.1 logarithmic cycles by the chitosan nanoCEL nanobiocomposite [38].

For use as packaging materials, Reddy and Rhim created nanoCEL (NC)-reinforced polymer nanocomposites. With a 5% weight percentage of NC in the agar matrix, the nanobiocomposite films demonstrated higher tensile strength and modulus; however, with a 10% weight percentage of NC reinforcement, the strength declined because of potential agglomeration in the biopolymer matrix [39].

To create CEL film (CF), the cotton linter-dissolving pulp was dissolved in an ionic solvent and then regenerated. Diadehyde CEL film (DACF) was oxidized to generate tannin CEL films (TCFs). The tannin's strong antioxidant and UV-blocking qualities came from its covalent bonding to the DACF surface. The CEL film's hydrophobicity and water vapor barrier qualities were enhanced by the addition of tannin. When compared to pure CEL film, the wet tensile strength and strain at break rose considerably. These TCFs' strong mechanical, UV-blocking, and antioxidant qualities make them very suitable for use in active food packaging [40].

CEL is tactful and has a strong hydrogen bond, making it difficult to dissolve in organic solvents like water. Although several solvent solutions have been developed to dissolve CEL, ionic liquids have garnered attention due to their greater dissolubility, low toxicity, thermal stability, and recyclability. To produce a novel pure paper-based composite material with high mechanical performance and barrier properties, research was conducted to cover paper with dissolved CEL. Ionic liquid (1-allyl-3-methylimidazolium chloride) was used as the solvent system to dissolve cotton linter. SEM, AFM techniques, and contact angle analysis were employed to describe the surface properties of the reinforced paper [58].

It is acknowledged that the most prevalent renewable polymeric material is CEL nanofibers. Therefore, using CEL nanofibers materials in packaging would cut down on environmental pollution and product packaging expenses. An efficient CEL nanofibers design may be a helpful tool for developing environmentally friendly packaging and for managing the environmental quality of the packaging materials [59].

9.5 Future Opportunities and Challenges

The U.S. National Energy Policy lays forth objectives to ensure energy supply by using a more diverse mix of locally sourced renewable energies. The production and consumption of downstream chemical and polymer products from oil refineries are experiencing the long-term consequences of this energy policy's implementation. Developing biomass technologies that allow for the more cheap and efficient production of bioenergy and bio based products is one obstacle to reaching this energy goal. As a result, regenerated CEL fibers are gaining popularity as the only type of artificial fiber manufactured from CEL biomass. But viscose rayon continues to be the most popular fiber. Its production uses a lot of hazardous chemicals, which has seriously harmed the environment. Microfiber rayon output is anticipated to decrease, whereas lyocell rayon production will steadily expand. Additional innovations may include the utilization of a more diversified CEL substrate and improvements to fiber efficiency for medical services, personal hygiene, and luxurious utility instances.

Wood pulp is now the most common CEL feedstock for the manufacturing of regenerated CEL fiber. Other CEL products or wastes, such as bamboo, bast fibers, cotton lintels, and sugarcane bagasse, may become more often utilized as CEL feedstock.

Military uniforms currently use viscose rayon fiber with enhanced flame retardancy for breathability, comfort, and burn prevention. Superior absorbency hygiene products also employ trilobal-shaped viscose rayon. Utilizing innovative spinning techniques, antimicrobial lyocell rayon fiber is created by loading copper or silver salt components. Healthcare fabrics, pants, sportswear and hygiene goods can all be made with this novel lyocell rayon fiber. Phase-change compounds such as paraffin have also been incorporated into lyocell rayon to create a thermal self-regulating property. Sportswear and bedding products are the end-user markets where this specialty lyocell rayon is being introduced.

9.6 Conclusion

Due to pollution from petroleum-based polymers and the depletion of non-renewable resources, the 21st century is seeing a transition toward environmentally friendly materials and sustainable methods. It is suggested to use CEL instead of petroleum-based plastic packaging; hydrogels and regenerated CEL sheets are in the development stage. For pharmaceuticals to be stable, efficacious, and safe, packaging is essential. Characteristics like child- and tamper-resistant packaging are essential for modern electronics. In Canada, food packaging makes up the majority of the 20% of household waste that is recycled. The renewable biopolymer CEL provides low-cost, non-toxic substitutes that are low in density. CEL is derived from food and agricultural waste and is utilized in packaging among other things. The value of plant waste for the synthesis of CEL is being studied. Regenerated CEL fibers come in four main varieties: viscose, lyocell, cupro, and acetate. With almost 93% of the market, viscose rayon is the most common kind. A new generation of rayon with enhanced characteristics and ecologically friendly production is called Lyocell. Wet-spinning CEL dissolved in a cuprammonium solution yields cupro rayon. Specialized businesses create ceriamonium rayon for use in artificial silk and medical filter materials. Food packaging made of regenerated CEL is an environmentally acceptable and sustainable substitute for conventional plastics. It has better barrier properties, less oxygen transmission, and may break down quickly in soil. The antimicrobial characteristics properties of nanoCEL-reinforced chitosan nanocomposite films have been shown to inhibit gram-positive as well as gram-negative bacteria, reducing the lactic acid bacteria population in ground beef. Cellulosic nanoparticles, nanocrystals, and bacterial nanoCEL are superior materials with mechanical properties, low densities, biodegradability, and numerous hydroxyl groups. They can be modified to suit various polymeric matrix materials.

AI disclosure statement: The author acknowledges the use of AI-based tools for minor language editing or grammatical correction. However, all intellectual contributions, analyses, and conclusions presented in this chapter are entirely the author's own.

References

1. Wang, S., Lu, A., Zhang, L., Recent advances in regenerated cellulose materials. *Prog. Polym. Sci.*, 53, 169–206, 2016.
2. Huang, K. and Wang, Y., Recent applications of regenerated cellulose films and hydrogels in food packaging. *Curr. Opin. Food Sci.*, 43, 7–17, 2022.
3. Amarji, B., Kulkarni, A., Deb, P.K., Maheshwari, R., Tekade, R.K., Package development of pharmaceutical products: Aspects of packaging materials used for pharmaceutical products, in: *Dosage Form Design Parameters*, pp. 521–52, Elsevier, 2018.
4. Zhao, G., Lyu, X., Lee, J., Cui, X., Chen, W.-N., Biodegradable and transparent cellulose film prepared eco-friendly from durian rind for packaging application. *Food Packag. Shelf Life*, 21, 100345, 2019.
5. Swicofil, A.G., Textile Services, 2011.
6. Kamide, K. and Nishiyama, K., Cuprammonium processes. *Regenerated Cellul. Fibers*, 88–155, 2001.
7. Brody, T., Food and Dietary Supplement Package Labeling—Guidance from FDA's Warning Letters and Title 21 of the Code of Federal Regulations. *Compr. Rev. Food Sci. Food Saf.*, 15, 1, 92–129, 2016.
8. Woodings, C., 3. the viscose process, in: *Regenerated Cellulose Fibres*, Woodhead Publishing, 2001.
9. Woodings, C., *Regenerated cellulose fibres*, Elsevier, 2001.
10. Hearle, J.W. and Morton, W.E., *Physical properties of textile fibres*, Elsevier, 2008.
11. McCorsley III, C.C., Process for shaped cellulose article prepared from a solution containing cellulose dissolved in a tertiary amine N-oxide solvent. Google Patents, 1981.
12. Chen, J., Synthetic textile fibers: Regenerated cellulose fibers, in: *Textiles and fashion*, pp. 79–95, Elsevier, 2015.
13. Feng, B., Ren, Y., Shi, L., Xue, Z., Li, X., Xiao, Z., *et al.*, Efficient anti-fibrillation of lyocell fabric by low-formaldehyde resin. *Ind. Crops Prod.*, 213, 118406, 2024.
14. Wang, X., Wang, S., Li, Y., Jin, X., Dong, C., Preparation and properties of low fibrillated antibacterial Lyocell fiber. *Arabian J. Chem.*, 105658, 2024.

15. Zhao, J., Jiang, L., Zuo, C., Tan, W., Ren, Y., Liu, X., Biomass-inspired fabrication of eco-friendly, durable flame retardant and ultraviolet resistant lyocell fabric. *Int. J. Biol. Macromol.*, 135332, 2024.
16. Jang, M., Choi, D., Kim, Y., Kil, H.-S., Kim, S.-K., Jo, S.M., *et al.*, Role of sulfuric acid in thermostabilization and carbonization of lyocell fibers. *Cellulose*, 30, 12, 7633–52, 2023.
17. Li, X., Fang, K., Shi, L., Xue, Z., Feng, B., Zheng, Q., *et al.*, Low-temperature and efficient crosslinking of lyocell knitted fabric with an eco-friendly diepoxy agent for fibrillation resistance. *Cellulose*, 31, 1, 717–35, 2024.
18. Li, X., Liu, J., Chen, F., Zhang, J., Li, H., Gu, X., *et al.*, Eco-friendly flame retardant coating based on ammonium polyphosphate and tyramine for lyocell fabrics. *Polym. Degrad. Stab.*, 218, 110572, 2023.
19. Collier, B.J., Effects of Finishing and Dyeing on Lyocell and Lyocell-Blend Fabrics. *Text. Chem. Color. Am. Dyest Rep.*, 1, 2, 1999.
20. Hongu, T., Takigami, M., Phillips, G.O., *New millennium fibers*, Elsevier, 2005.
21. Kayseri, GO., Bozdoğan, F., Hes, L., Performance properties of regenerated cellulose fibers. *Text. Apparel*, 20, 3, 208–12, 2010.
22. Hearle, J.W. and Woodings, C., Fibres related to cellulose, in: *Regenerated Cellulose Fibers*, pp. 156–73, Woodhead Publishing, Sawston, Cambridge, UK, 2001.
23. Joseph, M.L., Introductory textile science. (No Title), 1986.
24. Trovatti, E., The future of bacterial cellulose and other microbial polysaccharides. *J. Renewable Mater.*, 1, 1, 28–41, 2013.
25. Zhang, J., Luo, N., Zhang, X., Xu, L., Wu, J., Yu, J., *et al.*, All-cellulose nanocomposites reinforced with *in situ* retained cellulose nanocrystals during selective dissolution of cellulose in an ionic liquid. *ACS Sustainable Chem. Eng.*, 4, 8, 4417–23, 2016.
26. An, N., Qin, J., Zhou, X., Wang, Q., Fang, C., Guo, J., *et al.*, Recent progress in cellulose-based flexible sensors. *JRM*, 10, 2319–34, 2022.
27. Bayram, B., Ozkan, G., Kostka, T., Capanoglu, E., Esatbeyoglu, T., Valorization and application of fruit and vegetable wastes and by-products for food packaging materials. *Molecules.*, 26, 13, 4031, 2021.
28. Acharya, S., Liyanage, S., Parajuli, P., Rumi, S.S., Shamshina, J.L., Abidi, N., Utilization of cellulose to its full potential: A review on cellulose dissolution, regeneration, and applications. *Polymers.*, 13, 24, 4344, 2021.
29. B-x, Z., J-i, A., Uyama, H., Preparation and characterization of a transparent amorphous cellulose film. *RSC Adv.*, 5, 4, 2900–7, 2015.
30. Huang, L., Xu, H., Zhao, H., Xu, M., Qi, M., Yi, T., *et al.*, Properties of thermoplastic starch films reinforced with modified cellulose nanocrystals obtained from cassava residues. *New J. Chem.*, 43, 37, 14883–91, 2019.
31. Liu, X., Qin, Z., Ma, Y., Liu, H., Wang, X., Cellulose-Based Films for Food Packaging Applications: Review of Preparation, Properties, and Prospects. *J. Renewable Mater.*, 11, 8, 3203–25, 2023.

32. Verde, A.B., Alvim, I.D., Luccas, V., Marangoni, L., Alves, R.M.V., The influence of formulation and packaging material on the rheological properties of milk chocolate. *Appl. Food Res.*, 2, 2, 100199, 2022.
33. An, H., Lu, Z., Wang, Z., Abbas, M.Q., Du, Z., Safety assessment and quality control of regenerated cellulose food packaging in different processes. *Food Control*, 163, 110543, 2024.
34. Duncan, T.V., Applications of nanotechnology in food packaging and food safety: Barrier materials, antimicrobials and sensors. *J. Colloid Interface Sci.*, 363, 1, 1–24, 2011.
35. Carbone, M., Donia, D.T., Sabbatella, G., Antiochia, R., Silver nanoparticles in polymeric matrices for fresh food packaging. *J. King Saud Univ. Sci.*, 28, 4, 273–9, 2016.
36. Halagalimath R, R.J. and Kottom, N., Biodegradable Nano-Cellulose and its Composite Materials for Food Packaging Applications: A Review. *J. Mines Met. Fuels*, 70, 10A, 104–11, 2023.
37. He, M., Xu, M., Zhang, L., Controllable stearic acid crystal induced high hydrophobicity on cellulose film surface. *ACS Appl. Mater. Interfaces*, 5, 3, 585–91, 2013.
38. Dehnad, D., Emam-Djomeh, Z., Mirzaei, H., Jafari, S.-M., Dadashi, S., Optimization of physical and mechanical properties for chitosan–nanocellulose biocomposites. *Carbohydr. Polym.*, 105, 222–8, 2014.
39. Reddy, J.P. and Rhim, J.-W., Characterization of bionanocomposite films prepared with agar and paper-mulberry pulp nanocellulose. *Carbohydr. Polym.*, 110, 480–8, 2014.
40. Huang, X., Ji, Y., Guo, L., Xu, Q., Jin, L., Fu, Y., *et al.*, Incorporating tannin onto regenerated cellulose film towards sustainable active packaging. *Ind. Crops Prod.*, 180, 114710, 2022.
41. Jeevahan, J. and Chandrasekaran, M., Nanoedible films for food packaging: A review. *J. Mater. Sci.*, 54, 19, 12290–318, 2019.
42. Geyer, R., Production, Chapter 2-Production, use and fate of synthetic polymers. *Plast. Waste Recyl.*, 13–22, 2020.
43. Foroughi, F., Rezvani Ghomi, E., Morshedi Dehaghi, F., Borayek, R., Ramakrishna, S., A review on the life cycle assessment of cellulose: From properties to the potential of making it a low carbon material. *Materials.*, 14, 4, 714, 2021.
44. Akhtar, S., Gupta, A.K., Kumar, H., Manufacturing and Applications of Cellulosic Films in Packaging: An Alternative for Plastic Films. *Curr. Nutr. Food Sci.*, 20, 4, 384–400, 2024.
45. Grzybek, P., Dudek, G., van der Bruggen, B., Cellulose-based films and membranes: A comprehensive review on preparation and applications. *Chem. Eng. J.*, 495, 153500, 2024.
46. kimecopak, Cellulose Packaging: The Future Game-changer for Food Business, 2023.

47. Li, T., Chen, C., Brozena, A.H., Zhu, J., Xu, L., Driemeier, C., *et al.*, Developing fibrillated cellulose as a sustainable technological material. *Nature*, 590, 7844, 47–56, 2021.
48. Liu, Y., Ahmed, S., Sameen, D.E., Wang, Y., Lu, R., Dai, J., *et al.*, A review of cellulose and its derivatives in biopolymer-based for food packaging application. *Trends Food Sci. Technol.*, 112, 532–46, 2021.
49. Rol, F., Belgacem, M.N., Gandini, A., Bras, J., Recent advances in surface-modified cellulose nanofibrils. *Prog. Polym. Sci.*, 88, 241–64, 2019.
50. Saedi, S., Garcia, C.V., Kim, J.T., Shin, G.H., Physical and chemical modifications of cellulose fibers for food packaging applications. *Cellulose.*, 28, 14, 8877–97, 2021.
51. Ventura-Cruz, S. and Tecante, A., Nanocellulose and microcrystalline cellulose from agricultural waste: Review on isolation and application as reinforcement in polymeric matrices. *Food Hydrocolloids*, 118, 106771, 2021.
52. Wang, Y., Wang, X., Xie, Y., Zhang, K., Functional nanomaterials through esterification of cellulose: A review of chemistry and application. *Cellulose.*, 25, 3703–31, 2018.
53. Song, L., Wang, Z., Lamm, M.E., Yuan, L., Tang, C., Supramolecular polymer nanocomposites derived from plant oils and cellulose nanocrystals. *Macromolecules.*, 50, 19, 7475–83, 2017.
54. Farooq, M., Zou, T., Riviere, G., Sipponen, M.H., Osterberg, M., Strong, ductile, and waterproof cellulose nanofibril composite films with colloidal lignin particles. *Biomacromolecules.*, 20, 2, 693–704, 2018.
55. Guan, Q.-F., Yang, H.-B., Han, Z.-M., Zhou, L.-C., Zhu, Y.-B., Ling, Z.-C., *et al.*, Lightweight, tough, and sustainable cellulose nanofiber-derived bulk structural materials with low thermal expansion coefficient. *Sci. Adv.*, 6, 18, eaaz1114, 2020.
56. Jiang, Z. and Ngai, T., Recent Advances in Chemically Modified Cellulose and Its Derivatives for Food Packaging Applications: A Review. *Polymers.*, 14, 8, 1533, 2022.
57. Mondal, S., Nanocellulose reinforced polymer nanocomposites for sustainable packaging of foods, cosmetics, and pharmaceuticals, in: *Sustainable nanocellulose and nanohydrogels from natural sources*, pp. 237–53, Elsevier, 2020.
58. Zhu, R., Liu, X., Song, P., Wang, M., Xu, F., Jiang, Y., *et al.*, An approach for reinforcement of paper with high strength and barrier properties *via* coating regenerated cellulose. *Carbohydr. Polym.*, 200, 100–5, 2018.
59. Khalil, H.A., Davoudpour, Y., Saurabh, C.K., Hossain, M.S., Adnan, A., Dungani, R., *et al.*, A review on nanocellulosic fibres as new material for sustainable packaging: Process and applications. *Renewable Sustainable Energy Rev.*, 64, 823–36, 2016.

Part III

CELLULOSE IN BIOMEDICAL FIELD

10

Targeted Drugs and Bioactive Compounds Delivery by Regenerated Cellulose

Avik Kumar Dhar[1]*, Vijay Mohakar[1], Sneha Ghosh[2], Pravalika Irukulla[3], Niloy Chowdhury[4], Piyas Halder[5], Ivan Moldavchuk[1] and Md. Shamsuzzaman[6]

[1]Department of Textiles, Merchandising, and Interiors, University of Georgia, Athens, GA, USA
[2]Department of Pharmaceutical and Biomedical Sciences, College of Pharmacy, University of Georgia, Georgia, USA
[3]School of Chemical, Materials, and Biomedical Engineering, College of Engineering, University of Georgia, Athens, GA, USA
[4]Department of Yarn Engineering, Textile Engineering College, Begumganj, Noakhali, Bangladesh
[5]Department of Textile Engineering, Shyamoli Textile Engineering College, Dhaka, Bangladesh
[6]Department of Textile Engineering, World University of Bangladesh, Uttara, Dhaka, Bangladesh

Abstract
Regenerative cellulose (RC) is a cellulose product obtained by precipitation of cellulose polymer from solutions containing cellulose-rich sources like wood pulp, jute, ramie, cotton, etc. RC has the advantages of being both natural and synthetic, as it is derived from abundant renewable biomaterial sources. Due to its biodegradability, biocompatibility, low toxicity, tunable properties, self-pore forming features, unique encapsulation properties, tailored solubility, and cost-effectiveness, it has gained immense research interest in biomedical and pharmaceutical applications. These applications include targeted drugs and bioactive compounds delivery, wound dressings, biosensors, scaffolds for tissue engineering, etc. This chapter reviews different forms of RC materials (such as micro-nano fibers, membranes, beads, films, hydrogels, capsules, scaffolds, aerogels, microspheres, nanoparticles, composites, etc.) in the targeted delivery of drugs and bioactive compounds.

**Corresponding author*: avik.kumar@uga.edu

Mohd Shabbir (ed.) Regenerated Cellulose: A Renewable Alternative to Petroleum-Based Plastics, (267–310)

The cellulose regeneration techniques, the fabrication process of different RC materials, their performance properties, and the factors affecting the performance were critically analyzed in this chapter. Moreover, RC nanocomposites with different nanofillers for their application in drugs and bioactive compound delivery, drug loading techniques, and release kinetics will also be reviewed. Finally, potential challenges and limitations with the prospectives of RC-based materials, leveraging the excellent properties and applications in practice will be provided.

***Keywords*:** Regenerated cellulose, drug and bioactive compounds, controlled release, solvent systems, shapes, loading mechanisms, release kinetics

Abbreviations

RC	Regenerated Cellulose
ORC	Oxidized Regenerated Cellulose
R-BC	Regenerated Bacterial Cellulose
NMMO	N-Methylmorpholine N-Oxide
DDS	Drug Delivery System (drugs and bioactive compounds)
DMAc	N, N-dimethylacetamide
LiCl	Lithium Chloride
DMSO	Dimethyl Sulfoxide
PF	Paraformaldehyde
ILs	Ionic liquids
OES	Organic electrolyte solutions
BmimCl	1-Butyl-3-methylimidazolium chloride
BmimAc	1-butyl-3-methylimidazolium acetate
AmimCl	1-allyl-3-methylimidazolium chloride
EmimAc	1-ethyl-3-methylimidazolium acetate
MCC	Microcrystalline cellulose
TEAOH	Tetraethylammonium hydroxide
TBAOH	Tetrabutylammonium hydroxide
DP	Degree of polymerization
RCMSs	Regenerated Cellulose Microstructures
NPs	Nanoparticles
MB	Methylene blue
APIs	Active Pharmaceutical Ingredients
SEM	Scanning Electron Microscopy
TEM	Transmission Electron Microscopy
CBs	Cellulose Beads
CC	Cellulose Carbamate
CS	Chitosan

CG	Carrageenan
DMOG	Dimethyloxallyl Glycine
CA	Cellulose Acetate
CD	Cyclodextrin
CNC	Cellulose Nanocrystals
CM-DEAEC	Carboxymethyl-diethyl amino ethyl cellulose
OCBs	Oxidized RC Beads
FNB	Fenofibrate
IND	Indomethacin
OPEFB	Oil Palm Empty Fruit Bunches
SSD	Silver Sulfadiazine
EDA-CD	Ethylenediamine-Cyclodextrin
DN	Double Network
IPN	Interpenetrating Networks
ECH	Epichlorohydrin
NIPAM	N-Isopropylacrylamide
AM	Acrylamide
LCST	Low Critical Solution Temperature
PEC	Polyelectrolyte Complexes
PPy	Polypyrrole
TCH	Tetracycline Hydrochloride
G	Gentamicin
RSP	Riboflavin Sodium Phosphate
LiHCl	Lidocaine Hydrochloride Monohydrate
Thp	Anhydrous theophylline
HAp	Hydroxyapatite
VRP	Verapamil Hydrochloride
CDP	Clindamycin 2-phosphate
IBU	Ibuprofen
CMF	Cellulose micro-/nanofiber
RNCC	Regenerated Nanocellulose Composite

10.1 Introduction

The dual threats of nonrenewable energy depletion and environmental contamination produced by petrochemical-based polymers stimulate the use of naturally occurring polymers for fabricating novel materials [1]. Cellulose is one of the most plentiful and renewable natural polymers which is composed of glucose-repeating units connected by β-(1,4)-glucosidic bonds with each other. It can be obtained from numerous sources, such as the cell walls of wood and plants, some types of bacteria and algae, and tunicates. This natural abundance offers the path for exploring new applications for their versatile properties [2–4]. Consequently, this natural polysaccharide has been largely exploited in a wide range of applications, including textiles, drugs & bioactive compounds delivery, wastewater treatment, construction, pharmaceutical, packaging, agricultural, pulp and paper industries [5].

Nonetheless, cellulose utilization is constrained due to its intricate structure, which includes a wide intra- and intermolecular network of H-bonds that serve as the foundation for cohesion between cellulose molecules. Intramolecular H-bonds are anticipated to provide chain stiffness, whereas intermolecular H-bonds allow linear polymer molecules to combine in a sheet-like shape. As a result of its stable structure, cellulose doesn't melt and is insoluble in nearly all organic and inorganic solvents. So far various novel solvents have been proposed to address this issue and make cellulose more adaptable for targeted modifications [6–8].

Regenerated cellulose (RC) is a class of materials produced by converting natural cellulose (cellulose I) to a soluble cellulosic derivative (through dissolution) and then regeneration (cellulose II). Although the regeneration process does not chemically change the polymer, it does change the polymorphic state (thermodynamically more stable cellulose II form) of the source materials which results in more amorphous regions (Figure 10.1) [9]. A variety of technologies and solvents have been used to fabricate RC-based fibers, films, beads, hydrogels, aerogels, and composite structures, suggesting that these materials may find us in packaging, textiles, wastewater treatment, optical & electrical devices, biomedicine, and other fields [10]. Because of their improved biocompatibility, biodegradability, successful drug loading capacity, stability of matrices, controlled release properties, cost-effectiveness, functionalization potential, tunable properties, and diverse shapes (membrane/film, capsules, matrices, beads, hydrogels, micro/nanostructures, fibers, scaffold, and composites), RC-based materials have garnered a lot of research attention among various application areas, especially in the biomedical sectors as a carrier of drugs and bioactive compound molecules (protein, peptides, antibodies, enzymes, hormones, vaccines, growth factors, etc.) [11–14].

Therefore, this chapter focuses on the recent progress in the fabrication of RC-based materials for drug and bioactive compound delivery which includes the regeneration techniques, solvent system, structure-performance relationship, and factors affecting the drug loading and controlled release of the compounds. Furthermore, the mechanism of drug loading, stability, and release criteria in the administration pathway

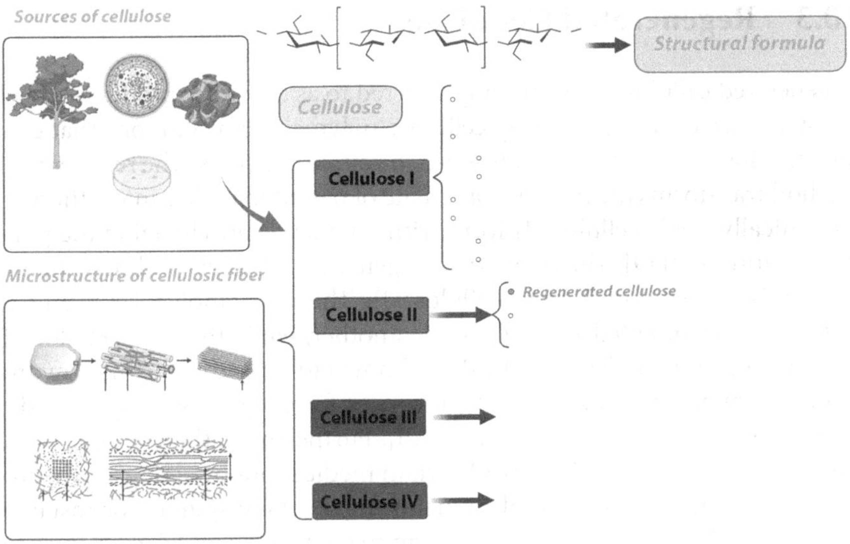

(oral, ocular, gastrointestinal, etc.) was comprehensively analyzed. Furthermore, the critical focus was to provide insight into future research and encourage future applications.

10.2 Cellulose

One of the most prevalent polysaccharides on Earth is cellulose, a semi-crystalline organic natural polymer having the formula $(C_6H_{10}O_5)_n$. In addition, it is the main structural element of the cell wall in oomycetes, algae, and plants. Interestingly, some bacterial strains are renowned for producing microbial cellulose, which is 20-100 nm in size [15, 16]. The crystallinity of cellulose varies according to its origin, extraction process, and preparation. For instance, cellulose derived from plants and wood usually has a degree of crystallinity between 40% and 60%, but cellulose from other sources (e.g., bacteria and tunicin) exhibits a higher degree of crystallinity ranging from 80% and 100% [17–19]. Because of its highly crystalline structure, wide availability, exceptional biodegradability, remarkable biocompatibility, and good mechanical qualities, tunable properties (due to the presence of extremely accessible and easily reactive –OH groups) cellulose has garnered a lot of interest from researchers looking at developing a variety of useful materials [20–22].

10.3 Regenerated Cellulose

Regenerated cellulose is commonly referred to as cellulosic products, which are synthesized by precipitating cellulose polymer from solutions that contain cellulose derivatives or cellulose-rich source materials. The regeneration method transforms the polymorphic state of parent materials into a thermodynamically stable cellulose II form without chemically changing the polymer (Figure 10.2) [9]. The H-bond arrangement in their crystalline structure distinguishes cellulose I from cellulose II. The macromolecular chains in cellulose I are oriented parallel to one another, while those in cellulose II are oriented antiparallel [23, 24]. RC is known to be one of the most hydrophilic polymers and indeed excellent carrier for drug delivery systems due to their controlled swelling behavior [25]. Furthermore, the presence of RC can increase the water solubility of certain medications by acting as a hydrophilic drug carrier. For example, a nano-microsphere system, for instance, was designed with cellulose and chitosan to control the body's ongoing production of the fat-soluble anti-inflammatory drug ketoprofen [26].

10.3.1 Oxidized RC

Oxidized RC (ORC) has been commercially available since the 1960s, and it has proven to be an effective hemostatic assistance in a variety of biomedical fields [27]. ORC with aldehyde groups has strong hemostatic and antibacterial capabilities, making it appropriate for use in sterile wound dressings with biocompatible features [28, 29]. But because of its high

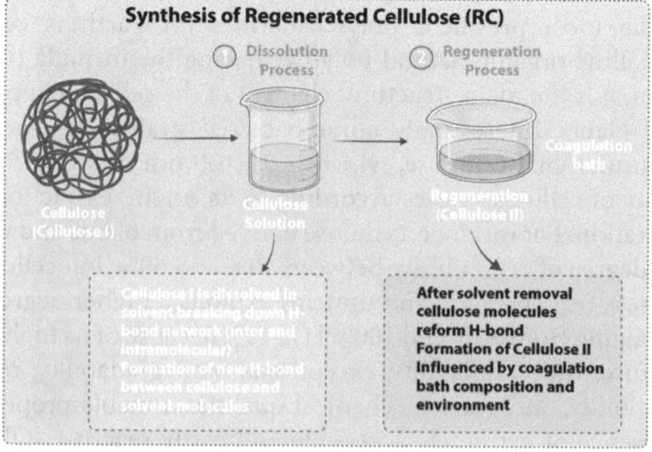

levels of crystallinity, natural cellulose finds it challenging to come into contact with the oxidant during the oxidation process [30]. A combination of $NaBH_4$-HCl formation and $NaIO_4$ oxidation can be used to further improve the oxidation of cellulose. The $NaBH_4$-HCl swelling and then the $NaIO_4$ oxidation pathway allow the preparation of ORC with a higher degree of oxidation. This technique uses diluted HCl and H_2 generated by $NaBH_4$ to cause cellulose to swell during regeneration, enhancing the specific surface area and porosity of cellulose structure. However, the degree of oxidation and ORC performance can be affected by several oxidation conditions, such as oxidation time, pH, and concentration of $NaIO_4$ solution [31].

10.4 Controlled Drug and Bioactive Compound Delivery

To trigger the intended therapeutic effect, the drug delivery system makes it possible for the drugs and active pharmaceutical components to be released. Tablets, capsules, syrups, ointments, and other traditional methods have poor bioavailability, fluctuate in plasma drug levels, and cannot provide sustained release. The entire therapy procedure may be ineffective in the absence of an effective delivery system. To attain the highest level of safety and efficacy, the medication must also be administered at a precisely controlled rate and the intended location. The development of controlled drug delivery systems aims to address the issues with traditional approaches [32].

Controlled drug delivery is an effective method for eventually increasing local therapeutic dosages, especially in hard-to-reach places [33]. Controlling the duration and quantity of drug release, delivering the drug to specific parts of the body, and overcoming cellular and tissue barriers are the ultimate objectives of this system. This makes it possible to maximize the therapeutic's effectiveness and minimize any potential negative effects (Figure 10.3) [34, 35].

The following procedures are typically involved in drug-release mechanisms from polymer matrices-

(i) Water entering the complex polymer matrices,
(ii) Swelling (matrices), and
(iii) Drug dissolution and diffusion as it moves outside the matrix

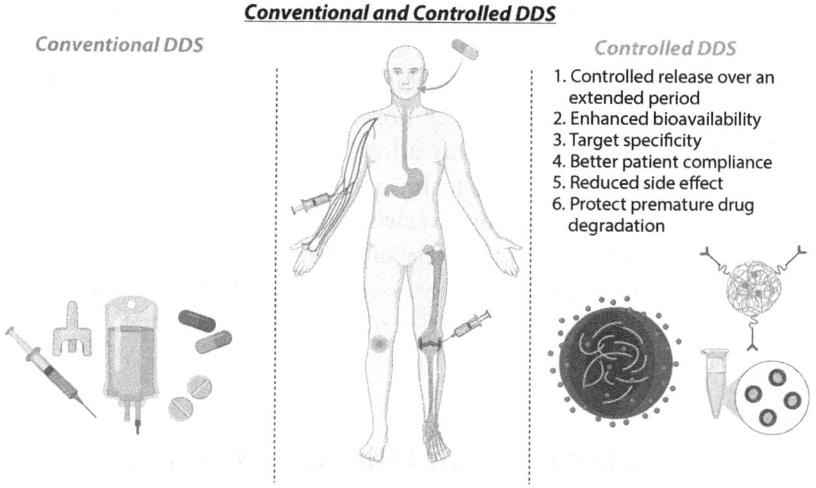

As a result, the degree of swelling and porosity are crucial factors in the drug release mechanism from the matrices. The water-soluble drug particles break quickly and create a large number of pores that allow other drug molecules to enter the release medium, accelerating the release rate [36]. In recent work, regenerated bacterial cellulose (R-BC) matrices were prepared using 50% NMMO (w/w) and loaded with famotidine and tizanidine (model drug). The drug matrices exhibited both chemical and thermal stability. Furthermore, 90% of the loaded drugs were released in the first half hour of the dissolution experiment (*in-vitro* drug release study) [12].

10.5 Solvent Systems for Cellulose Regeneration

Even though cellulose has several benefits over conventionally synthesized petroleum-based polymers, its full potential in its chemically unaltered state is not currently being fully utilized. Unlike thermoplastics, which are frequently utilized, cellulose is insoluble in the majority of common solvents and cannot be melted without structural degradation [7]. Disrupting the several intra- and intermolecular H-bond networks between the cellulose chains is the main obstacle here. Thus far, a few solvent systems have been developed, including such as $Cu(NH_3)_4(OH)_2$ (cupro process), NaOH and CS_2 (viscose process), NMMO (lyocell process), DMAc/LiCl, DMSO and PF solvent system, and some ionic liquids (ILs) [7, 14, 37–39].

However, before regeneration, the cellulose sources need to be purified to remove impurities (dirt, lignin, and hemicelluloses) [13, 40, 41]. Furthermore, the purification process (10% w/w at 90°C for 15 h) can reduce the molecular weight of the polymer [14]. Among different solvents, 7-9 wt% NaOH aqueous solution is one of the cheapest cellulose solvents in which cellulose can be dissolved near 4°C since intermolecular H-bonds are destroyed [42, 43]. Consequently, the viscose process became the most important method of regenerating & shaping cellulose where a soluble intermediate is formed by treating the cellulose source with alkali and carbon disulfide (CS_2). Finally, cellulose can be regenerated in an H_2SO_4/salt solution known as a coagulation bath [44]. However, because CS_2 is expensive, hazardous, and flammable, the viscose sector must use increasingly efficient methods to recover CS_2 and H_2S, a process byproduct [45].

The Lyocell technique is an additional method of cellulose regeneration where cellulose is dissolved immediately in NMMO monohydrate (contains approximately 13-23% water) at 80°-120°C. However, because NMMO is susceptible to thermal deterioration and can form homolytic and heterolytic cleavages at high temperatures (during dissolution and solvent recycling procedures), the Lyocell process is costly and unstable. In addition to causing the solvent to evaporate, this also degrades the cellulose and impairs the functionality of the product [46–48]. However, careful solvent selection is mandatory for dissolving & optimizing the properties of RC-based drug delivery carriers. So far, ILs, organic electrolyte solutions (OES), alkali/urea systems (organic/inorganic alkali), and acid hydrolysis procedures have been used to dissolve cellulose to fabricate drug-loading carriers.

10.5.1 Ionic Liquid System

Recently, room temperature ILs as new *green* solvents (BmimCl, BmimAc, AmimCl, EmimAc) have been widespread concern in the public due to their various advantages (Figure 10.4). Furthermore, it has also been reported that ILs exhibit outstanding dissolving capability for cellulose [41, 49–51]. For example, environment-friendly RC films were fabricated dissolving cellulose sources (cotton, pine, bamboo, and MCC) in EmimAc for their applications in food packaging and medical materials [40]. Voon *et al.* dissolved cellulose fibers (0.4 g) in AmimCl (10.0 g), preheated for 0.5 h at 80°C and stirred for 20 min at 500 rpm to obtain cellulose solution (4 wt%) to fabricate porous RC bead for potential drug delivery applications [13]. However, high production cost and processing difficulties (for high viscosities of ILs) during cellulose dissolution limit their wide range

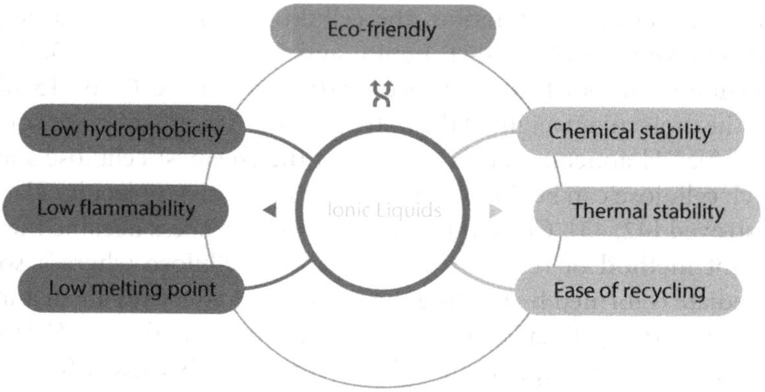

of applications. Furthermore, depending on their structure some ILs can be toxic to aquatic life and have slow biodegradation rates [52]. Table 10.1 summarizes the recent research on cellulose dissolution for fabricating drug-loading carriers.

10.5.2 Organic Electrolyte Solution

Organic electrolyte solution (OES) composed of a polar aprotic solvent and ILs have been studied recently to dissolve cellulose [53]. When compared to simple ionic liquids, OES solutions offer several benefits, such as reduced viscosity, high dissolution capacity at low temperatures, and adjustable physicochemical characteristics [54, 55]. Therefore, OES systems are used to fabricate RC film with excellent mechanical properties with minimum cellulose degradation [56]. Cheng *et al.* used an OES/ILs system (EmimAc and DMSO) to dissolve cotton linter at room temperature (Figure 10.5). Interestingly, the OES system (EmimAc molar fraction 0.2 to 0.4) rapidly dissolved cotton linters at room temperature, whereas ILs system (EmimAc control) required 80°C [57].

10.5.3 Inorganic Alkali Urea System

A new *green* cellulose solvent system known as the alkali (NaOH, LiOH)/urea/H_2O has attracted increasing interest. In this system, only very cheap chemicals, such as alkali and urea are adopted. Most significantly, it is regarded as a genuinely *green* cellulose-dissolving approach because no harmful compounds are employed or produced during the dissolving process [58, 59]. In aqueous alkali solutions, the solubility of cellulose is greatly enhanced by the addition of 2 to 4 wt% urea, and moderate urea is

Table 10.1 Recent research on cellulose dissolution in ILs for fabricating drug-loading carriers.

Cellulose sources	IL system	Concentration	References
Cotton linter, MCC, and raw cellulose (pine and bamboo)	EmimAc	Cellulose samples (5 g) in EmimAc (100 g), magnetic stirring for 0.5 h at 80°C	[40]
Paper wastes	AmimCl	4 wt% cellulose solution was prepared dissolving 0.4 g cellulose fibers in 10 g preheated AmimCl for 0.5 h at 80°C	[13]
Borassus fruit fibers	AmimCl	2 g cellulose was dissolved in 100 g AmimCl at 80°C for 120 min in an inert atmosphere and then degassed	[41]
Pine kraft pulp	BmimCl	2% and 3% (w/v) cellulose solutions were prepared dissolving cellulose in BmimCl for 1 h at 90°C (40-Hz mechanical agitation and vacuum conditions)	[51]
Denim waste	BmimCl BmimAc	100°C with magnetic stirring	[14]

important for both enhancing solubility and preventing the development of cellulose gel. The mechanism involved here NaOH swells the amorphous regions of cellulose and urea penetrates the cellulose structure and breaks the intermolecular H-bonds loading to enhance water solubility to form new H-bonds [43]. Cheng *et al.* used the freezing-thawing process for dissolving cellulose in LiOH/urea solution (4.6 wt%/15.0 wt%). After thawing at room temperature cellulose solution was centrifuged to eliminate the bubbles in the viscous solution [60].

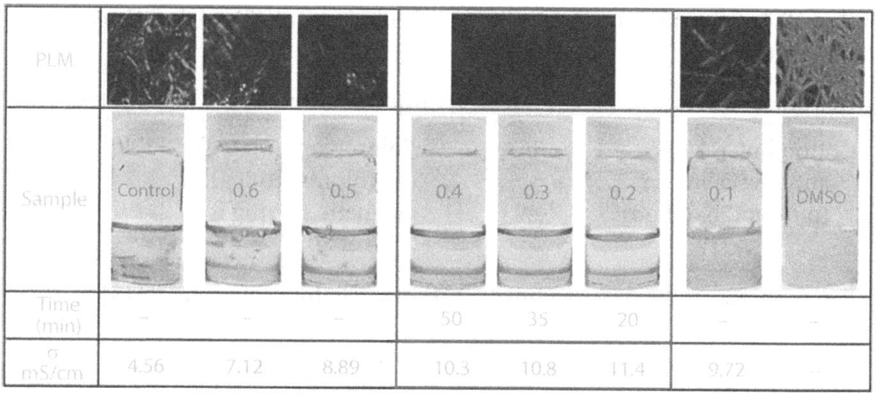

However, instead of urea, thiourea can also be used for cellulose dissolution. The thiourea-based system generally offers superior dissolution performance and allows for greater control over the properties of the RC materials [61–63]. The inability to dissolve high molecular weight cellulose is a drawback of thiourea-based alkali solvents, despite their superior cellulose solubility compared to alkali/urea solvents. Thus, to reduce molecular weight and eliminate hemicellulose and other contaminants, pretreatment techniques such as acid hydrolysis and hot-water treatment are employed [61, 64, 65].

10.5.4 Organic Alkali Urea System

Organic alkali/urea aqueous solvents (TBAOH, TEAOH) can readily dissolve cellulose at room temperature. Consequently, because of its lower cost, decreased viscosity, and simpler processing, its industrial-scale use is being further expanded [66, 67]. Urea is added to organic alkali to boost solubility by increasing the mixing period and decreasing solvent entropy losses during dissolution [68]. Xu *et al.* used a TEAOH/urea/H_2O system to dissolve hardwood bleached kraft pulp (HwBKP, Brazil). After 5 minutes, cellulose fibers swelled and dissociated rapidly, resulting in partial disintegration. After 60 minutes, the cellulose was almost totally dissolved, with only a few air bubbles and undissolved particles remaining. A clear cellulose solution was then obtained by centrifuging the air bubbles and undissolved material (Figure 10.6). The dissolved fraction was up to 99.6%, indicating excellent solubility in the TEAOH/urea/H_2O system [11].

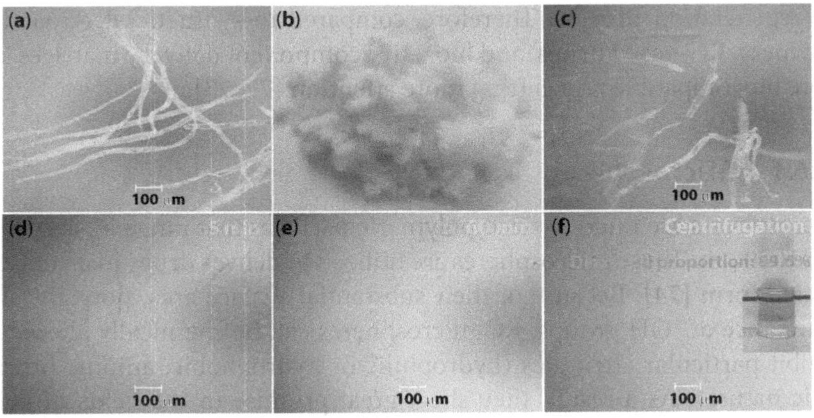

10.5.5 Acid Hydrolysis

Acid hydrolysis is carried out using dilute or concentrated inorganic acid. Among all, H_2SO_4 and HCL are mostly used for hydrolysis of lignocellulosic biomass [69]. Low-temperature H_2SO_4 treatment decreases cellulose DP very quickly by breaking ether linkage between glucose molecules at the primary stage. It begins to dissolve the cellulose to produce a stable cellulose solution as soon as the cellulose DP drops to a level low enough for dissolution. Consequently, it lowers the mechanical performance of RC [61]. In recent work, H_2SO_4 was added dropwise in the mixture of distilled water and lignocellulosic mixture to form a viscous solution. Then after ultrasonication, RC was precipitated in cold deionized water. The mixture was centrifuged to remove excess acid solution and dialyzed to maintain pH 7. Finally, the suspension was freeze-dried for further use (shaping in different forms) [70]. Although acid hydrolysis accomplishes the goal of decrystallization, large-scale usage of acid causes significant equipment deterioration and wastage of acid residue [71].

10.6 Form of RC for DDS

Because of its special processing abilities and structural characteristics, RC may be designed into a variety of shapes. Furthermore, it is feasible to develop RC products with precise shapes and characteristics for particular uses by carefully choosing the dissolving solvent and managing the regeneration process. Therefore, compared to synthetic electroactive polymers, RC-based drugs, and bioactive component delivery matrices are more biocompatible and getting more attention [72, 73].

10.6.1 Microspheres

Microspheres are micron-sized polymeric particles that range in size from 1 to 1000 μm. These microspheres are utilized to deliver drugs in an encapsulated form [74]. Because of their substantial surface area, porosity, and abundance of -OH groups, RC microspheres can be chemically altered to exhibit particular attributes (hydrophilic or hydrophobic, anionic or cationic nature). As a result, they show great promise in the fields of plastic additives, water purification, and drug delivery [75, 76]. Drug release can be controlled with the microsphere system, despite its lack of convenience compared to the nanoscale system. Additionally, by lowering the frequency of doses, microspheres can increase patient compliance [77, 78]. In a recent work, a simple emulsion–coagulation–oven-drying process was used to fabricate homogenous and spherical porous RCMSs for a variety of applications, including medication administration, electrolysis, biocatalysis, and the purification of water and proteins. This method involves dissolving cellulose and then regenerating it (cellulose–TEAOH/urea/H_2O solution). The mean size of the porous RCMSs can be adjusted between 280 and 103 μm. Furthermore, the RCMSs with cellulose II structure are distinguished by their good thermal stability and physical characteristics. Notably, the synthesis method is simple and environmentally benign [11].

10.6.2 Nanoparticles

Nanoparticles (NPs) are highly efficient at transporting macromolecules through the nasal, oral, tracheal, and ocular epithelium [79]. Because of their unique qualities, RC NPs have a lot of promise for use in a variety of industries, including food, electronics, personal care, and medicines. Cellulose can be hydrolyzed with concentrated H_2SO_4 for forming RC NPs, with average particle diameters between 30 and 120 nm. However, H_2SO_4 requires specific processing reactors because it is a powerful acid and a highly corrosive reagent. For the large-scale fabrication process, the synthesis method is therefore not promising. Additionally, the sizes of the cellulose NPs cannot be precisely controlled using this process [80, 81].

Another approach to fabricating RC NPs is to disperse RC (5 gL^{-1}) in deionized water through ultrasonication. Dispersed RC loses its crystallinity during regeneration (NaOH/urea system) and dispersion step.

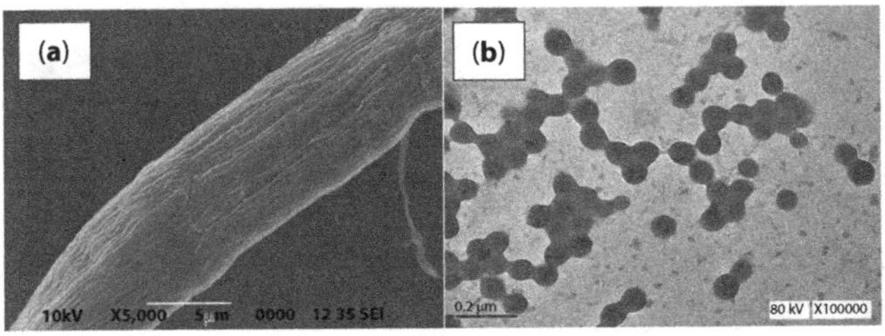

The dispersed cellulose may synthesize translucent, free-standing films and is a better material for drug administration and reinforcement in physiological environments [82]. Furthermore, in a recent study, spherically shaped RC NPs were fabricated from cotton fibers by a simple nanoprecipitation method. This method relies on the mechanism of precipitation where cellulose solution (1 mL) was added dropwise in ethanol (20 mL) under sonication. The average particle diameter of NPs was controlled by modulating the synthesis conditions such as concentration of cellulose solution and solvent/non-solvent volume ratio. Furthermore, methylene blue (MB) was loaded (as a model hydrophilic drug) onto cellulose NPs for their potential applications in drug delivery systems (Figure 10.7) [63].

10.6.3 Capsules

RC capsules exhibit potential as a controlled-release, biodegradable drug delivery method. Recently, a customized dip-coating process was used for fabricating RC capsules. The final capsule shell (Figure 10.8a) had a form similar to the halves of conventionally sized gelatin capsules. In the dry state following the last fabrication step, the RC membrane appeared monolithic, according to the microscopic evaluation of the membrane. However, when the membrane was exposed to an aqueous medium, the intruding water caused the polymer to swell, which resulted in the self-formation of porous routes for drug transport across them (Figures 10.8b and c). Potassium chloride, diphenhydramine hydrochloride, tramadol hydrochloride, niacinamide, acetaminophen, and ketoprofen were used as model APIs to assess the drug release properties from RC capsules. Compared to traditional encapsulated methods, RC capsules were more appropriate for osmotic drug delivery applications [38, 83–85].

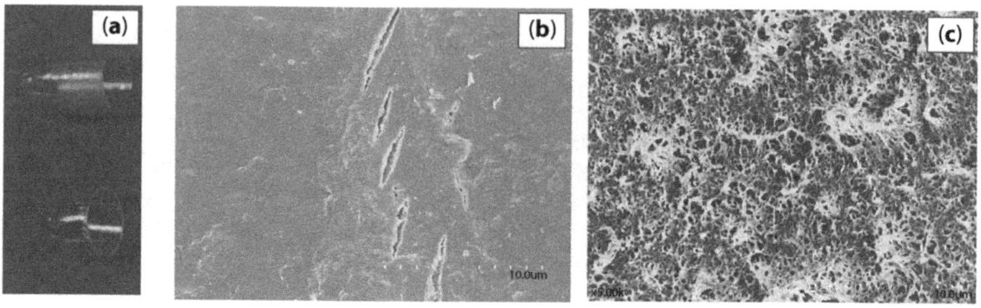

10.6.4 Beads

Beads are tiny, round particles that are used in many different fields of science and medicine. They could be anywhere from nanometers to millimeters in size. Beads can interact with biological materials, cells, or tissues because of their unique surface chemistry or functional structures. Their main roles include increasing cell separation, aiding in drug delivery, and serving as parts of scaffolds for tissue engineering or biosensors [86]. The biocompatibility of cellulose combined with its high surface area, porosity, and low density makes RC beads very attractive for usage in biomedical applications, such as drug molecule carriers for oral delivery [87]. RC beads can incorporate pharmacological compounds due to their porous structure. Furthermore, the CBs' size and shape may provide several technological advantages over conventional powders, such as increased flowability, less dust during processing, and particle size consistency. Furthermore, due to their micro-porous nature, RC beads can be loaded to contain small amounts of medication for use in low-dose products [88].

RC beads can be manufactured by dissolving cellulose in a solvent mixture and preparing the beads by coagulation of dissolved cellulose in ultrapure water or acid solution. However, the increase in coagulation bath temperature and molarity causes a rise in the volume and weight of water-swollen beads. Also, porosity changes little, since the volume and weight of the beads increase, simultaneously [13, 89]. Recently, RC of paper wastes was used for fabricating spherical RC beads with customized mean diameters (20–80 μm) and nanoscale (40–200 nm) ranges using water-in-oil microemulsion and nanoprecipitation procedures, respectively. By adjusting manufacturing parameters (stirring speed, surfactant & cellulose concentration, and reaction temperature) the mean diameter ranges of RC beads were precisely monitored [90].

10.6.4.1 *Effect of Nanoparticle Incorporation*

The incorporation of NPs during bead formation is a straightforward approach for tuning the mechanical, physical, and swelling behavior of RC beads, expanding their potential for numerous purposes such as separations and controlled release. RC beads, synthesized by a simple one-step *in situ* modification using CNC NPs and dropwise precipitation of cellulose/DMAc + LiCl solution resulted in a compacted structure with a rougher surface when the precipitation media contained higher concentrations of CNC. Nevertheless, the addition of CNCs considerably decreased the beads' ability to retain water [91].

10.6.4.2 *Smart Beads*

In the realm of stimulus-responsive drug release, smart beads with low costs, configurable release rates, and good biocompatibility have demonstrated considerable promise [92]. Anionic functional groups were introduced in RC bead by the TEMPO/NaClO/$NaClO_2$ mediated oxidation process (60°C, 5h) for pH-responsive drug delivery in the gastrointestinal tract. The swelling degree of the dried porous bead (OCBs) was regulated by (i) the hydrophilicity and (ii) the electrostatic repulsion between the RCOO- groups. However, OCBs become more hydrophilic as the degree of oxidation increases. Furthermore, higher electrostatic repulsion promoted water absorption into the beads and improved swelling capacity (Figure 10.9). Compared to unoxidized cellulose beads, OCBs swelled by 276 to 521%. The drug release trials with FNB & IND demonstrated a pH-responsive release ability of OCBs [93].

10.6.4.3 *Surface Coating*

Another method for developing RC beads is to coat hydrogel beads with RC. Recently, the chitosan (CS) hydrogel beads were coated with cellulose dissolved in EmimAc using a one-pot one-step procedure. Water molecules in the CS beads diffused outward when interacting with the cellulose-IL mixture, acting as an anti-solvent. This enabled the surface of the beads to be coated with RC (Figure 10.10). When the VRP (model drug)-impregnated beads were immersed in simulated stomach fluid (pH 1.2), the drug was released in a nearly perfect linear pattern which demonstrated the possibility of application of RC-coated CS beads as carriers for drug release control [94].

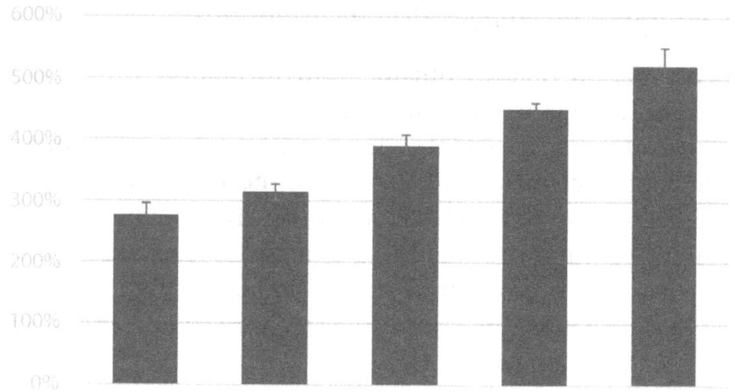

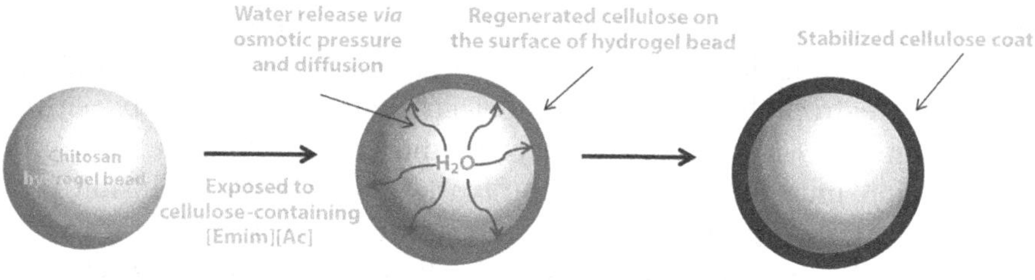

10.6.5 Hydrogel

Hydrogel is a three-dimensional structure of hydrophilic polymers that can hold and absorb huge amounts of bodily fluids or water. Because of their unusual mix of solid-like mechanical qualities and fluid retention, hydrogels are frequently employed in biomedical applications, particularly in drug delivery networks, wound dressings, tissue engineering scaffolds, and contact lenses [95, 96]. Furthermore, the spontaneous reaction to stimuli exerted on them makes hydrogel an appealing approach to overcome the limits of conventional drug delivery systems [97]. RC hydrogels are biocompatible and provide high oxygen permeability and low endotoxin levels. The results suggest that as cellulose concentration increases, the oxygen permeability decreases. Furthermore, the RC hydrogel can be modified by incorporating other natural-derived polysaccharides to increase water-holding capacity, tailorable physical properties, and non-toxic nature demonstrating a significant promise for use as a corneal bandage immediately following ocular trauma [39].

10.6.5.1 Composite Hydrogel

Composite hydrogels are innovative drug delivery systems that combine multiple components to enhance their properties and performance [98]. A thermo-responsive composite RC hydrogel was synthesized by incorporating OPEFB-extracted cellulose (3 w/v%) and Pluronic F127 polymer (20 w/v%). The composite hydrogel demonstrated high storage modulus (81 kPa), complex viscosity (9.6 kPa.s), excellent swelling ratio (4.22±0.70), and low degradation rate (31.3±5.9%), in addition to sustained release of SSD *in-vitro* drug [97]. However, purification dissolution, and extraction of cellulose are always challenging for their efficient use in drug delivery systems. Therefore, there is a great demand for easily dissolvable derivatives of cellulose. Among different cellulose derivatives, cellulose carbamate (CC) is readily dissolvable in numerous alkaline solutions and organic solvents [99]. Furthermore, CC can be successfully synthesized under hydrothermal conditions without any catalyst and organic solvent to fabricate RC products (hydrogels and aerogels) for drug delivery systems [59].

10.6.5.2 Oxidized Hydrogel

Recently, using ORC to fabricate a hydrogel system for wound healing has a fast-growing market. Sun *et al.* developed a multifunctional wound dressing with improved antibacterial characteristics and wound healing enhancement using ORC. The synthesis process included preparing ORC using a modified approach, synthesizing CS/Ag NPs using an *in-situ* reduction method, and finally preparing ORC/CS/Lys@Ag NPs hydrogels. This innovative approach enables homogeneous dispersion and minimum agglomeration of Ag NPs within the hydrogel, hence improving bacterial inhibition. Concurrently, the interaction between CS and Lys with the aldehyde group of ORC resulted in the formation of a Schiff base hydrogel. Importantly, the Schiff base bond between ORC and Lys, favored by the wound's weak acidity, facilitates the controlled release (degrade by 11.05 % ± 0.43 within 8 h) of Lys at the wound site, thereby enhancing the wound healing process (Figure 10.11) [29].

Nevertheless, increasing the level of oxidation of RC and fabricating matrices that are specifically sensitive to the wound environment are never effortless tasks. Herein, a highly oxidized aldehyde-based ORC was

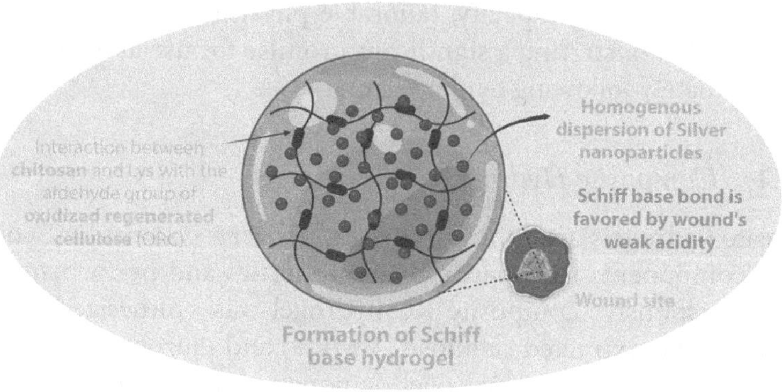

prepared by swelling ($NaBH_4$-HCl) and oxidation ($NaIO_4$) pathway. Later, CS and EDA-CD were incorporated with ORC to design pH-responsive hydrogel matrices capable of controlled drug release (ZnO NPs and IBU) for antimicrobial wound dressings [31].

10.6.5.3 Dual Network Composite Hydrogels

Double network (DN) hydrogel formation is a recent development in tough hydrogels. It is composed of two interpenetrating networks (IPN) with different structures: (i) a brittle, densely cross-linked network of low concentration and (ii) a ductile, lightly cross-linked network of high concentration [100, 101]. The exceptional mechanical qualities of DN hydrogels, such as their high-water content and impressive toughness and failure resistance, are among their most intriguing characteristics. Because of these benefits, they are more suitable for developing innovative drug delivery systems [102, 103]. The chemical cross-linking method (ECH) can be used to synthesize DN hydrogel, which is comprised of the CMC network and the copolymer networks of NIPAM and AM. Temperature responsiveness was added by NIPAM, which makes it easier to use for smart drug administration. Furthermore, the copolymerization of AM and NIPAM modulates the ratio of hydrophilic to hydrophobic groups in the hydrogel system, controlling its LCST and bringing it closer to the body's normal physiological temperature. The hydrogel exhibits good biocompatibility by *in vitro* cell compatibility test and demonstrates responsiveness to both the pH environment of the small intestine and the physiological temperature, thus it may be employed as the carrier of tetracycline for precisely treating intestinal bacterial infection [104].

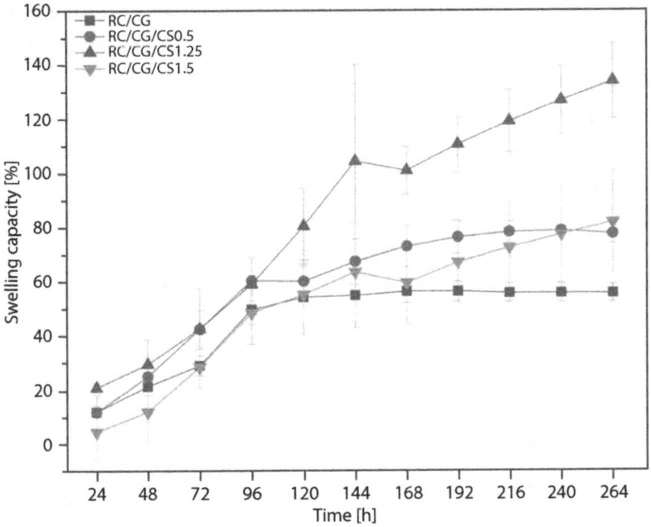

Figure 10.12 Swelling capacity of RC/CG, RC/CG/CS0.5, RC/CG/CS1.25, and RC/CG/CS1.5 in distilled water. OPEFBC concentration – 3 wt%, CG concentration – 2 wt%, ECH concentration – 10 wt%. Reproduced with permission from [107].

10.6.5.4 PEC Hydrogels

Polyelectrolyte complexes (PEC) serve as the electrostatic counterpart of anionic and cationic polyelectrolytes. The diverse applications afforded by PEC in a wide range of sectors such as biomedical have gained notable interest in recent years [105]. Furthermore, PEC hydrogel serves as a stimuli-responsive hydrogel and may be classified into two categories based on the physical and chemical stimuli triggers [106]. An ampholytic polyelectrolyte RC-hydrogel was fabricated *via* a facile chemical crosslinking technique for pH-responsive drug delivery systems. A homogeneous mixture of RC/CS/CG was obtained in the NaOH/urea solvent system without phase separation between the anionic and cationic polymer mixture. ECH was added to crosslink the mixture. However, the incorporation of CS imparted the highest swelling capacity and gel fraction (104.30% and 97.93% respectively) in the hydrogel matrix. Furthermore, the ampholytic RC/CG/CS hydrogels had commendable water uptake in low pH due to the macromolecular chain extension of CS (Figure 10.12) [107].

10.6.6 Nanofiber Mat

A novel technique called electrospinning uses electric force to pull charged fibers from polymer solutions to formulate nanofiber mats (2D structures).

This method develops nanofibers with favorable characteristics like high porosity, large surface areas, and bionic nanofiber structure [108, 109].

Additionally, a variety of drugs and bioactive substances can be transported using nanofiber mats [110]. For example, a dual-functional nanofiber mat was fabricated through an electrospinning technique. Initially, DMOG (wound healing drug) was packaged with β-CD and then combined with CA to prepare the spinning solution. Ag-NPs were directly synthesized on the drug-loaded CA nanofiber surface through an *in-situ* synthesis technique. CA and CA/β-CD/DMOG nanofibers had a diameter of 40-120 nm, while CA/DMOG/Ag-NPs nanofibers had diameters of 120-200 nm [111].

10.6.7 Composite Scaffold

Polymer scaffolds (3D structures) are implants that are frequently utilized to transfer cells, medicines, and genes into the body. Their regular porous structure provides the necessary support for cell adhesion, proliferation, differentiated function, and migration [112]. Among various techniques. Electrospinning and chemical crosslinking are generally established procedures for fabricating RC-based composite scaffolds for drug and bioactive compound delivery systems. For instance, electrospun cellulose acetate containing modified CNC (*via* physical adsorption of polyamidoamine) can perfectly control the release rate of tranexamic acid and improve the ultimate release amount of the gatifloxacin without needing any toxic reagent and chemical reaction [113]. In another study, a novel CM-DEAEC scaffold was synthesized through a chemical cross-linking method (2% ECH- v/v) and loaded with curcumin as a model drug (an important chemotherapeutic agent). A plethora of functional groups (–COOH and $-NH_2$) was incorporated into cellulose to make it more versatile as a non-toxic, antibacterial, and antifungal scaffold [114].

10.6.8 Film and Membrane

RC composite films provide a versatile and promising substrate for drug administration and can be designed using solvent casting, melt film extrusion, or layer-by-layer deposition processes [57, 70, 115, 116]. Furthermore, technologies for the controlled distribution of antibiotics have been designed that can lower drug dosages, thus reducing side effects and difficulties. The solution casting method was used to fabricate RC composite membranes loaded with antibiotic drugs, TCH (RC-TCH).

Table 10.2 Recent works on the development of RC film/membrane for drug delivery system.

Film/membrane	Method of synthesis	Potential applications	Refs.
TCH loaded RC	Solution casting	Wound dressing & medical uses	[117]
RC/polypyrrole/Ag NPs/IL	*In situ* chemical oxidation polymerization	Wound Healing	[119]
RC/collagen	Periodate oxidation, followed by collagen casting *via* Schiff base reaction	Tissue engineering scaffold	[60]
RC/poly(globalide)	Layer-by-layer casting	Skin tissue engineering	[70]
RC/citric acid	Solution casting	Food packaging & drug delivery	[116]
RC/zeolite	Solution casting	Biodegradable packaging & biomedical application	[121]
PLA/hydroxyapatite, and two types of RC (particulate and fibrous)	Melt extrusion	Biomaterial & tissue engineering	[115]
RC/Polyacrylic acid composite film	Free radical polymerization of polyacrylic acid and film formed in Petri plates	Drug release, antibacterial & antifungal wound dressing	[122]

The RC-TCH composites, more significantly, demonstrated good biocompatibility, and outstanding antibacterial activity, validating their prospective use in wound dressings and other medical applications [117].

Researchers are attempting to design conductive RC materials (magnetic and electric) for triggered drug delivery system [118, 119]. Furthermore, RC has a strong affinity to organic conductive polypyrrole (PPy) [120].

For instance, RC/γ-Fe_2O_3/PPy composite film was designed in a recent study, where γ-Fe_2O_3 were dispersed in RC to fabricate casted films. Then, RC/γ-Fe_2O_3 matrices were coated by PPy and doped with sulfosalicylic acid (an anti-inflammatory drug) by chemical synthesis process. The matrices showed a fast release of drug (> 60% within 1 h) for use in emergency care [73]. Table 10.2 summarizes recent work on the development of RC film/membrane for controlled drug delivery systems.

10.6.9 Biocomposites

To achieve an antimicrobial activity and to prevent wound infection NPs can be incorporated with RC to fabricate composite structures. For example, AgNPs were impregnated into RC-CS composite in a recent study. The composites containing RC and CS impregnated with AgNP with and without gentamicin (G) antibiotic were synthesized by the following procedure where 90 ml distilled water, 9.5 g of NaOH, and 4.5 g of Thiourea were dissolved and pre-cooled to −5°C and cellulose was dissolved in this solution. 2% CS solution was added and finally, the pH of the solutions was brought to 7.0. The resultant RC–CS composites were dried at 30°C. Later RC-CS composite was incorporated with AgNPs and gentamicin to fabricate RC–CS–Ag and RC–CS–Ag–G composites for drug delivery and wound healing [62].

10.7 Drug Loading Mechanism

Drug loading into the polymer matrices is the crucial step for a controlled drug delivery system. It is the process of incorporation of a drug into a carrier system. While dealing with drug loading system, two common terms are utilized in the literature [123]. These are loading capacity (in mg/g) and loading efficiency (in %) and they are defined as -

$$Loading\ capacity = \frac{Mass\ of\ drug\ onto\ the\ carrier}{Mass\ of\ the\ carrier}$$

$$Loading\ efficiency = \frac{Mass\ of\ drug\ onto\ the\ carrier}{Mass\ of\ the\ drug\ initially\ applied} \times 100\%$$

Table 10.3 Recent research on the drug loading mechanisms and maximum loading capacity of RC-based drug & bioactive compounds carrier.

Carrier materials	**Drug**	**Drug loading mechanism**	**Loading capacity (%) or mg**	**Refs.**
RC beads	Thp RSP and LiHCl	Absorption	Thp – 0.072 mg LiHCl- 0.396 mg RSP- 0.207 mg	[89]
ORC/CS/Lys@Ag NPs hydrogels	Ag and Lys	*In situ* loading	Lys-12%	[29]
CA/DMOG/Ag-NPs nanofibers	DMOG/β-CD inclusion complexes	*In situ* loading	0.175 g	[111]
TEMPO-oxidized RC beads	FNB and IND	Incipient wetness	20%	[93]
RC nanofibers	HAp and Ag NPs	*In situ* reduction	HAp-0.5 wt%,1 wt%,1.5wt%. Ag NP-3%,5%,7%	[132]
Thermo-responsive RC hydrogel	SSD	Absorption	100 mg	[97]
RC beads derived from waste textiles	LiHCl and Thp	Absorption	LiHCl- 43% and Thp- 10%	[14]
RC coated CS beads	VRP	Encapsulation	8.62 ± 0.07%,	[94]
RC/TiO_2/ZnO composite	Tramadol hydrochloride	Microencapsulation	RNCC5- 21.6%	[129]
Spherically shaped RC NPs	MB	*In situ* loading	89% (60 mg)	[63]
RC micro-nano fiber matrices	IBU on CMF matrix	*In situ* loading	6%	[51]
RC capsules	API powders	Capsule filling and sealing	KCl – 800 mg, Diphenhydramine -700 mg, Tramadol - 650 mg, Niacinamide – 550 mg, Acetaminophen - 450 mg, Ketoprofen – 450 mg	[85]
Porous RC beads	Curcumin	Absorption	88%	[13]

(Continued)

Table 10.3 Recent research on the drug loading mechanisms and maximum loading capacity of RC-based drug & bioactive compounds carrier. (*Continued*)

Carrier materials	Drug	Drug loading mechanism	Loading capacity (%) or mg	Refs.
CS NPs coated RC fibers	CDP	Encapsulation	-	[130]
Spherical RC beads	Curcumin	Absorption	~95%	[90]
RC–Ch–Ag Biocomposites	Gentamicin	Coupled by soaking	0.04% was coupled	[62]
RC/Poly (acrylic acid) composite films	Antibiotic drug miconazole nitrate	*In situ* loading	9.6 mg	[122]
RC/γ-Fe_2O_3/PPy composite matrices	5-Sulfosalicylic acid dihydrate	H-bonding	85.4% ±0.78%	[73]

So far different approaches have been taken to load drugs and bioactive compounds in RC-based carriers, such as *in situ* loading, dissolution-diffusion-absorption, encapsulation, and chemical interaction. However, the loading efficiency of the carrier depends on the material properties (surface area, porosity, functional properties, etc.), the nature of drugs (hydrophilicity), and the loading environment (temperature, pH). For example, the loading capacities of the RC beads were found positively correlated with their microporous characteristics (mainly the surface area and cumulative pore volume) [14]. Table 10.3 summarizes the recent research on the drug loading mechanism and loading capacity of RC-based drug carriers.

10.7.1 *In Situ* Loading

In situ drug loading greatly reduces preparation time and dramatically enhances the loaded amount and released fraction of the drug [124]. The advantages of *in situ* forming systems include the convenience of administration, straightforward fabrication, and less stressful production settings for delicate drug molecules [125]. In this technique, a precalculated amount of drug is added to the reaction mixture before the reaction is started [122]. Li *et al.* dissolved CA (10 wt%) in DMAc and acetone (2:1), stirred for 12 h at room temperature, and added DMOG/β-CD inclusion (2.5 wt%, relative to the polymer) to obtain the final spinning solution. Until a homogenous solution was generated, the resulting mixture was constantly stirred. After that, a plastic syringe was used to retrieve the fluid for electrospinning [111]. Sun *et al.* developed ORC/CS/Lys@Ag NPs hydrogels stepwise. The ORC solution and 1.00 mL of a specific concentration of Lys hydrochloride solution were slowly loaded into the CS/Ag NPs solution while stirring continuously. The solution was thoroughly mixed until homogeneous, and then it was poured into a mold for a designated period to form ORC/CS/Lys@Ag NPs hydrogel. The resulting hydrogels were undisturbed for 12 h and dialyzed in deionized water for 12 h [29].

Furthermore, during the nanoprecipitation process, the *in-situ* approach can also be used to load drugs into RC NPs. The effectiveness of drug loading is significantly impacted by the average particle sizes of cellulose NPs. For example, the highest MB loading efficiency (89%) in RC NPs was correlated with the smallest average particle diameter (70 nm). The loading efficiency of MB dropped dramatically from 89% to 56% when the average particle sizes of NPs increased from 70 to 365 nm. Additionally, NPs with smaller average particle sizes showed faster rates of solvent removal, which caused the drug molecules to become trapped more quickly [63]. This finding was supported by the literature [126].

10.7.2 Absorption

Absorption is the uptake of drug molecules into a porous structure or onto the surface of the carrier materials [127]. This procedure involves immersing the carriers in loading solutions and continuously agitating them. However, the selection of solvent is determined by the drug's solubility in that particular solvent. The drug-loaded carriers are withdrawn from the loading solution and dried at room temperature until no water remains in the materials. Loaded and dry carriers are stored at room temperature for future use. For example, readily water-soluble drugs, RSP and LiHCl (solubilities 112 mg/ml and 147 mg/ml, respectively), were loaded into the CBs by immersion and absorption technique. Drug compounds filled the pores of unloaded beads, and model pharmaceuticals did not exhibit crystals inside the beads. The quantity of entrapped drug in the beads might be increased by raising the drug loading concentration or the porosity of the unloaded beads. Additionally, the water solubility of the APIs had an impact on loading efficiency [89]. Furthermore, their morphological properties—specifically, porosity, mean diameter, and specific surface area—have a significant impact on the loading capacity. Later medication absorption may also be impacted by the surface area and fusing regions of the carriers [13, 51].

10.7.3 Encapsulation

The process of designing particles with sizes ranging from 1 to 1000 μm in which solid or liquid therapeutic compounds are enclosed either as a core encased in a polymeric shell (microcapsules) or as a distributed and/or dissolved in a polymer matrix (microspheres) is known as *microencapsulation* [128]. Recently, the microencapsulation method was used for loading tramadol hydrochloride drug in RC/TiO_2/ZnO Composite. The composite and the tramadol HCl drug were dissolved in CH_2Cl_2. The organic dispersion was emulsified at room temperature and agitated constantly until the dichloromethane evaporated. The drug microsphere was filtered, repeatedly washed, and dried overnight at 40°C [129]. The encapsulation effectiveness of the composite can be determined by the difference between the overall quantity of drug added during synthesis and the amount of free drug in the supernatant. A UV spectrophotometer is used to measure the amount of nonencapsulated drug at a particular wavelength [130].

10.7.4 Chemical Interaction

Drug-carrier linkage can be formed through chemical contact, which includes hydrogen bonding, ionic and covalent bonding, van der Waals force, hydrophobic and electrostatic interactions, and Pi–Pi stacking. However, altering the drug loading sites of the carrier materials can increase the loading capacity [131]. Sofi *et al.* used an *in-situ* reduction technique for fabricating RC nanofibers containing Hap (0.5, 1.0, and 1.5 wt%) and were treated with $AgNO_3$ (3, 5, and 7%) for 4 h to allow adsorption of Ag^+ on RC mats. Following the $AgNO_3$ treatment, the surplus solution was drained, and the fibers were treated with $NaBH_4$ (0.05 M) for the reduction of deposited Ag^+. The tinted white color of fibers turned pale yellow and then dark brown, indicating that the Ag^+ had been reduced. After 24 hours of air drying, the fibers were preserved for additional analysis [132]. In another study, sulfosalicylic acid was loaded into the RC/γ-Fe_2O_3/PPy matrices by chemical linkages. The H-bonding between the sulfonate groups of sulfosalicylic acid and with the -NH_2 group of PPy contributed to drug loading into the polymer matrices [73].

10.8 Drug Release Kinetics

In the field of controlled drug delivery, the use of mathematical models not only helps to understand the dynamics of drug release but also saves resources (cost and time) by helping to plan more effective studies [133]. The kinetics of drug release from RC-based carrier depend on the particle size, additives, pH, and matrix characteristics [134].

10.8.1 Higuchi Model

Drug release from a matrix in a diffusion-controlled fashion is explained by the Higuchi model. It is conducive in comprehending the gradual release of medications from polymeric systems, like the magnetic nano-composite matrices [135].

This well-known equation deals with the rate at which a solute—usually a drug—leaks into a surrounding fluid from a matrix, generally a polymer, when the solute's loading, A, is greater than its solubility, Cs, in the matrix. Two crucial processes are involved in Takeru Higuchi's analysis. This is the physical representation of splitting the matrix, as shown in Figure 10.13, into an outer region, $x < \xi$, where all the drug is dissolved (no particles), and an inner region, $x > \xi$, where undissolved particles exist. Fick's law controls the rate at which the solute is released into the surrounding fluid due to a gradient of concentration [136].

The Higuchi model describes the release of soluble and poorly soluble drugs. This model is based on the following assumptions:

- (i) The drug's initial dosage form concentration is greater than its solubility
- (ii) Drug only spreads in one dimension
- (iii) Diffusivity of drugs remains constant; and
- (iv) The receptor media reaches a sink condition, where the drugs can be dissolved with great efficiency [97].

The Higuchi equation is commonly phrased as follows and is obtained under the premise of a semi-infinite medium -

$$Q = K_H \sqrt{t}$$

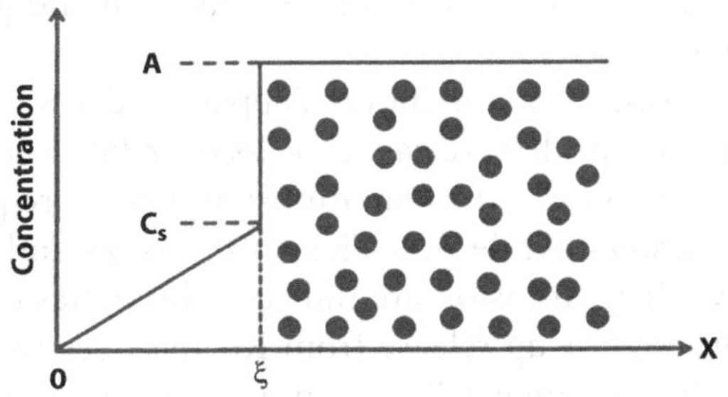

Figure 10.13 Schematic illustration of the Higuchi model. Reproduced with permission from [136].

where K_H is a Higuchi constant that is influenced by the drug's characteristics, the matrix, and the release circumstances, and Q is the quantity of drug released at time t [137].

The model may be used to forecast the drug release of RC-based carriers depending on loading and the cellulose matrix's physical characteristics. For example, the drug loading and microsphere size can have a big impact on the release kinetics. Generally, quicker release rates are achieved with smaller particles and bigger drug loadings. To understand how the medication diffuses through the cellulose matrix, researchers must know its porosity and tortuosity, which may be inferred from the release data [138].

10.8.2 Korsmeyer–Peppas Model

Korsmeyer–Peppas kinetic model, which belongs to the Fickian diffusion model and was primarily governed by the diffusion process. Iron oxide NPs and layered silicate magadiite were coupled to fabricate a drug carrier system [139]. Drug release from the polymeric system is described by the straightforward Korsmeyer–Peppas model which is sometimes referred to as the *Power law*. The Korsmeyer-Peppas model describes several release processes at once, including water migration into the matrix, matrix swelling, and matrix disintegration [140].

The model is expressed by

$$Q(t) = \frac{M_t}{M_{t0t}} = K_{kp}t^n$$

Here, M_t is the amount of drug at time t,

M_{tot} is the total amount of loaded drug, n is the release exponent (associated with drug release mechanism)

K_{kp} is the constant that depends on the structural and geometrical features of the systems [141].

The benefit of using the Korsmeyer-Peppas model was that it allowed for the description of the drug release process from liposomes by supplying the transport exponent, n. A Fickian diffusion process requires n = 0.5 in the absence of borders, while non-Fickian diffusion and boundary area effects on passive drug diffusion are indicated by values of n between 0.5 and 1 [142]. Although drug release from RC matrices may be effectively analyzed using the Korsmeyer-Peppas model, other models, such as the Higuchi model, may also be useful, especially *in situ*ations where diffusion is the main mechanism [143].

10.8.3 Zero-Order Kinetics Model

A set quantity of the medication is eliminated by the body per unit of time in zero-order kinetics. This indicates that the drug's bloodstream concentration has no bearing on the rate of removal. When the metabolic pathways that oversee drug removal become saturated, zero-order kinetics frequently happen [144]. Zero-order delivery is the process of a drug being released from a drug delivery device at a steady rate—that is, the same quantity of drug is released per unit of time regardless of how much drug is still in the delivery system. Mathematically, zero-order release can be described by the following equation -

$$\frac{M_t}{M_\infty} = K_0 t$$

M_t denotes the mass of the drug released over time t, M_∞ is the mass of the drug released when t approaches infinity (i.e., at equilibrium), and k_0 is a constant that depends on the nature of the delivery systems [145].

The following equation represents zero-order drug release in its most basic form-

$$C = C_0 + K_0 t$$

Here, C_0 is the initial drug amount in the solution,
C is the drug released amount, and
K_0 is the zero-order constant.

Zero-order drug release systems have a lot to offer in terms of patient compliance and drug delivery, but to make them more useful and successful in clinical settings, issues with material characteristics, biological variables, and device design need to be resolved [146].

Because of the weak connections between the medication and the carrier, it is especially difficult to achieve zero-order release for water-soluble medicines. Because of the weak connections between the medication and the carrier, it is especially difficult to achieve zero-order release for water-soluble medicines. This may make it more difficult to maintain a steady discharge rate [147].

10.8.4 First-Order Kinetics Model

Although it is challenging to theoretically conceive this process, this model has also been utilized to describe the absorption and/or elimination of some medications. The drug dissolution in pharmaceutical dosage forms, such as those that include water-soluble medicines in a porous matrix, may be described using this connection [148]. The release from a system where the release rate is concentration-dependent is described by the first-order equation expressed by the equation:

$$\frac{dC}{dt} = -Kt$$

where K is the first-order rate constant expressed in units of time^{-1}.

This equation can be expressed as:

$$Log\ C_t = \log C_0 - \frac{kt}{2.303}$$

where, C_0 is the initial concentration of the drug and C_t is the concentration of the drug in solution at time t.

The equation predicts a first-order dependence on the concentration gradient (Co-Ct) between the static liquid layer next to the solid surface and the bulk liquid [149]. First-order kinetics implies that the release rate of the medication declines with time, whereas zero-order kinetics shows that the release rate is constant [134].

10.8.5 Hixson-Crowell Model

The Hixson-Crowell system explains the change in drug release kinetics change as the polymer matrix's surface area changes (due to breakdown) [150]. In 1931, Hixon and Crowell proposed the Hixson-Crowell model. It explains how the drug is released (by dissolving) from dosage forms as the diameter and surface area of the particles or tablets vary. Hixon and Crowell provided the following equation after realizing that the particle's regular area is proportional to its cubic root of volume [151]. To represent the rate of dissolution for drugs with uniformly sized particles, Hixson and Crowell used the cube root of weight & radius of particles. The equation is used to express this.

$$M_0^{1/3} - M_t^{1/3} = kt$$

Here, M_0 is the initial drug amount in the pharmaceutical dosage, M_t is the remaining drug in pharmaceutical dosage at the time *t*, and *k* is the proportionality constant [152].

10.8.6 Weibull Model

The release rates of physiologically active substances (e.g., escin) from hydrogel matrices are examined using the Weibull model. It aids in comprehending how various formulations impact the active ingredient's release kinetics.

The Weibull equation is defined as

$$F(t) = 1 - e^{-(t/\tau)^{\beta}}$$

Here, F(t) is the drug released fraction at time *t*, τ is the scale parameter, and β is the shape parameter.

The Weibull model successfully captures these modifications in drug release behavior. The inclusion of hydrophilic polymers can modify the release kinetics [153]. Drug loading and release kinetics are strongly influenced by the size and shape of RC beads. For example, medication release patterns can be optimized by altering the average diameter of cellulose beads [13].

10.9 Conclusion

Due to its low toxicity, high biocompatibility, biodegradability, and commendable mechanical strength, RC is becoming more and more important in modern research of biomedical applications like tissue engineering, drug delivery, and biosensing. Furthermore, the presence of enormous hydroxyl groups helps to modify their functional properties for specific applications. This chapter focuses on the recent advancement of the solvent system for cellulose regeneration for their use in drugs and bioactive compound delivery. Among different architectures, microspheres, NPs, beads, films, membranes, hydrogels, capsules, scaffolds, capsules, and composites are widely used as carriers for controlled drugs and bioactive compound delivery. This chapter critically summarizes the fabrication techniques and their performance properties of those carriers. Furthermore, it will give an insight into the two most important aspects of an efficient drug delivery system—(i) drug loading techniques and (ii) release kinetics. However, promising areas for future research in this area include -

- *In vivo study:* Most of the research in this area is *in vitro*. Furthermore, these are in the initial stage of development. To better evaluate the efficacy of the treatment and to bolster the findings, larger and more specialized investigations are required. Furthermore, future research should focus on comprehensive studies of animal models to evaluate their *in vivo* behavior. In addition, long-term assessments of biocompatibility, cytotoxicity, and biodegradation can be done on living organisms to justify their practical applications.

- *Novel materials & applications*: Future studies on RC-based new carriers combined with various NPs may be fascinating. Furthermore, by simulating the environment (*in vitro*), the resulting materials' appropriateness for organ-specific applications can be validated.
- *Stimuli-responsive behavior:* Explore stimuli-responsive behavior of the developed materials for active targeting of specific organs.
- *Non-toxic and efficient cellulose dissolution system*: Application of non-toxic solvent and efficient regeneration of cellulose is still a challenge. However, when it comes to their application in biomedical (drug delivery systems), the nontoxicity of used solvents is mandatory. Therefore, future research should focus on the selection of biocompatible and industrially feasible (cost-effective) solvent systems for cellulose regeneration.
- *Biodegradation & release kinetics:* Biodegradation mechanism at different mediums *(in vitro & in vivo)* and kinetics were hardly explained in the literature.
- *Modeling & simulation:* Further research can be focused on modeling and simulation performance (drug release) to predict *in vivo* performance.

AI disclosure statement: The author confirms that no artificial intelligence (AI)-generated content or AI-assisted writing tools were used in the preparation of this chapter (including summaries and the abstract). All text, analysis, and conclusions are the result of the author's original work.

References

1. Wang, S., Lu, A., Zhang, L., Recent advances in regenerated cellulose materials. *Prog. Polym. Sci.*, 53, 169–206, 2016.
2. Lavoine, N., *et al.*, Microfibrillated cellulose–Its barrier properties and applications in cellulosic materials: A review. *Carbohydr. Polym.*, 90, 2, 735–764, 2012.
3. Crépy, L., *et al.*, Evaluation of a bio-based hydrophobic cellulose laurate film as biomaterial—Study on biodegradation and cytocompatibility. *J. Biomed. Mater. Res. Part B Appl. Biomater.*, 100, 4, 1000–1008, 2012.

4. Seddiqi, H., *et al.*, Cellulose and its derivatives: towards biomedical applications. *Cellulose*, 28, 4, 1893–1931, 2021.
5. Vo, L.T.T., *et al.*, Functionalisation of cellulosic substrates by a facile solventless method of introducing carbamate groups. *Carbohydr. Polym.*, 82, 4, 1191–1197, 2010.
6. Medronho, B. and Lindman, B., Brief overview on cellulose dissolution/regeneration interactions and mechanisms. *Adv. Colloid Interface Sci.*, 222, 502–508, 2015.
7. Liebert, T., Cellulose solvents–remarkable history, bright future, in: *Cellulose solvents: for analysis, shaping and chemical modification*, pp. 3–54, ACS Publications, Washington, DC, USA, 2010.
8. Kosan, B., Michels, C., Meister, F., Dissolution and forming of cellulose with ionic liquids. *Cellulose*, 15, 59–66, 2008.
9. Perez, S. and Samain, D., Structure and engineering of celluloses. *Adv. Carbohydr. Chem. Biochem.*, 64, 25–116, 2010.
10. Bhattacharjee, M., Dhar, A., Sikdar, P., Recent Advances in Regenerated Cellulosic Materials and Composites for Multifunctional Applications: A Review, in: *Regenerated Cellulose and Composites: Morphology-Property Relationship*, M. Shabbir (Ed.), pp. 37–78, Springer Nature Singapore, Singapore, 2023.
11. Xu, F. and Cho, B.-U., Preparation of porous regenerated cellulose microstructures *via* emulsion-coagulation technique. *Cellulose*, 29, 3, 1527–1542, 2022.
12. Badshah, M., *et al.*, Development and Evaluation of Drug Loaded Regenerated Bacterial Cellulose-Based Matrices as a Potential Dosage Form. *Front. Bioeng. Biotechnol.*, 8, 2020.
13. Voon, L.K., Pang, S.C., Chin, S.F., Porous Cellulose Beads Fabricated from Regenerated Cellulose as Potential Drug Delivery Carriers. *J. Chem.*, 2017, 1–11, 2017.
14. Zeng, B., Wang, X., Byrne, N., Cellulose Beads Derived from Waste Textiles for Drug Delivery. *Polymers*, 12, 7, 2020.
15. Popper, Z.A., *et al.*, Evolution and diversity of plant cell walls: From algae to flowering plants. *Annu. Rev. Plant Biol.*, 62, 1, 567–590, 2011.
16. Iguchi, M., Yamanaka, S., Budhiono, A., Bacterial cellulose—a masterpiece of nature's arts. *J. Mater. Sci.*, 35, 2, 261–270, 2000.
17. Müller, A., *et al.*, Bacterial nanocellulose with a shape-memory effect as potential drug delivery system. *RSC Adv.*, 4, 100, 57173–57184, 2014.
18. Avolio, R., *et al.*, A multitechnique approach to assess the effect of ball milling on cellulose. *Carbohydr. Polym.*, 87, 1, 265–273, 2012.
19. Moon, R.J., *et al.*, Cellulose nanomaterials review: Structure, properties and nanocomposites. *Chem. Soc. Rev.*, 40, 7, 3941–3994, 2011.
20. Xu, T., *et al.*, Advanced nanocellulose-based composites for flexible functional energy storage devices. *Adv. Mater.*, 33, 48, 2101368, 2021.

21. Sun, Z., *et al.*, Overview of cellulose-based flexible materials for supercapacitors. *J. Mater. Chem. A*, 9, 12, 7278–7300, 2021.
22. Zhang, Z., *et al.*, Bifunctional cellulose interlayer enabled efficient perovskite solar cells with simultaneously enhanced efficiency and stability. *Adv. Sci.*, 10, 8, 2207202, 2023.
23. Liu, H., *et al.*, Simulations reveal conformational changes of methylhydroxyl groups during dissolution of cellulose Iβ in ionic liquid 1-ethyl-3-methylimidazolium acetate. *J. Phys. Chem. B*, 116, 28, 8131–8138, 2012.
24. Molnár, G., *et al.*, Cellulose crystals plastify by localized shear. *Proc. Natl. Acad. Sci.*, 115, 28, 7260–7265, 2018.
25. Yamane, C., *et al.*, Two Different Surface Properties of Regenerated Cellulose due to Structural Anisotropy. *Polym. J.*, 38, 8, 819–826, 2006.
26. Abo-Elseoud, W., *et al.*, Use of cellulose nanocrystals in chitosan nanoparticles carrier system for the controlled release of ketoprofen. *OSP J. Nanomed. Nanotechnol.*, 1, 1, 1–11, 2020.
27. Petrov, D., *et al.*, 6 - Extracellular matrix-derived tissues for neurological applications, in: *Extracellular Matrix-derived Implants in Clinical Medicine*, D.L. Mooradian (Ed.), pp. 83–118, Woodhead Publishing, Duxford, UK, 2016.
28. Franceschini, G., Breast conserving surgery using oxidized regenerated cellulose as filling biomaterial: recommendations to improve clinical outcomes. *BMC Women's Health*, 21, 1, 281, 2021.
29. Sun, Q., *et al.*, Silver-infused lysine crosslinked hydrogel with oxidized regenerated cellulose for prospective advanced wound dressings. *Int. J. Biol. Macromol.*, 264, 2024.
30. Sirviö, J., *et al.*, Dialdehyde cellulose microfibers generated from wood pulp by milling-induced periodate oxidation. *Carbohydr. Polym.*, 86, 1, 260–265, 2011.
31. Jiang, L., *et al.*, Preparation of pH-responsive oxidized regenerated cellulose hydrogels compounded with nano-ZnO/chitosan/aminocyclodextrin ibuprofen complex for wound dressing. *Int. J. Biol. Macromol.*, 253, 126628, 2023.
32. Adepu, S. and Ramakrishna, S., Controlled Drug Delivery Systems: Current Status and Future Directions. *Molecules*, 26, 19, 5905, 2021.
33. Schilling, A.L., *et al.*, Advances in controlled drug delivery to the sinonasal mucosa. *Biomaterials*, 282, 121430, 2022.
34. Wang, M. and Duan, B., Materials and Their Biomedical Applications, in: *Encyclopedia of Biomedical Engineering*, R. Narayan (Ed.), pp. 135–152, Elsevier, Oxford, 2019.
35. Ginebra, M.P. and Montufar, E.B., 10 - Injectable biomedical foams for bone regeneration, in: *Biomedical Foams for Tissue Engineering Applications*, P.A. Netti (Ed.), pp. 281–312, Woodhead Publishing, UK, 2014.
36. Maderuelo, C., Zarzuelo, A., Lanao, J.M., Critical factors in the release of drugs from sustained release hydrophilic matrices. *J. Controlled Release*, 154, 1, 2–19, 2011.

37. Kulpinski, P., Cellulose nanofibers prepared by the N-methylmorpholine-N-oxide method. *J. Appl. Polym. Sci.*, 98, 4, 1855–1859, 2005.
38. Bhatt, B. and Kumar, V., Regenerated cellulose capsules for controlled drug delivery: Part I. Physiological characteristics of membrane formation and the influence of thermal annealing. *Cellulose*, 22, 5, 3237–3250, 2015.
39. Patchan, M.W., *et al.*, Evaluation of the biocompatibility of regenerated cellulose hydrogels with high strength and transparency for ocular applications. *J. Biomater. Appl.*, 30, 7, 1049–59, 2016.
40. Pang, J., *et al.*, Comparison of physical properties of regenerated cellulose films fabricated with different cellulose feedstocks in ionic liquid. *Carbohydr. Polym.*, 121, 71–8, 2015.
41. Reddy, K.O., *et al.*, Preparation and characterization of regenerated cellulose films using borassus fruit fibers and an ionic liquid. *Carbohydr. Polym.*, 160, 203–211, 2017.
42. Kamida, K., *et al.*, Study on the solubility of cellulose in aqueous alkali solution by deuteration IR and 13C NMR. *Polym. J.*, 16, 12, 857–866, 1984.
43. Zhou, J. and Zhang, L., Solubility of Cellulose in NaOH/Urea Aqueous Solution. *Polym. J.*, 32, 10, 866–870, 2000.
44. Woodings, C., *Regenerated cellulose fibres*, Elsevier, Woodhead Publishing, Cambridge, UK, 2001.
45. McCormick, C.L., Callais, P.A., Hutchinson Jr, B.H., Solution studies of cellulose in lithium chloride and N, N-dimethylacetamide. *Macromolecules*, 18, 12, 2394–2401, 1985.
46. Rosenau, T., *et al.*, Cellulose solutions in N-methylmorpholine-N-oxide (NMMO)–degradation processes and stabilizers. *Cellulose*, 9, 283–291, 2002.
47. Acharya, S., *et al.*, Utilization of Cellulose to Its Full Potential: A Review on Cellulose Dissolution, Regeneration, and Applications. *Polymers*, 13, 24, 4344, 2021.
48. Gao, Q., Shen, X., Lu, X., Regenerated bacterial cellulose fibers prepared by the NMMO· H2O process. *Carbohydr. Polym.*, 83, 3, 1253–1256, 2011.
49. Peng, X.-w., Ren, J.-l., Sun, R.-c., Homogeneous esterification of xylan-rich hemicelluloses with maleic anhydride in ionic liquid. *Biomacromolecules*, 11, 12, 3519–3524, 2010.
50. Liu, X., *et al.*, Regenerated cellulose film with enhanced tensile strength prepared with ionic liquid 1-ethyl-3-methylimidazolium acetate (EMIMAc). *Cellulose*, 20, 1391–1399, 2013.
51. Liu, Y., *et al.*, Regenerated cellulose micro-nano fiber matrices for transdermal drug release. *Mater. Sci. Eng. C*, 74, 485–492, 2017.
52. El Seoud, O.A., *et al.*, Cellulose in Ionic Liquids and Alkaline Solutions: Advances in the Mechanisms of Biopolymer Dissolution and Regeneration. *Polymers*, 11, 12, 1917, 2019.
53. Tian, X.-f., *et al.*, Pretreatment of microcrystalline cellulose in organic electrolyte solutions for enzymatic hydrolysis. *Biotechnol. Biofuels*, 4, 1–14, 2011.

54. Yang, B., *et al.*, Converting bleached hardwood kraft pulp to dissolving pulp by using organic electrolyte solutions. *Cellulose*, 28, 3, 1311–1320, 2021.
55. Clough, M.T., Organic electrolyte solutions as versatile media for the dissolution and regeneration of cellulose. *Green Chem.*, 19, 20, 4754–4768, 2017.
56. Gericke, M., *et al.*, Studies on the tosylation of cellulose in mixtures of ionic liquids and a co-solvent. *Carbohydr. Polym.*, 89, 2, 526–536, 2012.
57. Cheng, D., *et al.*, Facile preparation of regenerated cellulose film from cotton linter using organic electrolyte solution (OES). *Cellulose*, 24, 4, 1631–1639, 2017.
58. Geng, H., *et al.*, Characterisation of cellulose films regenerated from acetone/water coagulants. *Carbohydr. Polym.*, 102, 438–44, 2014.
59. Gan, S., *et al.*, Highly porous regenerated cellulose hydrogel and aerogel prepared from hydrothermal synthesized cellulose carbamate. *PLoS One*, 12, 3, e0173743, 2017.
60. Cheng, Y., *et al.*, The preparation, characterization and evaluation of regenerated cellulose/collagen composite hydrogel films. *Carbohydr. Polym.*, 107, 57–64, 2014.
61. Han, J. and Seo, Y., Comparison of Cellulose Dissolution Behavior by Alkaline and Sulfuric Acid Solvents and Their Films' Physical Properties. *Materials*, 14, 18, 5273, 2021.
62. Ahamed, M.I., *et al.*, Evaluation of biomaterial containing regenerated cellulose and chitosan incorporated with silver nanoparticles. *Int. J. Biol. Macromol.*, 72, 680–6, 2015.
63. Chin, S.F., Jimmy, F.B., Pang, S.C., Size controlled fabrication of cellulose nanoparticles for drug delivery applications. *J. Drug Delivery Sci. Technol.*, 43, 262–266, 2018.
64. Kumar, H. and Christopher, L.P., Recent trends and developments in dissolving pulp production and application. *Cellulose*, 24, 2347–2365, 2017.
65. Shi, Z., *et al.*, Dissolution of wood pulp in aqueous NaOH/urea solution *via* dilute acid pretreatment. *J. Agric. Food. Chem.*, 63, 27, 6113–6119, 2015.
66. Singh, N. and Prasad, K., Multi-tasking hydrated ionic liquids as sustainable media for the processing of waste human hair: a biorefinery approach. *Green Chem.*, 21, 12, 3328–3333, 2019.
67. Zhang, J., *et al.*, TBAH/Urea/H 2 O solvent for room temperature wet-spinning of cellulose and optimization of drawing process. *Cellulose*, 26, 6959–6977, 2019.
68. Walters, M.G., *et al.*, The role of urea in the solubility of cellulose in aqueous quaternary ammonium hydroxide. *RSC Adv.*, 10, 10, 5919–5929, 2020.
69. Niju, S., Swathika, M., Balajii, M., Chapter 10 - Pretreatment of lignocellulosic sugarcane leaves and tops for bioethanol production, in: *Lignocellulosic Biomass to Liquid Biofuels*, A. Yousuf, D. Pirozzi, F. Sannino (Eds.), pp. 301–324, Academic Press, California, USA, 2020.
70. Amaral, H.R., *et al.*, Synthesis of bilayer films from regenerated cellulose nanofibers and poly(globalide) for skin tissue engineering applications. *Carbohydr. Polym.*, 252, 2021.

71. Sun, B., *et al.*, Pretreatment by NaOH swelling and then HCl regeneration to enhance the acid hydrolysis of cellulose to glucose. *Bioresour. Technol.*, 196, 454–458, 2015.
72. Yun, S. and Kim, J., A bending electro-active paper actuator made by mixing multi-walled carbon nanotubes and cellulose. *Smart Mater. Struct.*, 16, 4, 1471, 2007.
73. Chowdhury, N.A., *et al.*, Fast Release of Sulfosalicylic Acid from Polymer Implants Consisting of Regenerated Cellulose/γ-Ferric Oxide/Polypyrrole. *J. Appl. Chem.*, 2014, 1–7, 2014.
74. Joshi, B. and Joshi, A., 10 - Ultrasound-based drug delivery systems, in: *Bioelectronics and Medical Devices*, K. Pal and (Eds.), pp. 241–260, Woodhead Publishing, UK, 2019.
75. He, M., *et al.*, Effects of addition method and fibrillation degree of cellulose nanofibrils on furnish drainability and paper properties. *Cellulose*, 24, 5657–5669, 2017.
76. Gericke, M., Trygg, J., Fardim, P., Functional cellulose beads: Preparation, characterization, and applications. *Chem. Rev.*, 113, 7, 4812–4836, 2013.
77. Gong, Y., *et al.*, pH-Responsive Cellulose-Based Microspheres Designed as an Effective Oral Delivery System for Insulin. *ACS Omega*, 6, 4, 2734–2741, 2021.
78. Uyen, N.T.T., *et al.*, Fabrication of alginate microspheres for drug delivery: A review. *Int. J. Biol. Macromol.*, 153, 1035–1046, 2020.
79. Dyer, A., *et al.*, Nasal delivery of insulin using novel chitosan based formulations: A comparative study in two animal models between simple chitosan formulations and chitosan nanoparticles. *Pharm. Res.*, 19, 998–1008, 2002.
80. Purkait, B.S., *et al.*, Isolation of cellulose nanoparticles from sesame husk. *Ind. Eng. Chem. Res.*, 50, 2, 871–876, 2011.
81. Zhang, J., *et al.*, Facile synthesis of spherical cellulose nanoparticles. *Carbohydr. Polym.*, 69, 3, 607–611, 2007.
82. Adsul, M., *et al.*, Facile approach for the dispersion of regenerated cellulose in aqueous system in the form of nanoparticles. *Biomacromolecules*, 13, 9, 2890–5, 2012.
83. Bhatt, B. and Kumar, V., Regenerated Cellulose Capsules for Controlled Drug Delivery, Part 2: Modulating Membrane Permeability by Incorporation of Depolymerized Cellulose and Altering Membrane Thickness. *J. Pharm. Sci.*, 104, 12, 4266–4275, 2015.
84. Bhatt, B. and Kumar, V., Regenerated cellulose capsules for controlled drug delivery: Part III. Developing a fabrication method and evaluating extemporaneous utility for controlled-release. *Eur. J. Pharm. Sci.*, 91, 40–9, 2016.
85. Bhatt, B. and Kumar, V., Regenerated cellulose capsules for controlled drug delivery: Part IV. *In-vitro* evaluation of novel self-pore forming regenerated cellulose capsules. *Eur. J. Pharm. Sci.*, 97, 227–236, 2017.
86. Shilpa, A., Agrawal, S., Ray, A.R., Controlled delivery of drugs from alginate matrix. *J. Macromol. Sci. Part C Polym. Rev.*, 43, 2, 187–221, 2003.

87. García-González, C.A., *et al.*, An opinion paper on aerogels for biomedical and environmental applications. *Molecules*, 24, 9, 1815, 2019.
88. Trygg, J., *et al.*, Physicochemical design of the morphology and ultrastructure of cellulose beads. *Carbohydr. Polym.*, 93, 1, 291–299, 2013.
89. Yildir, E., *et al.*, Tailored beads made of dissolved cellulose—Investigation of their drug release properties. *Int. J. Pharm.*, 456, 2, 417–423, 2013.
90. Voon, L.K., Pang, S.C., Chin, S.F., Optimizing Delivery Characteristics of Curcumin as a Model Drug *via* Tailoring Mean Diameter Ranges of Cellulose Beads. *Int. J. Polym. Sci.*, 2017, 1, 2581767, 2017.
91. Mystek, K., *et al.*, *In Situ* Modification of Regenerated Cellulose Beads: Creating All-Cellulose Composites. *Ind. Eng. Chem. Res.*, 59, 7, 2968–2976, 2020.
92. Singh, J. and Nayak, P., pH-responsive polymers for drug delivery: Trends and opportunities. *J. Polym Sci.*, 61, 22, 2828–2850, 2023.
93. Xie, F., *et al.*, TEMPO-Oxidized Cellulose Beads as Potential pH-Responsive Carriers for Site-Specific Drug Delivery in the Gastrointestinal Tract. *Molecules*, 26, 4, 2021.
94. Song, M.-H., Pham, T.P.T., Yun, Y.-S., Ionic liquid-assisted cellulose coating of chitosan hydrogel beads and their application as drug carriers. *Sci. Rep.*, 10, 1, 13905, 2020.
95. Vigata, M., *et al.*, Hydrogels as drug delivery systems: A review of current characterization and evaluation techniques. *Pharmaceutics*, 12, 12, 1188, 2020.
96. Li, J. and Mooney, D.J., Designing hydrogels for controlled drug delivery. *Nat. Rev. Mater.*, 1, 12, 1–17, 2016.
97. Al-Rajabi, M.M. and Teow, Y.H., Green synthesis of thermo-responsive hydrogel from oil palm empty fruit bunches cellulose for sustained drug delivery. *Polymers*, 13, 13, 2153, 2021.
98. Kass, L.E. and Nguyen, J., Nanocarrier-hydrogel composite delivery systems for precision drug release. *WIREs Nanomed. Nanobiotechnol.*, 14, 2, e1756, 2022.
99. Paunonen, S.I., Strength and barrier enhancements of cellophane and cellulose derivative films: A review, 2013.
100. Ahmed, S., *et al.*, Brittle–ductile transition of double network hydrogels: Mechanical balance of two networks as the key factor. *Polymer*, 55, 3, 914–923, 2014.
101. Gong, J.P., Why are double network hydrogels so tough? *Soft Matter*, 6, 12, 2583–2590, 2010.
102. Matsuda, T., *et al.*, Yielding criteria of double network hydrogels. *Macromolecules*, 49, 5, 1865–1872, 2016.
103. Haque, M.A., Kurokawa, T., Gong, J.P., Super tough double network hydrogels and their application as biomaterials. *Polymer*, 53, 9, 1805–1822, 2012.
104. Song, F., *et al.*, A robust regenerated cellulose-based dual stimuli-responsive hydrogel as an intelligent switch for controlled drug delivery. *Int. J. Biol. Macromol.*, 176, 448–458, 2021.

105. Ji, D.-Y., *et al.*, A novel injectable chitosan/polyglutamate polyelectrolyte complex hydrogel with hydroxyapatite for soft-tissue augmentation. *Carbohydr. Polym.*, 89, 4, 1123–1130, 2012.
106. Li, C.-P., Weng, M.-C., Huang, S.-L., Preparation and characterization of pH sensitive Chitosan/3-Glycidyloxypropyl Trimethoxysilane (GPTMS) hydrogels by sol-gel method. *Polymers*, 12, 6, 1326, 2020.
107. Zainul Armir, N.A., *et al.*, pH-responsive ampholytic regenerated cellulose hydrogel integrated with carrageenan and chitosan. *Ind. Crops Prod.*, 178, 2022.
108. Wang, X., Ding, B., Li, B., Biomimetic electrospun nanofibrous structures for tissue engineering. *Mater. Today*, 16, 6, 229–241, 2013.
109. Rather, A.H., *et al.*, Overview on immobilization of enzymes on synthetic polymeric nanofibers fabricated by electrospinning. *Biotechnol. Bioeng.*, 119, 1, 9–33, 2022.
110. Ramakrishna, S., *et al.*, Electrospun nanofibers: solving global issues. *Mater. Today*, 9, 3, 40–50, 2006.
111. Li, C., *et al.*, Antibacterial properties and drug release study of cellulose acetate nanofibers containing ear-like Ag-NPs and Dimethyloxallyl Glycine/beta-cyclodextrin. *Appl. Surf. Sci.*, 590, 153132, 2022.
112. Zielińska, A., *et al.*, Scaffolds for drug delivery and tissue engineering: The role of genetics. *J. Controlled Release*, 359, 207–223, 2023.
113. Mianehro, A., Electrospun Bioscaffold Based on Cellulose Acetate and Dendrimer-modified Cellulose Nanocrystals for Controlled Drug Release. *Carbohydr. Polym. Technol. Appl.*, 3, 100187, 2022.
114. Tarrahi, R., *et al.*, Development of a cellulose-based scaffold for sustained delivery of curcumin. *Int. J. Biol. Macromol.*, 183, 132–144, 2021.
115. Mahboubi Soufiani, A., *et al.*, Thermomechanical properties of poly(lactic acid) films reinforced with hydroxyapatite and regenerated cellulose microfibers. *J. Appl. Polym. Sci.*, 131, 20, 2014.
116. Soheilmoghaddam, F., *et al.*, Development of Regenerated Cellulose/Citric Acid Films with Ionic Liquids. *J. Polym. Environ.*, 30, 2, 613–621, 2021.
117. Shao, W., *et al.*, Tetracycline hydrochloride loaded regenerated cellulose composite membranes with controlled release and efficient antibacterial performance. *RSC Adv.*, 6, 4, 3068–3073, 2016.
118. Skotheim, T.A., *Handbook of conducting polymers*, CRC press, Boca Raton, FL, USA, 1997.
119. Chowdhury, N.A. and Al-Jumaily, A.M., Regenerated cellulose/polypyrrole/silver nanoparticles/ionic liquid composite films for potential wound healing applications. *Wound Med.*, 14, 16–18, 2016.
120. Huang, J., Ichinose, I., Kunitake, T., Nanocoating of natural cellulose fibers with conjugated polymer: hierarchical polypyrrole composite materials. *Chem. Commun.*, 13, 1717–1719, 2005.
121. Soheilmoghaddam, M., *et al.*, Bionanocomposites of regenerated cellulose/zeolite prepared using environmentally benign ionic liquid solvent. *Carbohydr. Polym.*, 106, 326–334, 2014.

122. Bajpai, M., Bajpai, S.K., Gautam, D., Investigation of Regenerated Cellulose/Poly(acrylic acid) Composite Films for Potential Wound Healing Applications: A Preliminary Study. *J. Appl. Chem.*, 2014, 1–9, 2014.
123. Judefeind, A. and de Villiers, M.M., Drug Loading into and *In Vitro* Release from Nanosized Drug Delivery Systems, in: *Nanotechnology in Drug Delivery*, M.M. de Villiers, P. Aramwit, G.S. Kwon (Eds.), pp. 129–162 Springer New York, New York, NY, 2009.
124. Wan, M.M., *et al.*, *In Situ* Loading of Drugs into Mesoporous Silica SBA-15. *Chem. - A Eur. J.*, 22, 18, 6294–6301, 2016.
125. Packhaeuser, C.B., *et al.*, *In situ* forming parenteral drug delivery systems: an overview. *Eur. J. Pharm. Biopharm.*, 58, 2, 445–455, 2004.
126. Sharma, N., Madan, P., Lin, S., Effect of process and formulation variables on the preparation of parenteral paclitaxel-loaded biodegradable polymeric nanoparticles: A co-surfactant study. *Asian J. Pharm. Sci.*, 11, 3, 404–416, 2016.
127. Sharma, R., *et al.*, Recent advances in biopolymer-based mucoadhesive drug delivery systems for oral application. *J. Drug Delivery Sci. Technol.*, 91, 105227, 2024.
128. Abadi, S.S.H., Moin, A., Veerabhadrappa, G.H., Fabricated microparticles: An innovative method to minimize the side effects of NSAIDs in arthritis. *Crit. Reviews™ Ther. Drug Carrier Syst.*, 33, 5, 2016.
129. Olatunji, G.A., *et al.*, Tramadol hydrochloride delivery by regenerated cellulose nanofiber-TiO2-ZnO composites. *Korean J. Chem. Eng.*, 35, 3, 784–791, 2018.
130. Ristić, T., *et al.*, Chitosan nanoparticles as a potential drug delivery system attached to viscose cellulose fibers. *Cellulose*, 24, 2, 739–753, 2016.
131. Li, Y. and Yang, L., Driving forces for drug loading in drug carriers. *J. Microencapsulation*, 32, 3, 255–272, 2015.
132. Sofi, H.S., *et al.*, Regenerated cellulose nanofibers from cellulose acetate: Incorporating hydroxyapatite (HAp) and silver (Ag) nanoparticles (NPs), as a scaffold for tissue engineering applications. *Mater. Sci. Eng. C Mater. Biol. Appl.*, 118, 111547, 2021.
133. Elmas, A., *et al.*, Mathematical modelling of drug release. *Res. Eng. Struct. Mater.*, 6, 4, 2020.
134. Najib, N. and Suleiman, M., The kinetics of drug release from ethylcellulose solid dispersions. *Drug Dev. Ind. Pharm.*, 11, 12, 2169–2181, 1985.
135. Jafari, S., Soleimani, M., Badinezhad, M., Application of different mathematical models for further investigation of *in vitro* drug release mechanisms based on magnetic nano-composite, in: *Polymer Bulletin*, pp. 1–18, 2022.
136. Paul, D., Elaborations on the Higuchi model for drug delivery. *Int. J. Pharm.*, 418, 1, 13–17, 2011.
137. Baggi, R.B. and Kilaru, N.B., Modulation of drug release mechanism by Higuchi model: Estimation of percent deviation. *Asian J. Pharm. Technol.*, 6, 4, 249–256, 2016.

138. Dubernet, C., *et al.*, Ibuprofen-loaded ethylcellulose microspheres: Release studies and analysis of the matrix structure through the Higuchi model. *J. Microencapsulation*, 7, 4, 555–565, 1990.
139. Ge, M., *et al.*, Preparation of organic-modified magadiite–magnetic nanocomposite particles as an effective nanohybrid drug carrier material for cancer treatment and its properties of sustained release mechanism by Korsmeyer–Peppas kinetic model. *J. Mater. Sci.*, 56, 25, 14270–14286, 2021.
140. Permanadewi, I., *et al.*, Modelling of controlled drug release in gastrointestinal tract simulation, in: *Journal of Physics: Conference Series*, IOP Publishing, Bristol, UK, 2019.
141. Rostamitabar, M., *et al.*, Cellulose aerogel micro fibers for drug delivery applications. *Mater. Sci. Eng. C*, 127, 112196, 2021.
142. Wu, I.Y., *et al.*, Interpreting non-linear drug diffusion data: Utilizing Korsmeyer-Peppas model to study drug release from liposomes. *Eur. J. Pharm. Sci.*, 138, 105026, 2019.
143. Altun, E., *et al.*, Kinetic release studies of antibiotic patches for local transdermal delivery. *Pharmaceutics*, 13, 5, 613, 2021.
144. Borowy, C.S. and Ashurst, J.V., *Physiology, zero and first order kinetics*, 2018.
145. Talevi, A. and Ruiz, M.E., Zero-Order Drug Release, in: *The ADME Encyclopedia: A Comprehensive Guide on Biopharmacy and Pharmacokinetics*, pp. 1–6, Springer, Switzerland, 2021.
146. Laracuente, M.-L., Marina, H.Y., McHugh, K.J., Zero-order drug delivery: State of the art and future prospects. *J. Controlled Release*, 327, 834–856, 2020.
147. Li, X., Li, Q., Zhao, C., Zero-order controlled release of water-soluble drugs using a marker pen platform. *ACS Omega*, 6, 21, 13774–13778, 2021.
148. Dash, S., *et al.*, Kinetic modeling on drug release from controlled drug delivery systems. *Acta Pol. Pharm.*, 67, 3, 217–223, 2010.
149. Paarakh, M.P., *et al.*, Release kinetics–concepts and applications. *Int. J. Pharm. Res. Technol. (IJPRT)*, 8, 1, 12–20, 2018.
150. Shurshina, A., Galina, A., Kulish, E., Kinetics of the release of a drug from a soluble or hydrolyzable polymer matrix. *Russ. J. Phys. Chem. B*, 10p, 1001–1006, 2016.
151. Raza, S.N. and Khan, N.A., Role of mathematical modelling in controlled release drug delivery. *Int. J. Med. Res. Pharm. Sci.*, 4, 5, 84–95, 2017.
152. Ramteke, K., *et al.*, Mathematical models of drug dissolution: A review. *Sch. Acad. J. Pharm.*, 3, 5, 388–396, 2014.
153. Ghosal, K. and Chandra, A., Statistical Modelling for Controlled Drug Delivery Systems and its Applications in HPMC based Hydrogels, in: *AIP Conference Proceedings*, American Institute of Physics, 2010.

11

Wound Dressing of Regenerative Cellulose

Rajdip Goswami[1], Meghamouli Rana[1], Biswajit Basu[1*], Swarnali Dutta[2], Sourav Ghosh[1] and Sudarshan Singh[3]

[1]*Department of Pharmaceutical Technology, School of Health and Medical Sciences, Adamas University, Barasat, Kolkata (WB), India*
[2]*Department of Pharmacology, Birla Institute of Technology, Mesra, Ranchi, Jharkhand, India*
[3]*Faculty of Pharmacy, Chiang Mai University, Chiang Mai, Thailand*

Abstract

Wound dressings are essential in managing skin lesions, providing a protective barrier while promoting healing. Regenerative cellulose, a biopolymer derived from cellulose, has emerged as a promising material for wound dressing applications due to its unique properties and versatility. Regenerative cellulose has emerged as promising material for wound dressing applications due to its biocompatibility, biodegradability, and versatility. Recent studies have demonstrated its effectiveness in promoting wound healing, and future advancements in its development and application are expected to further enhance its development and application are expected to further enhance its potential. Moreover, studies have focused on the use of regenerative cellulose in combination with other bioactive agents to accelerate wound healing. Therefore, the use of regenerative cellulose in wound dressings offers significant potential for future advancements. This chapter explore the development of novel regenerative cellulose-based dressings that can be tailored to specific wound types and healing needs. In addition, in this chapter various form of regenerative cellulose made dressing such as film, membranes, nanofibers, composite, and scaffolds that retain distinct properties and

**Corresponding author*: bbasu.pharma@gmail.com; ORCID: 0000-0001-8442-801X
Meghamouli Rana: ORCID: 0009-0002-6530-7081
Rajdip Goswami: ORCID: 0000-0003-3296-8799
Swarnali Dutta: ORCID: 0009-0008-4465-909X
Sourav Ghosh: ORCID: 0000-0003-4796-6016
Sudarshan Singh: ORCID: 0000-0002-7929-3322

Mohd Shabbir (ed.) Regenerated Cellulose: A Renewable Alternative to Petroleum-Based Plastics, (311–344) © 2026 Scrivener Publishing LLC

capable in delivering drug onsite for the management of wound have been taken in considerations.

***Keywords*:** Cellulose, biopolymer, dressings, regenerative cellulose, wound management

11.1 Introduction

The increasing interest in regenerative cellulose for biomedical applications, especially in wound management, is driven by the pressing need for materials that are not only effective but also sustainable and biocompatible. In recent years, there has been a significant shift in both the medical and materials science communities toward the development and use of products that align with environmental sustainability goals [1]. This shift is largely motivated by the growing awareness of the environmental impact of traditional petroleum-based plastics and synthetic polymers, which are often non-biodegradable and can lead to long-term ecological harm. Regenerative cellulose stands out as a promising alternative because it is derived from renewable natural resources, such as plants, bacteria, and marine organisms, which can be sustainably harvested or cultured. Unlike synthetic polymers, cellulose is biodegradable, meaning it can break down into non-toxic byproducts that are easily assimilated into the environment. This characteristic is particularly important in biomedical applications where the use of disposable materials is common, and the potential for waste accumulation is high [2]. In addition to its environmental benefits, regenerative cellulose is highly valued for its biocompatibility-its ability to interact with human tissues without eliciting adverse reactions. This is crucial in wound management, where the materials used must support healing while minimizing the risk of inflammation, infection, or other complications. Regenerative cellulose meets these requirements due to its inherent non-toxicity, ability to absorb and retain moisture, and structural similarity to the extracellular matrix, which is the natural scaffold for cells in the body. Furthermore, the versatility of regenerative cellulose allows it to be processed into various forms, such as films, hydrogels, nanofibers, and composites, each tailored to meet specific clinical needs. This adaptability, combined with its sustainable and biocompatible nature, makes regenerative cellulose an attractive option for a wide range of wound care products, from simple bandages to advanced dressings that deliver therapeutic agents directly to the wound site [3]. As the demand for more sustainable and effective biomedical materials continues to grow, the role of regenerative cellulose in wound management is likely to expand. Researchers and clinicians are increasingly recognizing its potential to not only improve patient outcomes but also reduce the environmental footprint of healthcare practices [4]. This convergence of ecological responsibility and medical innovation underscores the burgeoning interest in regenerative cellulose as a material of choice for the future of wound care and beyond [5].

Biomaterials are intricately engineered substances devised to interact symbiotically with biological systems, serving a wide spectrum of therapeutic or diagnostic purposes. These materials, which can be either derived from natural sources or synthesized through advanced chemical processes, are meticulously designed to repair, augment, or substitute tissue functions within the human body [6]. The application of biomaterials is vast and multidisciplinary, encompassing fields such as orthopaedic implants, dental restoratives, tissue engineering scaffolds, and controlled drug delivery systems. Each application demands a biomaterial that is not only biocompatible-meaning it can coexist with bodily tissues without eliciting an adverse reaction also non-toxic, ensuring that it does not introduce harmful substances into the body. Moreover, the ideal biomaterial must possess a suite of tailored properties that align with its intended function. For example, materials used in load-bearing implants, such as hip replacements, must exhibit high mechanical strength and fatigue resistance, whereas those designed for drug delivery might require biodegradability and the ability to release therapeutic agents in a controlled manner [7]. The rigorous demands placed on biomaterials necessitate a deep understanding of both biological and material sciences, enabling the design of substances that can effectively interact with the complex environment of the human body. As the field of biomaterials continues to evolve, the integration of nanotechnology, regenerative medicine, and personalized medicine is pushing the boundaries of what these materials can achieve, heralding a new era of medical innovation [8].

Cellulose is a pervasive biopolymer, a fundamental structural component of the cell walls of plants, and stands as the most abundant organic polymer on the planet. Composed of linear chains of β (1-4) linked glucose units, cellulose's molecular architecture imparts it with extraordinary tensile strength, making it a cornerstone of plant rigidity and structural integrity. Historically, cellulose has been utilized in myriad forms, from the papyrus of ancient civilizations to the textiles and paper products that remain ubiquitous today. Its natural abundance, coupled with its renewability and biodegradability, positions cellulose as a pivotal resource in the quest for sustainable material development. In recent years, cellulose has transcended its traditional roles, emerging as a critical material in the biomedical sector, particularly in the realm of wound care and tissue engineering. Its intrinsic properties—such as its ability to form hydrogels, its biocompatibility, and its ease of modification—render it an excellent candidate for a variety of medical applications [9]. Furthermore, the ongoing research into cellulose has revealed its potential for functionalization, where it can be chemically or physically altered to introduce new functionalities, such as antimicrobial properties or enhanced mechanical strength. As the world moves toward greener and more sustainable technologies, cellulose is poised to play an increasingly prominent role in the development of biomaterials that align with both environmental and clinical needs [10].

Regenerative cellulose refers to cellulose that has been chemically or mechanically processed to modify its native structure, thereby enhancing its properties for specific applications. Unlike native cellulose, which can be rigid, brittle, and difficult to process, regenerative cellulose can be engineered to exhibit improved flexibility, mechanical strength, and bioactivity, making it highly suitable for advanced medical applications, particularly in the development of wound dressings, tissue scaffolds, and drug delivery systems [11]. The process of regenerating cellulose typically involves dissolving native cellulose in a suitable solvent and then regenerating it through a controlled precipitation process, which allows for the manipulation of its molecular alignment and crystallinity [12]. This restructuring can impart the material with enhanced properties, such as increased surface area, porosity, and the ability to form films or fibers with specific mechanical characteristics [8]. These enhancements make regenerative cellulose not only more versatile but also more effective in biomedical applications [13]. For instance, in wound dressings, regenerative cellulose can provide a moist environment that is conducive to healing, while also serving as a barrier to protect the wound from external contaminants [8]. Moreover, regenerative cellulose retains the intrinsic benefits of its precursor, such as biocompatibility and biodegradability, which are crucial for any material intended for use in the human body. However, it goes a step further by offering superior performance characteristics that are tailored to meet the stringent demands of medical applications [10]. The ability to customize regenerative cellulose by incorporating bioactive molecules, such as antimicrobial agents or growth factors, further expands its utility, allowing for the creation of multifunctional materials that can actively participate in the healing process [11]. As research continues to advance, regenerative cellulose is expected to play a central role in the development of next-generation biomaterials that not only meet but exceed the requirements of modern medicine [14].

This chapter endeavors to explore the development and application of regenerative cellulose-based wound dressings. It will elucidate the unique properties of regenerative cellulose that render it suitable for wound care, including its biocompatibility, biodegradability, and versatility.

Furthermore, the chapter will delve into various forms of regenerative cellulose dressings, such as films, membranes, nanofibers, composites, and scaffolds, each tailored to specific wound types and healing needs [12]. The integration of regenerative cellulose with bioactive agents to expedite wound healing will also be examined. The overarching objective is to provide a comprehensive overview of the current advancements and future potential of regenerative cellulose in wound care applications [15].

11.2 Sources of Cellulose

Regenerated cellulose, derived from naturally occurring sources, offers a renewable alternative to synthetic polymers, thus aligning with the global shift toward eco-friendly materials [16]. The ability of cellulose to be sourced from diverse biological entities—from bacteria to plants, marine organisms, and even insects—adds to its versatility and accessibility. In this chapter, we explore the myriad sources of cellulose and their relevance in the context of developing advanced regenerative cellulose-based wound dressings [17].

11.2.1 Bacterial Cellulose

Bacterial cellulose (BC) is a form of cellulose produced by certain bacterial species, such as *Gluconaceto bacterxylinus*. Unlike plant-derived cellulose, BC is characterized by its high purity, absence of lignin and hemicellulose, and its ability to form highly hydrated, three-dimensional networks [18]. These features confer exceptional mechanical strength, flexibility, and water-holding capacity, making BC particularly suited for wound dressing applications. The unique nanofibrillar structure of bacterial cellulose not only provides an optimal environment for cellular proliferation but also facilitates the controlled release of therapeutic agents when used in composite dressings [19]. Recent advancements have seen the incorporation of antimicrobial agents, growth factors, and other bioactive compounds into BC matrices, further enhancing its wound-healing capabilities [20]. Its ease of production through fermentation processes, combined with its innate biocompatibility and biodegradability, positions bacterial cellulose as a leading candidate for next-generation wound care products [21].

11.2.2 Plant-Based Sources of Cellulose

Cellulose is most commonly derived from plant sources, where it serves as a structural component of the cell wall. Plant-based cellulose is abundant, renewable, and can be extracted from various sources, including cotton, wood pulp, flax, and hemp [22]. The extraction process involves the removal of lignin, hemicellulose, and other non-cellulosic components, yielding a pure form of cellulose suitable for regenerative applications [23]. The versatility of plant-derived cellulose is evident in its ability to be processed into various forms such as films, fibers, and hydrogels, each of which can be tailored for specific wound types. For instance, cellulose nanocrystals (CNCs) and cellulose nanofibrils (CNFs) derived from plants have been employed to reinforce the mechanical properties of wound dressings while maintaining their biocompatibility [24]. Additionally, the natural abundance of plant cellulose and its well-established processing techniques make it an economically viable option for large-scale production of regenerative cellulose-based products [25].

11.2.3 Marine Waste as a Source of Cellulose

Marine waste, particularly that derived from algae and crustacean exoskeletons, represents an underexplored yet promising source of cellulose. Seaweed, for example, is rich in cellulose and can be sustainably harvested from marine environments [26]. The extraction of cellulose from marine sources not only contributes to the circular economy by utilizing waste materials but also offers unique properties that are distinct from terrestrial sources [27]. Algal cellulose, for instance, exhibits a unique combination of gel-forming and water-retention properties, making it ideal for the formulation of hydrogels and other moist wound dressings [28]. Furthermore, the chitin found in crustacean shells can be deacetylated to produce chitosan, a biopolymer that, when combined with cellulose, enhances the antimicrobial properties and biodegradability of wound dressings [29]. This synergy between marine-derived cellulose and chitosan opens new avenues for the development of advanced wound care materials that are both sustainable and effective [30].

11.2.4 Animal or Insect Sources of Cellulose

While less common, cellulose can also be derived from certain animals and insects, particularly those that produce silk or other fibrous materials. For example, the silk fibroin from silkworms contains cellulose-like properties that can be harnessed for biomedical applications. Additionally, some insect species produce cellulose during their cocoon formation or as part of their exoskeletal structure [31]. These animal-derived celluloses often exhibit unique mechanical properties, such as high tensile strength and elasticity, which can be advantageous in creating flexible and durable wound dressings. Furthermore, the biocompatibility of these materials ensures that they integrate well with human tissues, reducing the risk of adverse reactions. The exploration of these unconventional sources of cellulose may yield novel materials with properties that surpass those of more traditional sources [32].

11.3 Application of Waste in the Development of Regenerative Cellulose

The utilization of waste materials in the development of regenerative cellulose represents a paradigm shift in sustainable material science, wherein refuse is transmuted into valuable resources [33]. This approach not only mitigates environmental degradation but also introduces novel pathways for the creation of high-performance biopolymers tailored for biomedical applications, particularly in wound management. Marine waste, such as discarded seaweed and crustacean shells, offers a rich reservoir of polysaccharides, including cellulose, which can be extracted and repurposed. Seaweed, abundant in marine environments, can be harvested and processed to yield algal cellulose, which is distinguished by its gel-forming and moisture-retention properties-attributes highly desirable in the formulation of hydrogels for wound dressings [34]. Similarly, chitin, derived from crustacean exoskeletons, can be deacetylated to produce chitosan, a biopolymer that, when combined with cellulose, exhibits enhanced antimicrobial activity and biocompatibility [35]. This synergistic blend of chitosan and cellulose is particularly effective in creating composite dressings that promote wound healing while providing a barrier against infections. Agricultural waste, including byproducts from cotton, flax, and hemp production, presents another viable source of cellulose. These lignocellulosic residues, often considered waste, can be processed to extract cellulose, which can then be regenerated into various forms such as nanofibers, films, and scaffolds. The regeneration process typically involves the dissolution of cellulose in a solvent system followed by its reformation into a desired structure [36]. This not only adds value to what would otherwise be discarded but also reduces the reliance on virgin materials, thus contributing to a circular economy. Urban waste, particularly paper and textile waste, also holds potential as a source of cellulose. Through recycling and reprocessing, cellulose fibers from discarded paper products or worn-out textiles can be regenerated and repurposed for medical applications [37]. The development of such upcycled cellulose materials not only diverts waste from landfills but also aligns with the principles of sustainability and resource efficiency. Incorporating waste-derived cellulose into regenerative applications underscores the importance of innovative waste management strategies [38]. By transforming waste into a resource, we can produce regenerative cellulose that is both environmentally sustainable and capable of meeting the demanding requirements of advanced wound care. The exploration of waste materials as a cellulose source represents a forward-thinking approach that integrates environmental stewardship with the advancement of biomedical technologies [39].

11.4 Anatomy of Skin

The integumentary system, epitomized by the skin, is an intricate and multifunctional organ that serves as the primary interface between the organism and its external milieu. This complex structure is stratified into three principal layers: the epidermis, the dermis, and the hypodermis, each layer contributing uniquely to the skin's protective and regulatory functions. The epidermis, the most superficial stratum, is a stratified squamous epithelium predominantly composed of keratinocytes, which undergo a meticulous process of differentiation as they migrate from the basal layer to the stratum corneum [40]. The basal layer, or stratum basale, harbors mitotically active keratinocytes and melanocytes, the latter being responsible for melanin synthesis, which confers photoprotection by absorbing ultraviolet radiation [41]. As keratinocytes ascend through the spinous and granular layers, they accumulate keratin filaments and lamellar bodies, culminating in the formation of the cornified envelope in the stratum corneum, a resilient barrier impermeable to water and most pathogens. Beneath the epidermis lies the dermis, a robust and elastic connective tissue layer that imbues the skin with its tensile strength and pliability [42]. The dermis is dichotomized into the papillary and reticular dermis. The papillary dermis, immediately subjacent to the epidermis, is interspersed with dermal papillae, which interdigitate with the epidermal ridges, enhancing the structural integrity between these layers [43]. This region is rich in capillaries and sensory nerve endings, facilitating nutrient exchange and tactile sensation. The reticular dermis, constituting the bulk of the dermal layer, is replete with dense irregular connective tissue, predominantly collagen type I and elastin fibers, which provide structural support and elasticity. Embedded within the dermis are adnexal structures, including hair follicles, sebaceous glands, and sweat glands, each playing a pivotal role in thermoregulation, excretion, and the maintenance of skin homeostasis [44].

The hypodermis, or subcutaneous tissue, forms the innermost layer, primarily composed of adipocytes ensconced within a loose matrix of collagen and elastin fibers [45]. This layer functions as a thermal insulator, cushioning the underlying musculature and bones from external mechanical forces. The adipose tissue also serves as an energy reservoir, releasing fatty acids during periods of caloric deficit. The vascularization of the skin is extensive, with a complex network of blood vessels that not only nourish the tissues but also facilitate thermoregulation through vasoconstriction and vasodilation. Lymphatic vessels, interwoven with the blood vasculature, play a crucial role in immune surveillance and the removal

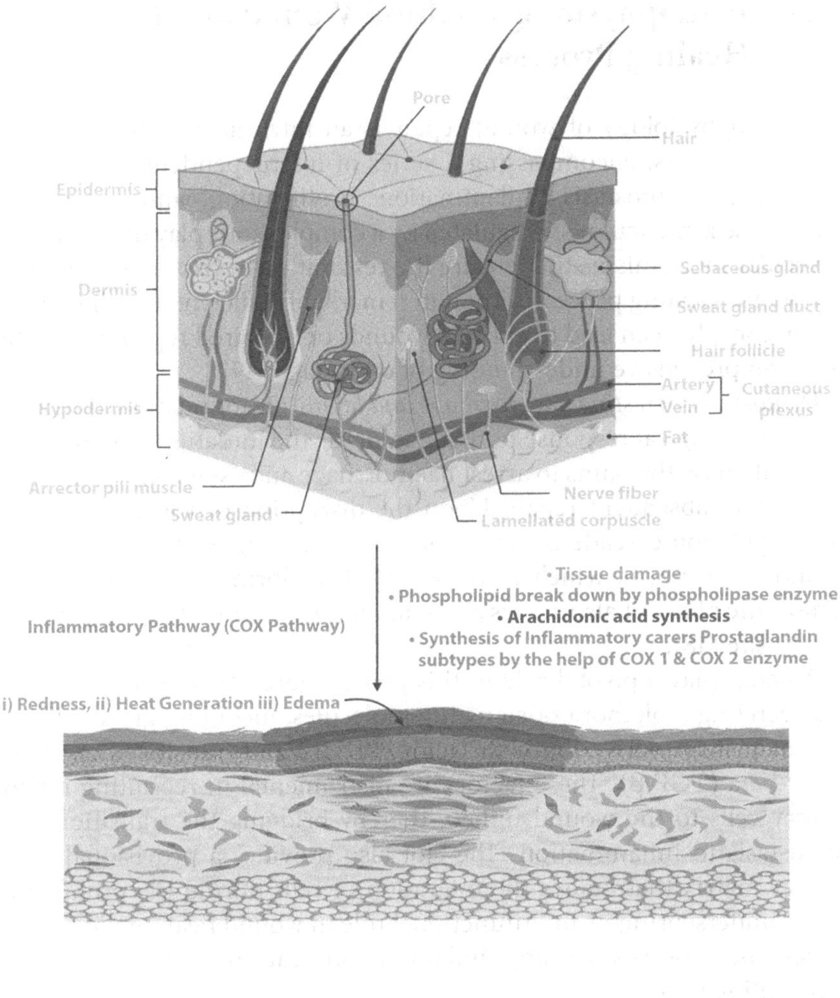

of interstitial fluid [46]. The skin's innervation is equally elaborate, with a myriad of sensory receptors, including mechanoreceptors, thermoreceptors, and nociceptors, enabling the perception of touch, temperature, and pain, thereby acting as a critical component of the somatosensory system. In its entirety, the skin's architecture is a testament to evolutionary ingenuity, intricately designed to perform a multitude of essential functions that are vital for the organism's survival [47]. Its complex interplay of cellular and extracellular components renders it not only a physical barrier but also a dynamic organ involved in homeostasis, sensation, and immune defence. Figure 11.1 depicts anatomy of skin and COX inflammatory pathway [48].

11.5 Pathophysiology Behind Wound and Its Natural Healing Process

The pathophysiology of wound repair is an intricately orchestrated and dynamic process, encompassing a series of interdependent and overlapping stages: hemostasis, inflammation, proliferation, and remodeling. Each phase is meticulously regulated by a complex interplay of cellular and molecular mechanisms that ensure the restoration of tissue integrity [49]. The wound healing process, although remarkably efficient, is vulnerable to disruptions that can lead to chronic wounds or impaired repair, necessitating a comprehensive understanding of each stage [49].

Hemostasis: arresting haemorrhage and laying the foundation: the immediate physiological response to tissue injury is the initiation of hemostasis, a critical phase that aims to arrest haemorrhage and establish a provisional matrix for subsequent repair. Upon the disruption of vascular integrity, the coagulation cascade is rapidly activated, leading to the conversion of fibrinogen to fibrin, which forms a clot. This fibrin clot is not merely a hemostatic plug but also serves as a scaffolding structure that stabilizes the wound site [48].

Platelets play a pivotal role in this phase, aggregating at the injury site and secreting a plethora of signaling molecules, including growth factors such as platelet-derived growth factor (PDGF) and transforming growth factor-beta (TGF-β). These factors are instrumental in recruiting inflammatory cells to the wound milieu, thereby bridging the transition from hemostasis to inflammation. The clot also provides a provisional extracellular matrix that guides the migration of cells during the proliferative phase, underscoring its multifunctional role in wound healing. Figure 11.2 depicts haemostasis causing inflammation, causing proliferation and remodelling [50].

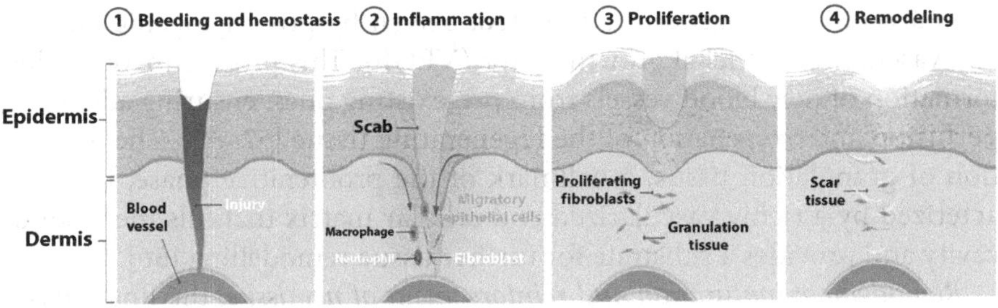

Inflammation: orchestrating the immune response: following hemostasis, the wound enters the inflammatory phase, a critical period characterized by the mobilization and infiltration of immune cells. Polymorphonuclear leukocytes (PMNs), predominantly neutrophils, are among the first responders, arriving at the wound site within hours of injury. Neutrophils perform essential functions, including the phagocytosis of necrotic debris, apoptotic cells, and microbial invaders. They also release reactive oxygen species (ROS) and proteolytic enzymes that help to sterilize the wound.

As the initial wave of neutrophils subsides, macrophages take center stage in the inflammatory process. These versatile cells not only continue the clearance of debris and pathogens but also secrete cytokines and growth factors that modulate the activity of other cells involved in healing. Macrophages exhibit a phenotypic shift during this phase, transitioning from a pro-inflammatory (M1) phenotype to a pro-reparative (M2) phenotype, which is crucial for the resolution of inflammation and the initiation of tissue repair [51]. This phase is pivotal, as a balanced inflammatory response is essential for proper wound healing. An excessive or prolonged inflammatory response can lead to tissue damage and chronic wounds, while an insufficient response may result in infection and impaired healing [52].

Proliferation: rebuilding tissue architecture: as the inflammatory phase resolves, the wound enters the proliferative phase, a period marked by vigorous cellular activity aimed at rebuilding the tissue architecture. Fibroblasts, which are recruited to the wound site by growth factors released during the earlier phases, play a central role in this stage [53]. These cells proliferate and synthesize extracellular matrix components, such as collagen, which provides structural support to the newly forming tissue [54]. Keratinocytes, the predominant cell type in the epidermis, migrate across the wound bed, re-epithelializing the surface and restoring the barrier function of the skin. This re-epithelialization is crucial for protecting the underlying tissue from environmental insults and microbial invasion [53].

Concurrently, angiogenesis is initiated, driven by pro-angiogenic factors like vascular endothelial growth factor (VEGF). This process involves the formation of new blood vessels from pre-existing ones, ensuring adequate perfusion and oxygenation of the regenerating tissue [52–55]. The formation of granulation tissue, a hallmark of the proliferative phase, is characterized by a richly vascularized and cellular matrix that fills the wound cavity and provides a substrate for further tissue remodelling [56].

Remodelling: maturation and reinforcement of neotissue: the final phase of wound healing is remodelling, a protracted and meticulous process that can span several months to years. During this phase, the extracellular matrix undergoes continuous reorganization, as the initially deposited collagen type III is gradually replaced by the more robust collagen type I. This collagen maturation process is essential for restoring tensile strength to the wound [57]. Fibroblasts differentiate into myofibroblasts, which generate contractile forces that reduce the wound size through a process known as wound contraction. Simultaneously, the newly formed blood vessels undergo pruning, and the cellularity of the wound decreases as the tissue reaches a more quiescent state. Despite the extensive remodeling, the tensile strength of the repaired tissue rarely matches that of uninjured skin, achieving approximately 70-80% of the original strength. This underscores the importance of proper wound management to minimize scarring and optimize functional recovery [58].

Perturbations in the healing process: implications for chronic wounds: any disruption or delay in the orderly progression of these healing phases can lead to pathological wound healing outcomes [59]. Factors such as diabetes, vascular insufficiency, and infection can impair the healing process, resulting in chronic wounds that are difficult to treat. Understanding the pathophysiological mechanisms underlying wound healing provides a foundation for developing targeted therapeutic interventions aimed at promoting efficient and complete tissue repair [60].

11.6 Enhancement in Wound Healing Management Through the Utilization of Cellulose-Based Dressings

Cellulose-based dressings have engendered a paradigm shift in the domain of wound management by proffering avant-garde solutions that surmount the limitations of conventional wound care substrates [60]. These dressings are meticulously engineered from regenerative cellulose, a biopolymer that is not only inherently biocompatible and biodegradable but also possesses the versatility to be tailored for various wound typologies. In contrast to traditional dressings, cellulose-based alternatives engender a moist wound environment, a sine qua non for optimal reparative processes [61]. This moisture retention sustains cellular activity and expedites re-epithelialization, thereby abbreviating the duration of healing. Moreover, cellulose-based dressings can be sophisticatedly designed to deliver bioactive agents such as antimicrobial peptides, trophic factors, or anti-inflammatory molecules directly to the wound bed [62]. This targeted administration amplifies the wound healing trajectory by addressing specific impediments, such as microbial colonization, excessive inflammatory responses, or impaired cellular proliferation. For instance, bacterial cellulose, with its distinctive nanofibrillar architecture, can be impregnated with silver nanoparticles to fabricate a dressing that not only accelerates healing but also exerts a potent antimicrobial effect [63]. The protean nature of cellulose permits its fabrication into diverse morphologies, including films, hydrogels, and nanofibers, each tailored for specific wound care exigencies. Hydrogels, for instance, are particularly efficacious in the management of burns and ulcers, where the maintenance of a moist microenvironment is paramount [64]. Conversely, nanofiber matrices can emulate the extracellular matrix, providing a scaffold that facilitates cellular migration and tissue regeneration. These innovations in cellulose-based dressings signify a substantial advancement in wound care, offering augmented healing outcomes and mitigating the risk of complications associated with recalcitrant wounds. As research in this field progresses, the integration of regenerative cellulose in wound management is poised to become a cornerstone of future therapeutic strategies [65].

11.7 Regenerated Cellulose: A Renewable Alternative to Petroleum-Based Plastics in Wound Dressings

Regenerated cellulose, derived from natural sources like wood pulp or cotton, offers a sustainable and eco-friendly alternative to petroleum-based plastics in the fabrication of wound dressings. Unlike conventional plastics, which are derived from non-renewable fossil fuels and pose environmental challenges due to their persistence in the environment, regenerated cellulose is biodegradable and renewable [66]. In wound dressings, regenerated cellulose provides several advantages. It has excellent biocompatibility, making it safe for direct contact with skin and tissues [67]. Its porous structure allows for effective moisture management, which is crucial for wound healing, as it helps maintain an optimal moisture balance while preventing the wound from becoming too dry or overly saturated [68]. Moreover, regenerated cellulose can be easily modified or combined with other bioactive materials to enhance its properties, such as antimicrobial activity or enhanced wound healing capabilities. This adaptability, coupled with its environmental benefits, positions regenerated cellulose as a promising material in the development of advanced wound dressings, offering a greener alternative to traditional petroleum-based materials [69]. Table 11.1 summarizes data on recently reported cellulose-based wound dressings.

Table 11.1 Summarize data on recently reported cellulose-based wound dressings.

Source of cellulose	Fabrication process	Critical attributes	References
Wood pulp	Regeneration *via* viscose process	Biodegradable, high tensile strength, effective moisture control	[70]
Cotton linters	Electrospinning of cellulose acetate	High surface area, good breathability, promotes cell adhesion	[71]
Bacterial cellulose	Static culture fermentation, regeneration	High purity, excellent water retention, conformability to wound	[72]
Agricultural waste	Alkaline treatment, followed by regeneration	Sustainable, cost-effective, antimicrobial properties	[73]
Wood Fibers	Ionic liquid treatment and regeneration	Environmentally friendly, customizable mechanical properties	[74]
Recycled paper	Dissolution in NaOH/urea, regeneration	Circular economy approach, good mechanical integrity	[75]

11.8 Application of Regenerated Cellulose in Wound Healing Management

Regenerated cellulose, derived from renewable biomass, has emerged as an indispensable material in the realm of wound healing, owing to its distinctive physicochemical attributes that position it as a superior and sustainable alternative to conventional petrochemical-derived polymers. Its intrinsic properties confer multifaceted benefits in wound care, elevating it as a pivotal component in contemporary therapeutic regimens [76].

Biocompatibility: The biocompatibility of regenerated cellulose is underpinned by its non-immunogenic nature, rendering it exceedingly amenable to integration with human tissue. This remarkable tolerance significantly mitigates the potential for hypersensitivity or adverse dermal responses, ensuring that wound dressings crafted from regenerated cellulose are both safe and efficacious in prolonged clinical use. Its inertness in biological systems also minimizes any risk of cytotoxicity, making it an optimal choice for chronic wound management [77].
Moisture management: a critical aspect of wound healing is the modulation of the wound bed's moisture milieu, a task at which regenerated cellulose excels. Its hygroscopic properties allow it to deftly equilibrate moisture levels, absorbing superfluous exudate while preserving a moist microenvironment conducive to cellular proliferation and tissue repair. This balanced moisture retention is paramount in preventing desiccation or excessive hydration of the wound site, both of which can impede the reparative processes and lead to complications such as maceration or infection [78].
Barrier protection: in the context of wound protection, regenerated cellulose serves as an impermeable yet breathable barricade, effectively precluding the ingress of pathogenic microorganisms and particulate contaminants. Its semi-permeable nature permits the exchange of essential gases, such as oxygen, while forming a robust defensive layer that diminishes the likelihood of infection. This protective quality is critical in maintaining the sterility of the wound environment, thereby fostering an optimal setting for uninterrupted healing [79].
Breathability: the microporous architecture of regenerated cellulose is engineered to facilitate adequate gaseous exchange, particularly the diffusion of oxygen, which is indispensable for cellular respiration and tissue regeneration. This breathability is instrumental in averting the phenomenon of skin maceration, a condition where excessive moisture leads to the breakdown of surrounding tissue, thereby exacerbating the wound. By promoting an

aerated yet hydrated environment, regenerated cellulose supports the natural reparative processes [80].

Sustainability: beyond its clinical merits, regenerated cellulose is lauded for its environmental stewardship. Sourced from renewable substrates, it embodies a biodegradable alternative to petrochemical plastics, thereby contributing to the reduction of medical waste and curtailing the environmental footprint of healthcare practices. The biodegradable nature of cellulose ensures that once disposed of, these wound dressings undergo natural decomposition, thus preventing the long-term ecological impact associated with synthetic polymers [81].

Customizability: the versatility of regenerated cellulose extends to its facile modification, enabling the incorporation of adjunctive therapeutic agents such as antimicrobial peptides, growth factors, or bioactive molecules. These functional enhancements can be tailored to specific wound types or healing phases, thereby augmenting the material's intrinsic properties with targeted therapeutic effects. The customizable nature of regenerated cellulose thus opens avenues for developing bespoke wound care solutions that address the nuances of individual clinical scenarios [82].

11.8.1 Cellulose-Based Film in the Management of Wound Healing

Regenerated cellulose films represent an innovative advancement in wound care, manifesting as diaphanous yet resilient layers that adhere directly to the wound bed. These films create a hermetic barrier against noxious external contaminants, while their inherent semi-permeability facilitates the diffusion of essential gases, particularly oxygen, which is indispensable for cellular respiration and tissue regeneration. The biocompatibility of these films is unparalleled, owing to the non-immunogenic nature of cellulose, which circumvents the risks of adverse inflammatory responses or allergenic complications that are often associated with synthetic polymers [83]. Moreover, the biodegradability of cellulose-based films renders them an ecologically benign alternative to conventional petrochemical-derived plastic films, aligning with the burgeoning demand for sustainable medical materials. A quintessential attribute of cellulose films is their proficiency in modulating the wound's moisture microenvironment [84]. By preserving a balanced hydration level, these films accelerate the re-epithelialization process, mitigate desiccation-induced tissue necrosis, and diminish the propensity for hypertrophic scar formation. This moisture-retentive capability is pivotal in creating an optimal milieu for the endogenous healing processes to transpire unimpeded. Furthermore, cellulose films can be ingeniously infused with a plethora of bioactive agents, such as antimicrobial peptides or pharmaceutical compounds, thus bestowing them with augmented therapeutic potential. This infusion can transform the film into a dynamic wound dressing that not only acts as a passive barrier but also delivers targeted pharmacological intervention to the wound site, mitigating infection risks and expediting the healing trajectory [85]. Cellulose-based films have been investigated for their potential to create a protective barrier over wounds while maintaining a moist environment conducive to healing. In a study published in MDPI's Materials journal, a cellulose-based film was developed with enhanced mechanical properties and antibacterial activity through the incorporation of silver nanoparticles. The study demonstrated accelerated wound closure and reduced infection rates in *in-vivo* models, highlighting the film's potential in wound care applications [86].

11.8.2 Cellulose-Based Membrane in the Management of Wound Healing

Cellulose-based membranes are engineered to serve as semi-permeable barriers that emulate the physiological characteristics of the human skin, providing a robust yet flexible protective cover for various types of wounds. These membranes, derived from regenerated cellulose, are designed with a nuanced understanding of the skin's natural functions, thereby enabling controlled transcutaneous moisture evaporation while simultaneously forestalling the ingress of pathogenic microorganisms [87]. This selective permeability is crucial in maintaining an ideal wound microenvironment, where the delicate balance between moisture retention and evaporation fosters optimal conditions for tissue repair. The mechanical robustness of cellulose-based membranes is another salient feature, endowing them with the durability required to protect wounds from mechanical stress and external perturbations. Their pliability ensures that they can conform to the contours of the wound, providing comprehensive coverage without compromising on comfort or mobility. This makes them particularly suitable for application over complex wound topographies, such as those encountered in burn injuries or post-surgical incisions. As an environmentally conscious alternative to traditional synthetic membranes, cellulose membranes decompose naturally, thereby obviating the environmental burden imposed by non-biodegradable wound care products. Their eco-friendly profile, coupled with their clinical efficacy, positions cellulose-based membranes as a cornerstone in the future of sustainable wound management [49].

Cellulose-based membranes offer a versatile platform for wound healing, providing a scaffold that supports cellular activities necessary for tissue regeneration. A recent publication in ACS Omega reported the development of a cellulose acetate membrane integrated with chitosan and honey, which exhibited excellent biocompatibility and promoted fibroblast proliferation. The membrane effectively facilitated wound closure in diabetic ulcer models, indicating its potential for chronic wound management [88].

11.8.3 Cellulose-Based Composite in the Management of Wound Healing

Cellulose-based composites represent a sophisticated convergence of regenerated cellulose with an array of bioactive adjuncts, engendering a material that is both functionally and structurally superior for wound healing applications. The amalgamation of cellulose with other biocompatible substances, such as collagen, chitosan, or metallic nanoparticles like silver, imbues these composites with a spectrum of enhanced properties that are tailored to address the multifaceted needs of wound management. For instance, the integration of collagen-a fundamental protein of the extracellular matrix-into cellulose composites augments their capacity to support cellular adhesion, proliferation, and differentiation, thereby expediting the wound healing process [89]. Chitosan, with its intrinsic antimicrobial properties, further elevates the composite's ability to stave off infections, making it particularly beneficial in the treatment of wounds susceptible to microbial invasion. The incorporation of silver nanoparticles imparts potent bactericidal activity, transforming the composite into a formidable agent against a broad spectrum of pathogens, including antibiotic-resistant strains [90]. The synergy between these materials and cellulose culminates in a composite that not only supports the structural integrity of the wound bed but also actively participates in the healing process by modulating the local biological environment. The versatility of cellulose-based composites also extends to their mechanical properties, which can be finely tuned to match the specific requirements of different wound types [91]. Whether the wound necessitates a pliant, elastic dressing or a more rigid, supportive scaffold, the properties of the composite can be adjusted accordingly, ensuring that it meets the demands of even the most challenging clinical scenarios. Composites combining cellulose with other biopolymers or nanoparticles have been shown to enhance the mechanical strength and bioactivity of wound dressings. A study published in Heliyon explored a cellulose-chitosan composite incorporated with curcumin, which exhibited potent anti-inflammatory and antioxidant properties. The composite dressing significantly improved wound healing in rat models by reducing inflammation and promoting collagen deposition [92].

11.8.4 Cellulose-Based Scaffolds in the Management of Wound Healing

Cellulose-based scaffolds are at the forefront of regenerative medicine, offering a three-dimensional architecture that closely mimics the extracellular matrix, thereby providing an optimal framework for tissue regeneration in wound healing. These scaffolds are meticulously engineered from regenerated cellulose, which endows them with a biocompatible and biodegradable profile, ensuring that they seamlessly integrate into the biological milieu without eliciting adverse immune responses or necessitating subsequent removal [93]. The porous morphology of these scaffolds is a key feature, as it permits the unhindered diffusion of nutrients and oxygen, both of which are critical for the sustenance of cellular activities and the promotion of tissue repair. This porosity also facilitates the efflux of metabolic waste products, thereby preventing their accumulation, which could otherwise hinder the healing process. Moreover, the scaffold's architecture is designed to support the ingrowth and migration of cells, particularly fibroblasts and endothelial cells, which are essential for the formation of new tissue and the establishment of a functional vascular network [94]. Cellulose-based scaffolds are particularly advantageous in the context of chronic wound management and skin grafting procedures, where the scaffold serves not only as a temporary matrix for tissue regeneration but also as a bioactive agent that can modulate the wound healing process [95]. By incorporating growth factors, cytokines, or other therapeutic molecules into the scaffold, it is possible to create a bioactive dressing that not only supports tissue regeneration but also actively promotes it. This ability to facilitate and enhance the body's natural healing processes makes cellulose-based scaffolds an invaluable tool in the management of complex wounds [96].

Cellulose-based scaffolds have been engineered to mimic the extracellular matrix, supporting cell attachment, proliferation, and differentiation. Research published in MDPI's Polymers journal described a novel scaffold composed of regenerated cellulose and gelatin, which enhanced the adhesion and proliferation of skin cells. The scaffold demonstrated promising results in wound healing assays, indicating its potential for use in skin tissue engineering [97].

11.8.5 Cellulose-Based Nanofibers in the Management of Wound Healing

Cellulose-based nanofibers represent a paradigm shift in advanced wound care, characterized by their ultra-fine, high-surface-area morphology, which is typically achieved through processes such as electrospinning. These nanofibers, crafted from regenerated cellulose, exhibit a unique set of properties that make them particularly well-suited for interacting with wound tissues at the molecular level [49, 88–92]. The high surface area-to-volume ratio of cellulose nanofibers significantly enhances their capacity to engage with the wound bed, facilitating the adsorption of exudates, the delivery of therapeutic agents, and the promotion of cellular activities such as adhesion and proliferation. Their fine-scale porosity ensures that they provide an ideal environment for cell growth, while also allowing for the controlled release of incorporated bioactive compounds, such as antibiotics, anti-inflammatory agents, or growth factors. This controlled release is crucial in providing sustained therapeutic effects, thereby reducing the frequency of dressing changes and minimizing disruption to the wound healing process. Moreover, the breathability of cellulose nanofibers, coupled with their moisture-retentive properties, creates a microenvironment that is conducive to wound healing, preventing the wound from becoming either too dry or overly saturated. This balance is critical in promoting optimal tissue regeneration while avoiding complications such as maceration or infection. The renewable origin and biodegradability of cellulose nanofibers further enhance their appeal as a sustainable alternative to petroleum-based nanomaterials, aligning with global efforts to reduce the environmental impact of medical waste. Nanofibers made from cellulose offer a high surface area-to-volume ratio, which is advantageous for drug delivery and wound healing applications. A study in ACS Omega investigated electrospun cellulose nanofibers loaded with aloe vera extract, which showed enhanced antibacterial properties and accelerated wound healing in *in vivo* models. The nanofibers were found to be particularly effective in promoting re-epithelialization and reducing scar formation [90, 91].

11.9 Regulatory Aspects of Cellulose-Based Wound Dressings

Regulatory approval is a critical step in the commercialization of cellulose-based wound dressings, particularly because they are intended for medical use. Given that these products are designed to be in direct contact with wounds and human tissues, they must meet stringent safety and efficacy standards set by regulatory bodies such as the U.S. Food and Drug Administration (FDA) and the European Medicines Agency (EMA) [23]. For cellulose-based wound dressings, which are considered medical devices, the regulatory process typically involves:

Biocompatibility testing: since regenerated cellulose is used as an alternative to petroleum-based plastics, it must undergo rigorous biocompatibility testing to ensure it does not cause adverse reactions when in contact with the skin or wound tissue [24].
Clinical trials: depending on the classification of the wound dressing (e.g., Class I, II, or III), clinical trials may be required to demonstrate its safety and effectiveness. These trials assess factors such as wound healing rates, infection control, and patient comfort [27].
Environmental impact: as cellulose-based wound dressings are biodegradable, they may also need to comply with environmental regulations that assess the product's lifecycle, including its disposal and impact on the environment [21].
Manufacturing standards: compliance with Good Manufacturing Practices (GMP) is essential to ensure that the wound dressings are consistently produced and controlled according to quality standards [6].

11.10 Patent Information on Cellulose-Based Wound Dressings

The development of cellulose-based wound dressings, as a renewable alternative to petroleum-based plastics, has led to numerous innovations, many of which are protected by patents. Patent information provides insight into the proprietary technologies and formulations used in these products [91–94].

Patented fabrication processes: many patents focus on the unique methods of regenerating cellulose and incorporating it into wound dressings. This includes processes like electro-spinning, film casting, or composite formation, which are designed to enhance the functional properties of the dressings [24–30].
Functional additives: patents also cover the incorporation of functional additives into cellulose-based wound dressings, such as antimicrobial agents, growth factors, or bioactive compounds. These additives are intended to improve wound healing, reduce infection risk, and provide additional therapeutic benefits [88].
Innovative applications: some patents are granted for the novel applications of regenerated cellulose in wound care. This includes specialized dressings for chronic wounds, burns, or surgical sites, where the unique properties of cellulose provide distinct advantages over traditional materials [94].
Sustainability claims: given the growing emphasis on sustainability, patents may also focus on the environmental benefits of using regenerated cellulose.

These patents highlight the biodegradable nature of the materials and the reduced environmental footprint compared to petroleum-based plastics [90].

11.11 Challenges Associated with Regenerated Cellulose in Wound Dressings

Scalability of production: One of the most formidable challenges is the scalability of production. The processes involved in the fabrication of regenerated cellulose, such as viscose processing or electrospinning, are inherently intricate and resource-intensive. Viscose processing, for example, involves the dissolution of cellulose in a chemical solvent to produce a viscous solution, which is then extruded into a coagulation bath to regenerate the cellulose. This method, while effective on a small scale, poses significant challenges when scaled up for commercial production. The complexity of controlling the various parameters-such as solvent concentration, temperature, and extrusion speed-on an industrial scale often results in variability in the quality of the final product. Moreover, the high cost of the chemicals and the need for extensive purification steps make it difficult to produce large quantities of regenerated cellulose at a price point that is competitive with synthetic alternatives. This challenge is further compounded by the need for specialized equipment and facilities, which require significant capital investment [59–61].

Material properties: while the inherent biocompatibility and biodegradability of regenerated cellulose make it an attractive material for wound dressings, optimizing its mechanical properties to meet the diverse requirements of different wound types presents a significant challenge. Wound dressings must possess a delicate balance of tensile strength, elasticity, and moisture retention to be effective. For instance, a dressing intended for use on a joint or other mobile area of the body must be sufficiently elastic to accommodate movement without tearing, while still maintaining adequate tensile strength to protect the wound. However, enhancing these mechanical properties often comes at the expense of the material's biodegradability [51–53]. The introduction of cross-linking agents or other chemical modifications to improve strength and elasticity can hinder the material's ability to break down naturally, undermining one of its key advantages. Thus, finding a balance between these properties without compromising biodegradability remains a critical challenge [13].

Integration of functional additives: the incorporation of bioactive agents, such as antimicrobial compounds or growth factors, into regenerated cellulose wound dressings is another area fraught with technical difficulties. These additives must be evenly distributed throughout the cellulose matrix to ensure consistent therapeutic efficacy [3]. However, achieving such uniformity can be challenging due to the chemical and physical interactions between the additives and the cellulose. Furthermore, the stability of these bioactive agents over the shelf life of the product is a major concern. Many of these compounds are sensitive to environmental factors such as light, temperature, and humidity, which can cause them to degrade or lose efficacy over time. Developing advanced formulation techniques that can protect these additives and ensure their stability and activity throughout the product's lifespan is essential for the success of regenerated cellulose-based wound dressings [29].

Regulatory hurdles: navigating the regulatory landscape for new wound dressing materials is a time-consuming and costly endeavor. Regenerated cellulose-based dressings must undergo extensive testing to demonstrate their safety, efficacy, and consistency before they can be approved for clinical use. These tests, which include biocompatibility assessments, mechanical testing, and long-term stability studies, are designed to ensure that the product performs as expected under real-world conditions. However, the rigorous nature of these tests, combined with the lengthy approval process, can significantly delay the entry of new products into the market. The cost of this regulatory process can also be prohibitive, particularly for smaller companies or startups with limited resources [37].

Market competition: the wound care market is saturated with well-established products, many of which are made from synthetic materials that have been used and trusted for decades. Introducing a new material such as regenerated cellulose requires substantial marketing and education efforts to convince healthcare providers and consumers of its benefits. Overcoming entrenched preferences for synthetic materials, which are often perceived as more reliable or cost-effective, is a significant challenge. To succeed, companies must not only demonstrate the superior performance of regenerated cellulose-based dressings but also effectively communicate their environmental and health benefits in a way that resonates with both clinicians and patients [50].

11.12 Future Prospects

Advanced functionalization: future developments in regenerated cellulose wound dressings may focus on advanced functionalization techniques, such as embedding nanomaterials or controlled-release systems, to enhance healing properties and provide targeted therapy [60].

Sustainability and circular economy: as the demand for sustainable medical products grows, regenerated cellulose wound dressings are well-positioned to become a preferred choice. Innovations in the sourcing and processing of cellulose, such as using agricultural waste or recycled paper, can further enhance their environmental benefits and align with circular economy principles [72].

Personalized wound care: with advancements in bioprinting and customization technologies, regenerated cellulose could be used to create personalized wound dressings tailored to the specific needs of individual patients. this could improve healing outcomes and reduce the risk of complications [83].

Integration with smart technologies: the future may see the integration of regenerated cellulose wound dressings with smart technologies, such as sensors that monitor wound healing in real-time. These "smart dressings" could provide data on wound conditions, allowing for more proactive and precise wound management [92].

Global market expansion: as awareness of the environmental impact of petroleum-based plastics increases, there is significant potential for regenerated cellulose wound dressings to expand into global markets. With ongoing research and development, these products could become a standard in wound care, particularly in regions with strong environmental regulations and a focus on sustainability [82].

11.13 Conclusion

Regenerated cellulose emerges as a potent alternative to petroleum-based plastics, especially within the realm of wound care, offering a blend of sustainability and high performance. Sourced from renewable natural materials such as wood pulp and cotton, regenerated cellulose represents a significant shift toward eco-friendly solutions in an industry increasingly focused on environmental stewardship. The biodegradability of regenerated cellulose underscores its potential to mitigate the environmental impact of medical waste, which is a growing concern in healthcare systems worldwide. The intrinsic properties of regenerated cellulose make it particularly advantageous for medical applications, specifically in wound dressings. One of the most compelling attributes of regenerated cellulose is its remarkable biocompatibility. As a naturally derived material, it is inherently non-toxic and well-tolerated by human tissues, which minimizes the risk of adverse reactions when used in direct contact with wounds. This makes it an ideal candidate for wound dressings, where maintaining a safe and conducive environment for healing is paramount. Moreover, regenerated cellulose excels in moisture management—a critical factor in wound healing. It has the capacity to absorb excess exudate while maintaining an optimal moisture level at the wound site, which is essential for accelerating the healing process. This balance prevents the wound from either drying out or becoming overly saturated, both of which can impede recovery and increase the risk of complications such as infection or scarring. In addition to its moisture-regulating properties, regenerated cellulose also provides a robust protective barrier against external contaminants. Its structural integrity and porosity allow it to act as a shield, preventing the ingress of bacteria and other harmful agents while still permitting the exchange of oxygen. This breathability is vital for tissue regeneration, as oxygen is a crucial component of cellular metabolism and repair processes. The porous nature of cellulose also helps prevent maceration of the surrounding skin, further supporting the healing process. The versatility of regenerated cellulose is another aspect that amplifies its potential in wound care. It can be readily functionalized with a variety of additives, including antimicrobial agents, growth factors, and other therapeutic compounds. This customizability enables the development of advanced wound care products tailored to specific clinical needs, such as dressings that promote faster healing or provide targeted protection against infections. The ability to integrate these functionalities into a single, biodegradable material represents a significant advancement over traditional wound care products, which often rely on multiple layers or components to achieve similar effects. However, despite these promising attributes, the widespread adoption of regenerated cellulose in wound care is not without its challenges. Scaling up production to meet commercial demand remains a significant hurdle, particularly given the complex and resource-intensive nature of cellulose regeneration processes. Additionally, navigating the regulatory landscape to secure approval for new cellulose-based products can be both time-consuming and costly, often delaying their entry into the market.

Market competition further complicates the landscape, as synthetic materials have long dominated the wound care sector. Overcoming the entrenched preferences for these established products requires not only demonstrating the superior performance of regenerated cellulose but also effectively communicating its environmental benefits to healthcare providers and consumers alike. Looking ahead, the future of regenerated cellulose in wound care is rife with opportunities for innovation. Advances in functionalization, such as the integration of smart technologies or the development of personalized medicine applications, are likely to expand its utility and appeal. As the global medical industry continues to shift toward greener alternatives, regenerated cellulose is poised to play a transformative role, offering a sustainable and efficacious option for next-generation medical materials. In conclusion, regenerated cellulose not only serves as an eco-friendly substitute for petroleum-based plastics but also holds the potential to revolutionize wound care through its unique combination of biocompatibility, functionality, and environmental consciousness. As the demand for sustainable medical products intensifies, regenerated cellulose stands ready to become a cornerstone of the future of medical materials, driving innovation and sustainability in equal measure.

AI disclosure statement: The author acknowledges the use of AI-based tools for minor language editing or grammatical correction. However, all intellectual contributions, analyses, and conclusions presented in this chapter are entirely the author's own.

References

1. Acharya, S., Liyanage, S., Abidi, N., Parajuli, P., Rumi, S.S., Shamshina, J.L., Utilization of cellulose to its full potential: A review on cellulose dissolution, regeneration, and applications. *Polym. (Basel)*, 13, 24, 2021.
2. Alven, S. and Aderibigbe, B.A., Chitosan and cellulose-based hydrogels for wound management. *Int. J. Mol. Sci.*, 21, 24, 1–30, 2020.
3. Amira, N., Armir, Z., Zulkifli, A., Gunaseelan, S., Palanivelu, S.D., Salleh, K.M., *et al.*, Regenerated cellulose products for agricultural and their potential: A review. *Polym. (Basel)*, 1–29, 2021. Available from: https://doi.org/10.3390/polym13203586.
4. Angele, M.K., Schneider, C.P., Chaudry, I.H., Bench-to-bedside review: Latest results in hemorrhagic shock. *Crit. Care*, 12, 4, 1–13, 2008.
5. Ansari, M.M., Heo, Y., Do, K., Ghosh, M., Son, Y.O., Nanocellulose derived from agricultural biowaste by-products–Sustainable synthesis, biocompatibility, biomedical applications, and future perspectives: A review. *Carbohydr. Polym. Technol. Appl.*, 8, June, 100529, 2024. Available from: https://doi.org/10.1016/j.carpta.2024.100529.
6. Arif, A., Azeem, F., Rasul, I., Siddique, M.H., Zubair, M., Muneer, F., *et al.*, Bioplastics from waste biomass of marine and poultry industries. *J. Biosci.*, 48, 2, 2023.
7. Arif, Z.U., The Role of Polysaccharide-based Biodegradable Soft Polymers in the Healthcare Sector. *Adv. Ind. Eng. Polym. Res.*, 2024. Available from: https://doi.org/10.1016/j.aiepr.2024.05.001.
8. Arif, Z.U., Khalid, M.Y., Sheikh, M.F., Zolfagharian, A., Bodaghi, M., Biopolymeric Sustainable Materials and their Emerging Applications Advancements in polymer science and engineering have helped the scientific

community to environmental impact of conventional synthetic plastics, in: *Biopolymers are environmentally Biomedical*, pp. 1–43, 2019.

9. Ates, B., Koytepe, S., Ulu, A., Gurses, C., Thakur, V.K., Chemistry, structures, and advanced applications of nanocomposites from biorenewable resources. *Chem. Rev.*, 120, 17, 9304–62, 2020.
10. ORCA – Online Research @ Cardiff Characterization of cellulose based sponges for wound dressings.
11. Regenerated Nanofibrous Cellulose-Based Adsorbent for Water Pollutants Master's Final Degree Project Regenerated Nanofibrous Cellulose-Based Adsorbent for Water Pollutants, 2023.
12. Benalaya, I., Alves, G., Lopes, J., Silva, L.R., A Review of Natural Polysaccharides: Sources, Characteristics, Properties, Food, and Pharmaceutical Applications. *Int. J. Mol. Sci.*, 25, 2, 2024.
13. Aziz, T., Ullah, A., Ali, A., Shabeer, M., Shah, M.N., Haq, F., *et al.*, Manufactures of bio-degradable and bio-based polymers for bio-materials in the pharmaceutical field. *J. Appl. Polym. Sci.*, 139, 29, 1–21, 2022.
14. Bashari, A., Rouhani Shirvan, A., Shakeri, M., Cellulose-based hydrogels for personal care products. *Polym. Adv. Technol.*, 29, 12, 2853–67, 2018.
15. Boulton, F., A hundred years of cascading - Started by Paul Morawitz (1879-1936), a pioneer of haemostasis and of transfusion. *Transfus. Med.*, 16, 1, 1–10, 2006.
16. Broda, M., Yelle, D.J., Serwańska-Leja, K., Biodegradable Polymers in Veterinary Medicine—A Review. *Molecules.*, 29, 4, 1–30, 2024.
17. Chakrapani, I.S., Rajasekhar, M., Bhaskar, L., Mohan, C., Smart Investigation on Nanocomposites Composed of Carbon Dioxide-Derived, Repeatable Biological Polymers – A review, July, 2024.
18. Cheng, H., Lijie, L., Wang, B., Feng, X., Mao, Z., Vancso, G.J., *et al.*, Multifaceted applications of cellulosic porous materials in environment, energy, and health. *Prog. Polym. Sci.*, 106, xxxx, 101253, 2020. Available from: https://doi.org/10.1016/j.progpolymsci.2020.101253.
19. Čolnik, M., Hrnčič, M.K., Škerget, M., Knez, Z., Biodegradable polymers, current trends of research and their applications, a review. *Chem. Ind. Chem. Eng. Q.*, 26, 4, 401–18, 2020.
20. Côto, T., Moura, I., de Sá, A., Vilarinho, C., Machado, A.V., Sustainable materials based on cellulose from food sector agro-wastes. *J. Renew. Mater.*, 6, 7, 688–96, 2018.
21. Das, P.P. and Chaudhary, V., Moving towards the era of bio fibre based polymer composites. *Clean Eng. Technol.*, 4, 100182, 2021. Available from: https://doi.org/10.1016/j.clet.2021.100182.
22. de Amorim, J.D.P., da Silva Junior, C.J.G., de Medeiros, A.D.M., do Nascimento, H.A., Sarubbo, M., de Medeiros, T.P.M., *et al.*, Bacterial Cellulose as a Versatile Biomaterial for Wound Dressing Application. *Molecules*, 27, 17, 2022.
23. De, S., James, B., Ji, J., Wasti, S., Zhang, S., Kore, S., *et al.*, Biomass-derived composites for various applications. *Adv. Bioenergy.*, 8, 145–96, 2023.

24. Digitalcommons@umaine D and Kugell, D., Nanocellulose Based Foams For Low-Cost Disposable Medical Nanocellulose Based Foams For Low-Cost Disposable Medical Applications Applications, 2023. Available from: https://digitalcommons.library.umaine.edu/etd/3866.
25. Elkaliny, N.E., Alzamel, N.M., Moussa, S.H., Elodamy, N.I., Madkor, E.A., Ibrahim, E.M., *et al.*, Macroalgae Bioplastics: A Sustainable Shift to Mitigate the Ecological Impact of Petroleum-Based Plastics. *Polym. (Basel)*, 16, 9, 2024.
26. Emenike, E.C., Iwuozor, K.O., Saliu, O.D., Ramontja, J., Adeniyi, A.G., Advances in the extraction, classification, modification, emerging and advanced applications of crystalline cellulose: A review. *Carbohydr. Polym. Technol. Appl.*, 6, June, 100337, 2023. Available from: https://doi.org/10.1016/j.carpta.2023.100337.
27. Farooq, A., Patoary, M.K., Zhang, M., Mussana, H., Li, M., Naeem, M.A., *et al.*, Cellulose from sources to nanocellulose and an overview of synthesis and properties of nanocellulose/zinc oxide nanocomposite materials. *Int. J. Biol. Macromol.*, 154, 1050–73, 2020. Available from: https://doi.org/10.1016/j.ijbiomac.2020.03.163.
28. Farshi, P., Salarian, R., Rabiee, M., Alizadeh, S., Gholipourmalekabadi, M., Ahmadi, S., *et al.*, Design, preparation, and characterization of silk fibroin/carboxymethyl cellulose wound dressing for skin tissue regeneration applications. *Polym. Eng. Sci.*, 62, 9, 2741–9, 2022.
29. Felgueiras, C., Azoia, N.G., Gonçalves, C., Gama, M., Dourado, F., Trends on the Cellulose-Based Textiles: Raw Materials and Technologies. *Front. Bioeng. Biotechnol.*, 9, March, 1–20, 2021.
30. Fouad, S. and Khalifa, G., Fabrication and characterization of antibacterial herbal drug- loaded polylactic acid/cellulose acetate composite nanofibers for wound dressing applications, November, 2015.
31. Galiano, F., Briceño, K., Marino, T., Molino, A., Christensen, K.V., Figoli, A., Advances in biopolymer-based membrane preparation and applications. *J. Memb. Sci.*, 564, 562–86, 2018.
32. Gamage, A., Thiviya, P., Mani, S., Ponnusamy, P.G., Manamperi, A., Evon, P., *et al.*, Environmental Properties and Applications of Biodegradable. *Polym. (Basel)*, 14, 1, 2022.
33. Garrison, T.F., Murawski, A., Quirino, R.L., Bio-based polymers with potential for biodegradability. *Polym. (Basel)*, 8, 7, 1–22, 2016.
34. George, M. and Montemagno, C., Cellulose Based Materials: in-Depth Property Survey and Assessment. *Int. Ref. J. Eng. Sci.*, 6, 5, 55–76, 2017. Available from: www.irjes.com.
35. Gómez-Gast, N., Cuellar, M.D.R.L., Vergara-Porras, B., Vieyra, H., Biopackaging Potential Alternatives: Bioplastic Composites of Polyhydroxyalkanoates and Vegetal Fibers. *Polym. (Basel)*, 14, 6, 1–24, 2022.
36. Goralnick, E., Ezeibe, C., Chaudhary, M.A., McCarty, J., Herrera-Escobar, J.P., Andriotti, T., *et al.*, Defining a Research Agenda for Layperson Prehospital

Hemorrhage Control: A Consensus Statement. *JAMA Netw. Open*, 3, 7, 1–14, 2020.
37. Gregory, D.A., Tripathi, L., Fricker, A.T.R., Asare, E., Orlando, I., Raghavendran, V., *et al.*, Bacterial cellulose: A smart biomaterial with diverse applications. *Mater. Sci. Eng. R Rep.*, 145, 2021.
38. Hachimi Alaoui, C., Réthoré, G., Weiss, P., Fatimi, A., Sustainable Biomass Lignin-Based Hydrogels: A Review on Properties, Formulation, and Biomedical Applications. *Int. J. Mol. Sci.*, 24, 17, 2023.
39. Helanto, K., Matikainen, L., Talj, R., Rojas, O.J., Bio-based polymers for sustainable packaging and biobarriers: A critical review. *BioResources.*, 14, 2, 4902–51, 2019.
40. Huang, Y., Kormakov, S., He, X., Gao, X., Zheng, X., Liu, Y., *et al.*, Conductive Polymer Composites from Renewable Resources: An Overview of Preparation, Properties, and Applications. *Polym. (Basel)*, 11, 2, 2019.
41. Ikumapayi, O.M., Laseinde, O.T., Adebayo, A.S., Oluwafemi, J.R., Ogedengbe, T.S., Akinlabi, S.A., *et al.*, An Overview on recent trends in Biopolymer Base Composites for Tissue Regeneration. *E3S Web Conf.*, 391, 1–9, 2023.
42. Imani, M., Carreras, I.M.V., Dimić-Mišić, K., Kostić, M., Barceló, E., García, M.A.C., *et al.*, Investigating waste mineral-filled cellulose sourcing in circular economy for regeneration into composite: Matching existing market volumes of oil-based plastics for packaging. *Clean Circ. Bioecon.*, 8, April, 2024.
43. Janaswamy, S., Yadav, M.P., Hoque, M., Bhattarai, S., Ahmed, S., Cellulosic fraction from agricultural biomass as a viable alternative for plastics and plastic products. *Ind. Crops Prod.*, 179, 1–26, 2022.
44. Jin, X., Open Research Online. *Choice Rev. Online*, 51, 06, 51-2973-51–2973, 2014.
45. Joseph, T.M., Unni, A.B., Joshy, K.S., Kar Mahapatra, D., Haponiuk, J., Thomas, S., Emerging Bio-Based Polymers from Lab to Market: Current Strategies, Market Dynamics and Research Trends. *C-Journal Carbon Res.*, 9, 1, 2023.
46. Kamal, T., Ul-Islam, M., Fatima, A., Ullah, M.W., Manan, S., Cost-Effective Synthesis of Bacterial Cellulose and Its Applications in the Food and Environmental Sectors. *Gels.*, 8, 9, 1–24, 2022.
47. Kiselioviene, S., Baniukaitiene, O., Harkavenko, V., Babenko, N.A., Liesiene, J., Cellulose hydrogel sheets for wound dressings. *Cellul. Chem. Technol. Cellul. Chem. Technol.*, 50, 9, 915–23, 2016.
48. Kubicek, C.P., Synthetic biopolymers. *Synth. Biol.*, 307–335, 2015.
49. Vasile, C. and Baican, M., Lignins as Promising Renewable Biopolymers and Bioactive Compounds for High-Performance Materials. *Polym. (Basel)*, 15, 15, 1–52, 2023.
50. Kucińska-Lipka, J., Gubanska, I., Janik, H., Bacterial cellulose in the field of wound healing and regenerative medicine of skin: recent trends and future prospectives. *Polym. Bull.*, 72, 9, 2399–419, 2015.

51. Latifi, S., Saoiabi, S., Loukili, E.H., Azzaoui, K., Hammouti, B., Abidi, N., *et al.*, Preparation of cellulose-hydroxyapatite composites using 3D printing for biomedical applications. *Morrocan J. Chem.*, 12, 2, 884–914, 2024.
52. Lawson, L., Degenstein, L.M., Bates, B., Chute, W., King, D., Dolez, P.I., Cellulose Textiles from Hemp Biomass: Opportunities and Challenges. *Sustain.*, 14, 22, 1–16, 2022.
53. Magalhães, S., Alves, L., Medronho, B., Fonseca, A.C., Romano, A., Coelho, J.F.J., *et al.*, Brief overview on bio-based adhesives and sealants. *Polym. (Basel)*, 11, 10, 2019.
54. Mahmood, H., Moniruzzaman, M., Yusup, S., Welton, T., Ionic liquids assisted processing of renewable resources for the fabrication of biodegradable composite materials. *Green Chem.*, 19, 9, 2051–75, 2017.
55. MZ Al, M., MD, I., Mobarak, M.H., The Development of Eco-Friendly Biopolymers for Use in Tissue Engineering and Drug Delivery. *J. Nanomater.*, 2023, 2023.
56. Mao, L., Hu, S., Gao, Y., Wang, L., Zhao, W., Fu, L., *et al.*, Biodegradable and Electroactive Regenerated Bacterial Cellulose/MXene (Ti3C2Tx) Composite Hydrogel as Wound Dressing for Accelerating Skin Wound Healing under Electrical Stimulation. *Adv. Healthc. Mater.*, 9, 19, 1–13, 2020.
57. Masek, A. and Kosmalska, A., Technological limitations in obtaining and using cellulose biocomposites. *Front. Bioeng. Biotechnol.*, 10, August, 1–17, 2022.
58. Meghana, M.C., Nandhini, C., Benny, L., George, L., Varghese, A., A road map on synthetic strategies and applications of biodegradable polymers, in: *Polymer Bulletin*, vol. 80, pp. 11507–11556, Springer Berlin Heidelberg, Berlin, Heidelberg, 2023, Available from: https://doi.org/10.1007/s00289-022-04565-9.
59. Mikhailidi, A., Ungureanu, E., Belosinschi, D., Tofanica, B., Volf, I., Cellulose-Based Metallogels — Part 3: Multifunctional Materials, 2023.
60. Mishra, R.K., Ha, S.K., Verma, K., Tiwari, S.K., Recent progress in selected bio-nanomaterials and their engineering applications: An overview. *J. Sci. Adv. Mater. Devices*, 3, 3, 263–88, 2018. Available from: https://doi.org/10.1016/j.jsamd.2018.05.003.
61. Mohan, D., Teong, Z.K., Bakir, A.N., Sajab, M.S., Kaco, H., Extending cellulose-based polymers application in additive manufacturing technology: A review of recent approaches. *Polym. (Basel)*, 12, 9, 2020.
62. Moradpoor, H., Mohammadi, H., Safaei, M., Mozaffari, H.R., Sharifi, R., Gorji, P., *et al.*, Recent Advances on Bacterial Cellulose-Based Wound Management: Promises and Challenges. *Int. J. Polym. Sci.*, 2022, 2022.
63. Mori, R., Replacing all petroleum-based chemical products with natural biomass-based chemical products: a tutorial review. *RSC Sustain.*, 1, 2, 179–212, 2023.
64. Mujtaba, M., Fernandes Fraceto, L., Fazeli, M., Mukherjee, S., Savassa, S.M., Araujo de Medeiros, G., *et al.*, Lignocellulosic biomass from agricultural waste to the circular economy: A review with focus on biofuels, biocomposites and

bioplastics. *J. Clean Prod.*, 402, March, 136815, 2023. Available from: https://doi.org/10.1016/j.jclepro.2023.136815.
65. Mukheem, A., Hossain, M., Shahabuddin, S., Muthoosamy, K., *Bioplastic Polyhydroxyalkanoate (PHA): Recent Advances in Modification and Medical Applications*, pp. 1–37, Preprints, Basel, Switzerland, 2018, Available from: https://www.preprints.org/manuscript/201808.0271/v1.
66. Muñoz-Bonilla, A., Echeverria, C., Sonseca, A., Arrieta, M.P., Fernández-García, M., Bio-based polymers with antimicrobial properties towards sustainable development. *Mater. (Basel)*, 12, 4, 2019.
67. Omidian, H., Akhzarmehr, A., Chowdhury, S.D., Advancements in Cellulose-Based Superabsorbent Hydrogels, 2024.
68. Paulauskiene, T., Teresiute, A., Uebe, J., Tadzijevas, A., Sustainable Cross-Linkers for the Synthesis of Cellulose-Based Aerogels: Research and Application. *J. Mar. Sci. Eng.*, 10, 4, 2022.
69. Portela, R., Leal, C.R., Almeida, P.L., Sobral, R.G., Bacterial cellulose: A versatile biopolymer for wound dressing applications. *Microb. Biotechnol.*, 12, 4, 586–610, 2019.
70. Radu, C.D., Verestiuc, L., Ulea, E., Lipsa, F.D., Vulpe, V., Munteanu, C., *et al.*, Evaluation of keratin/bacterial cellulose based scaffolds as potential burned wound dressing. *Appl. Sci.*, 11, 5, 1–17, 2021.
71. Ramezani, G., Stiharu, I., van de Ven, T.G.M., Nerguizian, V., Advancements in Hybrid Cellulose-Based Films: Innovations and Applications in 2D Nano-Delivery Systems. *J. Funct. Biomater.*, 15, 4, 2024.
72. Raza, S., Zhang, J., Ali, I., Li, X., Liu, C., Recent trends in the development of biomass-based polymers from renewable resources and their environmental applications. *J. Taiwan Inst. Chem. Eng.*, 115, 293–303, 2020.
73. Ribul, M., Regenerative textiles: A framework for future materials circularity in the textile value chain. *Sustain.*, 13, 24, 2021.
74. Samir, A., Ashour, F.H., Hakim, A.A.A., Bassyouni, M., Recent advances in biodegradable polymers for sustainable applications. *npj Mater. Degrad.*, 6, 1, 2022.
75. Sariboga, R. and Sarioglu, O.F., Applications of Cellulose-Based Nanomaterials for Sustainability and Therapeutics: A Review. *ChemBioEng Rev.*, 4, 1–15, 2024.
76. Satchanska, G., Davidova, S., Petrov, P.D., Natural and Synthetic Polymers for Biomedical and Environmental Applications. *Polym. (Basel)*, 16, 8, 2024.
77. Selvamurugan Muthusamy, M. and Pramasivam, S., Bioplastics – An Eco-friendly Alternative to Petrochemical Plastics. *Curr. World Environ.*, 14, 1, 49–59, 2019.
78. Shabbir, M., Regenerated Cellulose and Composites: Morphology-Property Relationship, 2023.
79. Sharif, A., Renewable Resource-Based Polymers Renewable Resource-Based Polymers, May, 2019.

80. Sikhosana, S.T., Gumede, T.P., Malebo, N.J., Ogundeji, A.O., Motloung, B., Medicinal plants as a cellulose source for the fabrication of poly(lactic acid) composites: A mini-review. *Polym. Renew. Resour.*, 14, 1, 44–57, 2023.
81. Singh, P., Duarte, H., Alves, L., Antunes, F., Le, M.N., Dormanns, J., *et al.*, From Cellulose Dissolution and Regeneration to Added Value Applications - Synergism Between Molecular Understanding and Material Development, in: *To cite this version: HAL Id: hal-03080667 World's largest Science, Technology & Medicine Open Access book pu*, 2020.
82. Sukhodub, L., Kumeda, M., Sukhodub, L., Tsyndrenko, O., Petrenko, O., Prokopiuk, V., *et al.*, The effect of carbon and magnetic nanoparticles on the properties of chitosan-based neural tubes: Cytotoxicity, drug release, *In Vivo* nerve regeneration. *Carbohydr. Polym. Technol. Appl.*, 8, June, 100528, 2024. Available from: https://doi.org/10.1016/j.carpta.2024.100528.
83. Taokaew, S., Recent Advances in Cellulose-Based Hydrogels Prepared by Ionic Liquid-Based Processes. *Gels*, 9, 7, 2023.
84. Tayeb, A.H., Amini, E., Ghasemi, S., Tajvidi, M., Cellulose nanomaterials-binding properties and applications: A review. *Molecules.*, 23, 10, 1–24, 2018.
85. Thakur, M., Chandel, M., Rani, A., Sharma, A., Introduction to Biorenewable Nanocomposite Materials: Methods of Preparation, Current Developments, and Future Perspectives. *ACS Symp. Ser.*, 1411, 1–24, 2022.
86. Thomas, B., Raj, M.C., Athira, B.K., Rubiyah, H.M., Joy, J., Moores, A., *et al.*, Nanocellulose, a Versatile Green Platform: From Biosources to Materials and Their Applications. *Chem. Rev.*, 118, 24, 11575–625, 2018.
87. Ul-Islam, M., Khan, T., Khattak, W.A., Park, J.K., Bacterial cellulose-MMTs nanoreinforced composite films: Novel wound dressing material with anti-bacterial properties. *Cellulose.*, 20, 2, 589–96, 2013.
88. Veeman, D., Sai, M.S., Sureshkumar, P., Jagadeesha, T., Natrayan, L., Ravichandran, M., *et al.*, Additive Manufacturing of Biopolymers for Tissue Engineering and Regenerative Medicine: An Overview, Potential Applications, Advancements, and Trends. *Int. J. Polym. Sci.*, 2021, 2021.
89. Visage, L., Chaparro, A.M., Press, A., Documents Education, (2023) Science Direct. Biomaterials, Introduction to the Chemistry of Life. Charles Molnar and Jane Gair (2021) Concepts of Biology, Cole, L., Kramer, P. R. Macronutrients (2016). *Hum. Physiology Biochem. Bas.*, 1, 2023.
90. Wang, Y.W., Liu, C.C., Cherng, J.H., Lin, C.S., Chang, S.J., Hong, Z.J., *et al.*, Biological effects of chitosan-based dressing on hemostasis mechanism. *Polym. (Basel)*, 11, 11, 1–14, 2019.
91. Wanjuu, C.W., Open PRAIRIE: Open Public Research Access Institutional Repository and Information Exchange Biodegradable Cellulose Films as Alternatives to Plastics, 2020.
92. Xiong, R. and Eng, M., Cellulose-Based Biodegradable Blends and Composites Regenerated from Ionic Liquids, 2012.

93. Xu, A., Wang, Y., Xu, X., Xiao, Z., Liu, R., A Clean and Sustainable Cellulose-Based Composite Film Reinforced with Waste Plastic Polyethylene Terephthalate. *Adv. Mater. Sci. Eng.*, 2020, 2020.
94. Yang, M., Ward, J., Choy, K.L., Nature-Inspired Bacterial Cellulose/Methylglyoxal (BC/MGO) Nanocomposite for Broad-Spectrum Antimicrobial Wound Dressing. *Macromol. Biosci.*, 20, 8, 2020.
95. Zhong, C., Industrial-Scale Production and Applications of Bacterial Cellulose. *Front. Bioeng. Biotechnol.*, 8, December, 1–19, 2020.
96. Zorah, M., Mudhafar, M., Naser, H.A., Mustapa, I.R., the Promises of the Potential Uses of Polymer Biomaterials in Biomedical Applications and Their Challenges. *Int. J. Appl. Pharm.*, 15, 4, 27–36, 2023.
97. Zwawi, M., A review on natural fiber bio-composites, surface modifications and applications. *Molecules*, 26, 2, 2021.

12

Bacterial Cellulose for Advanced Wound Healing

Mrunal Patil, Nikita Shinde and Shivaji Kashte*

Department of Stem Cell and Regenerative Medicine and Medical Biotechnology, Centre for Interdisciplinary Research, D. Y. Patil Education Society (Deemed to be University), Kolhapur (MS), India

Abstract

Cellulose is a commonly found biopolymer in plant cell walls and certain bacteria and archaea. When produced by bacteria, it is known as "bacterial cellulose" (BC). Alginate, gelatin, and collagen are well-known natural polymers appreciated for biocompatibility but often lack mechanical strength. On the other hand, synthetic polymers like poly-lactic acid and polyethylene offer significant mechanical resilience but compromise biocompatibility. BC is outstanding due to its remarkable hydrophilic properties, biocompatibility, porosity, water retention, gas exchange capacity, and thermal insulation abilities. The superior properties of BC make it a versatile candidate for wound healing biomaterial in medical and hygiene industries. BC can potentially reduce inflammation and fight microbial infections, making it a promising material for treating various types of wounds, such as chronic, acute, burn, and diabetic ulcers. Therefore, utilizing BC for wound healing can leverage its multifaceted benefits to improve patient outcomes and expedite healing.

Keywords: Bacterial cellulose, wound healing, wound dressings, biocompatibility, biodegradability

12.1 Introduction

Thousands of deaths arise due to road accidents and surgical processes globally. As per Road Traffic Crash and Injury Data published officially by

**Corresponding author*: ShivajiKashte.cir@dypgroup.edu.in

Mohd Shabbir (ed.) Regenerated Cellulose: A Renewable Alternative to Petroleum-Based Plastics, (345–386) © 2026 Scrivener Publishing LLC

the Ministry of Road Transport and Highways, India (MoRTH), a total of 371,884 people were injured and 153,972 persons lost lives in road accidents in the year 2021, which corresponds to a mortality of 11.3 deaths per 100,000 populations. The primary reason for these deaths is uncontrolled bleeding, hemorrhage occurring immediately after the accident, and the risk of infection arising due to natural as well as hospital environment within 24-72 hrs after the accident. Therefore, in such cases, there is a need to execute proper wound care management, which will help to reduce mortality [1, 2].

Traditionally, sutures, staples, and cotton wound dressings were used to cover the wound openings by maintaining a dry environment, and blotting exudates and preventing infections were the only considerations. These methods might show side effects like bleeding, sepsis, fissures, fistula formations, and, in the worst conditions, organ damage (Food and Drug Administration, US). Nowadays, the expectancy for wound dressing is transformed as scientists are experimenting to produce materials that show not only basic expectations but advanced features like skin regeneration where the dressing material promotes the skin regeneration process actively. Therefore, it's pivotal to update traditional ways with multifunctional dressings that resist infections as well as promote the healing process [3].

Polymer-based biomaterials are practiced to manage wounds through dressings or regenerative scaffolds [4]. The polymers are of 2 types: natural and synthetic. Natural polymers are derived from natural sources such as mammals, plants, and microorganisms. The examples of natural polymers are cellulose (plant or animal-derived polysaccharide), chitosan (animal-based polysaccharide), hyaluronic acid (polysaccharide sourced from animals), sodium alginate (polysaccharide obtained from seaweeds, brown algae), carrageenan (polysaccharide sourced from red algae) and collagen (protein derived from mammals), silk fibroin (protein obtained from a silkworm), keratin (protein sourced from mammals). Choosing naturally produced products delivers several advantages, such as biocompatibility for prolonged use without adverse effects, biodegradability for self-removal after work has been done, versatility for use in various biomedical applications, sustainability as derived from renewable sources, and fictionalization ability for modification that may be used to enhance properties [5]. Natural polymers offer a range of properties, but some properties have limitations, such as mechanical strength and antimicrobial activity. Therefore, synthetic polymers are developed, giving mechanical strength, antibacterial properties, and biodegradability. Examples are polyethylene glycol (PEG), polyurethane (PU), polycaprolactone (PCL), polyvinyl alcohol (PVA), polyethylene oxide (PEO), polyvinyl pyrrolidone (PVP). Synthetic polymers are different from ECM. Therefore, they may show toxicity or inflammation, and the cost is high. Therefore, there is a need to opt for the materials by combining natural and synthetic polymers to manage desired properties [6]. Therefore, bacterial cellulose and its composite materials could be the best alternative to existing materials for wound healing.

12.2 Wound Healing

Being the first barrier, the skin is the most vulnerable part of the body that can easily be wounded due to mechanical injury, ulcers, or surgical processes. The skin has the intrinsic healing property for small-diameter wounds (measuring 1-2 cm). The healing process may be delayed if the wound has a larger diameter. This extended exposure of the inner dermal layer to the outer environment increases the risk of infection. Therefore, proper wound management is essential in these situations [3].

The wound healing process comprises four distinct phases (Figure 12.1);

12.2.1 Hemostasis

Immediately after injury, the body's remarkable natural healing process kicks in. The injured site is identified, and platelets gather at the wound site to form a blood clot, effectively stopping the bleeding. This process is orchestrated by various factors in the Extracellular Matrix (ECM), such as Platelet-Derived Growth Factors (PDGF), Epidermal Growth Factor (EGF), Fibroblast Growth Factor (FGF), and cytokines like Interleukin-1β (IL-1β) and transforming Growth Factor (β-TGF), begins releasing growth factors after 6 hours of injury.

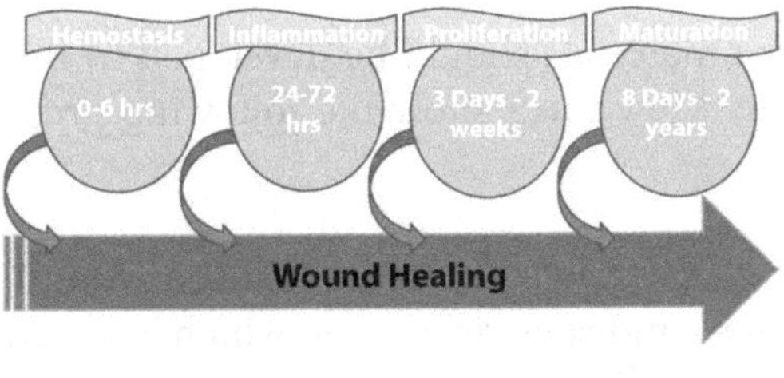

12.2.2 Inflammation

The second phase of wound healing, inflammation, is a meticulous process. Inflammatory cells are activated to clear necrosis products like dead tissues and opportunistic bacteria. Macrophages migrate toward the wound area and secrete cytokines and growth factors, further facilitating healing. Neutrophils migrate to the wound site during the initial stage of inflammation (24-36 hrs), and macrophages arrive at the site during the secondary stage (48-72 hrs). This process includes angiogenesis and wound contraction, all of the body's intricate healing mechanisms.

12.2.3 Proliferation

After the inflammation stage, the proliferation stage (from 3 days to 2 weeks) starts. During this stage, fibroblasts migrate toward the wound site, angiogenesis occurs, and re-epithelialization occurs to build new ECM, followed by neovascularization to form blood vessels.

12.2.4 Remodeling and Maturation

The remodeling or maturation phase begins after the 8th day of injury, which may last up to 2 years in adults. This stage helps in the generation of collagen network and to mature scar tissue [4, 7].

On the basis of the above wound healing pattern, the wounds are classified into two categories;

a. **Acute Wounds:** In this type, the wound healing process follows distinct organized phases (Natural Process) that may overlap: hemostasis, Inflammation, Proliferation, and remodeling. It recovers in 2-4 weeks.
b. **Chronic Wounds:** In this type, the natural wound healing pattern of phases is not followed as they halt in one phase and do not show healing for up to 4 weeks [8–11].

The wound care materials should be developed to improve the quality of the patient's life. They must show the following properties to manage wounds properly;

a. Provide Moist Environment: Advanced wound care material retains a moist environment, which is beneficial for cell migration, proliferation, and tissue regeneration.

b. Minimizes Rate of Infection: Advance wound care materials reduce the chances of infection by their inherent antibacterial activity or by adding antimicrobial agents.
c. Promotes Wound Healing: The advanced wound care materials are supplemented with various factors that promote wound healing.
d. Painless Delivery: Advanced wound care materials (injection, patch, spray, etc.) are easier to apply than traditional methods like sutures and staples.
e. Other Advantages: The advanced wound care materials bear ideal features like Biocompatibility, Hemocompatibility, Swellability, Gaseous exchange capacity, Angiogenic property, and Biodegradability [8, 9].

The natural polymer cellulose is a material that can provide these properties for wound management.

12.3 Cellulose

Cellulose is one of the most abundant and readily available polysaccharides on earth. It can be sourced from wood plants (cell wall is cellulosic), some bacterial species, and tunicates (Only animal source). The cellulose showed various morphological forms: fibers, nanofibrils, etc. These varieties are found either due to the inherent diversity of organic material sources or conditions maintained during the biosynthesis process [14, 15].

12.3.1 Structure of Cellulose

The cellulose is a high molecular weight, homopolysaccharide made up of glucose units linked by beta-1, 4 glycosidic bonds. It is arranged in organized parallel layers to form elementary fibrils. This parallel structure is again linked with weak forces like van der Waals forces and hydrogen bonds. Cellulose is semi-crystalline, and its degree of crystallinity changes with the source. In plant-based material, it is 40-60%, while in bacteria and tunicin, it is 80-100%. The crystalline cellulose may present in allomorphs: cellulose I, cellulose II, cellulose III, and cellulose IV. The presence of allomorphs varies from source to source. The crystalline structure of cellulose has a hydrogen-bonded network, which makes it durable with no definite melting point. This helps to make it stable thermally and eventually to maintain consistency in properties, e.g., rigidity. Reactive functional groups (e.g., OH group) make cellulose chemically active and hence suitable for various biochemical applications [14, 16–18].

12.4 Bacterial Cellulose

The Bacterial cell produces a natural nanomaterial, an exopolysaccharide called bacterial cellulose (BC). BC is chemically equivalent to plant cellulose but has a high degree of crystallinity and purity.

12.4.1 Biosynthesis of BC

The Biosynthesis pathway of BC (Figure 12.2) starts with precursors such as Glucose or Fructose hexoses (6-carbon sugars), Pyruvate, Glycerol, Dihydroxyacetone Trioses (3 Carbon sugar), or dicarboxylic acid like tetroses (4-Carbon sugar). The trioses and tetroses are intermediate compounds from Kreb's or TCA Cycle. These sugars are first converted to Uridine diphosphoglucose (UDP Glucose) as UDP Glucose is the building block of Cellulose structure. The precursors are first converted to Glucose, which is then phosphorylated by the enzyme glucokinase to form Glucose 6

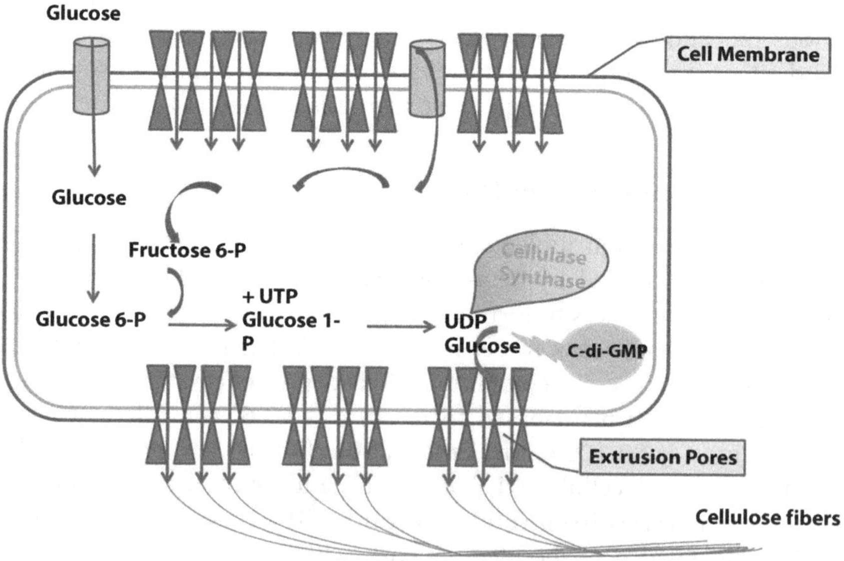

phosphate (Glucose 6-P). Then, The Glucose 6 - P is converted to Glucose 1 Phosphate (Glucose 1-P) by the enzyme Phosphoglucomutase. With the help of one UTP molecule, Glucose 1-P is converted to UDP Glucose, a monomer that polymerizes into a Cellulose polymer. Here, the major role played by bis- (3',5')- cyclic diguanylic acid (c-di- GMP), which acts as a secondary messenger to activate an enzyme cellulose synthase complex which brings about a polymerization reaction [19, 20].

The BC can be synthesized by various bacterial species, including *Sarcina ventriculi* (Gram-Positive) and *Acetobacter Azotobacter*, Gluconacetobacter, Rhizobium, Agrobacterium, Pseudomonas, Salmonella, Alcaligenes (Gram-Negative) [21]. *Komagataeibacter xylinus* (*K. xylinus*), can produce four allomorphs (I–IV) of BC by fermenting sugars extracellularly. The cellulose biosynthesis pathway starts with utilizing glucose molecules and the subsequent formation of 1, 4 linear glucan chains with the help of the cellulose synthase enzyme. The BC produces an oxygen-rich surface. Therefore, it forms a membrane at the air-liquid surface in static culture conditions at 30o C, pH 4-6, and dissolved oxygen (DO) concentration of 3.1 ppm [22].

Yuan *et al.* produced wet nanocellulose membranes using *K. xylinus* strains, namely ATCC 23770, DHU-ATCC-1, and DHU-WX-1, by employing static culture to examine the efficiency of full-thickness wound healing. The produced nanocellulose by these strains differs in fiber density, fiber diameter, and mechanical properties. The DHU-WX-1 showed the best performance due to slow water loss, which helps the wound to stay moist for a longer period, ultimately resulting in fast healing Within 12 days, while DHU-ATCC-1 in 15 days and ATCC 23770 in 17 days [23].

Saleh *et al.* isolated *Lactiplantibacillus plantarum* from rotten fruit samples with 16s rRNA confirmation and optimized the Yamanaka medium components for the fermentation procedure. They found that 15g/l glucose, 13 g/l yeast extract, 1 g/l MgSO4, 4 g/l KH2PO4, and 7 ml/l ethanol, at pH 7.2, 30°C, for 8 days, with an inoculum size of 11%, increases the yield of BC from 2.7 g/L to 4.51 g/L. [24].

Various fruit peels are used to design culture media for sustainable production of BC. Hasanin *et al.* used Strawberry, Pineapple, Citrus orange, Apple, Pomegranate, and Tomato to isolate new BC producer *Achromobacter* S33 among which Strawberry is a potent producer. The researcher also screened various fruit peel waste to culture BC producers, among which Mango Peel Waste (MPW) showed a considerable yield of 1.22 g/l. The BC produced has 79% crystallinity, 10 nm diameter, and 200 nm fiber length [25].

The scientist also conducted a co-culture method of fermentation to evaluate the effect on BC production. Nazarova *et al.* co-cultured

Komagataeibacter sucrofermentans with *Leuconostoc mesenteroids* and *Xanthomonas campestris* to check whether competition or antagonism within a mixed culture helps to increase output significantly or not. They found that *K. sucrofermentans* with *L. mesenteroids* show increased yield (5.99 ± 0.02 g/l) two and a half fold compared to a monoculture of BC producer (2.25 ± 0.05 g/l). While with *X. campestris*, there is no difference in BC production [26].

12.4.2 Factors Affecting the Production of BC

The various physiochemical factors affecting BC synthesis are yield, structural variation, and physicochemical properties (Table 12.1). The first primary consideration is the type of fermentation method. If the static culture (The synthesis container/fermenter is placed in static condition), the synthesis of BC film takes place at the air-liquid interface as the bacteria are considered to move in the direction of the oxygen-rich surface. Chen *et al.* used *Acetobacter xylinum* ATCC 23767 for fermentation with media components glucose, yeast extract, peptone, and citric acid, maintaining optimal conditions of 28°C, pH 6 for 9 days. The optimization with static methods yields 8.5g/l BC with dense three Dimensional network structure with uniform pores with good hydrophilicity and structural stability [27].

The synthesis path depends on the starting raw material, the Carbon source. Digel *et al.* cultured *Medusomyces gisevii* with different carbon sources, glucose, and sucrose concentrations. The yield of BC with 10 g/l sugar concentration and 18-20 days of incubation is higher in glucose (173 g L^{-1}) than sucrose (163.5g/l) as it can be directly incorporated into the cellulose synthesis pathway. But the density, crystallinity, and surface roughness are more in the presence of sucrose, which is advantageous, but it also possesses large pore size up to 2 μm, which is not ideal for wound healing applications [28].

Another factor to be considered is the pH of the culture media; while fermenting BC, gluconic acid is formed as a by-product, which over accumulation reduces the yield of BC. To overcome this, Hishammuddin *et al.* used 0.2 M acetate buffered medium with a static condition, which maintains pH value to optimized level at 3.8- 4.0and increases yield of BC (5.66 g/l) as compared to Hestrin and Schramm (HS) medium (2.56 g/l) and control group (2.72 g/l) [29].

Pham *et al.* varied three parameters while fermenting BC using pineapple waste solution as a carbon source in the presence of *Acetobacter xylinum*. The researchers varied three parameters and evaluated the obtained structural changes in BC with the help of SEM and XRD. The fermentation

Table 12.1 Bacterial cellulose producing species and yield.

Sr. no.	Bacterial species	Method	Yield	References
1.	*Komagataeibacter sucrofermentans* B-11267 with *Xanthomonas campestris* (Xanthan producer) and *Leuconostoc mesenteroids* (Dextran producer)	Co-culture Method	*Komagataeibacter sucrofermentans* B-11267 with *Xanthomonas campestris* - 5.99 ± 0.02g/l *Komagataeibacter sucrofermentans* B-11267 with *Leuconostoc mesenteroids*- 2.25 + 0.05 g/l	[26]
2.	*Lactiplantibacillus plantarum* AS.6	Static Culture	4.7 g/l	[24]
3.	*Acetobacter xylinum* ATCC 23,767	Static Culture	8.5 g/L	[27]
4.	*Medusomyces gisevii*	Static Culture	Glucose (Carbon Source) - 173 g L^{-1} Sucrose (Carbon Source) - 163.5 g L^{-1}	[28]
5.	*Acetobacter xylinum* 0416	Static Culture	5.66 g/L (Acetate buffer pH: 4 – 4.5)	[29]
6.	*Medusomyces gisevii* Sa-12	static culture with aeration	9.1 %	[30]
7.	*Achromobacter* S33	Static Culture	1.22 g/L (Mango peel waste (MPW) hydrolysate)	[25]
8.	Rhizobium sp. (Isolated from bean root)	Static Culture	2.5 g/L	[31]
9.	*Acetobacter senegalensis* MA1	Static Culture	469.83 g/ L	[32]
10.	*Acetobacter xylinum*	Static Batch Culture	44.2 g at 4th cycle	[33]

was carried out using a temperature range varying from 25°C to 35°C, with bacterial inoculum ranging from 5-15% for 5-10 days, which resulted in crystallinity ranging from 40.6 to 83.4. The optimal crystallinity of 82.2% was observed when 14% of inoculum fermented for 13 Days at 30°C. The pineapple wastewater source of culture media makes the process more sustainable with low cost [34].

BC producers are strictly aerobic. Therefore, aeration affects the rate of production. Shavyrkina *et al.* optimized the static culture of *Medusomyces gisevii* Sa-12 in a binder-400 climate chamber with aeration to enhance the yield by 25%. XRD pattern of produced PC showed a Crystallinity index of 89% and Transmission geometry of 92%. However, there was a decrease in elastic modulus by 12.6% and the degree of polymerization by 12%. The study proves the importance of gaseous oxygen in fermentation is foremost [30].

12.5 Properties of BC for Wound Healing

BC stands out in biomedical applications due to its exceptional properties compared to plant cellulose. It showed the highest thermal stability (glass transition temperature of 191°C) and 84% - 89% crystallinity. The microfiber structure of BC showed a resemblance with natural tissue properties such as robustness and flexibility. This property makes BC highly suitable for biomedical application [18]. The BC bears various inherent properties that render the wound healing process (Figure 12.3).

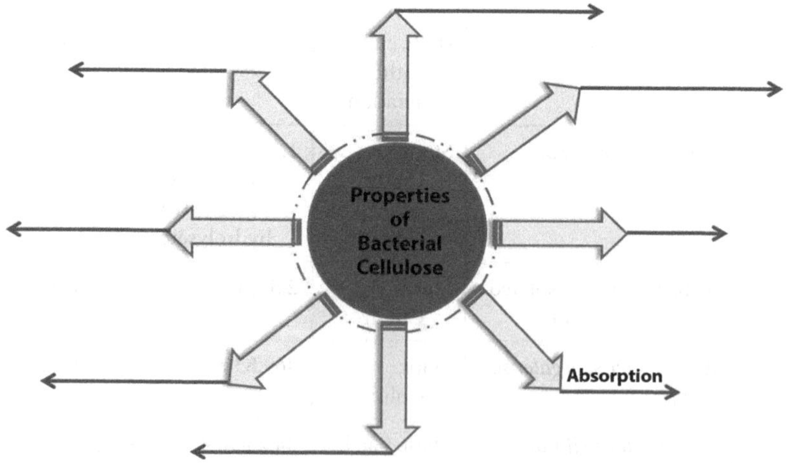

12.5.1 Three-Dimensional Network Structure

BC is a natural nanomaterial with a 3-D network structure. This property gives BC its porous structure and enhances its mechanical Strength, making it suitable for wound healing.

Wahid *et al.* designed a dressing with BC and ε-polylysine (ε-PL) dopamine with a self-crosslinking strategy (Figure 12.4). When the BC membrane is treated with dopamine, the monomers diffuse through BC's three-dimensional network structure. Dopamine then self-crosslinked, forms a coating over the BC membrane, and converts white material to black, confirming polymerization. Subsequently, the ε-PL was conjugated on polymerized BC/PDA through the Michael Addition Reaction. The prepared dressing demonstrated antibacterial activity against *Escherichia coli (E. coli)* and *Staphylococcus aureus (S. aureus)*. It demonstrated 95% viability when treated with NIH3T3 cells. The dressing was hemostatic and showed 1.7% hemolysis, which comes under the safe value. *In vivo* studies of full-thickness wound models were carried out in Sprague-Dawley rats. On the 14th day of the experiment, the inflammatory cells decreased with re-epithelialization and the complete dermal and epidermal layer regeneration with smooth and thick surfaces compared to the control group. This makes the dressing preferable choice [35].

Zhou *et al.* engineered a multifunctional wound dressing specifically for treating infected wounds with BC. The advantage of the 3-D network structure of BC is that it is used to incorporate the bioactive component collagen I and the antimicrobial agent hydroxypropyltrimethyl ammonium chloride chitosan (HACC). The BC incorporation gave a porous nature with >89% porosity to the dressing with a high-water absorption

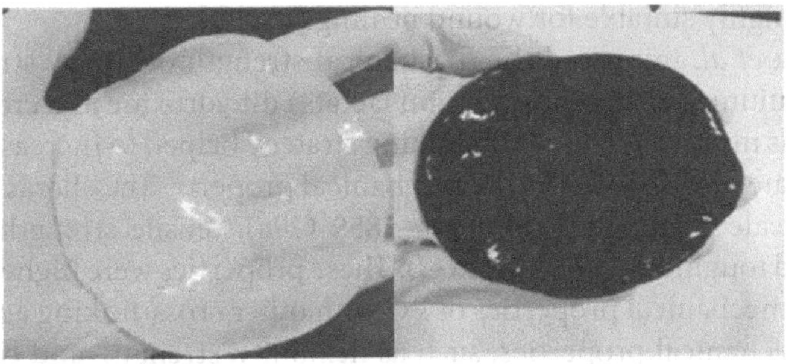

Figure 12.4 (a) before polymerization, (b) after polymerization. (Reproduced from Wahid *et al.* [35], Copyright 2021 American Chemical Society).

capacity of 35.78 ± 2.45 g/g that helped to absorb wound exudates and maintain moist environments, which is necessary for faster wound healing. It showed a Water Vapor Transmission Rate of 35.78 ± 2.45 g/g and 3084 ± 56 g m−2·day−1. The BC-based dressing showed high cytocompatibility by inducing the proliferation rate of NIH-3T3 cells and HUVECs [36].

Yang *et al.* developed a high-performance bacterial-cellulose-based dressing with silver nanowires (AgNWs). To overcome the cytotoxicity of silver, they effectively used a 3-D network structure of BC to entrap silver nanoparticles. The electrostatic interaction between silver and Col-I helped silver aggregate on the BC surface, which was confirmed by SEM analysis. The dressing obtained showed improved mechanical properties with good thermostability, absorption, and water retention capacity. The fast release of silver showed cytotoxicity. Therefore, the entrapping method helped control the release of silver, which enhanced cytocompatibility with NIH-3T3 cell proliferation. The dressing also showed an antibacterial effect on strains ATCC 25923 (*E.coli*) and ATCC 25922 (*S. aureus*) and promoted wound healing in 8 days with collagen deposition at a wound site, with neovascularization for blood vessel formation and hair follicle growth [37].

12.5.2 Mechanical Strength

The mechanical strength of traditional suturing and stapling is considerably higher than biomaterials. For example, the Storage Modulus of hydrogel-like material falls within the range of 1 to 100 kPa, which is comparatively lower than cotton dressing or gauze, which is 1MPa- 1GPa; still, it is suitable per human skin requirements. The wound healing material should have appropriate mechanical strength to ensure structural rigidity, adaptation to the pressure of the wound environment, and for patients' comfort. BC showed 240 MPa tensile strength and 10 GPa Young Modulus, which is highly suitable for wound healing [38, 39].

Andrea *et al.* improved the mechanical strength of BC by strategically using titanium (IV) bis-(ammonium lactate) dihydroxide as a crosslinking agent. This multivalent ion crosslinking strategy helped to increase the production rate of BC as well as the mechanical property. The characterization study revealed Young's Modulus (~1859 GPa), tensile strength (~43670 MPa), and toughness (~681 MJ m$-^{3}$). These properties were higher than the reported mechanical properties of BC with other cross-linking agents [40].

The mechanical properties, such as flexibility, strength, and conductivity, can be increased by modifying BC. Fei *et al.* developed BC membrane/paper, which can be used efficiently for modern wound dressings and give real-time monitoring of wound status. The BC was modified with epoxy, then polyaniline (PANI) was grafted using a chemical method. This assembly PANI-g-EBC3 followed by the addition of polyacrylamide (PAM). The resulting PANI-g-EBC3/PAM nanocomposite was mechanically strong, flexible, and rigid, with a strongly interconnected three-dimensional network structure [41].

Wu *et al.* prepared a high-strength ultrathin BC membrane, which can be used to design wearable smart wound dressings. The strategy applied for synthesis is concluded under three steps: one-step solvent-assisted drawing, hot-pressing drying, and top-down peeling. Here, Hydrogen bonds were first weakened by solvent; then, the nanofibers were stretched 20-30% by the hot press, followed by peeling of the ultrathin layer of BC. Further, the twisting ultrathin films, flexible, strong, and tough fibers with a strength of 954.2 MPa and toughness of 93.2 MJ m−3, were obtained [42].

The BC membranes can maintain structural integrity under mechanical stress or body movements, describing their flexibility for wound dressing application. Sriplai *et al.* designed a magnetic and flexible BC membrane using a simple diffusion of ferrofluid solution. The BC pellicles were obtained from a commercial source through the fermentation of *Gluconacetobacter xylinum* (strain TISTR 975). To enhance compatibility, the biopolymer cellulose (BC) was substituted with alcohol (referred to as BC-N) or freeze-dried (referred to as BC-F) and then immersed in ferrofluid. Following the immersion, Fe3O4 nanoparticles were uniformly distributed throughout the material. The mechanical properties of the BC-F sample, including tensile strength and Young's modulus, were more significant due to its more compact structure than BC-N. These membranes, when deformed, effectively attained their original shape without causing any damage [43].

12.5.3 Porosity

Due to its porous nature, BC resembles the selective permeability of human skin. It allows oxygen, water, and gaseous molecules to pass through its porous structure. This promotes fluid absorption and helps maintain a moist environment during wound healing. Proliferation also helps to entrap molecules and seed living cells. This pore should not lose mechanical stability and affect the pliability of the dressing [44, 45].

Pasaribu *et al.* developed bioactive wound dressing for burn wound treatment with BC implanted *in situ* with collagen and *ex-situ* with Chitosan. The three-dimensional porous structure provides a backbone to impregnate supportive polymers chitosan and collagen. Collagen helps with cellular regeneration at the wound site, while Chitosan is an antibacterial component that adheres to red blood cells and helps in hemostasis. The resulting wound dressing BC/Collagen/Chitosan showed 95.5% moisture content, 76.8% porosity, and 1.51% hemocompatibility. The cell survival rate was 90% after 24 hours of dressing application tested by MTT assay on the HeLa cell line. The dressing material showed water thermal degradation at around 205°C to 303°C. Therefore, it can sustain autoclaving treatment of sterilization before the practical application of dressing, which was carried out at 110°C to 120°C. For *in vivo* study, a second-degree burn wound model was developed in *Rattus norvegicus.* Thick granular tissue and re-epithelization were observed on the 21st day, confirming wound healing [46].

Petrova *et al.* developed a variety of composites by combining BC synthesized by *Komagataeibacter rhaeticus* with negatively charged hyaluronan (HA), sodium alginate (ALG), or K-carrageenan (CAR), as well as with positively charged chitosan (CS). The polysaccharide impregnation in BC formed a polyelectrolyte complex, resulting in morphological changes in the BC network. The BC-ALG-CS showed the highest porosity, resulting in the highest swelling property, which was utmost for wound management. The thickness of the composite was 0.13 mm. The Mesenchymal Stem Cells were used for biocompatibility testing, which showed adherence of cells with 50-75% confluence [47].

12.5.4 Water and Exudate Absorption Capacity

Before the wound healing process can begin, components of the wound exudate, such as fibrous tissue and other debris, must be removed through autolytic debridement. This ensures that the hydro balance is maintained. Moisture or wet dressings are used throughout the procedure to maintain optimum moisture. The BC, comprised of functional hydroxyl groups, also showed a 3-D network porous structure. Therefore, the water and wound exudate absorption capacity is more significant. Consequently, it can help to maintain hydro balance, which is the characteristic property of wound healing [48].

Ciecholewska-jusko *et al.* employed *ex-situ modification to design a* superabsorbent dressing of BC after drying. Here, the researcher cross-linked BC obtained from *K. xylinus* strain (ATCC 53524) with Citric acid using different catalysts, namely, disodium phosphate, sodium bicarbonate, ammonium bicarbonate, and lastly, the mixture of all. They found promising results with amalgamation of disodium phosphate and sodium bicarbonate as catalysts, water holding capacity 5 times more than native BC, and water release ability decreased 6-fold than native BC. The cytotoxicity assay on fibroblast L-929 cell lines showed>85% viability. Therefore, this dressing provides higher absorption and maintains a moist condition, which is a prominent characteristic required for wound healing [49].

In another study, Ciecholewska-jusko *et al.* developed dressing material especially for infectious, highly exuding wounds by crosslinking BC and impregnating octenidine dihydrochloride (OCT) to control infection of common pathogens. The material was spongy, layered, and non-translucent. BC's exudate absorption capacity helped maintain the proper balance of moisture and dryness, which was necessary for rapid healing. OCT showed antimicrobial activity against *S. aureus* and *Pseudomonas aeruginosa (P. aeruginosa)*, inhibiting both planktonic (69-93%) and biofilm growth (84-100%) continuously for up to 7 days. The cell viability with the L929 Fibroblast cell line was below 70%, assessed by both extract and direct contact assay, which is a threshold for cytotoxicity given by ISO 10993–5 [50].

12.5.5 Thermal Stability

The thermal stability of BC helps it withstand higher temperatures without compromising mechanical integrity. At 200°C, BC only showed a rearrangement of Hydrogen Bonds, and above 400°C actual mass loss starts [51].

Mamouri *et al.* designed a thermostable BC-based biocomposite loaded with Savory essential oil for wound dressing. BC, when combined with polyvinyl alcohol and Barhang seed gum, showed characteristics improved. The prepared composite was nontoxic and soft and had antimicrobial activity. Thermal stability played a crucial role while storing and applying dressing material as the samples were stored in low temperatures but applied on natural skin [52].

12.5.6 Hemostatic Effect

Bleeding is the primary cause of accidental injuries, where controlling bleeding is crucial. The BC showed a 3-D network structure, which helps in the entrapment of platelets, red blood cells, etc. [53]. Tran *et al.* fabricated hemostatic dressing by NO2-based oxidation of BC. Here, the oxidation formed the COOH group to the chains of BC, which helped to interact with components of blood and promote healing [54].

Cao *et al.* developed a photothermal antibacterial injectable cryogel for rapid hemostasis and to repair deep, irregular wounds. They modified BC by oxidation with $NaIO_4$ and polymerized with polydopamine to form DBC/PDA cryogel. This prepared porous cryogel showed quick absorption of blood and stopped bleeding [55].

Bian *et al.* constructed a hemostatic OBNC-DFO sponge using oxidized bacterial nanocellulose grafted with a desferrioxamine drug through an amide bond. The sponge showed high porosity and specific surface area, which gives it high water and blood absorption capacity and promotes blood clotting. It showed 1.34% hemolysis, considered a safe value; the blood clotting index decreased with time and showed rapid clot formation. The sponge not only showed hemostasis but also promoted vascularization in damaged tissue [56].

12.5.7 Biocompatibility

Evaluation of cytotoxicity is a crucial step before entering animal or clinical studies. The BC is compatible with living tissue and helps cellular adhesion, proliferation, and migration without causing adverse effects or inflammation. Therefore, it is highly suitable for wound healing application [57].

The impurities found in fermented BC, like endotoxins, are of significant consideration while evaluating biocompatibility. The *Gluconacetobacter xylinus* produces high Lipopolysaccharide, interfering with the dehydrogenation method as it adheres to the BC membrane. In this case, methods like dry heat dehydrogenation and organic solvent washing are used to remove endotoxin impurities, ultimately increasing biocompatibility [58].

Fatih Ciftci *et al.* synthesized BC/AgNP composite film by incorporating silver nanoparticles on BC fibrils. The Biocompatibility of the film was examined by cell staining and subsequent fluorescence visualization. Composite biofilm treated with L929 mouse Fibroblast Cells resulted in adherence and growth of epidermal cells without any cytotoxic effect. This biocompatible nature is foremost for wound healing applications. Further, this bioactive material may be used to lower inflammation in skin barriers and heal a wound [59].

Khan *et al.* formulated a composite of BC and turmeric for therapeutic wound dressing. The excellent biocompatibility of BC and wound exudate absorption helped dress the wound without showing any toxic or allergic effects. The Biocompatibility testing expressed fibroblast cell attachment and proliferation when treated with prepared dressing. The physiochemical and biological properties of therapeutic turmeric dressing also promoted tissue regeneration, hair growth as well as wound healing [60].

12.5.8 Biodegradability

Biodegradability is a property where the biomaterial implanted should break down into small metabolites utilized by the body's metabolism to maintain homeostasis. This helps to avoid secondary surgery and reduces expenses and time. The process is time-dependent and affected by various parameters such as pH, molecular weight, chemical composition, and crystallinity. The natural process of biodegradation can be influenced by external and internal stimuli, which can be processed through various mechanisms, including hydrolysis, enzymatic activity, oxidation, and physical degradation [5, 61].

The BC immersed in Phosphate Buffered Saline (PBS) absorbs liquid and swells over incubated time. The swelling of BC fibers stretches C-O bonds, eventually weakening bond breaks due to the influence of water and ion molecules in the surroundings. The degradation can be improved using ionic liquids such as BMIM HSO_4 and PY-HSO_4. Munir *et al.* experimented with degrading BC membrane and found significant results over 28 28-day studies. BMIM HSO_4 degraded BC by 56% while PY-HSO_4 by 36% [62].

Yuan *et al.* created a nanocomposite, referred to as OBC/CS/COL, by combining oxidized bacterial cellulose (OBC) and chitosan (CS) with collagen (COL). The OBC/CS/COL nanocomposite demonstrated approximately 98% degradation in phosphate-buffered saline (PBS) within 30 days. Additionally, it exhibited antimicrobial properties against *E. coli, S. aureus*, and *K. xylinus*. Also, it promoted adhesion and proliferation of fibroblast cells, confirmed with calcein AM staining on the 3rd Day, declaring its biocompatibility. The nanocomposite showed hemostasis by clotting blood in 86 seconds [63]. The BC has properties that researchers can use to develop formulations for wound healing.

12.6 Commercially Available BC Products

Dermafill® is a semi-opaque, translucent Membrane for wound management Biosynthesized using *Xylinum spp*. It is a biocompatible, nontoxic membrane provided in a sterile package. The structure of Dermafill® microscopically resembles the Extracellular Matrix (ECM). It gives a high Water Vapor Transfer Rate and fluid balance. The transparent nature allows visualization of the wounds without removing the dressing. Therefore, one can monitor healing throughout in real-time, so there is no need to change the dressing. It will shed off itself spontaneously. The product cannot be used on infected wounds directly before treatment. The clinical trial (NCT00137163) (Table 12.2) carried out over 18 years and in old adult men and women showed a 75% reduction in wound size after the application of the product in 11 chronic wounds and epithelization in 81 days [64, 65].

Table 12.2 Clinical trials.

Sr. no.	Name of product	Clinical trial ID	Sponsor	Year of completion	References
1	oxidized regenerated cellulose (ORC)	NCT00947089	Catholic University of the Sacred Heart	Ongoing Phase 4	[71]
2	Bacterial cellulose-monolaurin hydrogel	NCT05079763	University of the Philippines	Ongoing Phase 2	[72]
3	GATT-Patch	NCT05900037	GATT Technologies BV	Completed (2024)	[73]
4	Hemopatch	NCT05103956	*Foundation of the Andalusian Public Health System*	Completed (2021)	[74]
5	Oxidized cellulose strip	NCT01637025	Baxter Healthcare Corporation	Completed (2013)	[75]

Xcell® is a cellulose-based product for chronic wound treatment. The *Acetobacter Xylinum* cellulose is used, which has high biocompatibility, superior mechanical strength, and both moisture absorption and donation ability. The product removes dead tissue and promotes autolytic debridement. It also helps to cleanse wound margins, which promote epithelial migration and wound size reduction. The clinical trial (NCT00446823) subjected patients between 18 and 90 years of age to the secondary stage of chronic venous insufficiency. The patients are observed for 12 weeks or till wound healing. 75% granulation within 43 days and 50% re-epithelization seen within 57 days [66]. Cellulose dressing is in phase 3 of clinical trial. The 18 year and old patients with Split Thickness Donor Sites Burn and wound were observed. The dressing can absorb wound exudate and blood exudate; the application is easy due to adherence and can be removed atraumatic. The >95% epithelization was seen in the 14-day observation [67].

EpiProtect® is biologically synthesized from *Acetobacter Xylinum* wound dressing designed explicitly for full-thickness wounds in the pediatric population. The product provides sustainable mechanical strength, facilitates exudate drainage through a mesh-like structure, provides pain relief effectively, and provides transparency and permeability, providing ease of antibiotic application from surfaces without changing the dressing.

The flexibility and adherence properties help to apply dressing easily without requiring a skillful person to undergo anesthesia. For clinical trials, <5 years old patients were observed, and results showed that Median healing time in Superficial partial thickness was 13 days, the mid partial thickness was 13 days, the deep partial thickness was 124 days [68, 69].

Bionext® is BC based wound care product used for biological wound treatment and can be used as skin replacement in case of burns or traumatic wounds. It serves antibacterial activity. The remarkable selective permeability of Bionext® allows gaseous exchange but limits the entry of other liquids and external agents. The Bionext® adheres to wounded areas and provides mechanical support [70].

12.7 Enhancements and Modifications of BC

Some characteristics are missing in BC that limit the potency of wound healing. It lacks inherent antimicrobial and antioxidant activity, which is paramount for wound healing as the wound exposes the dermal layer to the environment, increasing the risk of infection. The second important consideration is about storage of BC. It remains in its native structure in moist (wet) conditions only. As soon it gets dry, it will lose its structural integrity [76].

Various modifications and composites have been developed to overcome lacunas and improve BC's inherent characteristics, making BC an ideal wound dressing (Table 12.3). These are categorized into four groups: physical modification, chemical modification, composite preparation, and incorporation of stem cells.

12.7.1 Physical Modifications

Physical modification of BC boosts its properties for wound healing applications. Various studies showed that these modifications improve mechanical strength, swelling behavior, and antimicrobial activity, making BC a versatile biomaterial for wound dressing.

Deng *et al.* synthesized a modified form of BC with quaternized chitosan (QCS) through a chemical process, creating a dehydrated dialdehyde bacterial cellulose/quaternized chitosan (DBC/QCS) hydrogel. This hydrogel demonstrated a remarkable swelling capacity exceeding 1000% and exhibited 80.8% and 81.3% antimicrobial activity against *E. coli* and *S. aureus*, respectively. In a study involving wound healing in rats over 14 days, it was observed that the hydrogel, when infused with pirfenidone (PFD),

Table 12.3 Different bacterial cellulose materials used for wound healing.

Sr. no.	Name of material	Bacterial species source	Properties	References
1	Bacterial Nanocellulose Membrane	*K. xylinus* ATCC 23770, DHU-ATCC-1, DHU-WX-1	Crystallinity index (CI): ATCC 23770 - 85.42%, DHU-ATCC - 1 - 87.31, DHU-WX - 1- 89.72% Young's moduli: ATCC 23770 – 0.45 + 0.11 MPa, DHU-ATCC-1 – 1.36 + 0.16 MPa, DHU-WX-1 - 7.82+.13 MPa Antibacterial Activity: Against *E. coli* and *S. aureus* Preventable Efficiency- ~100% (For all membranes) Cytocompatibility on L-929 cell line by Calcein AM/PI staining: ATCC 23770 – Highest Viability, DHU-ATCC-1 – Moderate Viability, DHU-WX-1- Poor Viability Wound Size Reduction: On 9th Day (Sprague-Dawley rats), ATCC 23770 – 58.7%, DHU-ATCC-1 – 74.9%, DHU-WX-1- 90.3% Wound Closure: ATCC 23770 – 17 days, DHU-ATCC-1 – 15 days, DHU-WX-1- 12 days	[23]

(*Continued*)

Table 12.3 Different Bacterial cellulose materials used for wound healing. (*Continued*)

Sr. no.	Name of material	Bacterial species source	Properties	References
2	BC-based dressings containing ε-polylysine (ε-PL), cross-linked by mussel-inspired polydopamine (PDA)	*K. xylinus* (CGMCC no. 2955)	Antibacterial Activity: *In vitro* - *E. coli, S. aureus*- 100% (For all membranes) *In vivo* - *S. aureus* BCP@ε-PL1 - 96.9% BCP@ε-PL3 - 99.83% *In vitro* Cell Viability on mouse embryonic fibroblast cells NIH3T3– 95-100% Hemolysis – 0.18- 1.70% *In vivo* full thickness wound in Rat model (male Sprague-Dawley rats) 14th Day observation- complete Re-epithelialization, wound healed with Dermis and epidermis regeneration with smooth surface	[35]

(*Continued*)

Table 12.3 Different Bacterial cellulose materials used for wound healing. (*Continued*)

Sr. no.	Name of material	Bacterial species source	Properties	References
3	bacterial cellulose, collagen I, hydroxypropyltrimethyl ammonium chloride chitosan	*Acetobacter xylinum*	For CHBC dressings: water absorption rate - 35.78 ± 2.45 g/g water vapor transmission rate - 3084 ± 56 g m−2·day−1 Porosity: 77.96 ± 1.82%; Decomposition: 310 to 400°C Absorption Capacity: 310 to 400°C Antibacterial Activity –*S. aureus* Cytocompatibility: NIH3T3 cell and HUVEC used and observed by SEM and CLSM- complete spread of proliferation and exhibited a polygonal and fusiform morphology *In vivo* wound healing- Full thickness round wound *in Kunming* mice: wound healing – ~100% within 8 days and re-epithelization as well as new hair follicles seen	[36]

(*Continued*)

Table 12.3 Different Bacterial cellulose materials used for wound healing. (*Continued*)

Sr. no.	Name of material	Bacterial species source	Properties	References
4	bacterial cellulose, Octenidine dihydrochloride (OCT)	*K. xylinus* (ATCC 53524)	Swelling Ratio – 300% Antimicrobial activity - Biofilm reduction by MTT *S. aureus*- 89-93% *P. aeruginosa*- 69-86% Cytotoxicity on ATCC CCL-1 (L929) murine fibroblasts- – <70% (threshold level as per ISO 10993–5)	[50]
5	Oxidized Bacterial Nanocellulose (OBNC), Desferrioxamine (DFO)	*K. xylinus*	Tensile Strength – 40.32 + 2.88 kPa; Young's Modulus – 2.73 + 1.01 kPa; Hemolysis – 1.34%; BCI - 85.96% at 3 min, 73.71% at 6 min, 29.21 at 12 min; Proliferation of HUVECs - upto 5 th Day *In vivo* Hemostasis - rat tail amputation model Surface blood infiltration area – 81.78 %; Blood loss – 200.26 mg liver trauma model, Surface blood infiltration area - 68.59%, the blood loss - 110.37 mg *In vivo* Rat model – Sprague-Dawley rats 21st to 28th Day - collagen deposition and remodeling, as well as facilitating skin appendage regeneration	[56]

(*Continued*)

Table 12.3 Different Bacterial cellulose materials used for wound healing. (*Continued*)

Sr. no.	Name of material	Bacterial species source	Properties	References
6	Bacterial Cellulose (BC) Silver Nanoparticles (Ag) Graphitic Carbon Nitride (gCN) Eucalyptus Extract (EE) Polyvinyl Alcohol (PVA)	*G. xylinus*	Antibacterial Activity – For Ag/gCN/EE composite *E. coli*- 98.2 ± 1.2 %; *S. aureus*- 98.5 ± 0.8% size of the BC fibers: BC/Ag/gCN/EE nanocomposite - 138.7 ± 9 tensile strength: BC/Ag/gCN/EE - 6.99, Young Modulus - 121.37 Biodegradation (After 96 hrs): PVA/BC/Ag/gCN/EE - 44% -PVA/BC/Ag/gCN/EE - 79% Cell viability on NIH 3T3 fibroblast cells by MTT assay: BC/Ag/gCN/EE nanocomposite: 89 ± 2.31% PVA/BC/Ag/gCN/EE nanocomposite: 96 ± 3.28% Scratch Wound Assay 24 hrs: BC/Ag/gCN/EE composite: Reduced scratch area by 18.69% PVA/BC/Ag/gCN/EE composite: Reduced scratch area by 23.97%	[102]

(*Continued*)

Table 12.3 Different Bacterial cellulose materials used for wound healing. (*Continued*)

Sr. no.	Name of material	Bacterial species source	Properties	References
7	Dehydrated Bacterial Cellulose (DBC), Quartinized Chitosan (QCS), Pirfenidone (PFD)	*Gluconacetobacter xylinus*	Swelling ratio- 1000% Antimicrobial Activity *Escherichia coli* - 81.3 %; *S. aureus* - 80.8% Excellent Biocompatibility on L-929 Cells with proliferation *In vivo* rat model, 14th day – Addition of PFD promotes Collagen Deposition, Accelerate wound healing with Scar prevention	[77]
8	Bacterial Cellulose, 1 –butyl- 3 -methylimidazolium hydrogen sulfate (BMIM-HSO4), Pyridinium hydrogen sulphate (Py-HSO4)	*Scoby Culture (Both Bacteria and Yeast co-cultured)*	Contact angle - 46°C (Py-HSO4 treated), 30° C (BMIM-HSO4 treated) Swelling Ratio (24 hrs) – 996% (Py-HSO4 treated), 1113% (BMIM-HSO4 treated) Biodegradation (28 Days) – 36% (Py-HSO4 treated), 56.1% (BMIM-HSO4 treated) Antibacterial Activity - *E. coli, S. aureus* *In vitro* Biocompatibility on Fibroblast NIH3T3 cells – 100% (Both treatments)	[62]

facilitated collagen deposition and sped up the healing process while also aiding in scar prevention [77].

Han *et al.* developed a technique to covalently crosslink γ-polyglutamic acid (γ-PGA) with chitosan oligosaccharides (COS) within BC. The mechanical analysis of the BC films indicated that incorporating 2.25% COS during incubation periods of 0.5, 1.5, and 2 days resulted in significant improvements, increasing Young's modulus and yield by 5.65 and 1.42 times, respectively. Conversely, there was a notable reduction in swelling behavior, with the modified films exhibiting a swelling capacity of 1774%, which is just 46% of that of unmodified BC. This method successfully balanced improved mechanical properties and reduced swelling by optimizing hydrogen bonding, electrostatic interactions, and amido bonds. The resulting modified bacterial cellulose composite films (BCCF) demonstrated effective transport capabilities for benzalkonium chlorides, indicating their potential for drug delivery. Consequently, this approach shows promise for enhancing BC for various applications, including wound dressings, artificial skin, and scar-reducing patches [78].

Khan *et al.* created composite films combining bacterial cellulose and turmeric powder (BC-TP) to function as wound dressings, leveraging the healing properties of turmeric. Research on their swelling and water release rates found that these composite films exhibit excellent water absorption and retention abilities. Studies examining antibacterial effectiveness against gram-positive and gram-negative bacteria showed that turmeric powder significantly boosts the antibacterial properties when integrated into a bacterial cellulose framework. The BC-TP composites achieved over 90% wound healing by the seventh day, compared to untreated wounds, which only displayed approximately 24% healing in the same timeframe. This research suggests that BC-TP composite films could provide effective alternatives to conventional wound dressing options [60].

12.7.2 Chemical Modifications

Chemical modifications of BC include varying its structure to enhance functionality, including antibacterial properties, biocompatibility, and mechanical strength. Commonly used methods for modifying BC include oxidation, esterification, and etherification. This enables the introduction of various functional groups, which can help to develop the interaction with cells and increase the maintenance and release of therapeutic agents [79, 80].

Various methods have been used for the chemical modification of BC. Surface-induced atom transfer radical polymerization (ATRP) is a notable technique, much like conventional synthetic methods and cross-linking with silane coupling agents [81]. Additionally, a green- -chemistry method focuses on the derivations of BC's surface hydroxyl groups using epoxides that contain quaternary ammonium groups and alkyl chains, contributing to its antibacterial properties [79]. Many chemical reactions, such as esterification, oxidation, acetylation, and graft copolymerization, are still utilized to improve the characteristics of BC. The esterification of BC is an acetylation process that modifies the hydroxyl (OH) groups using carboxylic acid as the acetylating agent [82].

Ngai *et al.* developed an eco-friendly, straightforward surface esterification method to produce hydrophobic films using BC sourced from kombucha tea. They used alkenyl succinic anhydrides with different chain lengths as esterifying agents, introducing pendant carboxylic groups. This modification created multiple cross-linking sites, enhancing the overall strength of the BC structure. Consequently, the resulting hydrophobic BC films showed considerable resistance to water, ultraviolet light, oxygen, and bacteria [83].

Sufiandi *et al.* chose the immersion method to incorporate citric acid (CA) as the cross-linking agent and polyethylene glycol 600 (PEG-600) as the plasticizer to incorporate into the BC film. The addition of CA facilitated the establishment of covalent bonds with the hydroxyl (OH) groups present on both the BC and PEG-600 through the process of esterification. This led to enhanced tensile strength and tensile modulus of the material [84].

The *in-situ* oxidation process of BC caused very little harm to the fibers. This led to carboxyl-functionalized ionogels with impressive tensile strength, strong adhesion, and favorable biodegradability. Additionally, nitrogen dioxide serves as another widely used oxidizing agent, effectively transforming the primary hydroxyl (OH) group of BC into a carboxyl group [85].

BC dressings cushion the wound and keep the environment moist, facilitating bacterial growth and delaying wound healing. Therefore, researchers functionalized BC with ε-poly-L-Lysine (ε-PLL), a non-toxic, broad-spectrum antimicrobial polymer by carbodiimide chemistry, to modify BC to render an antimicrobial nature without hampering its mechanical properties. This dressing showed antimicrobial activity against *S. epidermidis* and was very low for *E. coli*. This may happen due to the concentration of ε-PLL. The Human Dermal Fibroblast (HDF) cell line was utilized to assess cytocompatibility, demonstrating normal cell morphology and cell proliferation after 13 days of culture [86].

12.7.3 Composite Preparations (With Polysaccharides, Proteins, and Drugs)

Bacterial cellulose (BC) is a versatile biopolymer demonstrating significant potential in wound healing applications. Various modifications using polysaccharides, proteins, and drugs have been employed to improve its properties.

12.7.3.1 Modification with Polysaccharides

In BC, combining polysaccharides, such as collagen, chitosan, hyaluronic acid, and gelatin, has been widely used to improve its wound healing applications. These modifications control the natural biocompatibility and biodegradability of polysaccharides and boost the BC-based dressings' structural integrity and functionality. Collagen, being a significant component of the extracellular matrix, can improve cell adhesion and proliferation, while gelatin offers excellent moisture retention properties, making it valuable for effective wound management [39, 48].

12.7.3.2 Modification with Proteins

BC can also be modified with proteins to improve its wound-healing properties. For example, the addition of antimicrobial peptides or growth factors can significantly boost the antibacterial response of BC dressing, promoting faster healing and reducing infection risk [48]. These proteins can be combined into the BC matrix through covalent bonding or physical entrapment, allowing sustainable release while maintaining the integrity of dressing materials [39].

12.7.3.3 Incorporation of Drugs

One of the most promising methods for increasing BC's effectiveness in wound healing is the incorporation of drugs. Antibiotics such as ciprofloxacin, ceftriaxone, and tetracycline have been implanted in BC to provide localized antimicrobial action and stimulate healing. This drug loading not only helps in controlling infections but also allows for a controlled release of medications, ensuring prolonged therapeutic effects [87].

12.7.3.4 Benefits of Composite Preparations

Composite preparations using BC, polysaccharides, proteins, and drugs exhibit numerous benefits in wound healing. Combining these materials enhances moisture retention, improved mechanical properties, and a favorable cell growth and migration microenvironment. Also, this synergy can lead to better absorption of wound exudates and a reduction in infection rates, making BC-based composites particularly suitable for chronic wound management [39].

Zmejkoski *et al.* prepared composite hydrogel by encapsulating nano chitosan dots (nChiD) into the BC matrix. The nChiD was developed using gamma radiations. The encapsulation of nChiD helped to increase antioxidant and antibacterial activity, which was negligible to absent in natural BC [88]. In another study, Zmejkoski *et al.* incorporated Graphene Quantum Dots (GQDs) into BC to develop a biocompatible and antibacterial CQDs-BC hydrogel composite. The composite showed impressive wound fluid absorption as well as water retention. The CQDs-BC hydrogel composite showed considerable activity against *S. aureus, Staphylococcus agalactia, E. coli, P. aeruginosa,* and Methicillin-resistant *S. aureus*. Therefore, the prepared composite satisfies wound dressing properties [89].

Mohaghegh *et al.* prepared multilayer dressing to overcome the limits of one-layer dressings for full-thickness wound healing. The gelatin/alginate hydrogel and polycaprolactone loaded with Ciprofloxacin, an antibiotic, were added to the BC matrix. The BC provided a significant role in mechanical support, and wound exudate absorption maintained an environment for rapid healing. The other components ensure potent antibacterial activity with a slow release of antibiotics from multilayer dressing [90]. Sajjad *et al.* developed a nanocomposite of Montmorillonite-BC to replace burn tissue and regenerate tissue. The modified Montmorillonite (Cu-MMT, Ca-MMT, Na-MMT) was impregnated *via* particle impregnation. The MMT's loaded BC showed enhanced wound healing activity with antibacterial activity, so it can be used as an artificial substitute for burned or damaged skin or as a scaffold for tissue engineering [91].

12.7.4 Incorporation of Stem Cells

BC is a natural polymer with exceptional properties, including high tensile strength, flexibility, and water retention capacity, making it suitable for wound dressing. Incorporating stem cells, specifically Mesenchymal Stem Cells (MSCs) or Endothelial Progenitor Cells (EPCs), into biocompatible (BC) scaffolds creates a favorable environment for tissue regeneration. This integration not only supports the healing process by providing a stable matrix but also enhances the functionality of the dressing by enabling the controlled release of essential growth factors from the stem cells [92]. Incorporating BC and stem cells promotes wound healing by addressing various biological functions. MSCs can differentiate into multiple cell types, including fibroblasts and keratinocytes, essential for skin repair and regeneration [93]. They also secrete cytokines and growth factors that help with angiogenesis, collagen deposition, and tissue renewal. BC's 3D network structure significantly helps maintain moisture at the wound site, which is critical for effective healing [92].

Wang *et al.* conducted a study in which Bone Marrow Stem Cells (BMSTs) were applied to Bioactive Collagen (BC), providing a three-dimensional structure that supports cell adherence. In this experiment, six-week-old male Balb/C mice were divided into three groups: a control group, a group treated with BC Membrane (BCM), and a group receiving Bioactive Wound Dressing (BWD). Full-thickness skin defects were created using a tissue punch on the backs of the mice, and gauze infused with either BCM or BWD was placed on the wounds. The findings revealed that by day 14, the wounds treated with BCM and BWD showed significant healing, with the BWD exhibiting the most notable improvement. On day 7, the full-thickness skin from the scars was removed for further analysis, which included fixation, dehydration, and embedding in paraffin for sectioning. Hematoxylin, eosin, and Masson staining were performed to assess blood vessel formation and collagen production. The BWD group displayed more closely packed fibroblasts, well-organized collagen fibers, and an increased presence of new capillaries and hair follicles. Additionally, the epidermal gaps in the BWD group were smaller on days 7 and 14 after the injury, indicating a quicker re-epithelialization process. Overall, these results suggest that the Bioactive Wound Dressing significantly enhances the healing process of wounds [94].

Researchers led by Souza developed a membrane dressing composed of nanostructured cellulose, gellan, xyloglucan, and lysozyme infused with Mesenchymal Stem Cells (MSCs) to treat deep second-degree burn wounds effectively. The study utilized Wistar *R. norvegicus* rats, each weighing 300 grams, to evaluate the performance of the cellulose membrane treatments. The *in vivo* experiments' findings indicated a 40% to 60% reduction in acute and chronic inflammatory infiltrates. Additionally, a remarkable 96% decrease in the injury area and increased vascular proliferation, enhanced collagen deposition, and complete epithelialization were observed after a 30-day period. The presence of MSCs was confirmed in the burned tissue, indicating that these cells successfully migrated and multiplied in the affected area [95].

Radu and colleagues created a composite biomaterial using BC derived from *Komagataeibacter xylinum* and a keratin-based scaffold. They introduced adipose-derived stem cells from rabbits into this scaffold. The stem

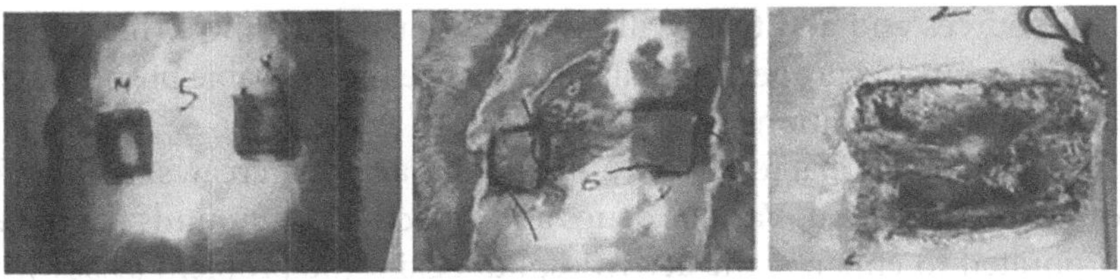

Figure 12.5 (a) The burnedwounds, (b) The burned and grafted wounds, (c) Image after 14 days of the grafted wound (Figure reproduced from [96], Creative Common License 4.0).

cells stimulated tissue growth with a satisfactory level of viability. This composite was subsequently applied to the burn wounds of rabbits of a standard breed. The healing process in the treated wounds was notably more effective than in the control group. The recovery involved the removal of dead tissue and the rebuilding of damaged areas caused by thermal injury, occurring at a significantly faster rate. This highlights the vital contributions of keratin, cellulose, and stem cells in the healing process. Additionally, cellulose and keratin provided essential protection and support for the cells engaged in tissue repair and regeneration (Figure 12.5) [96].

12.8 Limitations and Future Directions

The BC is selectively used and frequently opted polysaccharide among researchers due to its remarkable properties addressing wound healing. Still, there is scope for budding researchers to make it better or idea material. The demand for pure BC is high compared to the production yield, and the cost of production is high. Fermentation is a central research area as the fulfilling supply is necessary. The optimization of parameters such as Bacterial Strain, Media Components (mainly carbon and nitrogen sources), pH, temperature, aeration, etc., is the point of dwelling for upcoming researchers [92, 97].

The hydrogen bonding in BC delivers a hierarchical structure to BC, which makes the polymer chain fold into a fibrous structure. This structure allows BC to participate in various Biomedical applications, including wound healing, drug delivery, etc. This structure can be affected by cultural conditions such as carbon and nitrogen sources, physical and chemical environment, culture movement (static, agitation), and culture additives. When low temperature and static conditions are maintained, a stable amorph structure of BC will be produced, while agitation with alternative carbon sources and additives will give BC thermal stability. This ensures that BC has tunable properties that can enhance the desirable properties by altering culture conditions [98].

The lack of antibacterial or antioxidant activity creates a gap for researchers to work on. Various drugs and bioactive compounds have been incorporated to enhance the activities, but again, the problem arises with drug-resistant bacterial species. Therefore, there is a need to explore antibacterial activity [99, 100].

When exposed to the environment, BC quickly dehydrates as water molecules leave the place from BC structure through pores, ultimately affecting mechanical properties. Some researchers used alternative solvents with high evaporation temperatures, such as polyethylene glycol (PEG) and glycerol, to overcome this. These solvents resemble hydrogen bonding interactions with BC, effectively increasing Young's modulus (130-fold). Still, the network between this solvent and BC keeps bacteria from surviving within a few weeks. Therefore, there is still a need to look for other solvents or BC modifications in the future [101].

12.9 Conclusion

The BC is a breakthrough for various biomedical applications. It is an exclusive choice of researchers to compose wound dressing materials. BC does not show any adverse or toxic effect when contacting living cells or tissue. It also showed compatibility with blood cells without harm or hemolysis. In case of accidental bleeding or hemorrhage, it helps to clot blood by entrapping cells with its dynamic three-dimensional fibrous architecture. The mechanical strength of BC is enough to support and adhere to living cells, which is necessary for normal body cells to proliferate and repair the wounded area. Overall inherent properties, including Biocompatibility, Hemocompatibility, mechanical Strength, and Three-Dimensional network structure, are satisfactory to fulfill the above-mentioned four phases of wound healing, namely Hemostasis, Inflammation, Proliferation, remodeling, and Maturation, making it suitable for wound healing applications. Though it completes basic needs for wound healing, it lacks some equally important accessory characteristics, such as antibacterial and antioxidant activity. Therefore, there is a need to modify BC by employing various physical and chemical methods. This may help to achieve the best results, making BC an ideal component for wound management and healing.

AI disclosure statement: The author acknowledges the use of AI-based tools for minor language editing or grammatical correction. However, all intellectual contributions, analyses, and conclusions presented in this chapter are entirely the author's own.

Bibliography

1. Tiwari, G., Goel, R., Bhalla, K., Road Safety in India: Status Report 2023, in: *Transp. Res. Inj. Prev. Centre, Indian Inst. Technol. Delh. [Internet]*, pp. 1–83, 2023, Available from: https://tripc.iitd.ac.in.
2. Ashoobi, M.T., Asgary, M.R., Sarafi, M., Fathalipour, N., Pirooz, A., Jafaryparvar, Z., *et al.*, Incidence rate and risk factors of surgical wound infection in general surgery patients: A cross-sectional study. *Int. Wound J.*, 20, 2640–8, 2023.
3. Britto, E.J., Nezwek, T.A., Popowicz, P., Robins, M., Wound Dressings, in: *Tech. Small Anim. Wound Manag. [Internet]*, pp. 127–54, 2024 [cited 2024 Sep 11]. Available from: https://www.ncbi.nlm.nih.gov/books/NBK470199/.
4. Oliveira, C., Sousa, D., Teixeira, J.A., Ferreira-Santos, P., Botelho, C.M., Polymeric biomaterials for wound healing. *Front. Bioeng. Biotechnol.*, 11, 1–19, 2023.
5. Zhang, H., Lin, X., Cao, X., Wang, Y., Wang, J., Zhao, Y., Developing natural polymers for skin wound healing. *Bioact. Mater.*, 33, 355–76, 2024.
6. Mir, M., Ali, M.N., Barakullah, A., Gulzar, A., Arshad, M., Fatima, S., *et al.*, Synthetic polymeric biomaterials for wound healing: A review. *Prog. Biomater. [Internet]*, 7, 1–21, 2018. Available from: https://doi.org/10.1007/s40204-018-0083-4.
7. Ansari, M. and Darvishi, A., A review of the current state of natural biomaterials in wound healing applications. *Front. Bioeng. Biotechnol.*, 12, 1–26, 2024.
8. Bowers, S. and Franco, E., Chronic wounds: Evaluation and management. *Am. Fam. Physician*, 101, 159–66, 2020.
9. Grubbs, H. and Manna, B., Wound Physiology, in: *StatPearls [Internet]*, 2023, [cited 2024 Sep 11]; Available from: https://www.ncbi.nlm.nih.gov/books/NBK518964/.
10. Wallace, H.A., Basehore, B.M., Zito, P.M., Wound Healing Phases, in: *StatPearls [Internet]*, 2023, [cited 2024 Sep 11]; Available from: https://www.ncbi.nlm.nih.gov/books/NBK470443/.
11. Drahansky, M., Paridah, M., Moradbak, A., Mohamed, A., Owolabi, F., Asniza, M., *et al.*, We are IntechOpen, the world's leading publisher of Open Access books Built by scientists, for scientists TOP 1%, in: *Intech [Internet]*, vol. i, p. 13, 2016, Available from: https://www.intechopen.com/books/advanced-biometric-technologies/liveness-detection-in-biometrics.

12. Sheokand, B., Vats, M., Kumar, A., Srivastava, C.M., Bahadur, I., Pathak, S.R., Natural polymers used in the dressing materials for wound healing: Past, present and future. *J. Polym Sci.*, 61, 1389–414, 2023.
13. Gounden, V. and Singh, M., Hydrogels and Wound Healing: Current and Future Prospects. *Gels*, 10, 43, 2024 [Internet]. 2024 [cited 2024 Sep 18];10:43. Available from: https://www.mdpi.com/2310-2861/10/1/43/htm.
14. Seddiqi, H., Oliaei, E., Honarkar, H., Jin, J., Geonzon, L.C., Bacabac, R.G., *et al.*, Cellulose and its derivatives: towards biomedical applications [Internet], in: *Cellulose*, Springer, Netherlands, 2021, Available from: https://doi.org/10.1007/s10570-020-03674-w.
15. Hasanin, M.S., Cellulose-Based Biomaterials: Chemistry and Biomedical Applications. *Starch - Stärke [Internet]*, 74, 2200060, 2022 [cited 2024 Aug 15]. Available from: https://onlinelibrary.wiley.com/doi/full/10.1002/star.202200060.
16. Zhong, C., Industrial-Scale Production and Applications of Bacterial Cellulose. *Front. Bioeng. Biotechnol.*, 8, 1–19, 2020.
17. Verma, J., Petru, M., Goel, S., Cellulose based materials to accelerate the transition towards sustainability. *Ind. Crops Prod. [Internet]*, 210, 118078, 2024, doi: 10.1016/j.indcrop.2024.118078.
18. Naomi, R., Idrus, R.B.H., Fauzi, M.B., Plant-vs. Bacterial-derived cellulose for wound healing: A review. *Int. J. Environ. Res. Public Health*, 17, 1–25, 2020.
19. Mishra, S., Singh, P.K., Pattnaik, R., Kumar, S., Ojha, S.K., Srichandan, H., *et al.*, Biochemistry, Synthesis, and Applications of Bacterial Cellulose: A Review. *Front. Bioeng. Biotechnol. [Internet]*, 10, 780409, 2022 [cited 2024 Sep 16]. Available from: www.frontiersin.org.
20. Lahiri, D., Nag, M., Dutta, B., Dey, A., Sarkar, T., Pati, S., *et al.*, Bacterial Cellulose: Production, Characterization, and Application as Antimicrobial Agent. *Int. J. Mol. Sci. [Internet]*, 22, 12984, 2021 [cited 2024 Sep 16]. Available from: /pmc/articles/PMC8657668/.
21. Wang, J., Tavakoli, J., Tang, Y., Bacterial cellulose production, properties and applications with different culture methods – A review. *Carbohydr. Polym. [Internet]*, 219, 63–76, 2019. Available from: https://doi.org/10.1016/j.carbpol.2019.05.008.
22. Swingler, S., Gupta, A., Gibson, H., Kowalczuk, M., Heaselgrave, W., Radecka, I., Recent advances and applications of bacterial cellulose in biomedicine. *Polym. (Basel)*, 13, 1–29, 2021.
23. Yuan, H., Chen, L., Hong, F.F., Evaluation of wet nanocellulose membranes produced by different bacterial strains for healing full-thickness skin defects. *Carbohydr. Polym.*, 285, 119218, 2022.
24. Saleh, A.K., El-Gendi, H., Soliman, N.A., El-Zawawy, W.K., Abdel-Fattah, Y.R., Bioprocess development for bacterial cellulose biosynthesis by novel Lactiplantibacillus plantarum isolate along with characterization and antimicrobial assessment of fabricated membrane. *Dent. Sci. Rep. [Internet]*,

12, 2181, 2022 [cited 2024 Sep 3]. Available from: https://typeset.io/papers/bioprocess-development-for-bacterial-cellulose-biosynthesis-58za5m1p.
25. Hasanin, M.S., Abdelraof, M., Hashem, A.H., El Saied, H., Sustainable bacterial cellulose production by Achromobacter using mango peel waste. *Microb. Cell Fact [Internet]*, 22, 1–12, 2023 [cited 2024 Sep 16]. Available from: https://microbialcellfactories.biomedcentral.com/articles/10.1186/s12934-023-02031-3.
26. Nazarova, N.B., Liyaskina, E.V., Revin, V.V., The Production of Bacterial Cellulose by Co-Cultivation of Komagataeibacter sucrofermentans with Dextran Producers Leuconostoc mesenteroides and Xanthan Xanthomonas campestris, in: *Вестник Томского государственного университета [Internet]*, pp. 23–42, 2022, [cited 2024 Sep 3]. Available from: https://typeset.io/papers/the-production-of-bacterial-cellulose-by-co-cultivation-of-21elo8d0.
27. Chen, J., Hong, F., Zheng, H., Zheng, L., Du, B., Using Static Culture Method to Increase the Production of Acetobacter Xylinum Bacterial Cellulose. *J. Nat. Fibers [Internet]*, 21, 2024. Available from: https://doi.org/10.1080/15440478.2023.2288286.
28. Digel, I., Akimbekov, N., Rogachev, E., Pogorelova, N., Bacterial cellulose produced by Medusomyces gisevii on glucose and sucrose: Biosynthesis and structural properties. *Cellul. [Internet]*, 30, 11439–53, 2023. Available from: https://doi.org/10.1007/s10570-023-05592-z.
29. Hishammuddin, N., Radzun, K.A., Syafiq, M.H., Rahman, S.A., Bahari, S.A., Osman, S., *et al.*, Green environment: Effects of acetate buffer on cellulose production by Acetobacter xylinum 0416 in fermented static cultivation. *IOP Conf. Ser. Earth Environ. Sci.*, 951, 2022.
30. Shavyrkina, N.A., Skiba, E.A., Kazantseva, A.E., Gladysheva, E.K., Budaeva, V.V., Bychin, N.V., *et al.*, Static Culture Combined with Aeration in Biosynthesis of Bacterial Cellulose, 2021 [cited 2024 Sep 6]; Available from: https://doi.org/10.3390/polym13234241.
31. Almihyawi, R.A.H., Musazade, E., Alhussany, N., Zhang, S., Chen, H., Production and characterization of bacterial cellulose by Rhizobium sp. isolated from bean root. *Sci. Rep. [Internet]*, 14, 1–17, 2024. Available from: https://doi.org/10.1038/s41598-024-61619-w.
32. Aswini, K., Gopal, N.O., Uthandi, S., Optimized culture conditions for bacterial cellulose production by Acetobacter senegalensis MA1. *BMC Biotechnol.*, 20, 1–16, 2020.
33. Khanchezar, S. and Babaeipour, V., Bacterial Cellulose Production Enhancement in Repeated Static Batch Culture of Acetobacter xylinum in Bench-Scale. *J. Chem. Chem. Eng. Res. Artic.*, 41, 2022, 2022.
34. Pham, T.T. and Tran, T.T.A., Evaluation of the Crystallinity of Bacterial Cellulose Produced From Pineapple Waste Solution By Using Acetobacter Xylinum. *ASEAN Eng. J.*, 13, 81–91, 2023.
35. Wahid, F., Zhao, X.J., Zhao, X.Q., Ma, X.F., Xue, N., Liu, X.Z., *et al.*, Fabrication of Bacterial Cellulose-Based Dressings for Promoting Infected Wound Healing. *ACS Appl. Mater. Interfaces*, 13, 32716–28, 2021.

36. Zhou, C., Yang, Z., Xun, X., Ma, L., Chen, Z., Hu, X., *et al.*, De novo strategy with engineering a multifunctional bacterial cellulose-based dressing for rapid healing of infected wounds. *Bioact. Mater.*, 13, 212–22, 2022.
37. Yang, Z., Ma, L., Han, X., Xun, X., Li, T., Duan, K., *et al.*, A facile, biosynthetic design strategy for high-performance multifunctional bacterial cellulose-based dressing. *Compos. Part B Eng.*, 238, 109945, 2022.
38. Liu, X., Cai, Z., Pei, M., Zeng, H., Yang, L., Cao, W., *et al.*, Bacterial Cellulose-Based Bandages with Integrated Antibacteria and Electrical Stimulation for Advanced Wound Management. *Adv. Healthc. Mater. [Internet]*, 13, 2302893, 2024 [cited 2024 Aug 24]. Available from: https://onlinelibrary.wiley.com/doi/full/10.1002/adhm.202302893.
39. Portela, R., Leal, C.R., Pedro, L., Minireview Bacterial cellulose: A versatile biopolymer for wound dressing applications, 2020.
40. Knöller, A., Widenmeyer, M., Bill, J., Burghard, Z., Fast-Growing Bacterial Cellulose with Outstanding Mechanical Properties *via* Cross-Linking by Multivalent Ions, pp. 1–8, 2020.
41. Fei, G., Wang, Y., Wang, H., Ma, Y., Guo, Q., Huang, W., *et al.*, Fabrication of Bacterial Cellulose/Polyaniline Nanocomposite Paper with Excellent Conductivity, Strength, and Flexibility. *ACS Sustain. Chem. Eng.*, 7, 8215–25, 2019.
42. Wu, Z., Chen, S., Wu, R., Sheng, N., Zhang, M., Ji, P., *et al.*, Top-down peeling bacterial cellulose to high strength ultrathin fi lms and multifunctional fi bers. *Chem. Eng. J. [Internet]*, 123527, 2019. Available from: https://doi.org/10.1016/j.cej.2019.123527.
43. Sriplai, N., Mongkolthanaruk, W., Eichhorn, S.J., Pinitsoontorn, S., Magnetically responsive and flexible bacterial cellulose membranes. *Carbohydr. Polym. [Internet]*, 192, 251–62, 2018. Available from: https://doi.org/10.1016/j.carbpol.2018.03.072.
44. Farzaneh Jabbari, V.B., Bacterial cellulose as a potential biopolymer for wound care. A review. *Int. J. Polym. Mater. Polym. Biomater.*, 73, 455–77, 2024.
45. Homaeigohar, S. and Boccaccini, A.R., Antibacterial biohybrid nanofibers for wound dressings. *Acta Biomater.*, 107, 25–49, 2020.
46. Pasaribu, K.M., Ilyas, S., Tamrin, T., Radecka, I., Swingler, S., Gupta, A., *et al.*, Bioactive bacterial cellulose wound dressings for burns with collagen *in-situ* and chitosan *ex-situ* impregnation. *Int. J. Biol. Macromol.*, 230, 2023.
47. Petrova, V.A., Khripunov, A.K., Golovkin, A.S., Mishanin, A.I., Gofman, I.V., Romanov, D.P., *et al.*, Bacterial cellulose (Komagataeibacter rhaeticus) biocomposites and their cytocompatibility. *Mater. (Basel) [Internet]*, 13, 1–18, 2020 [cited 2024 Sep 8]. Available from: https://www.researchgate.net/publication/346218482_Bacterial_Cellulose_Komagataeibacter_rhaeticus_Biocomposites_and_Their_Cytocompatibility.
48. Zahel, P., Beekmann, U., Eberlein, T., Schmitz, M., Werz, O., Kralisch, D., Bacterial Cellulose—Adaptation of a Nature-Identical Material to the Needs of Advanced Chronic Wound Care. *Pharmaceuticals.*, 15, 2022.

49. Ciecholewska-Juśko, D., Żywicka, A., Junka, A., Drozd, R., Sobolewski, P., Migdał, P., *et al.*, Superabsorbent crosslinked bacterial cellulose biomaterials for chronic wound dressings. *Carbohydr. Polym.*, 253, 117247, 2021.
50. Ciecholewska-Juśko, D., Junka, A., Fijałkowski, K., The cross-linked bacterial cellulose impregnated with octenidine dihydrochloride-based antiseptic as an antibacterial dressing material for highly-exuding, infected wounds. *Microbiol. Res.*, 263, 127125, 2022.
51. Torgbo, S. and Sukyai, P., Biodegradation and thermal stability of bacterial cellulose as biomaterial: The relevance in biomedical applications. *Polym. Degrad. Stab. [Internet]*, 109232, 2020. Available from: https://doi.org/10.1016/j.polymdegradstab.2020.109232.
52. Mamouri, K.S., Rahaiee, S., Zare, M., Kenari, M.N., Mirzakhani, N., Physicochemical and thermal characterization, and evaluation of a bacterial cellulose/Barhang gum-based dressing for wound healing. *Int. J. Biol. Macromol.*, 242, 1–5, 2023.
53. SZ, B., JL, A., SC, A., XZ, C., JM, A., JH, D., NOxidized cellulose-based hemostatic materialso Title. *Carbohydr. Polym. [Internet]*, 230, 115585, 2020. Available from: https://www.sciencedirect.com/science/article/abs/pii/S0144861719312536.
54. Hoang-Bao Tran, T., Minh Nguyen, T., Huu Nguyen, H., Nguyen, B.G., KL-MT& THN, Fabrication and Characterization of Oxidizedbacterial Cellulose Membrane as a Potential Hemostatic Dressing. *9th Int. Conf. Dev. Biomed. Eng. Vietnam*, 95, 206–218, 2023.
55. Cao, S., Zhang, K., Li, Q., Zhang, S., Chen, J., Injectable and photothermal antibacterial bacterial cellulose cryogel for rapid hemostasis and repair of irregular and deep skin wounds. *Carbohydr. Polym.*, 320, 121239, 2023.
56. Bian, J., Bao, L., Gao, X., Wen, X., Zhang, Q., Huang, J., *et al.*, Bacteria-engineered porous sponge for hemostasis and vascularization. *J. Nanobiotechnol. [Internet]*, 20, 1–20, 2022. Available from: https://doi.org/10.1186/s12951-022-01254-7.
57. Pandey, A., Singh, M.K., Singh, A., Bacterial cellulose: A smart biomaterial for biomedical applications. *J. Mater. Res. [Internet]*, 39, 2–18, 2024. Available from: https://doi.org/10.1557/s43578-023-01116-4.
58. Girard, V. and Chaussé, J., Bacterial cellulose: A comprehensive review, pp. 1–35, 2024.
59. Ciftci, F., Bioadhesion, antimicrobial activity, and biocompatibility evaluation bacterial cellulose based silver nanoparticle bioactive composite films. *Process Biochem.*, 137, 99–110, 2023.
60. SK, A., HC, B., SW, B., MI, C., SU, C., MU-I, A., *et al.*, Bacterial cellulose-turmeric powder composites as potential therapeutic wound dressings. *Ind. Crops Prod.*, 211, 118237, 2024.
61. Moradpoor, H., Mohammadi, H., Safaei, M., Mozaffari, H.R., Sharifi, R., Gorji, P., *et al.*, Recent Advances on Bacterial Cellulose-Based Wound Management: Promises and Challenges. *Int. J. Polym. Sci.*, 2022, 2022.

62. Munir, M., Nosheen, S., Muhammad, N., Uroos, M., Mustafa, W., Khan, R., *et al.*, Ionic liquid treated bacterial cellulose sheets as prospective biodegradable implant materials. *Cellulose*, 31, 2327–40, 2024.
63. Yuan, H., Chen, L., Hong, F.F., A Biodegradable Antibacterial Nanocomposite Based on Oxidized Bacterial Nanocellulose for Rapid Hemostasis and Wound Healing. *ACS Appl. Mater. Interfaces*, 12, 3382–92, 2020.
64. *Dermafill - Cellulose Membrane Dressing*, Xylinum Cellulose [Internet], [cited 2024 Sep 9]. Available from: https://www.dermafill.com/.
65. *Microbial cellulose wound dressing in the treatment of nonhealing lower extremity ulcers*, PubMed [Internet], [cited 2024 Sep 18]. Available from: https://pubmed.ncbi.nlm.nih.gov/25904579/.
66. Frankel, V.H., Serafica, G.C., Damien, C.J., Development and testing of a novel biosynthesized XCell for treating chronic wounds. *Surg. Technol. Int.*, 12, 27–33, 2004.
67. A Trial Comparing a Cellulose Dressing to Two Standard of Care Dressings in Treating Split Thickness Donor Sites in Burn and Wound Patients, in: *Researcher View*, ClinicalTrials.gov [Internet], [cited 2024 Sep 19]. Available from: https://clinicaltrials.gov/study/NCT05499104?cond=wound&term=Cellulose&page=2&rank=19&tab=table.
68. Shanks, L.A., Cronshaw, A., Alexander, K.S., Davies, J.A., O'Boyle, C.P., Evaluation of EpiProtect® microbial cellulose burns dressings in young children. *Scars Burn Heal.*, 6, 2020. 205951312094050.
69. Evaluation of EpiProtect® microbial cellulose burns dressings in young children [Internet]. [cited 2024 Sep 9]. Available from: https://www.researchgate.net/publication/343794738_Evaluation_of_EpiProtectR_microbial_cellulose_burns_dressings_in_young_children.
70. Bionext Biocellulose Wound Dressing [Internet]. [cited 2024 Sep 10]. Available from: https://www.bennetthealth.net/bionext/#Properties.
71. Role of Oxidized Regenerated Cellulose (ORC) Applied to "Dirty" Surgical Wounds, in: *Researcher View*, ClinicalTrials.gov [Internet], [cited 2024 Sep 19]. Available from: https://clinicaltrials.gov/study/NCT00947089?cond=dirty surgical wounds&term=Oxidized regenerated cellulose&rank=1&tab=table.
72. Bacterial Cellulose-monolaurin Hydrogel for Acute Radiation Dermatitis, in: *Researcher View*, ClinicalTrials.gov [Internet], [cited 2024 Sep 19]. Available from: https://clinicaltrials.gov/study/NCT05079763?cond=wound&term=Cellulose&page=3&rank=26&tab=table.
73. Efficacy of Buccal Pad of Fat, Advanced Platelet Rich Fibrin, Fibrin Glue and Oxidized Cellulose Plug in Management of Oro-Antral Communication, Comparative Clinical Study, in: *Study Details*, ClinicalTrials.gov [Internet], [cited 2024 Sep 19]. Available from: https://clinicaltrials.gov/study/NCT06220877?term=Oxidized cellulose&rank=2.
74. Hemostatic Patch/Collagen Sealant and Polyethylene Glycol (Hemopatch) in the Postoperative of the Thyroidectomy, in: *Study Details*, ClinicalTrials.gov [Internet], [cited 2024 Sep 19]. Available from: https://clinicaltrials.gov/

study/NCT05103956?term=Oxidizedcellulose&page=2&rank=11#study-record-dates.

75. Oxidized Cellulose hEmostAsis evaluatioN, in: *Study Details*, ClinicalTrials.gov [Internet], [cited 2024 Sep 19]. Available from: https://clinicaltrials.gov/study/NCT01637025?term=Oxidized cellulose&page=4&rank=39.
76. Ou, K., Liu, Y., Deng, L., Chen, S., Gu, S., Wang, B., Covalently grafting polycation to bacterial cellulose for antibacterial and anti-cell adhesive wound dressings. *Int. J. Biol. Macromol.*, 269, 132157, 2024.
77. Deng, L., Li, F., Han, Z., Qu, X., Li, J., Zhou, Z., *et al.*, Bacterial cellulose-based hydrogel with regulated rehydration and enhanced antibacterial activity for wound healing. *Int. J. Biol. Macromol.*, 267, 1–5, 2024.
78. Yu, C., Han, Z., Sun, H., Tong, J., Hu, Z., Wang, Y., *et al.*, Balancing mechanical property and swelling behavior of bacterial cellulose film by *in-situ* adding chitosan oligosaccharide and covalent crosslinking with γ-PGA. *Int. J. Biol. Macromol.*, 267, 2024.
79. Orlando, I., Basnett, P., Nigmatullin, R., Wang, W., Knowles, J.C., Roy, I., Chemical Modification of Bacterial Cellulose for the Development of an Antibacterial Wound Dressing. *Front. Bioeng. Biotechnol.*, 8, 1–19, 2020.
80. Sulaeva, I., Henniges, U., Rosenau, T., Potthast, A., Bacterial cellulose as a material for wound treatment: Properties and modifications: A review. *Biotechnol. Adv.*, 33, 1547–71, 2015.
81. Liang, S., Advances in drug delivery applications of modified bacterial cellulose-based materials. *Front. Bioeng. Biotechnol.*, 11, 1–14, 2023.
82. Qian, H., Liu, J., Wang, X., Pei, W., Fu, C., Ma, M., *et al.*, The state-of-the-art application of functional bacterial cellulose-based materials in biomedical fields. *Carbohydr. Polym.*, 300, 120252, 2023.
83. Jiang, Z., Shi, Z., Cheung, K.M., Ngai, T., Robust, Flexible, and High-Barrier Films from Bacterial Cellulose Modified by Long-Chain Alkenyl Succinic Anhydrides. *ACS Sustain. Chem. Eng.*, 11, 2486–98, 2023.
84. Pratama, R.A., Saepudin, E., Sufiandi, S., The effect of polyethylene glycol-citric acid (PEG-Ca) addition on mechanical properties of dry bacterial cellulose film, in: *AIP Conf Proceding*, 2022.
85. Wahid, F., Huang, L.-H., Zhao, X.-Q., Li, W.-C., Wang, Y.-Y., Jia, S.-R., *et al.*, Bacterial cellulose and its potential for biomedical applications. *Biotechnol. Adv.*, 53, 2021.
86. Fürsatz, M., Skog, M., Sivlér, P., Palm, E., Aronsson, C., Skallberg, A., *et al.*, Functionalization of bacterial cellulose wound dressings with the antimicrobial peptide ε-poly-L-Lysine. *Biomed. Mater. [Internet]*, 13, 025014, 2018 [cited 2024 Sep 19]. Available from: https://iopscience.iop.org/article/10.1088/1748-605X/aa9486.
87. Varaprasad, K., Jayaramudu, T., Kanikireddy, V., Toro, C., Sadiku, E.R., Alginate-based composite materials for wound dressing application: A mini review. *Carbohydr. Polym.*, 236, 116025, 2020.

88. Zmejkoski, D.Z., Marković, Z.M., Budimir, M.D., Zdravković, N.M., Trišić, D.D., Bugárová, N., *et al.*, Photoactive and antioxidant nanochitosan dots/biocellulose hydrogels for wound healing treatment. *Mater. Sci. Eng., C*, 122, 2021.
89. Zmejkoski, D.Z., Marković, Z.M., Mitić, D.D., Zdravković, N.M., Kozyrovska, N.O., Bugárová, N., *et al.*, Antibacterial composite hydrogels of graphene quantum dots and bacterial cellulose accelerate wound healing. *J. Biomed. Mater. Res. Part B Appl. Biomater. [Internet]*, 110, 1796–805, 2022 [cited 2024 Sep 4]. Available from: https://onlinelibrary.wiley.com/doi/full/10.1002/jbm.b.35037.
90. Mohaghegh, H., Assadi, Z., Derakhshan, A., Masaeli, E., Accelerating Full-Thickness Wound Healing with Bacterial Cellulose-Based Multilayer Composites. *J. Pharm. Sci.*, 113, 754–63, 2024.
91. WS, A., TK, B., MU-I, C., RK, D., ZH, A., AK, A., *et al.*, Development of modified montmorillonite-bacterial cellulose nanocomposites as a novel substitute for burn skin and tissue regeneration. *Carbohydr. Polym.*, 206, 548–56, 2019.
92. Zheng, L., Li, S., Luo, J., Wang, X., Latest Advances on Bacterial Cellulose-Based Antibacterial Materials as Wound Dressings. *Front. Bioeng. Biotechnol.*, 8, 1–15, 2020.
93. Kanji, S. and Das, H., Advances of Stem Cell Therapeutics in Cutaneous Wound Healing and Regeneration. *Mediators Inflamm.*, 2017, 2017.
94. Wang, X., Zhao, J., Wang, X., Zhang, J., Wang, Y., Wang, X., *et al.*, Bacterial cellulose membrane combined with BMSCs promotes wound healing by activating the notch signaling pathway, pp. 1–14, 2023.
95. Costa De Oliveira Souza, C.M., De Souza, C.F., Mogharbel, B.F., Irioda, A.C., Franco, C.R.C., Sierakowski, M.R., *et al.*, Nanostructured cellulose-gellan-xyloglucan-lysozyme dressing seeded with mesenchymal stem cells for deep second-degree burn treatment. *Int. J. Nanomed.*, 16, 833–50, 2021.
96. Radu, C.D., Verestiuc, L., Ulea, E., Lipsa, F.D., Vulpe, V., Tamas, C., *et al.*, Evaluation of Keratin/Bacterial Cellulose Based Scaffolds as Potential Burned Wound Dressing. *Appl. Sci.*, 2021.
97. Behera, B., Laavanya, D., Balasubramanian, P., Techno-economic feasibility assessment of bacterial cellulose biofilm production during the Kombucha fermentation process. *Bioresour. Technol.*, 346, 126659, 2022.
98. van Zyl, E.M. and Coburn, J.M., Hierarchical structure of bacterial-derived cellulose and its impact on biomedical applications. *Curr. Opin. Chem. Eng.*, 24, 122–30, 2019.
99. Echeverría, C., Arnaldos-Pérez, I., Barranco, R., Muñoz-Bonilla, A., Fernández-García, M., Synthesis and characterization of cellulose derivatives for enhanced antimicrobial activity. *Cellul. [Internet]*, 31, 1729–40, 2024 [cited 2024 Sep 10]. Available from: https://link.springer.com/article/10.1007/s10570-023-05655-1.

100. Dawwam, G.E., Al-Shemy, M.T., El-Demerdash, A.S., Green synthesis of cellulose nanocrystal/ZnO bio-nanocomposites exerting antibacterial activity and downregulating virulence toxigenic genes of food-poisoning bacteria. *Sci. Rep. [Internet]*, 12, 1–18, 2022 [cited 2024 Sep 10]. Available from: https://www.nature.com/articles/s41598-022-21087-6.
101. Caro-Astorga, J., Lee, K.Y., Ellis, T., Increasing bacterial cellulose compression resilience with glycerol or PEG400 for robuster engineered living materials. *Carbohydr. Polym. Technol. Appl. [Internet]*, 4, 100245, 2022. Available from: https://doi.org/10.1016/j.carpta.2022.100245.
102. Sorourian, M., Pourmadadi, M., Yazdian, F., Rashedi, H., Nigjeh, M.N., Sorourian, G., *et al.*, Engineered bacterial cellulose-based Ag nanoparticles/g-C3N4/Eucalyptus extract nanocomposites for wound dressing: *In vitro* evaluation. *Eur. J. Med. Chem. Rep.*, 12, 100190, 2024.

13

Bacterial Cellulose for Bone Tissue Engineering

Nikita A. Shinde, Mrunal S. Patil and Shivaji B. Kashte*

Department of Stem Cell and Regenerative Medicine and Medical Biotechnology, Centre for Interdisciplinary Research, D. Y. Patil Education Society (Deemed to be University), Kolhapur, India

Abstract

Bacterial cellulose (BC) has shown unique structural and mechanical properties and biocompatibility, making it suitable for the regeneration or repair of bone tissue. Since BC lacks lignin and hemicellulose, it reduces the likelihood of inflammatory reactions compared to plant-derived cellulose. Furthermore, the porous structure of BC can be engineered to enhance cell adhesion, growth, and differentiation, which are crucial for efficient bone regeneration. Current materials used for bone tissue engineering, such as synthetic polymers, autografts, and allografts, have notable drawbacks. Although autografts are biocompatible, they require additional surgery and are not widely available. Allografts carry risks of immunological rejection and disease transmission. While synthetic polymers are versatile, they often lack the mechanical and biological properties needed for effective bone repair. In contrast, BC offers numerous advantages over these conventional materials. Its natural origin and the simplicity of bacterial fermentation make manufacturing scalable and sustainable. BC's osteoinductive properties can be enhanced by functionalizing it with bioactive substances. BC composites support osteoblast adhesion and proliferation, thus facilitating bone tissue formation. Therefore, BC presents a promising and innovative biomaterial for bone tissue engineering. However, further research into BC-based composites and their *in vivo* performance is crucial to fully realizing their potential in bone tissue engineering applications.

***Keywords*:** Bacterial cellulose, cellulose scaffolds, bone defects, bone regeneration, bone tissue engineering

**Corresponding author*: ShivajiKashte.cir@dypgroup.edu.in

Mohd Shabbir (ed.) Regenerated Cellulose: A Renewable Alternative to Petroleum-Based Plastics, (387–420) © 2026 Scrivener Publishing LLC

13.1 Introduction

Bone is a rigid connective tissue that forms the skeleton of vertebrates, providing structural support and protection for vital organs. It consists of a matrix containing collagen fibers and inorganic minerals, primarily calcium phosphate, which give strength and rigidity to the bone. Bone plays various roles in the human body. They provide structural support for the body, helping movement through the control of muscles. About 80% of bone mass comprises cortical bone, which gives the skeleton mechanical strength. Its metabolic activity includes the continuous regeneration of the bone structure through the coordinated actions of osteoblasts and osteoclasts. Osteoblasts are responsible for the formation of new bone, while osteoclasts are involved in the resorption of bone tissue. Trabecular bone (spongy) is less dense and has higher metabolic activity than cortical bone, with osteoblasts being greater in trabecular bone than cortical bone. It is essential for load transfer and creates a space for blood cell production [1]. An estimated 1.5 million people experience fractures as a result of bone disease each year globally [2]. Bone defects or fractures, often caused by high-energy trauma or conditions like osteoporosis and cancer, can also result from surgical procedures such as tumor resections. Surgical methods like distraction osteogenesis or the Masquelet operation are often employed to encourage bone regeneration. Treatment options vary and usually consist of immobilization surgery or bone transplantation. Because of their healing features and compatibility, autografts—derived from the patient's body—are considered the best option for bone repair. However, their availability is limited, and extra surgery is necessary. Although allografts, or donors from the same species, avoid the need for a second round of treatment, there is a higher risk of rejection and a longer time to recover [3]. The field of tissue engineering seeks to create biological substitutes to replace or repair damaged tissues. In particular, bone tissue engineering (BTE) uses hydrogels and scaffolds to support various cell types during their development and repair of broken bones. The fundamental goal of BTE treatments is to incorporate these components to encourage bone tissue regeneration and healing successfully [4].

BTE scaffolds are of various biomaterials like ceramics and natural or synthetic polymers. Biomaterials are engineered to mimic the hierarchical structure of bone, providing an optimal environment for tissue integration and cell proliferation. Ceramics like hydroxyapatite are used for scaffolds to increase osteoconductive properties and boost the framework of scaffolds [5]. In BTE, synthetic polymers that boost mechanical properties and degradability include polycaprolactone, polyvinyl pyrrolidone, and poly (lactic-co-glycolide). In BTE, natural polymers, including collagen, gelatin, chitosan [6], and cellulose, are primarily used to improve biocompatibility and decrease immunogenic reactions. The most common polymer found on Earth is cellulose. The primary source of cellulose is the walls of plant cells. However, other substances, such as lignin, pectin, and hemicelluloses, are mixed in with this type of cellulose. Additional purification is necessary before the cellulose can be utilized for practical applications. Furthermore, as a byproduct of their regular metabolic activities, cellulose is produced by many algal and bacterial taxa, including Aerobacter, Salmonella, Escherichia, Acetobacter, Rhizobium, Agrobacterium, Achromobacter, and Sarcina [7, 8]. The purest type of cellulose is bacterial cellulose (BC), initially extracted and described by Brown in 1886 [9]. From a structural perspective, BC comprises porous sheets formed by a dense, orderly, three-dimensional network of fibrils. Compared to plant cellulose, BC exhibits significantly higher mechanical properties, thermal stability, and crystallinity, which can be attributed to its well-organized structure and high purity [10]. Therefore, BC can be a remarkable natural polymer for applications.

13.2 Biosynthesis of Bacteriological Cellulose

Numerous bacterial species can produce cellulose, but *Gluconacetobacter xylinus (G. xylinus)* and *Komagataeibacter xylinus (K. xylinum)* are unique organisms for researching this biopolymer. The biosynthesis of bacterial cellulose (BC) is intricate and involves multiple steps involving multiple proteins, with cellulase synthase being the central enzyme. It starts with the polymerization of glucose molecules through the linkage of β-1, 4-glucan units, forming a chain structure. Cellulose microfibrils are more organized structures that result from synthesizing these cellulose chains. Various forces, including electrostatic interactions, van der Waals forces, and hydrogen bonds, play a crucial role in stabilizing BC microfibrils [11, 12]. Cellulose produced by *K. xylinus* possesses remarkable features, such as multidirectional polarity and variable thickness. Two forms of cellulose are produced during the microfibril crystallization process in *K. xylinus*: cellulose I, which consists of microfibrils commanded in a parallel pattern, and cellulose II, which comprises microfibrils arranged antiparallel [13]. The oxidation of organic acids, functional carbohydrates, and the Krebs and pentose metabolic cycles are necessary for *K. xylinus* to produce cellulose. Since *K. xylinus* is an aerobic bacterium, it lacks the activity of the enzyme phosphofructokinase-1, which catalyzes the synthesis of fructose-6-phosphate to fructose-1, 6-bisphosphate. The glycolytic intermediate glucose-6-phosphate is the first step in synthesizing cellulose from glucose [14, 15]. The enzyme glucokinase phosphorylates glucose (C6) as the initial step in the path, resulting in glucose-6-phosphate. Then, this molecule is isomerized by the enzyme phosphoglucomutase (EC 5.4.2) to produce glucose-1-phosphate (G-1-P). UTP-glucose-1-phosphate uridylyltransferase (EC 2.7.7.9) catalyzes the rate-limiting step in the chain reaction between glucose-1-phosphate and UTP to create uridine-5′-diphosphate-α-D-glucose (UDP-glucose). In *K. xylinus*, UDP-glucose is a crucial intermediate sugar nucleotide essential for cellulose synthesis [16]. Ultimately, cellulose synthase transfers glucosyl residues from UDP-glucose to a developing β-1, 4-glucan chain. The BC synthase complex then polymerizes this chain to produce BC. The cellulose synthase complex in the cytoplasmic membrane is the most crucial enzyme in producing cellulose, as shown in Figure 13.1 [17].

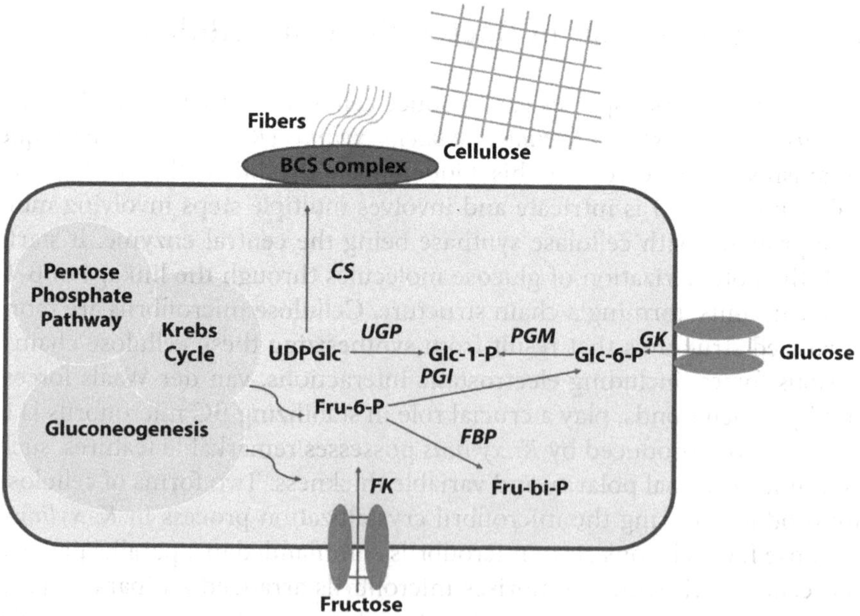

Figure 13.1 Biosynthesis of bacteriological cellulose [Figure adapted from [18]].

13.3 Properties of Scaffolds for Bone Tissue Engineering

13.3.1 Crystallinity

The crystallinity of BC is crucial in applications related to bone tissue engineering (illustrated in Figure 13.2). BC is characterized by high crystallinity, which contributes mechanical strength and hydrophilicity, which is suitable for mimicking the extracellular matrix in tissue engineering applications. The crystallinity index of BC can be affected by its modifications. The addition of hydroxyapatite nanoparticles reduced the crystallinity from 82.5% to 62% [19]. The mineralization process in BC scaffolds also impacts crystallinity, as calcium phosphate deposition can boost the properties of the scaffold for bone regeneration application [20].

13.3.2 Porosity

Porosity is referred to as the percentage of void space in a solid (shown in Figure 13.2). Scaffolds that help cellular activities are essential for bone regeneration. Porosity encourages mesenchymal cell migration, proliferation, osteoblast proliferation, and vascularization. Pores play an essential role in the development of new bone tissue. The porosity is crucial to the

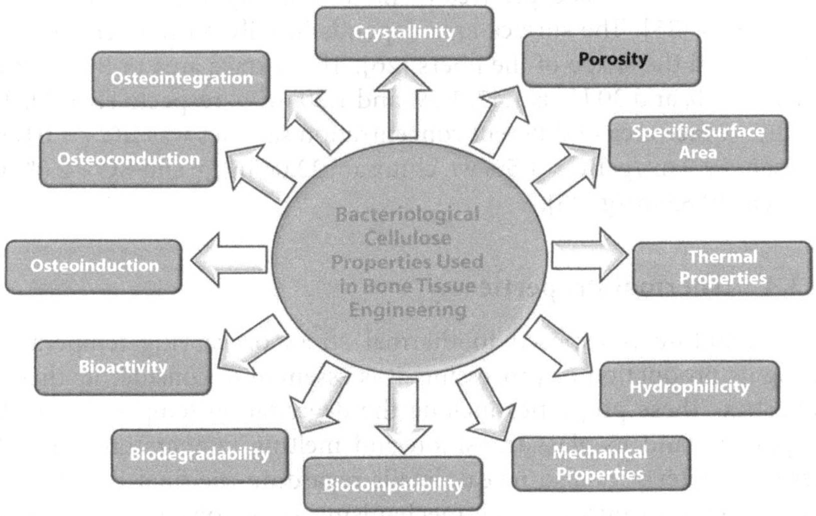

effective movement of nutrients and oxygen, which are required for cell survival, healthy growth, migration, and the enhanced three-dimensional (3D) organization required for vascularization and ECM synthesis [21]. Macropores (>100 μm) are crucial for cellular infiltration and vascularization and promoting new bone formation. Micropores (1–10 μm) enhance cell adhesion and nutrient absorption and are fundamental for osteogenic differentiation [22]. The pore size of the bacterial cellulose-gelatin/hydroxyapatite scaffold (BC-GEL/HAp) was between 0.2 and 0.5 μm [23]. Nanocomposite scaffolds that incorporate hydroxyapatite embedded within a cellulose network (PAA-g-Cellulose/HAp) showed a pore size of 72 to 125 μm and a porosity of 85.7%, showing their potential for bone tissue engineering [24].

13.3.3 Specific Surface Area

The specific surface area of the scaffold is crucial in bone tissue engineering, as it enhances cell adhesion, promotes osteogenic differentiation, and increases cell proliferation. High-specific surface area scaffolds develop a favorable microenvironment for bone regeneration, facilitating nutrient exchange and cellular interactions. The surface area of cellulosic materials can be determined by applying the Brunauer–Emmett–Teller (BET) method combined with nitrogen adsorption. These investigations showed the surface area in the 0.58 to 200 m^2/g range, suggesting that variables including sample choice, pre-treatment, and drying techniques affect the surface area [25]. The surface area depends heavily on the percentage of cellulose and the shape of the fibers [26]. The surface area of BC micronized at 10, 20, and 30 Hz is 1.47, 1.59, and 1.20 m^2/g, respectively [27]. The specific surface area of different concentration samples was BC + 1.0% wt. chitosan, 29.5 m^2/g; BC + 1.5% wt. chitosan, 22.06 m^2/g; and BC + 2.0% wt. chitosan, 19.85 m^2/g [28].

13.3.4 Thermal Properties

When cellulose is exposed to thermal stress or varying temperatures during its production or processing, it is essential to consider its thermal properties. These properties include the degradation temperature of BC components and its glass transition and melting temperatures [29, 30]. These characteristics also reveal details regarding the final BC product's composition and purity. Three mechanisms are involved in BC thermal degradation, together with mass losses. First, the evaporation of moisture or leftover water in the matrix, particularly in the amorphous areas, is thought to cause mass loss before reaching 200°C [29, 31]. While eliminating this water influences hydrogen bonding, the material structure is not significantly affected. Within this temperature range, residual proteinaceous debris left over from fermentation can break down at approximately 159°C. A 70% to 80% of the mass loss at up to 150°C reported by Thermogravimetric analysis (TGA) can be attributed to the second step, which is associated with the degradation of BC fibers [32]. This includes removing crystalline areas through depolymerization, which leads to the breakdown of the process of the glycosyl units and dehydration (of bound water) [33]. Either esterification of the hydroxyl groups or direct catalysis can speed up this process. After degradation is finished, oxidation produces a charred residue. After that, the remaining carbonaceous matter decomposes to produce gaseous low-molecular-weight compounds [34]. BC-derived nanocrystals were created by treating BC with enzymes, resulting in the formation of BC nanocellulose crystals. It showed the following thermal properties: T_d at 379°C (TGA), T_g at 66.7°C (DSC), T_m at 203.3°C (DSC), and ΔH_m at 83.2 J/g (DSC) where, T_d is decomposition temperature; T_g, glass transition temperature; T_m, melting temperature; and ΔH_m, enthalpy of melting [35].

13.3.5 Hydrophilicity

Hydrophilicity refers to the affinity of material interacting with water, essential for cell adhesion, nutrient transport, and biocompatibility (shown in Figure 13.2). BC scaffolds show significant hydrophilicity, increasing their suitability for bone regeneration applications. The contact angle of BC scaffolds varies from 30° to 50°, showing hydrophilic properties, which simplify cell attachment and proliferation of cells [36, 37]. BC underwent treatment with TEMPO-mediated oxidation (resulting in TO-BC) and maleic acid (resulting in MA-BC) to develop uniform BC dispersions suitable for creating 3D-printed scaffolds aimed at bone regeneration. The water contact angle for MA-BC was 80°, significantly higher than the 33° for TO-BC. This variation can be explained by the ester bonds formed during the esterification process between the hydroxyl groups in BC and the carboxyl groups in maleic acid [38]. The scaffolds BC and b-glucan (BgC) BgC-1.1, BgC-1.2, BgC-1.3, and BgC-1.4 had water contact angles of 118.40°, 103.50°, 92.20°, and 79.70° at 10 minutes, and 95.60°, 82.00°, 77.20°, and 64.10° at that time, respectively. In addition to boosting the interactions among the oxygenated functionalities and the polymer, the increased amount of graphene oxide (GO) additionally altered the surface hydrophilicity to hydrophobicity by regulating the porosity of the scaffolds [39].

13.3.6 Mechanical Properties

The mechanical property of scaffolds in BTE is essential for their performance and integration within the body. These properties ensure scaffolds can bear physiological loads while stimulating cell growth and tissue regeneration. To represent the material's mechanical properties, the ultimate tensile or fracture strength, or stress at breaking down. Young's modulus shows the elastic area slopes of the strain and stress curve and the material's stiffness, which is usually expressed in GPa. However, MPa or KPa is typically employed for calculating extremely elastic materials like hydrogels, foams, and scaffolds. The cortical bone has a Young's modulus of 15 to 20 GPa, but the cancellous bone has a Young's modulus of 0.1 to 2 GPa. Cortical bone has a 100 to 200 MPa strength, while cancellous bone has a compressive strength of 2 to 20 MPa [40]. Tensile testing of BC reveals that the initial stretching step modifies the interactions between the polymer and water and the structure of the polymer network.

As a result, until the ribbons straighten and finally fracture, the layers align parallel to the direction of tension. Research has shown that aligning fibers in an ordered manner decreases resistance to deformation. Therefore, the most important structural parameters influencing BC tensile characteristics are anticipated to be cellulose content and fiber orientation/structural shape [41, 42]. BC scaffolds have tensile strengths that vary from about 10 MPa to more than 15000 MPa, depending on how they are hydrated and specifically formulated [43]. The structural integrity, mechanical stability, and possible applications of BC can be better understood by investigating its compression properties. In both wet and dried phases, the concentration and configuration of the cellulose fibril network significantly impact BC stiffness [41]. There are some differences in the three-phase behavior of hydrated BC. The second phase is not a plateau because the BC sample loses water and decreases in volume when stress is applied. The water, which is partially attached to the fibers, contributes to the material's overall stiffness. As this water is removed, it increases the stiffness, which means that the exit of the incompressible water significantly influences the sample's compression behavior. Furthermore, the third and final phase, which is related to the densification of the material, occurs at around 75% strain. In contrast, the second phase, characterized by fiber entanglement and a denser network, occurs at approximately 50% strain [41]. The mechanical compression for BC, BC-HAp, oxBC30, oxBC60, oxBC30-HAp, and oxBC60-HAp tissue scaffold were 190 kPa, 840 kPa, 265 kPa, 312.5 kPa, 745 kPa, and 865 kPa, respectively [44].

13.3.7 Biocompatibility

Biocompatibility is critical as it determines how healthy material interacts with biological systems. It encompasses the capacity of a material to effectively carry out its intended functions without causing opposing reactions. The importance of biocompatibility in BTE is featured by its influence on cell behavior, scaffold integration, and the overall healing process. An essential requirement for producing suitable biomedical materials is their capacity to prevent damage to living tissues and encourage the appropriate host response [45]. Recent research has focused on BC implants due to their remarkable biocompatibility and capacity to promote cell attachment, differentiation, and tissue ingrowth. The excellent biocompatibility of BC is often shown by *in vivo* research [46]. After transplanting, the host tissues usually accept BC membranes or scaffolds without adverse effects. Inflammation reactions were generally found to be negligible or below tolerable limits, which is a common sign of biocompatibility. The biocompatibility could be assessed in both the short and long terms because the implantation times differ between the studies [47]. *In vitro* toxicity tests showed that oxidized-BC membranes with strontium apatite (SrAp) had over 70% viability in the MC3T3 mouse cell line and were less toxic than calcium hydroxyapatite (CaHA) membranes [48].

13.3.8 Biodegradability

Biodegradability is the capacity of a material to decompose through biological processes. Biodegradability is a crucial characteristic of BC scaffolds in BTE, influencing their performance and integration within biological systems. This property is enhanced for BC through modifications, such as crosslinking treatments, which can improve its structural integrity while maintaining its degradable nature [49, 50]. The composite scaffold of BC and gelatin showed more biodegradability (20.8% to 44.4%), which is suitable for tissue engineering [50]. Gamma irradiation was used to modify BC nanofibrils to enhance their biodegradability. In comparison to a non-irradiated BC membrane that resists degradation, those that have undergone gamma irradiation at doses of 25 kGy and 50 kGy experienced weight losses of 18% and 25%, respectively, after about six months of being in a simulated bodily fluid (SBF) solution. The tensile strength of the irradiated membranes is notably affected, especially after exposure to the SBF. This phenomenon is primarily attributed to the loss of smaller molecular weight components, which results in a more flexible fibril network [51].

13.3.9 Bioactivity

"Bioactivity" refers to a biomaterial's capacity to interact with biological systems to promote cell adhesion, proliferation, and differentiation, especially in osteoblasts for bone repair. This idea is essential to developing scaffolds that have the same characteristics as actual bone [52]. BC has significant bioactivity, making it a promising scaffold material for BTE. Bioactive materials, including hydroxyapatite and growth factors, enhance their osteogenic properties by promoting the more natural utilization of bone tissue. In addition, modifications like phosphate crosslinking enhance BC's biomineralization, which is essential for producing apatite that mimics bone [37]. In BC/HAp composite hydrogel, BC combined with HAp and platelet-rich plasma significantly increased cell differentiation and bone healing *in vitro* [53]. Integrating MXene with BC enhances mechanical properties and promotes osteogenic gene expression, indicating its potential for effective bone regeneration [54].

13.3.10 Osteoinduction

Osteoinduction is defined as the scaffold's capacity to recruit progenitor cells to the healing site and encourage osteogenic differentiation in those cells through bimolecular signaling [55]. Studies show that the scaffold's textured surface and nanoscale features facilitate the conversion of stem cells into osteoblasts, a process crucial for bone formation. Furthermore, it has been found that implants designed with a lower oxygen partial pressure can encourage pericytes in blood vessels to transform into cells that contribute to bone production [56].

13.3.11 Osteoconduction

Osteoconduction is the process by which bone grows on the surface of scaffolds or implant materials [29]. Adding other materials boosts BC's osteoconductive properties by providing an optimal environment for attachment and growth of cells, which is crucial to increase the bone regeneration process [19]. The BC combined with calcium phosphate boosts its mechanical properties, promotes osteoblastic activity, and is essential for new bone formation (evident in Table 13.1) [57].

13.3.12 Osteointegration

Osteointegration is the strong connection between living bone and the surface of a load-bearing implant. This connection is important for keeping

Table 13.1 Application of bacteriological cellulose with various materials in bone tissue engineering.

Scaffolds composition	Source of BC	Physico-chemical–mechanical properties	*In vitro* findings	Osteoblast differentiation and bone regeneration	Ref.
BC/ PHB	*A. xylinum*	Pore size, 5 to 50 µm. Young's modulus, 15 ± 1.0 MPa and 1400 ± 101 MPa	Cell viability and proliferation of 3T3-L1 mice preadipocytes, exhibiting a nontoxic nature	Osteoblastic differentiation increased osterix (OSX) expression and early ALP activity; calvarial defects regenerated after 20 weeks	[57]
BC/Fe_3O_4/ HA	*Gluconace- tobacter xylinus*	Porosity, 81.1% compressive strength, 9.87 MPa	MTT assay using L929 cells and incubated for 24 hr. Showed 80% cell viability	Promoted cell attachment, differentiation, and proliferation	[43]
BC/GEL/BHAp	*G. xylinum*	Pore size, 45 to 210 µm Porosity, 68.49–80.94% Young's modulus, 9.61 to 11.33 MPa compressive strength, 35.02 to 94.9 MPa	-	Saos-2 cells attachment, differentiation, and proliferation and migration within BHA-containing scaffolds at a significantly higher level at day 14	[60]
BC-GELPCL/ HA	*Glucanoaceto-bacter xylinus*	Pore size, 314.14 ± 23.2 µm	150% cell viability after 3 to 7 days of scaffold incubation with Human osteoblast cells	Osteoblast cell attachment was observed after 7 days	[61]

(*Continued*)

Table 13.1 Application of bacteriological cellulose with various materials in bone tissue engineering. (*Continued*)

Scaffolds composition	Source of BC	Physico-chemical–mechanical properties	*In vitro* findings	Osteoblast differentiation and bone regeneration	Ref.
BC/HA/PDA	*Komagataeibacter xylinus*	-	70% cell viability after 24 hrs of scaffold incubation with MC3T3-E1 cells	Excellent adherence of MC3T3-E1 cells, followed by 24 hours of cell culture time, with preosteoblasts.	[62]
BC/Au	*G. xylinum*	Total pore volume, 17.9165 ml/g an average pore area, 1.3781 m^2/g Young's modulus, ~5.5 MPa	Nontoxic nature with hBMSCs	Enhanced osteogenic differentiation. Bone regeneration in a rabbit femoral defect model, with significant bone formation observed within 2 months	[63]
mNBC	*Komagataeibacter europaeus*	Pore sizes, 20 to 1500 μm	MTT assays and live/dead staining indicated high cell viability and biocompatibility	Electron microscopy confirmed strong attachment and fibroblast-like morphology of murine MSCs on mNBC scaffolds	[65]
BC-g-PNCl	-	Mechanical strength, 57.3 MPa	Viable L929 fibroblasts for 7 days	Pro-osteogenic potential, as confirmed by high ALP activity and calcium nodule formation in BMSCs. *In vivo*, this material promoted bone regeneration in rat calvarial defect models.	[66]

(*Continued*)

Table 13.1 Application of bacteriological cellulose with various materials in bone tissue engineering. (*Continued*)

Scaffolds composition	Source of BC	Physico-chemical–mechanical properties	*In vitro* findings	Osteoblast differentiation and bone regeneration	Ref.
BC/HAP@PRP	*Acetobacter xylinum*	-	Non toxic	Enhanced mineralized nodule formation and promoted MC3T3-E1 cell differentiation after 3 weeks.	[53]
BC/PVA/hBN	*Gluconacetobacter xylinus*	Viscosity, 81.3 ± 4.5 mPa.s, surface tension- 65.3 ± 5 mN/m, density, 1.07 ± 0.02 g/mL, tensile strength, 0.1 ± 0.05 MPa, elongation at break, 93 ± 23%.	~150% cell viability after 72 hours with human osteoblasts	After 72 hours of incubation, the morphological structures of the human osteoblasts attached to the scaffolds appeared to have a rounded morphology	[67]
TO-BC-MA-BC	-	Tensile strength, 61.2 MPa, elongation at break, 2.8%; Young's modulus, 2.1 GPa.	Promoted cell viability at 100 μg/mL, with TO-BC maintaining 143% and MA-BC decreasing from 137% to 107% as concentration increased.	ALP activity and mineralized nodule formation, assessed at 7–14 days and 28 days, respectively, showed enhanced osteogenic differentiation. *In vivo,* improved bone regeneration.	[38]

(*Continued*)

Table 13.1 Application of bacteriological cellulose with various materials in bone tissue engineering. (*Continued*)

Scaffolds composition	Source of BC	Physico-chemical–mechanical properties	*In vitro* findings	Osteoblast differentiation and bone regeneration	Ref.
MgONPs-BC	*Acetobacter xylinum*	Porosity, 89.2 ± 4.85%. Compressive strength, 53.09 ± 5.02 MPa Young's modulus, 191.19 ± 15.10	The MTT assay was used to assess viability of MG-63 cells on days 7, 14, and 21 showed ~150% cell viability.	After 14 and 21 days, MG63 cells on MgONPs-BC showed enhanced ALP activity.	[68]
3D BNC	*Komagataeibacter medellinensis*	Crystal size, 25.12 to 35.9 nm	95% cell viability using BM-MSCs	Alkaline phosphatase, osteocalcin, and osteopontin indicated osteogenic differentiation, strong cellular adhesion of BM-MSCs	[69]
BC/SF/BGNP	-	Lower rate of breakdown and swelling pore size, 205 μm	Cytotoxicity of the was evaluated using hASCs for 3, 7, and 21 days showed ~150% cell viability	BC/SF/BGNP scaffolds exhibited the highest levels of osteocalcin, ALP, RUNX2 expression, ALP activity, and mineral formation.	[70]

(*Continued*)

Table 13.1 Application of bacteriological cellulose with various materials in bone tissue engineering. (*Continued*)

Scaffolds composition	Source of BC	Physico-chemical–mechanical properties	*In vitro* findings	Osteoblast differentiation and bone regeneration	Ref.
COL/BC/BMP-2	-	Specific surface area, 123.4 m^2/g, pore volume, 0.59 cm^3/g; average pore diameter, 198.5 nm; and particle sizes, 8–12 microns; crystallinity, 11.92 ± 1.21%	Cytotoxicity tests using MC3T3-E1 cells showed ~105% cell viability	Promoting cell adhesion, proliferation, and osteogenic differentiation	[73]
BC-GEL/HAp DN	-	Fracture stress, 0.28 MPa; elastic modulus, 0.27 MPa	-	rBMSCs cultured on the BC-GEL/HAp composite had better adhesion and higher proliferation and differentiation	[23]
MBG/BC	-	-	hBMSCs onto scaffolds, their impacts on biocompatibility	Scaffolds demonstrated higher hBMSCs proliferation, elevated ALP activity, and enhanced osteogenic gene expression	[75]
CS/OBC/nHAP	-	-	-	Rat skull defect model, the scaffold was shown to promote bone tissue formation	[76]

the implant stable and ensuring it works well over the long term. It includes a time-dependent healing process where bone cells bond to the implant, ensuring adequate mechanical stability necessary for load-bearing functions [58]. BC is familiar for its beneficial properties, such as biocompatibility and inherent structure that mimic collagen fibers found in natural bone [59]. Its unique microporous characteristic allows for good cell adhesion and vascularization, making it an effective scaffold material for bone tissue engineering. Research shows that combining BC with calcium phosphate and other biomaterials improves its ability to support bone growth and helps it integrate with bone tissue [59].

13.4 Scaffolds for Bone Tissue Engineering

Various scaffolds were prepared by using BC and studied for bone tissue engineering. The composites were prepared from hydroxyapatite (HA) on BC and magnetite (Fe3O4) using ultrasonic radiation. The composite scaffold showed 81.1% porosity, 9.87 MPa compressive strength, and 1.85 Ga stiffness. The scaffold's swelling ability was 19±4.0%. The addition of nanoparticles reduced its crystallinity and enhanced the scaffold's biodegradation property. Biocompatibility and cell viability analysis of scaffolds showed a nontoxic nature and also promoted cell attachment, differentiation, and proliferation, showing potential for BTE [43].

Bacterial cellulose was used to reinforce scaffolds made of boron-doped hydroxyapatite and gelatin (BC-GEL/BHAp). These scaffolds were created using a freeze-drying method. The pore size and porosity of the scaffold were 45 to 210 μm and 68.49% to 80.94%, respectively. BC-GEL/BHAp scaffolds showed Young's modulus of 9.61 to 11.33 MPa and compressive strength ranging from 35.02 to 94.9 MPa. Scaffold degradation analysis reveals a weight loss of 5.91% after four weeks. BC-GEL/BHAp scaffolds showed attachment of human osteosarcoma cell line (SAOS-2), migration, and proliferation within scaffolds, which was observed by 14 days significantly [60]. Bacterial cellulose/gelatin/polycaprolactone/hydroxyapatite (BC-GEL-PCL/HA) composite scaffolds were fabricated using 3D printing. The pore size of this composite scaffold was 314.14 ± 23.2 μm. The mechanical properties include tensile strength varying from 1.08 ± 0.34 MPa to 1.58 ± 0.19 MPa and strain at break from 3.29 ± 1.41% to 3.97 ± 2.27%. These scaffolds showed 150% cell viability after 3 to 7 days of scaffold incubation with Human osteoblast cells. Osteoblast cell attachment was observed after 7 days using DAPI staining [61].

Composite scaffolds (BC/HA/PDA) were prepared using sugarcane filter cake and polydopamine-modified BC by the freeze-drying method.

These scaffolds showed above 70% cell viability. The scaffolds promoted excellent adherence of MC3T3-E1 cells, followed by 24 hours of cell culture time, which showed their biocompatibility with these preosteoblasts. Protein adsorption analysis was carried out *in vitro*, which showed that the BC/HA/PDA performed better in adsorption efficiency than the unmodified scaffolds [62].

Gold nanoparticles (GNPs) were incorporated into BC hydrogels, forming Au/BC hydrogels. The hydrogels were prepared *via in-situ* fermentation, resulting in a total pore volume of 17.9165 ml/g and an average pore area of 1.3781 m^2/g. These scaffolds showed Young's modulus of ~5.5 MPa and improved degradability from 21% to 12.7%. *In vitro* results showed enhanced osteogenic differentiation, confirmed by increased ALP activity and more calcium depositions in hBMSCs cultured with Au/BC hydrogels. *In vivo*, the hydrogels demonstrated effective bone regeneration in a rabbit femoral defect model, with significant bone formation observed within 2 months [63].

Bacterial cellulose (BC) membranes were initially formed through static culture fermentation and subsequently underwent oxidation to produce 2, 3-dialdehyde bacterial cellulose (DABC). The fabrication of calcium phosphate-BC oxidized membranes was achieved by repeatedly immersing the membranes in precursor solutions, accompanied by ultrasonic irradiation. SEM of freeze-dried BC and DABC revealed that while DABC retained BC's fibrillar network, structural alterations were evident, including fewer tiny pores (0.5–1 μm) and larger pores (4–6 μm). TGA showed polymer degradation at around 135°C, with more significant mass loss in DABC due to oxidation. *In vitro* degradation studies over 28 days demonstrated a higher mass loss in DABC compared to a higher mass loss in DABC than native BC. Fluorescence microscopy and SEM for biological evaluation indicated improved adhesion and proliferation of cells on BC, with star-like, extended networks. At the same time, DABC showed lower cell density, with cells remaining rounded and less adherent. DABC samples suggested limited cell migration and adhesion, although longer culture times may promote more favorable cell morphology (Figure 13.3) [64].

3-D macro- and microporous nanofibrous bacterial cellulose (mNBC) scaffolds were prepared using freeze-drying. These scaffolds showed pore sizes of 20 to 1500 μm, with pores >300 μm enhancing osteogenic potential. Physicochemical characterization revealed scaffold degradation starting at ~285°C, with 33.19% mass remaining at 350 °C. In contrast, native BC membranes remained stable up to ~295°C with 43.69% residual mass. *In vitro*, mNBC scaffolds incubated in PBS showed complete degradation within 60 days. Electron microscopy confirmed strong attachment and

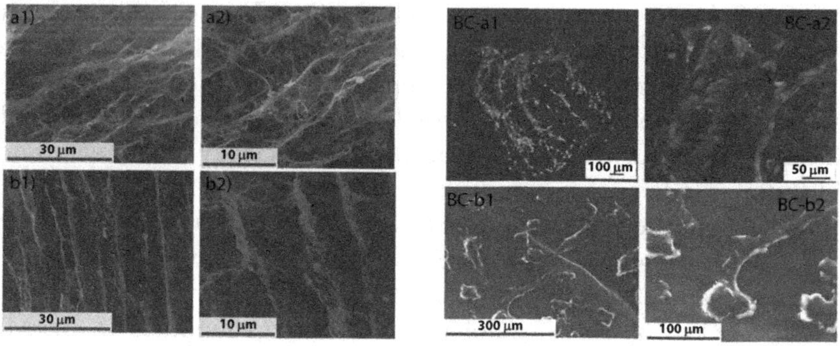

(a) (b)

fibroblast-like morphology of murine MSCs on mNBC scaffolds. MTT assays and live/dead staining indicated high cell viability and biocompatibility, with minimal cell death [65].

Polyacrylamide (PAM) was grafted on BC nanofibrils to create a PAM-grafted BC membrane (BC-g-PAM). Then, this membrane was chlorinated to create N-halamine functionalized BC, called BC-g-PNCl. This was combined with chitosan (CS) and hydroxyapatite (HAP) to form a bilayer membrane (BC-g-PNCl/CS-HAP) *via* lyophilization. It showed a wet-state mechanical strength of 57.3 MPa and degradation of 31% after 12 weeks in PBS. *In vitro*, BC-g-PNCl/CS-HAP demonstrated excellent cytocompatibility and pro-osteogenic potential, as confirmed by high ALP activity and calcium nodule formation in BMSCs. *In vivo*, this material promoted bone regeneration in rat calvarial defect models, effectively preventing connective tissue and bacterial infiltration while stimulating new bone formation [66].

BC/hydroxyapatite composite hydrogel scaffolds prepared *via* biomimetic mineralization, with varied mineralization times and ion concentrations. Platelet-rich plasma (PRP) was added to enhance growth factor release. *In vitro*, PRP-loaded BC/HAP@PRP scaffolds showed significant biocompatibility, enhanced mineralization, and promoted MC3T3-E1 cell differentiation after 3 weeks, indicating strong potential for bone tissue repair [53].

Polyvinyl Alcohol (PVA), Hexagonal Boron Nitride (hBN), and Bacterial Cellulose (BC) were used to fabricate 3D-printed scaffolds with various concentrations. The 12 wt% PVA/0.25 wt% hBN/0.5 wt% BC composite showed the best results. The viscosity of the composite scaffold revealed 81.3 ± 4.5 mPa.s, a surface tension of 65.3 ± 5 mN/m, and a density of 1.07 ± 0.02 g/mL. It showed 0.1 ± 0.05 MPa tensile strength and 93 ± 23% elongation at break. *In vitro*, the composite scaffold showed ~150% cell viability after 72 hours with human osteoblasts, indicating non-cytotoxicity and good cell adhesion. Overall, the 3D-printed PVA/hBN/BC scaffolds are suitable as an extracellular matrix for bone cells [67].

3D-printed scaffolds prepared from BC-treated TEMPO-mediated oxidation (TO-BC) and maleic acid (MA-BC). Physicochemical characterization revealed crystallinity indices of 89 for TO-BC and 92 for MA-BC. Neat BC demonstrated 61.2 MPa tensile strength, 2.8% elongation at break, and 2.1 GPa Young's modulus. Following treatment, the tensile strength reduced to 52.2 MPa for TO-BC and 36.9 MPa for MA-BC. *In vitro*, TO-BC and MA-BC dispersions promoted cell viability, with TO-BC maintaining 143% and MA-BC decreasing from 137% to 107% as concentration increased. ALP activity and mineralized nodule formation, assessed at 7–14 days and 28 days, respectively, showed enhanced osteogenic differentiation. MA-BC exhibited superior mineralization effects. *In vivo*, using a rat skull defect model, 3D-printed BC, TO-BC, and MA-BC scaffolds showed improved bone regeneration. Micro-CT analysis revealed that TO-BC and MA-BC scaffolds filled the defect areas more effectively than the control. MA-BC scaffolds showed the highest bone mineral density (0.223 ± 0.030 g/cm^2), indicating superior bone regeneration without adverse effects on surrounding tissues [38].

A 3D nanocomposite framework has been produced using BC nanofibers and eco-friendly magnesium oxide nanoparticles (MgONPs). Compared to pure BC (62.9 ± 3.49%), the porosity of the MgONPs-BC composite was higher at 89.2 ± 4.85%. The MgONPs-BC demonstrated improved mechanical properties compared to pure BC, including a compressive strength of 53.09 ± 5.02 MPa and Young's modulus of 191.19 ± 15.10 MPa. According to antibacterial testing, MgONPs at 1% and 5% w/v effectively inhibited methicillin-resistant *S. aureus*. MgONPs-BC scaffolds significantly improve *In vitro* cell survival and mineralization, as evidenced by elevated expressions of ALP and OCN at 21 days; however, expression of Runx2 remained constant [68].

Bacterial nanocellulose (BNC) carries a negative surface charge in physiological conditions, which enhances calcium ions' adsorption and triggers calcium phosphate formation. This study investigated mineralization methods for three-dimensional microporous BNC to mimic natural bone. During the fermentation of BNC with *Komagataeibacter medellinensis*, porogen particles were added. Calcium phosphates were applied to the BNC scaffolds through five immersion cycles involving calcium and phosphate salts. SEM showed pore sizes from 70 to 350 μm, with interconnectivity influenced by biomineralization. The BNC surface exhibited rod-shaped crystals, with the calcium-to-phosphate ratio increasing from 1.13 to 1.6 and crystal size from 25.12 to 35.9 nm over cycles. Alkaline phosphatase, osteocalcin, and osteopontin indicate osteogenic differentiation, high cell viability (up to 95%), and strong cellular adhesion of human bone marrow mesenchymal stem cells *in vitro* [69].

Human adipose-derived stem cells have been seeded on composite scaffolds composed of bioactive glass nanoparticles (BGNPs), silk fibroin (SF), and BC to encourage osteogenesis. The BGNPs were studied using scanning electron microscopy, demonstrating that they were spherical and varied in size from 15 to 30 nm. The pore width of SF scaffolds was reduced to approximately 210 ± 10 μm and 205 ± 10 μm, respectively, by including BC and BGNP. Consequently, their modulus and compressive strength increased. A Fourier transform infrared (FTIR) spectroscopy study confirmed the presence of BGNPs, BC, and SF in the scaffolds. Surface markers for hASCs have been verified by flow cytometry, and BC/SF/BGNP scaffolds demonstrated a lower rate of breakdown and swelling. Gene expression studies have demonstrated the osteogenic potential of BGNPs in scaffolds. The BC/SF/BGNP scaffolds showed the highest levels of osteocalcin, alkaline phosphatase (ALP), and Runx2 expression, along with increased ALP activity and mineral formation. These findings suggest that, due to their mechanical properties, osteogenic activity, and varying degradation rates, BC/SF/BGNP scaffolds hold promise for application in BTE [70].

This study investigated the effects of TEMPO-oxidized BNC scaffolds and bacterial nanocellulose (BNC) on calcium phosphate mineralization to assess the impact of carboxylate and hydroxyl groups on forming apatite crystals. Two mineralization methods were used to prepare scaffolds with interconnected pores and regulated microporosity (50–350 μm) for both BNC and TEMPO-oxidized BNC scaffolds. After 60 minutes of treatment, TEMPO oxidation induced changes in the porous morphology while maintaining structural integrity. Microcrystals were distributed uniformly in unmodified BNC scaffolds, whereas the distribution of crystals was uneven in TEMPO-oxidized scaffolds because of fiber fractures. Both types of scaffolds exhibited suitable cellular attachment and vitality in *in vitro* experiments [71].

Hydrogel scaffolds created using BC and infused with varying ratios of calcium phosphate—specifically, 0:00, 10:90, 20:80, 40:60, 50:50, and 60:40—were assessed for their swelling and rheological characteristics. These scaffolds incorporated polyvinylpyrrolidone (PVP), glycerin, agar, polyethylene glycol, and a combination of BC-PVP, β-tricalcium phosphate and hydroxyapatite (BC-PVP-β-TCP/HA). The BC-PVP-β-TCP/HA_20:80 and BC-PVP-β-TCP/HA_50:50 scaffolds demonstrated significant swelling and elasticity. SEM confirmed that the BC-PVP-β-TCP/HA_50:50 scaffolds had the highest surface roughness, emphasizing their porous structure. The BC-PVP-β-TCP/HA_20:80 and BC-PVP-β-TCP/HA_50:50 scaffolds exhibited beneficial mechanical and swelling properties, making them suitable candidates for applications in BTE [72].

To develop nanocomposite scaffolds for bone tissue, β-glucan (β-G) and BC were synthesized using freeze-drying and free-radical polymerization techniques. The BgC scaffolds exhibited maximum compression strengths between 3.90 ± 1.1 MPa for BgC-1.1 and 12.13 ± 2.4 MPa for BgC-1.4. Young's modulus increased for BgC-1.1 and BgC-1.4; the former measured 82.59 ± 1.5 MPa, while the latter recorded 394.87 ± 2.8 MPa. *In vitro* studies with MC3T3-E1 cells demonstrated significant cell adhesion and proliferation, with BgC-1.4 showing maximum viability of 93%, compared to 100% for gelatin. As GO content increased, the form of the cell shifted from thread-like to cylindrical. BgC-1.4 revealed higher antibacterial activity against *P. aeruginosa* (17.7 mm), *S. aureus* (21.5 mm), and *E. coli* (24 mm) [39].

A 3D porous microsphere scaffold containing bone morphogenetic protein 2 (BMP-2) using an initial approach and reverse-phase suspension regeneration was developed by chemically combining collagen (COL) and BC using Malaprade and Schiff-base reactions. The microspheres bore rigid surfaces using a specific surface area of 123.4 m^2/g, pore volume of 0.59 cm^3/g, average pore diameter of 198.5 nm, and particle sizes of 8 to 12 microns. In BC, the crystallinity was 11.92 ± 1.21%, while in COL, it was 80.37 ± 4.85%. Compared to controls, MC3T3-E1 cells grew at a rate of ~105% during cytotoxicity tests, which revealed no adverse effects. The cell adhesion rate for COL/BC microspheres increased significantly to 90.7%. ALP, COL I, OCN, OPN, Runx2, and OSX were all expressed more often when BMP-2 was present, indicating that it improved osteogenic differentiation [73].

The bacterial poly (3-hydroxybutyrate) (PHB) and BC were used to prepare PHB-BC scaffolds by salt leaching. The PHB-BC scaffold's pore size ranged from 5 to 50 μm. Its tensile strength and Young's modulus were 15 ± 1.0 MPa and 1400 ± 101 MPa, respectively. Over 5 days, the PHB-BC scaffold maintained the viability and proliferation of 3T3-L1 mice preadipocytes and showed biocompatibility exhibiting nontoxic qualities. Alkaline phosphate (ALP) activity showed a significant decrease over 4 weeks. The PHB-BC scaffold enhanced osteoblastic differentiation, as indicated by increased osterix (OSX) expression and early ALP activity. X-ray and histological/histomorphometric analyses of calvarial defects after 20 weeks following implantation indicated new bone formation by the PHB/BC scaffold (Figure 13.4) [57].

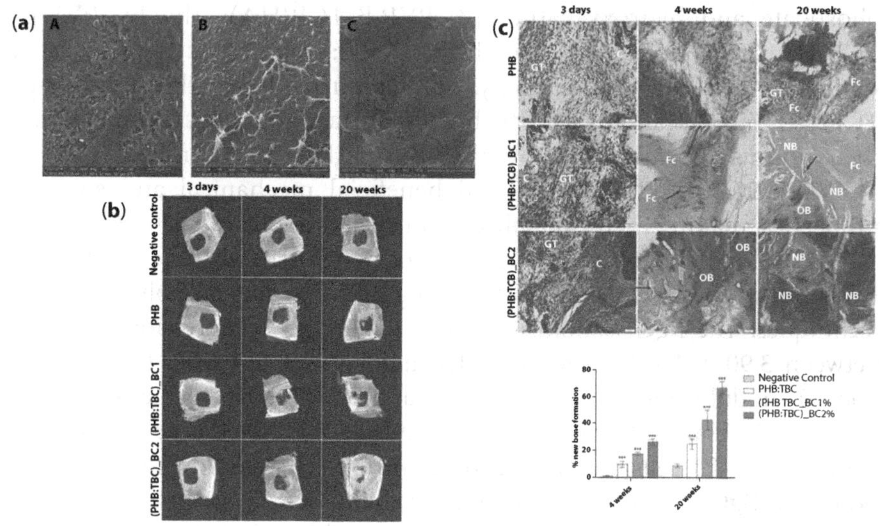

A non-resorbable guided tissue regeneration membrane was developed by combining bacterial cellulose membranes (BCMs) and human osteopontin (p-rhOPN). Polyacrylic acid (PAA) particles were attached to BCMs to provide several carboxyl groups for the conjugation of p-rhOPN, resulting in the development of p-rhOPN-BCM. The quantity of p-rhOPN was confirmed *via* ELISA. *In vitro* studies using human periodontal ligament stem cells demonstrated enhanced osteogenic differentiation, confirmed by immunocytochemistry, qRT-PCR, and mineralization assays. p-rhOPN-BCM showed comparable bioactivity to BCMs conjugated with mammalian-derived rhOPN. The cytotoxicity test showed no significant cytotoxicity, and the material showed potential for promoting bone tissue regeneration [74].

Using the freeze-drying method, a unique multi-component organic or inorganic composite scaffold, the BC/hydroxyapatite/gelatin double-network (BC-GEL/HAp DN), was created. The BC-GEL/HAp hydrogel, due to its multi-component double-network structure, demonstrated notably higher fracture stress (0.28 MPa) and elastic modulus (0.27 MPa) compared to BC/GEL hydrogel, which had fracture stress of 0.12 MPa and elastic modulus of 0.22 MPa. After 19 days of degradation, the weight loss for the BC-GEL/HAp composite reached 66.2%. Additionally, *in vitro* studies showed that rat bone marrow stem cells (rBMSCs) grown on the BC-GEL/HAp composite had enhanced adhesion and superior proliferation and differentiation capabilities compared to those grown on the BC/GEL composite [23].

A nanocomposite scaffold combining mesoporous bioactive glass and bacterial BC (MBG/BC) was created using an *in situ* membrane–liquid interface culture method. In this nanocomposite, the mesoporous bioactive glass was uniformly distributed within the three-dimensional matrix of BC. Tests conducted to assess mineralization *in vitro* showed that adding MBG significantly improved the bioactive properties of the BC scaffolds. Human bone marrow stromal cells (hBMSCs) were cultured on the MBG/BC scaffolds to investigate the scaffolds' effects on biocompatibility and bone regeneration further. Various factors were measured, including osteogenic gene expression, cell morphology, cell viability, and alkaline phosphatase activity. The findings indicated that the MBG/BC nanocomposite scaffolds promoted a more significant proliferation of hBMSCs compared to the BC scaffolds alone, along with increased alkaline phosphatase activity and enhanced expression of osteogenic genes [75].

A new, cost-effective method has been developed to create a composite scaffold called CS/OBC/nHAP, which utilizes entirely naturally derived materials. This scaffold combines oxidized bacterial cellulose (OBC) and chitosan (CS) with nano-hydroxyapatite (nHAP) and is evenly distributed throughout the structure. When comparing the CS/OBC/nHAP scaffold to a CS/nHAP scaffold, the results revealed that the latter had significantly poorer mechanical properties, reduced water retention, and a more consistent breakdown rate. In cell-based studies, the CS/OBC/nHAP scaffold demonstrated excellent biocompatibility and significantly promoted the growth of MC3T3-E1 cells. The scaffold effectively promoted bone tissue formation in a rat model with skull defects. H&E staining further highlighted the *in vivo* stability of the CS/OBC/nHAP scaffold and its ability to enhance bone tissue development [76].

To produce a cryogel, BC nanofibers were combined with sodium alginate and MXene, a 2D nanomaterial having excellent mechanical and electrical signaling abilities. Cross-linking significantly improved the mechanical characteristics and porosity of the cryogel. Effective reinforcement was indicated by the significant improvements in water absorption (852–1446%) and retention (692–973%) as well as the increases in compressive stress (0.85–1.43 MPa) and modulus (0.22–1.17 MPa) following the incorporation of MXene. The results of Osteogenic gene expression supported biological evaluations that demonstrated that MXene enhanced cell proliferation and promoted osteogenic differentiation. The macroporous MXene-BC cryogel facilitated cellular proliferation, and the quantitative polymerase chain reaction showed a strong osteogenic effect at lower MXene concentrations [54].

Chitosan (C), poly (vinyl alcohol) (P), nano bioactive glass (B), and nanocellulose (NC) were used to create the organic–inorganic hybrid CPBNC bionanocomposite. The bionanocomposite CPBNC showed electrostatic interactions between its functional groups. The mechanical examination revealed enhanced tensile strength (59.7 ± 1.5 MPa), enhanced compressive strength (4.9 ± 0.2 MPa), and Young's modulus (174.9 ± 6.1 MPa). Equivalent to the cancellous bone in terms of porosity (61–79%), the bionanocomposite exhibited excellent swelling and decreased degradation properties, providing it suitable for the seeding of cells. Biomimetic mineralization demonstrated hydroxyapatite nucleation. Further, the CPBNC bionanocomposite showed significant antibacterial activity against *S. aureus* and *E. coli* and was blood-compatible. Therefore, the CPBNC bionanocomposite appears to be a high-performance scaffold with the potential for bone tissue engineering [77].

13.5 Challenges and Future Directions

13.5.1 Challenges

Despite its advantages in bone tissue engineering, BC has shown some limitations. Key challenges include its structural inadequacies and lack of inherent osteogenic activities, which delay its effectiveness as a scaffold.

13.5.1.1 Mechanical Properties

BC has good flexibility and tensile strength for applications involving soft tissues, yet it does not possess the compressive strength needed for sustaining loads in bone tissues. The transplantation of bones in weight-bearing regions becomes difficult because of BC's mechanical limitations, whereas natural bone demonstrates both tensile and compressive strength [78].

13.5.1.2 Lack of Osteoconductivity

BC needs osteoinductive and osteoconductive qualities; it lacks the natural to regenerate bone. The scaffold's ability to enhance osteoblasts' adhesion and growth is called osteoconductivity. BC cannot come together in bone-forming cells or stimulate the production of new bone without functionalization or surface modifications [76, 78].

13.5.1.3 *Limited Degradation and Bioactivity*

Its inherent resistance to degradation is a significant obstacle that makes it more challenging to adapt for improved biodegradability without affecting its structural integrity and effectiveness. BC cannot accumulate in cells that generate new bone or stimulate the formation of new bone without functionalization or surface modifications [78, 79].

13.5.1.4 *Challenges in Vascularization*

Developing blood vessels within the scaffold, or vascularization, is essential for integrating massive tissue constructs and maintaining them functional. The formation of blood vessels, or angiogenesis, may not have enough surface area due to the thick fibrillar network of BC scaffolds. The implanted scaffold may necrotize over time as a result of inadequate vascularization, which will make bone regeneration more difficult [80].

13.5.1.5 *Complex Functionalization Requirements*

More advanced functionalization methods, such as mineralization utilizing hydroxyapatite or calcium phosphate, are required to enhance BC's mechanical characteristics, osteoconductivity, and bioactivity. Scaling up for clinical applications becomes more challenging due to these extra stages that complicate and raise the expense of the scaffold production process. Even with these adjustments, the outcomes are frequently not as favorable as with other scaffolds made with bone regeneration [76, 78].

13.5.1.6 *Difficulty in 3D Scaffold Fabrication*

Natural bone's anisotropic and porous characteristics must be simulated to develop 3D scaffolds based on BC. Compared to creating 3D structures with interconnected pores that facilitate cell movement and nutrient transport, the development of 2D BC membranes is less complex. The structural stability of the scaffold may be compromised by methods like freeze-drying or templating, even if they can increase porosity [78].

13.5.1.7 *Potential for Immune Response*

An immune reaction may still happen, even though BC is generally regarded as biocompatible if impurities remain from the production process. Toxins and bacterial cells must be eliminated to prevent inflammatory responses. Still, complete purification may be complex. Reactions of the foreign body caused by leftover elements could hinder the healing process [76, 78].

13.5.1.8 Scale-Up Challenges

Even though BC can be produced in large quantities in the lab, the amount produced can be increased in response to medical needs. Maintaining uniform quality, ensuring uniformity in scaffold properties, and expediting production processes for broad usage are challenges that need to be solved [78].

13.5.1.9 Lack of Clinical Data

Most of the research on BTE using BC has occurred *in vitro* or with animal models. However, few human studies or clinical trials utilize BC-based scaffolds, which limits our knowledge of their efficacy in practical medicinal applications. One of the main obstacles preventing BC scaffolds from being widely used in orthopedic surgery is the lack of clinical evidence [78].

13.5.2 Future Directions

The future direction of BC uses in BTE is promising. Research focuses on enhancing its properties through various modification techniques, composite materials, and advanced manufacturing technologies. Researchers also aim to address recent limitations, such as insufficient pore size and biodegradability, to improve BC's ability as a scaffold material for bone regeneration.

13.5.2.1 Engineering Advancements

Future research may emphasize BC's engineering through both *in situ* and *ex situ* modifications. This method aims to enhance the properties of BC, like porosity, mechanical stability, and bioactivity, permitting better integration with bone tissue regeneration [81]. The exploration of 3D printing will further facilitate the construction of patient-specific scaffolds with customizable designs [82].

13.5.2.2 Composite Materials

Combining BC with other biopolymers, such as collagen and hydroxyapatite, has shown the potential to enhance scaffolds' mechanical and biological properties. Ongoing research will focus on optimum ratios and processing methods to prepare composite materials that enhance osteogenesis and successfully support the healing process [81].

13.5.2.3 *Supercritical Carbon Dioxide Technology*

Supercritical carbon dioxide ($scCO_2$) technology in BC scaffolds could significantly increase their properties, including porosity and surface characteristics. Future research will discover how to integrate scCO2 technology with 3D printing to fabricate scaffolds that are not only biocompatible but also physically and chemically enhanced for the application of bone tissue engineering [83].

13.5.2.4 *Addressing Biodegradability Issues*

Future approaches will be needed to address the challenges surrounding BC'3s biodegradability. Research may focus on the addition of biodegradable materials or modification of BC's structure to facilitate controlled degradation while maintaining scaffold integrity during the tissue regeneration process [81].

13.5.2.5 *Preclinical and Clinical Studies*

The importance of translating this research into therapeutic applications will increase as the knowledge of the characteristics and uses of BCs increases through more preclinical studies. To determine the safety and efficacy of BC-based scaffolds in human patients, especially in the context of bone regeneration and repair, more preclinical and clinical trials are necessary [81].

13.6 Conclusion

In conclusion, BC has become a very suitable material for BTE due to its unique characteristics and versatility for scaffold development. A comprehensive investigation of BC's thermal, mechanical, biocompatible, and biosynthetic characteristics reveals its prospective for bone tissue engineering. BC-based scaffolds showed the ideal porosity, mechanical strength, thermal stability, and biocompatibility. BC is an acceptable choice for BTE applications. However, other materials, such as hydroxyapatite, must enhance their osteoinductive and osteoconductive properties.

Additional preclinical and clinical studies are necessary to explore the significant potential of BC-based scaffolds in BTE.

AI disclosure statement: The author acknowledges the use of AI-based tools for minor language editing or grammatical correction. However, all intellectual contributions, analyses, and conclusions presented in this chapter are entirely the author's own.

References

1. Yazdanpanah, Z., Sharma, N.K., Raquin, A., Cooper, D.M.L., Chen, X., Johnston, J.D., Printing tissue-engineered scaffolds made of polycaprolactone and nano-hydroxyapatite with mechanical properties appropriate for trabecular bone substitutes. *Biomed. Eng. Online [Internet]*, 22, 1–14, 2023, Available from: https://doi.org/10.1186/s12938-023-01135-6.
2. Choi, I.A., Umemoto, A., Mizuno, M., Park-Min, K.-H., Bone metabolism – an underappreciated player. *npj Metab. Heal. Dis. [Internet]*, 2, 1–10, 2024, Available from: http://dx.doi.org/10.1038/s44324-024-00010-9.
3. Schmidt, A.H., Autologous bone graft: Is it still the gold standard? *Inj. [Internet]*, 52, S18–22, 2021, Available from: https://doi.org/10.1016/j.injury.2021.01.043.
4. Amini, A.R., Laurencin, C.T., Nukavarapu, S.P., Bone tissue engineering: Recent advances and challenges. *Crit. Rev. Biomed. Eng.*, 40, 363–408, 2012.
5. Saurav, S., Sharma, P., Kumar, A., Tabassum, Z., Girdhar, M., Mamidi, N., *et al.*, Harnessing Natural Polymers for Nano-Scaffolds in Bone Tissue Engineering: A Comprehensive Overview of Bone Disease Treatment. *Curr. Issues Mol. Biol.*, 46, 585–611, 2024.
6. Manoukian, O.S., Sardashti, N., Stedman, T., Gailiunas, K., Ojha, A., Penalosa, A., *et al.*, Biomaterials for tissue engineering and regenerative medicine [Internet], in: *Encycl. Biomed. Eng.*, Elsevier Inc, 2019, Available from: http://dx.doi.org/10.1016/B978-0-12-801238-3.64098-9.
7. Ullah, M.W., Ul Islam, M., Khan, S., Shah, N., Park, J.K., Recent advancements in bioreactions of cellular and cell-free systems: A study of bacterial cellulose as a model. *Korean J. Chem. Eng.*, 34, 1591–9, 2017.
8. Seo, C., Lee, H.W., Suresh, A., Yang, J.W., Jung, J.K., Kim, Y.C., Improvement of fermentative production of exopolysaccharides from Aureobasidium pullulans under various conditions. *Korean J. Chem. Eng.*, 31, 1433–7, 2014.
9. Shi, Z., Gao, X., Ullah, M.W., Li, S., Wang, Q., Yang, G., Electroconductive natural polymer-based hydrogels. *Biomaterials*, 111, 40–54, 2016.
10. Gao, X., Shi, Z., Kuśmierczyk, P., Liu, C., Yang, G., Sevostianov, I., *et al.*, Time-dependent rheological behaviour of bacterial cellulose hydrogel. *Mater. Sci. Eng., C*, 58, 153–9, 2016.

11. Raghavendran, V., Asare, E., Roy, I., Bacterial cellulose: Biosynthesis, production, and applications [Internet], in: *Adv. Microb. Physiol.*, 1st ed, Elsevier Ltd, 2020, Available from: http://dx.doi.org/10.1016/bs.ampbs.2020.07.002.
12. Ullah, M.W., Manan, S., Kiprono, S.J., Ul-Islam, M., Yang, G., Synthesis, structure, and properties of bacterial cellulose. *Nanocellulose From Fundam. Adv. Mater.*, 81–114, 2019.
13. Mahmoud, Y.A.G., El-Naggar, M.E., Abdel-Megeed, A., El-Newehy, M.H., Recent advancements in microbial polysaccharides: Synthesis and applications. *Polym. (Basel)*, 13, 1–21, 2021.
14. Gullo, M., La China, S., Falcone, P.M., Giudici, P., Biotechnological production of cellulose by acetic acid bacteria: current state and perspectives. *Appl. Microbiol. Biotechnol.*, 2018.
15. Omadjela, O., Narahari, A., Strumillo, J., Mélida, H., Mazur, O., Bulone, V., *et al.*, BcsA and BcsB form the catalytically active core of bacterial cellulose synthase sufficient for *In vitro* cellulose synthesis. *Proc. Natl. Acad. Sci. U.S.A.*, 110, 17856–61, 2013.
16. Mehta, K., Pfeffer, S., Brown, R.M., Characterization of an acsD disruption mutant provides additional evidence for the hierarchical cell-directed self-assembly of cellulose in Gluconacetobacter xylinus. *Cellulose*, 22, 119–37, 2015.
17. Sunagawa, N., Fujiwara, T., Yoda, T., Kawano, S., Satoh, Y., Yao, M., *et al.*, Cellulose complementing factor (Ccp) is a new member of the cellulose synthase complex (terminal complex) in Acetobacter xylinum. *J. Biosci. Bioeng. [Internet]*, 115, 607–12, 2013, Available from: http://dx.doi.org/10.1016/j.jbiosc.2012.12.021.
18. Srivastava, S. and Mathur, G., Bacterial Cellulose: A Multipurpose Biomaterial for Manmade World. *Curr. Appl. Sci. Technol.*, 23, 1–19, 2023.
19. Raut, M.P., Asare, E., Syed Mohamed, S.M.D., Amadi, E.N., Roy, I., Bacterial Cellulose-Based Blends and Composites: Versatile Biomaterials for Tissue Engineering Applications. *Int. J. Mol. Sci.*, 24, 2023.
20. Nurlidar, F., Kobayashi, M., Yunus, A.L., Yunus, M.Y., Puspitasari, T., Darwis, D., Gamma-Irradiated Bacterial Cellulose as a Three-Dimensional Scaffold for Osteogenic Differentiation of Rat Bone Marrow Stromal Cells. *Indones. J. Chem.*, 22, 816–26, 2022.
21. Logeart-Avramoglou, D., Anagnostou, F., Bizios, R., Petite, H., Engineering bone: Challenges and obstacles. *J. Cell. Mol. Med.*, 9, 72–84, 2005.
22. Jeyachandran, D. and Cerruti, M., Glass, Ceramic, Polymeric, and Composite Scaffolds with Multiscale Porosity for Bone Tissue Engineering. *Adv. Eng. Mater.*, 25, 2023.
23. Ran, J., Jiang, P., Liu, S., Sun, G., Yan, P., Shen, X., *et al.*, Constructing multi-component organic/inorganic composite bacterial cellulose-gelatin/hydroxyapatite double-network scaffold platform for stem cell-mediated bone tissue engineering. *Mater. Sci. Eng. C [Internet]*, 78, 130–40, 2017, Available from: http://dx.doi.org/10.1016/j.msec.2017.04.062.

24. Beladi, F., Saber-Samandari, S., Saber-Samandari, S., Cellular compatibility of nanocomposite scaffolds based on hydroxyapatite entrapped in cellulose network for bone repair. *Mater. Sci. Eng. C [Internet].*, 75, 385–92, 2017, Available from: http://dx.doi.org/10.1016/j.msec.2017.02.040.
25. Guo, J. and Catchmark, J.M., Surface area and porosity of acid hydrolyzed cellulose nanowhiskers and cellulose produced by Gluconacetobacter xylinus. *Carbohydr. Polym. [Internet].*, 87, 1026–37, 2012, Available from: http://dx.doi.org/10.1016/j.carbpol.2011.07.060.
26. Asadullah, M., Rahman, M.A., Motin, M.A., Sultan, M.B., Preparation and adsorption studies of high specific surface area activated carbons obtained from the chemical activation of jute stick. *Adsorpt. Sci. Technol.*, 24, 761–70, 2006.
27. Celestino, M.F., Lima, L.R., Fontes, M., Batista, I.T.S., Mulinari, D.R., Dametto, A., *et al.*, 3D Filaments Based on Polyhydroxy Butyrate—Micronized Bacterial Cellulose for Tissue Engineering Applications. *J. Funct. Biomater.*, 14, 2023.
28. Li, G., Nandgaonkar, A.G., Habibi, Y., Krause, W.E., Wei, Q., Lucia, L.A., An environmentally benign approach to achieving vectorial alignment and high microporosity in bacterial cellulose/chitosan scaffolds. *RSC Adv.*, 7, 13678–88, 2017.
29. Torgbo, S. and Sukyai, P., Biodegradation and thermal stability of bacterial cellulose as biomaterial: The relevance in biomedical applications. *Polym. Degrad. Stab. [Internet]*, 179, 109232, 2020, Available from: https://doi.org/10.1016/j.polymdegradstab.2020.109232.
30. Stumpf, T.R., Pértile, R.A.N., Rambo, C.R., Porto, L.M., Enriched glucose and dextrin mannitol-based media modulates fibroblast behavior on bacterial cellulose membranes. *Mater. Sci. Eng. C [Internet]*, 33, 4739–45, 2013, Available from: http://dx.doi.org/10.1016/j.msec.2013.07.035.
31. Mohite, B.V. and Patil, S.V., Physical, structural, mechanical and thermal characterization of bacterial cellulose by G hansenii NCIM 2529. *Carbohydr. Polym. [Internet]*, 106, 132–41, 2014, Available from: http://dx.doi.org/10.1016/j.carbpol.2014.02.012.
32. Mohammadkazemi, F., Azin, M., Ashori, A., Production of bacterial cellulose using different carbon sources and culture media. *Carbohydr. Polym. [Internet]*, 117, 518–23, 2015, Available from: http://dx.doi.org/10.1016/j.carbpol.2014.10.008.
33. Jia, Y., Wang, X., Huo, M., Zhai, X., Li, F., Zhong, C., Preparation and characterization of a novel bacterial cellulose/chitosan bio-hydrogel. *Nanomater. Nanotechnol.*, 7, 1–8, 2017.
34. George, J., Ramana, K.V., Sabapathy, S.N., Jagannath, J.H., Bawa, A.S., Characterization of chemically treated bacterial (Acetobacter xylinum) biopolymer: Some thermo-mechanical properties. *Int. J. Biol. Macromol.*, 37, 189–94, 2005.

35. George, J., Ramana, K.V., Bawa, A.S., Siddaramaiah. Bacterial cellulose nanocrystals exhibiting high thermal stability and their polymer nanocomposites. *Int. J. Biol. Macromol. [Internet]*, 48, 50–7, 2011, Available from: http://dx.doi.org/10.1016/j.ijbiomac.2010.09.013.
36. Basu, P., Saha, N., Alexandrova, R., Andonova-Lilova, B., Georgieva, M., Miloshev, G., *et al.*, Biocompatibility and biological efficiency of inorganic calcium filled bacterial cellulose based hydrogel scaffolds for bone bioengineering. *Int. J. Mol. Sci.*, 19, 1–16, 2018.
37. Suneetha, M., Kim, H., Han, S.S., Bone-like apatite formation in biocompatible phosphate-crosslinked bacterial cellulose-based hydrogels for bone tissue engineering applications. *Int. J. Biol. Macromol.*, 256, 128364, 2024.
38. Wang, X., Tang, S., Chai, S., Wang, P., Qin, J., Pei, W., *et al.*, Preparing printable bacterial cellulose based gelatin gel to promote *in vivo* bone regeneration. *Carbohydr. Polym. [Internet]*, 270, 118342, 2021, Available from: https://doi.org/10.1016/j.carbpol.2021.118342.
39. Umar Aslam Khan, M., Haider, S., Haider, A., Izwan Abd Razak, S., Rafiq Abdul Kadir, M., Shah, S.A., *et al.*, Development of porous, antibacterial and biocompatible GO/n-HAp/bacterial cellulose/β-glucan biocomposite scaffold for bone tissue engineering. *Arab. J. Chem. [Internet]*, 14, 102924, 2021, Available from: https://doi.org/10.1016/j.arabjc.2020.102924.
40. Olszta, M.J., Cheng, X., Jee, S.S., Kumar, R., Kim, Y.Y., Kaufman, M.J., *et al.*, Bone structure and formation: A new perspective. *Mater. Sci. Eng. R Rep.*, 58, 77–116, 2007.
41. Chen, S.Q., Lopez-Sanchez, P., Wang, D., Mikkelsen, D., Gidley, M.J., Mechanical properties of bacterial cellulose synthesised by diverse strains of the genus Komagataeibacter. *Food Hydrocoll. [Internet]*, 81, 87–95, 2018, Available from: https://doi.org/10.1016/j.foodhyd.2018.02.031.
42. Krystynowicz, A., Czaja, W., Wiktorowska-Jezierska, A., Gonçalves-Miśkiewicz, M., Turkiewicz, M., Bielecki, S., Factors affecting the yield and properties of bacterial cellulose. *J. Ind. Microbiol. Biotechnol.*, 29, 189–95, 2002.
43. Torgbo, S. and Sukyai, P., Fabrication of microporous bacterial cellulose embedded with magnetite and hydroxyapatite nanocomposite scaffold for bone tissue engineering. *Mater. Chem. Phys. [Internet]*, 237, 121868, 2019, Available from: https://doi.org/10.1016/j.matchemphys.2019.121868.
44. Öz, Y.E., Bingül, N.D., Morçimen, Z.G., Şendemir, A., Hameş, E.E., Fabrication of porous bone scaffolds using degradable and mouldable bacterial cellulose. *Cellul. [Internet]*, 31, 2921–35, 2024, Available from: https://doi.org/10.1007/s10570-024-05771-6.
45. Huang, J. and Gu, Y., Self-assembly of various guest substrates in natural cellulose substances to functional nanostructured materials. *Curr. Opin. Colloid Interface Sci. [Internet]*, 16, 470–81, 2011, Available from: http://dx.doi.org/10.1016/j.cocis.2011.08.004.

46. Torgbo, S. and Sukyai, P., Bacterial cellulose-based scaffold materials for bone tissue engineering. *Appl. Mater Today [Internet]*, 11, 34–49, 2018, Available from: http://dx.doi.org/10.1016/j.apmt.2018.01.004.
47. Girard, V.D., Chaussé, J., Vermette, P., Bacterial cellulose: A comprehensive review. *J. Appl. Polym. Sci.*, 141, 1–35, 2024.
48. Luz, E.P.C.G., Soares, AL.B., Souzade, F.F.P., FK, A., Castro-Silva, I.I., Rosa M de, F., *et al.*, Implantable matrixes of bacterial cellulose and strontium apatite: Preclinical analysis of cytotoxicity and osteoconductivity. *Mater. Communication*, 33, 2022.
49. Rzhepakovsky, I., Piskov, S., Avanesyan, S., Sizonenko, M., Timchenko, L., Anfinogenova, O., *et al.*, Composite of bacterial cellulose and gelatin: A versatile biocompatible scaffold for tissue engineering. *Int. J. Biol. Macromol.*, 256, 128369, 2024.
50. Xun, X., Li, Y., Ni, M., Xu, Y., Li, J., Zhang, D., *et al.*, Calcium crosslinked macroporous bacterial cellulose scaffolds with enhanced *in situ* mineralization and osteoinductivity for cranial bone regeneration. *Compos. Part B Eng.*, 275, 111277, 2024.
51. Gorgieva, S., Bacterial cellulose as a versatile platform for research and development of biomedical materials. *Processes*, 8, 2020.
52. Dukle, A. and Sankar, M.R., 3D-printed polylactic acid scaffolds for bone tissue engineering: Bioactivity enhancing strategies based on composite filaments and coatings. *Mater. Communication*, 40, 109776, 2024.
53. Wang, X., Yang, X., Xiao, X., Li, X., Chen, C., Sun, D., Biomimetic design of platelet-rich plasma controlled release bacterial cellulose/hydroxyapatite composite hydrogel for bone tissue engineering. *Int. J. Biol. Macromol.*, 269, 132124, 2024.
54. Hu Pengfei Cai, T. and Xia, C., MXene reinforced microporous bacterial cellulose/sodium alginate dual crosslinked cryogel for bone tissue engineering. *Biomed. Mater*, 19, 2024.
55. Wennerberg, A., The importance of surface roughness for implant incorporation. *Int. J. Mach. Tools Manuf.*, 38, 657–62, 1998.
56. Habibovic, P., Yuan, H., Van Der Valk, C.M., Meijer, G., Van Blitterswijk, C.A., De Groot, K., 3D microenvironment as essential element for osteoinduction by biomaterials. *Biomaterials*, 26, 3565–75, 2005.
57. Codreanu, A., Balta, C., Herman, H., Cotoraci, C., Mihali, C.V., Zurbau, N., *et al.*, Bacterial cellulose-modified polyhydroxyalkanoates scaffolds promotes bone formation in critical size calvarial defects in mice. *Mater. (Basel)*, 13, 2020.
58. Hall, T.A.G., Theodoridis, K., Kohli, N., Cegla, F., van Arkel, R.J., Active osseointegration in an ex vivo porcine bone model. *Front. Bioeng. Biotechnol.*, 12, 1–16, 2024.
59. Aykora, D. and Uzun, M., Bone tissue engineering for osteointegration: Where are we now? *Polym. Bull. [Internet]*, 81, 8595–605, 2024, Available from: https://doi.org/10.1007/s00289-024-05153-9.

60. Atila, D., Karataş, A., Evcin, A., Keskin, D., Tezcaner, A., Bacterial cellulose-reinforced boron-doped hydroxyapatite/gelatin scaffolds for bone tissue engineering. *Cellulose*, 26, 9765–85, 2019.
61. Cakmak, A.M., Unal, S., Sahin, A., Oktar, F.N., Sengor, M., Ekren, N., *et al.*, 3D Printed Polycaprolactone/Gelatin/Bacterial Cellulose/Hydroxyapatite Composite Scaffold for Bone Tissue Engineering. 2020.
62. Boyetey, M.J.B., Sukyai, P., Kamonsutthipaijit, N., Nijpanich, S., Chanlek, N., Fabrication and Characterization of a Polydopamine-Modified Bacterial Cellulose and Sugarcane Filter Cake-Derived Hydroxyapatite Composite Scaffold. *ACS Omega*, 8, 43295–303, 2023.
63. Huang, C., Ye, Q., Dong, J., Li, L., Wang, M., Zhang, Y., *et al.*, Biofabrication of natural Au/bacterial cellulose hydrogel for bone tissue regeneration *via in-situ* fermentation. *Smart Mater. Med. [Internet]*, 4, 1–14, 2023, Available from: https://doi.org/10.1016/j.smaim.2022.06.001.
64. Busuioc, C., Isopencu, G., Banciu, A., Banciu, D.D., Oprea, O., Mocanu, A., *et al.*, Bacterial Cellulose Hybrid Composites with Calcium Phosphate for Bone Tissue Regeneration. *Int. J. Mol. Sci.*, 23, 2022.
65. Dubey, S., Mishra, R., Roy, P., Singh, R.P., 3-D macro/microporous-nanofibrous bacterial cellulose scaffolds seeded with BMP-2 preconditioned mesenchymal stem cells exhibit remarkable potential for bone tissue engineering. *Int. J. Biol. Macromol.*, 167, 934–46, 2021.
66. Wu, S., Luo, S., Cen, Z., Li, Q., Li, L., Li, W., *et al.*, All-in-one porous membrane enables full protection in guided bone regeneration. *Nat. Commun*, 15, 1–13, 2024.
67. Aki, D., Ulag, S., Unal, S., Sengor, M., Ekren, N., Lin, C.C., *et al.*, 3D printing of PVA/hexagonal boron nitride/bacterial cellulose composite scaffolds for bone tissue engineering. *Mater. Des.*, 196, 2020.
68. Ghanbari, E., Khazaei, M., Mehdipour, A., Khoshfetrat, A.B., Niknafs, B., Green Synthesized Magnesium Oxide Nanoparticles Reinforce Osteogenesis Properties of Bacterial Cellulose Scaffolds for Bone Tissue Engineering Applications: An *In vitro* Assessment. *Cell J.*, 25, 483–95, 2023.
69. Añas-GutiérreCz, A., Toro, L., Fornaguera, C., Borrós, S., Osorio, M., Castro-Herazo, C., *et al.*, Biomineralization in Three-Dimensional Scaffolds Based on Bacterial Nanocellulose for Bone Tissue Engineering: Feature Characterization and Stem Cell Differentiation. *Polym. (Basel)*, 15, 2023.
70. Niknafs, B., Meskaraf-asadabadi, M., Hamdi, K., Ghanbari, E., Incorporating bioactive glass nanoparticles in silk fibroin/bacterial nanocellulose composite scaffolds improves their biological and osteogenic properties for bone tissue engineering applications. *Int. J. Biol. Macromol. [Internet]*, 266, 131167, 2024, Available from: https://doi.org/10.1016/j.ijbiomac.2024.131167.
71. Cañas-Gutiérrez, A., Martinez-Correa, E., Suárez-Avendaño, D., Arboleda-Toro, D., Castro-Herazo, C., Influence of bacterial nanocellulose surface modification on calcium phosphates precipitation for bone tissue engineering. *Cellulose*, 27, 10747–63, 2020.

72. Basu, P., Saha, N., Saha, P., Swelling and rheological study of calcium phosphate filled bacterial cellulose-based hydrogel scaffold. *J. Appl. Polym. Sci.*, 137, 1–14, 2020.
73. Zhang, W., Wang, X.C., Li, X.Y., le, Z.L., Jiang, F., A 3D porous microsphere with multistage structure and component based on bacterial cellulose and collagen for bone tissue engineering. *Carbohydr. Polym. [Internet]*, 236, 116043, 2020, Available from: https://doi.org/10.1016/j.carbpol.2020.116043.
74. Klinthoopthamrong, N., Chaikiawkeaw, D., Phoolcharoen, W., Rattanapisit, K., Kaewpungsup, P., Pavasant, P., *et al.*, Bacterial cellulose membrane conjugated with plant-derived osteopontin: Preparation and its potential for bone tissue regeneration. *Int. J. Biol. Macromol. [Internet]*, 149, 51–9, 2020, Available from: https://doi.org/10.1016/j.ijbiomac.2020.01.158.
75. Xiao, J., Wei, Q., ZL, J.X., Li, Z., Zhou, Z., Chen, F., *et al.*, Mesoporous bioactive glass/bacterial cellulose composite scaffolds for bone support materials. *Colloids Surf. A Physicochem. Eng. Asp.*, 642, 128693, 2022.
76. Cao, S., Li, Q., Zhang, S., Liu, K., Yang, Y., Chen, J., Oxidized bacterial cellulose reinforced nanocomposite scaffolds for bone repair. *Colloids Surf. B Biointerfaces*, 211, 112316, 2022.
77. Narmatha, C., Khaleel, B., Sugantha, K., Multifunctional organic and inorganic hybrid bionanocomposite of chitosan/poly(vinyl alcohol)/nanobioactive glass/nanocellulose for bone tissue engineering. *J. Mech. Behav. Biomed. Mater.*, 2022.
78. Malandain, N., Sanz-Fraile, H., Farré, R., Otero, J., Roig, A., Laromaine, A., Cell-Laden 3D Hydrogels of Type I Collagen Incorporating Bacterial Nanocellulose Fibers. *ACS Appl. Bio Mater.*, 6, 3638–47, 2023.
79. Sandhu, S., Arpa, A., Chen, X., Kumar, R., Jaworski, J., Engineering Bacterial Cellulose for Diverse Biomedical Applications. *Recent Prog. Mater.*, 1, 1–1, 2019.
80. Owida, H.A., Al-Ayyad, M., Turab, N.M., Al-Nabulsi, J.I., Recent Biomaterial Developments for Bone Tissue Engineering and Potential Clinical Application: Narrative Review of the Literature. *Int. J. Online Biomed. Eng.*, 19, 4–33, 2023.
81. Khan, S., Siddique, R., Huanfei, D., Shereen, M.A., Nabi, G., Bai, Q., *et al.*, Perspective Applications and Associated Challenges of Using Nanocellulose in Treating Bone-Related Diseases. *Front Bioeng. Biotechnol*, 9, 1–19, 2021.
82. Iravani, S. and Varma, R.S., Cellulose-Based Composites as Scaffolds for Tissue Engineering: Recent Advances. *Molecules*, 27, 2022.
83. Iglesias-Mejuto, A., Magariños, B., Ferreira-Gonçalves, T., Starbird-Pérez, R., Álvarez-Lorenzo, C., Reis, C.P., *et al.*, Vancomycin-loaded methylcellulose aerogel scaffolds for advanced bone tissue engineering. *Carbohydr. Polym.*, 324, 2024.

Part IV

INDUSTRIAL APPLICATIONS

14

Regenerated Cellulose in Textiles

Sandra Varnaitė-Žuravliova* **and Julija Baltušnikaitė-Guzaitienė**

Center for Physical Sciences and Technology, Department of Textile Technologies, Kaunas, Lithuania

Abstract

Most fibers used in textiles nowadays are petroleum-based polymers, such as polyester and polyamide. However, regenerated cellulose fibers offer a semi-synthetic, more sustainable alternative. These fibers are created by converting natural cellulose from sources like bamboo, wood pulp, and cotton linters into soluble derivatives, which are then regenerated into fibers. Common examples include viscose, modal, lyocell, acetate, and cupro. These fibers imitate the softness, comfort, and breathability of natural fibers like cotton and silk, making them popular in apparel and home textiles.

Despite their eco-friendly reputation, the production of regenerated cellulose fibers involves chemical processes that can harm the environment if not managed properly. Advances in production, such as closed-loop systems and adherence to standards like the Global Recycle Standard (GRS), have mitigated some environmental concerns. Overall, regenerated cellulose fibers represent a significant step towards more sustainable textile manufacturing by combining the benefits of natural fibers with renewable resources, offering a viable alternative to petroleum-based synthetics.

Keywords**:** Regenerated cellulose, fibers, rayon, viscose, lyocell, eco-friendly

14.1 Introduction

In recent years, the definition of "textile" has broadened significantly beyond traditional fabrics, now encompassing a diverse array of materials utilized across various sectors. This transformation has been driven by advancements in material science, increasing emphasis on sustainability,

**Corresponding author*: sandra.varnaite.zuravliova@ftmc.lt

Mohd Shabbir (ed.) Regenerated Cellulose: A Renewable Alternative to Petroleum-Based Plastics, (423–462) © 2026 Scrivener Publishing LLC

and the rising demand for textiles with specialized functions in fields such as technology, construction, architecture, agriculture, and medicine [1–3]. Notably, regenerated cellulose fibers (RCFs) have gained considerable attention due to their eco-friendly attributes and adaptability for a wide range of applications, making them a crucial part of this evolving textile landscape. The inclusion of high-performance materials, such as composites, nanofibers, and smart textiles, underscores the ongoing shift toward textiles designed for specific technical and functional purposes. RCFs, in particular, exemplify the movement toward more sustainable materials, which have not only expanded the role of textiles beyond traditional uses in clothing and furnishings but have also made them critical components in industrial, technological, and environmental applications [4]. Their ability to balance performance and environmental responsibility has positioned them at the forefront of modern textile innovation.

As sustainability becomes a central focus across industries, there has been a re-evaluation of material choices, with renewed interest in bio-based alternatives [5]. Regenerated cellulose fibers, derived from renewable sources like wood pulp and produced using environmentally responsible methods—such as the closed-loop lyocell process—offer a viable solution. These fibers significantly contribute to the circular economy by reducing the environmental footprint associated with traditional petroleum-based synthetic fibers, aligning with global sustainability targets and the push for greener manufacturing practices [6–9].

A key benefit of cellulose-based textiles is their biodegradability, which is particularly advantageous in industries where waste and pollution are pressing concerns [10]. RCFs can be effectively used in applications ranging from disposable medical products to agricultural coverings and construction materials. Their ability to degrade naturally without leaving harmful residues offers a sustainable alternative to conventional plastics, further supporting efforts to reduce plastic waste and environmental pollution [11]. The increased adoption of biodegradable and renewable materials like regenerated cellulose fibers highlights the textile industry's commitment to fostering a sustainable future. These materials not only address ecological challenges but also meet the evolving functional demands of modern industrial and consumer applications. As the industry continues to innovate and adapt to global sustainability imperatives, the role of fibers like RCFs will become even more pivotal in shaping the future of textiles across multiple sectors.

Cellulose-based fibers, a type of polysaccharide, have become highly valued in the textile industry for their renewability, abundance, and diverse functional properties. As the most prevalent organic compound on Earth, cellulose forms the structural foundation of plant cell walls and is also synthesized by certain microbes. Its natural biopolymer structure, featuring β-1,4-glycosidic bonds that create durable chains of D-glucose units, gives cellulose remarkable strength, biodegradability, and compatibility, making it an essential material for sustainable textile advancements [12–14]. Cellulose is naturally derived from various sources, with cotton being nearly pure at around 90% cellulose content. In contrast, woody biomass contains approximately 40% to 50% cellulose, making both essential raw materials for cellulose-based fiber production [15]. These sources are increasingly harnessed to produce advanced textile materials like hydrogels, aerogels, fillers, and films, that find applications in a wide array of industries beyond medical fields [16, 17]. In the textile sector, cellulose's properties are utilized to create sustainable fabrics and fibers that are both environmentally friendly and functional [18–20].

A major advancement in cellulose fiber technology is the regeneration process, which transforms cellulose I into cellulose II. This modification improves crystallinity while increasing the presence of amorphous regions, enhancing the fiber's structural and functional properties. This modification improves flexibility and adaptability, making regenerated cellulose fibers, such as viscose, lyocell, and modal, suitable for a wide range of applications requiring specific performance characteristics like moisture management, durability, and comfort [21]. Regenerated cellulose fibers address the increasing demand for sustainable and eco-friendly materials across industries, including fashion, construction, and agriculture. These fibers are biodegradable and exhibit mechanical properties comparable to synthetic fibers, positioning them as a viable alternative in the transition towards greener materials [22–24]. Their versatility supports the development of innovative products such as breathable fabrics, eco-friendly insulation, and biodegradable agricultural covers.

Beyond textiles, regenerated cellulose materials show potential in diverse fields, including biomedicine, energy storage, and packaging, as they can be formed into films, membranes, hydrogels, aerogels, filaments, microspheres, and bioplastics. Importantly, these materials can biodegrade in soil and oceans, helping to mitigate environmental pollution [25]. This movement towards using renewable, biodegradable cellulose fibers aligns with global sustainability efforts, promoting the use of natural substances to reduce environmental impact. By shifting away from petroleum-based, non-biodegradable fibers, the textile industry actively contributes to a circular economy, helping to minimize waste and pollution across various sectors. With ongoing advancements, regenerated cellulose fibers are becoming key in the development of durable and eco-friendly materials.

Their versatility, combined with the industry's growing commitment to sustainability, ensures they will continue to shape the future of textile production. Research into cellulose solvents, dissolution techniques, and methods for enhancing fiber performance will drive further progress and address existing challenges.

14.2 Classification of Regenerated Cellulose Fibers

Regenerated cellulose fibers are traditionally grouped into five main types: viscose rayon, lyocell, cuprammonium (cupro), cellulose acetate, and modal. Their classification is determined by the specific production processes used to transform natural cellulose into textile fibers.

- Viscose Rayon fibers

Viscose fibers come from natural cellulose, usually sourced from wood pulp or cotton linters. The process of turning cellulose into viscose involves several chemical steps—soaking, shredding, and treating with xanthate—which help create fibers that resemble the softness and appearance of cotton or silk. Due to its adaptability and environmentally friendly nature, viscose is widely used in textile production [12]. Viscose fibers were first developed in the late 19th century as scientists aimed to produce a material with characteristics similar to natural silk. In 1892, British chemists Charles Cross, Edward Bevan, and Clayton Beadle secured a patent for the viscose process, laying the groundwork for its commercial use in the early 20th century [26].

Viscose rayon, one of the first and most widely used regenerated cellulose fibers, is produced by dissolving cellulose in sodium hydroxide and carbon disulfide, forming cellulose xanthate. This solution is then spun into fibers and solidified in an acidic bath [27, 28]. Due to its high absorbency and soft texture, viscose rayon is ideal for medical uses like wound dressings, bandages, and surgical swabs. Its ability to be effectively sterilized further reinforces its usefulness in healthcare settings [29, 30].

Viscose fibers typically have a round or slightly irregular cross-section, with their shape and size influenced by the coagulation process during spinning. Studies show they often appear circular or oval, depending on the exact conditions of their formation [31, 32]. Viscose fibers usually feature a hollow core, called a central lumen, surrounded by a cellulosic wall with varying degrees of fibrillation. This lumen forms as the cellulose solution rapidly solidifies during spinning, trapping air or gas within the fiber—an identifying trait of regenerated cellulose fibers [33]. The cellulosic wall around the lumen consists of aligned cellulose microfibrils, and their crystallinity and orientation influence the fiber's mechanical characteristics [34]. While viscose rayon's cross-section usually looks smooth and uniform, slight irregularities can emerge due to the production process. Factors like temperature and solvent concentration during coagulation can affect the fiber's shape, sometimes resulting in a less consistent appearance [35, 36]. Differences in the fiber's shape can stem from fluctuations in the extrusion flow rate, affecting its overall structure.

When seen lengthwise, viscose rayon fibers have a smooth surface with delicate longitudinal striations or fibrils. These features develop during the stretching and drawing steps of production, where cellulose chains align to create their unique fibrillar texture [31]. The surface of viscose fibers can sometimes show slight waviness or undulations, resulting from mechanical drawing and relaxation during spinning. Although generally smooth, their diameter may vary due to changes in spinning conditions. Factors such as pressure, temperature, and draw ratio during extrusion and coagulation influence these fluctuations, impacting the fiber's mechanical properties and its potential applications [32].

Recent progress in viscose rayon technology has aimed at enhancing fiber durability while minimizing environmental effects. Developments in the manufacturing process have resulted in more eco-friendly production techniques and improved fiber characteristics [37, 38].

- Lyocell

Lyocell fibers belong to the regenerated cellulose family and are recognized for their environmentally friendly production and excellent performance. Sourced mainly from wood pulp, lyocell is a type of rayon; however, unlike viscose, it is manufactured using a non-toxic solvent, N-methylmorpholine N-oxide (NMMO), making the process more sustainable. Since its introduction in the 1990s, lyocell has gained widespread popularity for its eco-conscious nature and high-quality properties [39, 40]. Lyocell was created in response to the demand for a more environmentally friendly substitute for conventional rayon fibers. Courtaulds fibers introduced the first commercial version, known as Tencel, in the early 1990s [32].

Lyocell fibers stand out due to their distinct structural and mechanical characteristics, differentiating them from other regenerated cellulose fibers like viscose. Their cross-section is typically round with a smooth, even surface. Unlike viscose, which often contains a central lumen, lyocell fibers possess a denser and more compact cellulosic structure with little to no lumen. This compact formation is a result of the solvent-spinning method, in which cellulose is dissolved in NMMO, then extruded and regenerated under controlled conditions [39, 41]. The consistent, round cross-section enhances lyocell fibers' tensile strength and smooth texture, making them ideal for applications requiring durability and a soft feel. Their compact structure, with minimal internal voids, combined with the alignment of cellulose microfibrils during spinning, results in fibers that are both strong and resistant to deformation—essential qualities for textiles requiring reliable performance [32].

Lyocell fibers are distinguished by their consistently smooth and uniform texture, visible in both cross-sectional and longitudinal perspectives. The spinning technique used in their production minimizes surface imperfections, unlike viscose fibers, which tend to display more noticeable striations and irregularities. This smooth surface not only contributes to the luxurious feel of lyocell fabrics but also enhances their dyeing properties, allowing for vibrant and uniform coloration [42]. In longitudinal views, lyocell fibers are typically straight, smooth, and highly uniform. This straightness, combined with the fiber's inherent strength, minimizes the likelihood of pilling and wear, issues commonly encountered with less uniform fibers. Thanks to their impressive mechanical properties—particularly high tensile strength and durability—lyocell fibers are well-suited for diverse applications, ranging from clothing and fashion textiles to specialized technical fabrics [43].

The NMMO process used to produce lyocell is considered more environmentally friendly than the viscose process [40, 44]. In this process, cellulose is dissolved in NMMO solvent before being extruded and regenerated into fibers. Lyocell fibers are known for their remarkable tensile strength, efficient moisture regulation, and biocompatibility, making them ideal for various applications such as surgical gowns, drapes, and wound dressings [12, 45].

Advancements in lyocell production continue to refine fiber properties while prioritizing sustainability. Ongoing research aims to improve functionality and minimize environmental effects, leading to more efficient and eco-friendly manufacturing methods [40]. Recent innovations in lyocell production focus on managing fibrillation and enhancing fiber durability for high-performance applications like medical textiles. Methods such as cross-linking agents and enzymatic treatments have greatly reduced fibrillation, improving lyocell's effectiveness in wound dressings and surgical fabrics [46, 47]. Research continues to focus on making lyocell production more environmentally sustainable, with advancements aimed at improving solvent recovery efficiency and exploring alternative cellulose sources for raw materials [48, 49].

- Modal

Modal fibers represent a refined version of rayon, engineered to surpass the qualities of conventional viscose rayon. Derived from beech tree pulp, they undergo specialized chemical treatments that enhance softness, strength, and shrinkage resistance. Recognized for their luxurious texture and durability, modal fibers were developed to offer a more resilient and high-performing alternative. Austrian company Lenzing AG played a pivotal role in their advancement, coining the term "modal" to define these sophisticated fibers [50, 51].

Modal fibers are manufactured using a process akin to viscose rayon production, but with adjustments that improve their durability and flexibility. These modifications result in fibers that are stronger, more resilient, and better able to maintain their shape, making modal a preferred choice for textiles requiring enhanced performance [49]. These refinements give modal fibers enhanced softness, strength, and shrinkage resistance—key attributes that make them particularly valuable for reusable medical textiles such as hospital bed linens and patient garments. Their durability and comfort support hygiene and longevity in healthcare environments [45]. Modal fibers possess a slightly oval cross-section, resembling viscose but with enhanced structural integrity and uniformity due to refined manufacturing techniques. Their cross-sectional shape maintains greater consistency in density and form compared to standard viscose fibers. Additionally, modal fibers exhibit a smoother, more even surface in both cross-sectional and longitudinal perspectives. Their longitudinal structure appears straight with minimal surface irregularities and fewer striations than viscose fibers, contributing to their smooth texture and enhanced performance [52].

- Cuprammonium Rayon (Cupro)

Cupro fibers, also known as cuprammonium rayon, are regenerated cellulose fibers derived from cotton linter, a byproduct of the cotton industry. Their production process involves dissolving cellulose in a copper-ammonium solution, resulting in fibers with a fine, smooth texture and excellent drapability [12, 53]. This process yields fibers that are silky, lustrous, and hypoallergenic, making them an excellent choice for applications requiring soft, skin-friendly materials, such as advanced wound care products. Their smooth texture and gentle nature enhance comfort and usability in sensitive environments [54, 55].

The cuprammonium method dissolves cellulose in a copper-ammonium solution, producing cupro fibers with distinct properties. This method produces fibers known for their fine texture, silk-like sheen, and excellent breathability, making them a popular choice in luxury textiles and high-performance applications [56]. These fibers are prized for their silky touch, excellent moisture absorption, and smooth finish, making them ideal for premium textiles such as luxury fashion and upscale home fabrics.

Recent improvements in the cuprammonium process have strengthened the fibers and broadened their use in contemporary textiles. Innovations have allowed cupro fibers to maintain their softness and sheen while being integrated into durable fabrics suitable for a variety of uses. Cupro fibers typically feature a round or slightly irregular cross-section with a dense, smooth internal structure and minimal lumen, giving them a refined, high-quality texture. Their cross-sectional appearance is sleek and uniform, emphasizing their luxurious nature. When viewed lengthwise, cupro fibers present an exceptionally smooth surface with minimal striations or irregularities, creating a straight form with a distinct sheen and softness [55, 57]. Beyond their silk-like texture and excellent moisture absorption, cupro fibers are biodegradable, making them highly attractive in the sustainability-driven textile industry. Their growing use in eco-friendly innovations reflects the global shift toward more sustainable and biodegradable materials. Their combination of luxury, functionality, and sustainability positions cupro fibers as a key material in both fashion and technical textile applications.

- Cellulose Acetate

Acetate fibers, or cellulose acetate fibers, originate from cellulose obtained from wood pulp or cotton linters. Their production involves acetylation, where cellulose reacts with acetic anhydride or acetyl chloride in the presence of a catalyst. This method creates fibers that blend natural and synthetic qualities, providing unique benefits for textile applications [12, 58].

Acetate fibers stand out among regenerated cellulose fibers due to the acetylation process, which imparts a smooth texture, glossy finish, and low moisture absorption. These attributes make them ideal for applications where elegance and comfort are key, such as fashion textiles, linings, and decorative fabrics. Their ability to mimic the sheen and feel of silk has contributed to their popularity in luxury and formal wear [59]. The smooth surface of acetate fibers helps reduce irritation when in contact with the skin, making them an excellent choice for fabrics that prioritize comfort.

Acetate fibers are valued for their soft, silky texture, making them a popular choice in luxury textiles like high-end fashion and home furnishings. Their hypoallergenic nature also makes them ideal for medical textiles, particularly for sensitive skin. They are used in bandages and patient garments, offering both comfort and a gentle, non-abrasive surface.

Cellulose acetate fibers are also valued for their versatility in dyeing, as they can achieve vibrant colors with ease, especially when used with disperse dyes. This makes them a popular choice for decorative fabrics, linings, and formalwear. Moreover, their inherent resistance to shrinking and wrinkling adds to their utility in various textile applications. Recent advancements in the production of acetate fibers have also focused on enhancing their biodegradability, making the a more sustainable choice in the increasingly eco-conscious textile industry [59, 60].

Regenerated cellulose fibers offer a compelling alternative to synthetic fibers, tackling environmental concerns while preserving the beneficial traits of natural fibers. According to the *Materials Market Report* 2023 by Textile Exchange, RCFs constituted approximately 6% of the global fiber market in 2022, amounting to 7.3 million tons [61]. Viscose fibers dominated RCF production, accounting for 80% of total output with around 5.8 million tons. Acetate fibers represented 13%, while lyocell, modal, and cupro fibers comprised 4%, 3%, and 0.2%, respectively. Recycled regenerated cellulose fibers made up 0.5% of the market, reflecting a growing commitment to sustainability and fiber recycling.

14.3 Methods of Production for Regenerated Cellulose Fibers

RCFs have garnered significant attention due to their ability to be efficiently processed into man-made fibers using bio-based feedstocks, a process that is more straightforward compared to natural fibers [62]. Several methods are available for producing high-quality RCFs through dissolution and regeneration processes [6, 63]. These fibers are commonly produced using fiber spinning, with wet spinning being a widely adopted technique. In this process, a cellulose solution is extruded through a spinneret into an acid bath, triggering coagulation. The fibers then undergo thorough washing—first with hot water, followed by distilled water—to remove any residual salts. Finally, they are air-dried at room temperature, ensuring the formation of the desired regenerated cellulose structure [64].

The properties of RCFs are highly dependent on the solvent systems used during the dissolution process [62]. The choice of solvent system plays a crucial role in determining the solubility and spinnability of cellulose, directly influencing the final fiber properties. Several factors affect how effectively cellulose dissolves, including the solvent's ability to interact with cellulose and its viscosity. Additionally, the raw material's polymerization degree, molecular structure, and chemical composition all contribute to the dissolution process. Key processing conditions, such as temperature, pressure, and duration, also significantly impact fiber formation and overall quality [65, 66].

The structural and mechanical properties of RCFs can be adjusted by modifying factors like draw ratio, spinning dopes, and the spinning process. Research indicates that RCFs created at lower draw ratios often develop irregular shapes with noticeable surface grooves, while those spun at higher draw ratios achieve a smoother surface and more rounded form, enhancing their mechanical performance [64, 68].

For high-quality fiber production, advanced purification methods like enzymatic treatments and bleaching help eliminate impurities and improve biocompatibility. These processes are especially crucial for applications with strict biological safety requirements, ensuring the fibers meet the necessary standards for medical and technical uses [67–70].

14.3.1 Methods of Production for Viscose Fibers

The production of viscose fibers involves several key steps, beginning with cellulose pulp preparation, followed by its dissolution, regeneration, and final finishing. Each stage plays a crucial role in defining the fiber's final properties, ensuring it achieves the desired texture, strength, and performance for various applications. Through advancements in processing techniques, viscose production continues to evolve, focusing on sustainability and improved fiber quality. Each step is critical in determining the quality and characteristics of the final fiber. The process begins with extracting cellulose from sources such as wood pulp or cotton linters. The cellulose pulp is purified to remove impurities like hemicellulose, lignin, and other residual compounds [71–73]. Following purification, the cellulose is treated with sodium hydroxide (NaOH) to form alkali cellulose, which increases its reactivity for subsequent chemical treatments [74].

During production, alkali cellulose undergoes controlled aging, usually for 1 to 2 days, to lower its degree of polymerization. This step is crucial for achieving the desired viscosity of the viscose solution, ensuring optimal fiber formation and processing efficiency. By carefully regulating aging conditions, manufacturers can refine the solution's characteristics for improved spinnability and fiber quality [75]. After aging, the xanthation process occurs, in which the alkali cellulose reacts with carbon disulfide (CS_2), forming cellulose xanthate. This reaction enables the cellulose to become soluble in dilute sodium hydroxide, creating the viscose solution. The viscose solution, an orange-yellow viscous liquid, is left to ripen, allowing the cellulose chains to rearrange and reach the required spinnability for fiber formation [28].

For maintaining fiber uniformity and high quality, the ripened viscose solution is carefully filtered to eliminate undissolved particles and degassed to remove air bubbles. These processes are essential in ensuring the final fibers have a smooth, consistent texture, enhancing their overall performance and durability. Optimizing these steps contributes to producing viscose fibers with reliable structural integrity and a refined appearance [76]. At this stage, the viscose solution is pushed through a spinneret into a sulfuric acid bath, where acid precipitation and coagulation convert the dissolved cellulose into solid fibers. This transformation is crucial, as it defines the final physical characteristics of the viscose fibers, influencing their strength, texture, and overall performance in textile applications. Careful control of these conditions ensures fibers with desirable qualities such as smoothness, uniformity, and durability [77, 78]. Viscose rayon is manufactured in two primary forms: continuous filament and staple fiber. While both types offer versatility, staple fibers are the most widely used, particularly in textiles, due to their adaptability in blending with other fibers and enhancing fabric softness and breathability. These characteristics make viscose staple fibers a popular choice across various applications, from apparel to home textiles.

After regeneration, the fibers undergo multiple treatments such as washing, bleaching, and finishing. These processes improve their whiteness, durability, and softness, ensuring they meet the quality standards required for diverse textile uses [63, 79].

Viscose fiber production involves chemicals like sodium hydroxide and carbon disulfide, which can present environmental and health concerns. However, modern technological advancements are working to minimize these risks through more efficient chemical recovery systems and improved waste management strategies, making the process more sustainable [78, 80]. These innovations are essential for making the production of viscose fibers more sustainable while maintaining their high quality and desirable properties for textile applications.

14.3.2 Methods of Production for Lyocell Fibers

Lyocell fiber manufacturing follows a multi-step process, including cellulose extraction, dissolution, spinning, and post-treatment. A key aspect of this method is the use of N-methylmorpholine N-oxide (NMMO), a safe and reusable solvent that contributes to the eco-friendly nature of lyocell production.

Lyocell fibers are primarily derived from wood pulp harvested from responsibly managed forests. The pulp undergoes a purification process to eliminate lignin and hemicellulose, resulting in high-purity cellulose that can be directly dissolved [81]. Unlike conventional rayon production, the lyocell process eliminates the need for chemical derivatization. Instead, purified cellulose dissolves directly in an aqueous NMMO solution, making the process both more efficient and environmentally sustainable [44, 82, 83].

The cellulose-NMMO mixture, known as the spinning dope, undergoes filtration to remove any undissolved particles and air bubbles. After filtration, it is extruded through a spinneret into a coagulation bath—usually water or a diluted NMMO solution—where the cellulose fibers form through regeneration [84, 32]. Following regeneration, lyocell fibers go through post-treatment processes such as washing to eliminate remaining NMMO, drying, and finishing. These procedures are essential for improving the fibers' mechanical strength and overall visual quality [82].

A major advantage of the lyocell production process is the efficient recovery and reuse of NMMO. Typically, more than 99% of the solvent is recovered and recycled, which significantly reduces the environmental impact and lowers production costs [41, 85]. This closed-loop system, combined with the non-toxic nature of NMMO, makes lyocell production far more environmentally friendly than other types of rayon.

In addition to being biodegradable, lyocell fibers are produced from renewable resources, making them highly aligned with the principles of sustainable development [22, 63, 86]. The combination of eco-friendly production processes and the fiber's biodegradability has led to widespread recognition of lyocell as a sustainable fiber choice within the textile industry.

14.3.3 Methods of Production for Modal Fibers

Modal fibers are derived from beech wood pulp, which is treated to eliminate lignin and hemicellulose. This refinement process ensures the cellulose is highly pure and ready for fiber manufacturing [49, 87]. Purified cellulose undergoes an initial treatment with sodium hydroxide, resulting in alkali cellulose. This then reacts with carbon disulfide to form cellulose xanthate. While derived from the traditional viscose method, this refined process enhances fiber strength, durability, and overall quality, making it more suitable for high-performance textile applications [88–90].

Once cellulose xanthate is formed, it is dissolved in a dilute sodium hydroxide solution, resulting in a thick spinning mixture. This solution undergoes meticulous processing to maintain uniformity and achieve optimal viscosity for efficient fiber formation. The spinning dope is then filtered and degassed to remove impurities and air bubbles, ensuring smooth fiber production. Finally, it is extruded through a spinneret into a sulfuric acid bath, where the cellulose solidifies and regenerates into fibers [32, 91].

Following the spinning process, modal fibers go through multiple post-treatment steps. These include washing to eliminate leftover chemicals, bleaching to attain the desired whiteness, and finishing treatments to enhance their softness, strength, and durability [47].

Modal fiber production is regarded as more sustainable than conventional viscose rayon due to advancements in manufacturing techniques. Companies like Lenzing utilize a closed-loop system that recovers and reuses chemicals such as sodium hydroxide and carbon disulfide, significantly lowering waste and environmental impact [92, 93]. Furthermore, modal fiber cellulose is usually obtained from sustainably managed beech wood forests, guaranteeing a renewable and responsibly harvested source of raw materials.

The closed-loop production system efficiently recovers and reuses chemicals, enhancing its sustainability and reducing environmental impact [82]. This approach reduces harmful emissions and chemical waste, making modal fiber production a more sustainable option within the textile industry.

While modal fibers are primarily used in textiles for their softness and durability, the process can be adapted to include additional purification steps to remove residual chemicals for specialized applications. Although not necessarily required for standard textile applications, the process for creating high-purity modal fibers can involve more rigorous washing and bleaching, ensuring cleaner fibers for high-performance uses [94].

14.3.4 Methods of Production for Cupro Fibers

Cupro fibers are derived from cotton linters, the short fibers left on cotton seeds after the longer fibers are removed. These linters undergo purification to eliminate impurities, yielding high-purity cellulose ideal for fiber manufacturing [71].

The purified cellulose is dissolved in a cuprammonium solution, composed of copper(II) hydroxide and aqueous ammonia. This unique dissolution process, exclusive to cupro fiber production, forms a viscous solution suitable for extrusion into fibers. Unlike other cellulose regeneration methods, such as the viscose process, this technique specifically relies on copper and ammonia to break down cellulose [52, 95]. The cellulose solution is pushed through a spinneret into a coagulation bath, usually containing a dilute sulfuric acid solution. Within this bath, the cellulose undergoes regeneration, forming solid fibers as the sulfuric acid precipitates it from the solution.

After fiber formation, they undergo posttreatment steps such as washing to eliminate residual chemicals, bleaching for optimal whiteness, and further treatments to enhance their characteristics. Stretching during these processes helps align cellulose molecules, boosting the fibers' strength, uniformity, and overall quality.

Cupro fiber production relies on the careful handling of copper and ammonia, which are essential to the process. To reduce environmental impact, contemporary manufacturing includes recovery and recycling systems for these chemicals, promoting a more sustainable approach [96]. While cupro fibers are known for their softness, sheen, and high moisture absorption, the production process must be carefully managed to ensure that environmental and safety standards are met. For medical use, the manufacturing process incorporates advanced purification techniques to eliminate any remaining copper ions or other potentially hazardous compounds, ensuring the fibers meet strict safety and biocompatibility standards. This ensures the fibers meet stringent safety and quality standards for specialized applications [97].

14.3.5 Methods of Production for Acetate Fibers

Acetate fibers are derived from purified wood pulp and cotton linters, with lignin, hemicellulose, and other impurities removed to obtain high-purity cellulose suitable for fiber production [98]. The purified cellulose undergoes acetylation, reacting with acetic acid and acetic anhydride with sulfuric acid as a catalyst. This process swaps the hydroxyl groups in cellulose for acetyl groups, transforming it into cellulose acetate [99].

Cellulose acetate is dissolved in a solvent, usually acetone, forming a dense solution known as the spinning dope. This mixture is then used in fiber spinning, where it is extruded through a spinneret to produce fibers [100]. Acetate fibers are produced using either wet or dry spinning. In wet spinning, they are extruded into a coagulation bath to remove the solvent and solidify. In dry spinning, the fibers harden as the solvent naturally evaporates in the air [101].

After spinning, acetate fibers go through several post-treatment steps, including washing to remove residual solvents, stretching to improve polymer alignment for enhanced strength, and finishing treatments that boost dyeability and luster. Because their production involves solvents and chemicals, careful handling and disposal help minimize environmental effects.

Advances in solvent recovery and recycling have improved sustainability, alongside eco-friendly approaches like responsibly sourced cellulose and safer solvents and catalysts in acetylation and dissolution [68].

Contemporary acetate fiber manufacturing integrates solvent recovery and reuse systems, particularly for acetone, to minimize waste and decrease environmental impact. Moreover, advancements in biodegradable acetate fibers contribute to greater sustainability, offering a more eco-friendly option for the textile sector [102, 103]. For specialized uses like high-performance textiles, acetate fiber production involves thorough purification to remove any remaining acetic acid or other chemicals. Additionally, treatments may be applied to enhance characteristics such as improved dye absorption or antimicrobial properties, supporting the increasing demand for acetate fibers in environmentally conscious markets [11].

14.4 New Methods of Production for RCFs

Recently, researchers have explored alternative solvents for dissolving cellulose, with ionic liquids (ILs) emerging as a promising option. ILs are recognized as a sustainable and eco-friendly option, demonstrating significant potential in producing regenerated cellulose fibers. Their ability to efficiently process cellulose makes them a promising alternative to conventional solvents, contributing to greener fiber production [104]. ILs are salts that remain in a liquid state below 100°C and exhibit unique properties such as low vapor pressure, high thermal stability, and the ability to dissolve both organic and inorganic compounds. They enable cellulose dissolution by disrupting hydrogen bonds, eliminating the need for chemical derivatization. Common ILs used for this purpose consist of cations like imidazolium, pyridinium, or ammonium, paired with anions such as chloride or acetate. After dissolution, cellulose can be regenerated into fibers using a coagulation bath. Due to their non-volatile nature and reusability, ILs are often referred to as "green solvents," although their cost-effectiveness and environmental impact can vary [105].

The urea/alkali method provides an alternative approach for dissolving cellulose, utilizing a blend of urea and sodium hydroxide. In this process, urea serves as a hydrogen bond acceptor, breaking cellulose's hydrogen bonding network. Conducted at subzero temperatures, it enhances energy efficiency and offers environmental benefits due to its non-toxic solvents. However, fine-tuning process parameters is critical for improving the mechanical properties of the resulting fibers. While promising, large-scale implementation of this method faces efficiency and scalability hurdles compared to ionic liquids [106].

The CarbaCell method dissolves cellulose using organic carbonates, primarily dimethyl carbonate (DMC). This approach is valued for its low toxicity and eco-friendly nature, as DMC is biodegradable and presents fewer risks than traditional solvents like carbon disulfide (CS_2) used in viscose production. In this process, organic carbonates activate cellulose's hydroxyl groups, breaking intermolecular hydrogen bonds. The resulting cellulose solution is then regenerated into fibers through wet or dry-jet wet spinning [52].

Each cellulose dissolution technique has distinct advantages, with the CarbaCell process standing out for its sustainability. It functions under mild conditions, relying on biodegradable, low-toxicity organic carbonates that can be effectively recovered, making it a promising option for eco-conscious fiber manufacturing [106]. ILs are highly efficient for dissolving cellulose and offer the advantage of recyclability. However, their cost can be significant, and their environmental impact varies based on the specific IL employed. While they excel in processing, their biodegradability and ecological footprint require thorough assessment to determine their long-term sustainability [107]. The urea/alkali method provides a cost-effective and environmentally friendly alternative for cellulose processing. Although its low toxicity contributes to sustainability, it struggles with scalability and efficiency compared to ILs and the CarbaCell method. Further refinements are necessary to enhance its viability for large-scale industrial applications [108].

14.5 Characteristics of Regenerated Cellulose Fibers

Regenerated cellulose fibers originate from natural cellulose, a glucose-based polymer sourced from materials like wood pulp or cotton linters. Through chemical processing, these fibers gain enhanced properties suited for diverse textile applications, combining a natural base with refined performance. Structurally, cellulose fibers vary significantly. Natural cellulose fibers exhibit tightly packed crystalline fibrillar formations arranged in distinct helical patterns, while regenerated cellulose fibers have a less structured arrangement with noticeable variations. Rayon, the earliest regenerated cellulose fiber, emerged in the late 19th century and exists in multiple forms, including viscose rayon, modal, lyocell (Tencel), cellulose acetate, and cuprammonium rayon. Each of these fibers is produced through distinct chemical processes, resulting in different properties and characteristics (see Table 14.1). This structural diversity allows regenerated cellulose fibers to be versatile, making them valuable in a variety of textile applications [51].

Table 14.1 Properties of the most common regenerated cellulose and cotton fibers [47, 57, 63, 71, 109–111].

Characteristic \ Fiber	Cotton	Viscose rayon	Modal	Lyocell	Tencel	Cupro	Acetate
Density, g/cm^3	1.52	1.46-1.54	1.53	1.5	1.5	1.5	1.29-1.33
Tenacity, cN/tex	19.5-35	8.8-24	34	30-45	36-44	9-28	12.8
Elongation, %	7-14	17-30	12	12-18	16-18	6-25	24
Moisture regain, %	7-8	11-14	11.8	10-13	11	11-12.5	6-7
Cristallinity, %	60-70	30-40	40-48	50-65	50-65	30-40	20-30
Fiber cross-section circularity	0.5-0.8	<1	closer to 1	closer to 1	closer to 1	0.6-0.8	Lower circularity

Note: *A value of 1 corresponds to a perfectly round shape, indicating complete circular symmetry.

The chemical characteristics of regenerated cellulose depend on the level of substitution and the specific method used for regeneration. For instance, viscose rayon typically has a lower degree of polymerization and is more prone to degradation, while lyocell offers superior mechanical and chemical properties, and modal fibers are known for their enhanced strength and durability [41].

These fibers have a strong affinity for water due to the hydroxyl groups (-OH) present on the cellulose chain, allowing them to efficiently absorb and retain moisture. This feature is particularly beneficial in medical uses such as wound dressings and surgical sutures, where controlling moisture is essential. Regenerated cellulose fibers exhibit moderate resistance to chemicals, being sensitive to strong acids and alkalis while maintaining stability in physiological conditions. While they remain insoluble in water, these fibers can dissolve in select solvents such as cuprammonium hydroxide and N-Methylmorpholine N-oxide (NMMO), making them viable for fiber production. Their limited solubility in biological fluids further enhances their suitability for medical applications [22, 111–114].

These fibers naturally decompose in the environment, making them environmentally friendly [114, 115]. These fibers break down when exposed to temperatures ranging from 200°C to 300°C, with lyocell generally displaying greater thermal resistance than viscose. The glass transition temperature (Tg) of regenerated cellulose fibers is around 250°C, though it varies based on molecular weight and crystallinity [116–118].

Regenerated cellulose fibers are known for their softness, breathability, and absorbency. They can mimic the texture and feel of natural fibers like silk, cotton, or wool, depending on their treatment and processing.

Properties of Viscose Rayon Fibers

Viscose rayon, a regenerated cellulose fiber, exhibits lower crystallinity than natural cellulose fibers like cotton. The manufacturing and post-treatment processes disturb cellulose's highly ordered crystalline structure, leading to a more amorphous fiber. As a result, viscose rayon is generally weaker and less durable than cotton, particularly when wet, making it more susceptible to mechanical and environmental stress [71, 119, 120].

The lower crystallinity of viscose rayon imparts several notable characteristics [51, 56, 119, 121]:

- Softness and drapability: the more amorphous structure provides viscose rayon softer and gives it a fluid drape. Viscose is well-suited for garments and textiles that need a graceful drape, offering a soft, smooth texture reminiscent of silk.
- Breathability: breathable and allows good air circulation, enhancing comfort.
- Improved ability to absorb dyes: the greater presence of amorphous areas facilitates deeper dye infiltration, leading to rich, consistent coloration.
- Moisture absorption: viscose rayon efficiently wicks moisture, much like cotton, contributing to greater comfort in apparel. Its ability to absorb and release moisture helps regulate body temperature and maintain breathability.
- Strength: moderate strength when dry but weaker when wet, requiring careful handling during washing.
- Versatility and luster: can have a high sheen. The versatile characteristics of viscose rayon make it ideal for a wide range of uses, including stylish apparel and home furnishings.

Properties of Modal Fibers

Modal is a regenerated cellulose fiber akin to viscose rayon, but its manufacturing process is refined to increase crystallinity, resulting in superior characteristics. This modification enhances strength, durability, and softness, making modal an excellent choice for high-performance textiles.

The increased crystallinity of modal fibers offers multiple advantages, enhancing their strength, durability, and resistance to wear compared to other regenerated cellulose fibers [122–124]:

- Enhanced strength and longevity: The higher crystallinity enhances tensile strength, making modal fibers more

resilient and long-lasting compared to viscose rayon, whether dry or wet.
- Luxurious texture: Although modal fibers exhibit high crystallinity, they maintain an exceptionally smooth and soft texture, making them a top choice for premium textiles and apparel. Their refined structure enhances durability while preserving a luxurious feel.
- Superior shape retention: Modal fibers maintain their dimensions well, ensuring they keep their form even after repeated washing and drying.
- Exceptional moisture management: These fibers are highly effective at wicking moisture, providing outstanding breathability and comfort, surpassing both cotton and viscose in performance.
- Long-lasting color vibrancy: Their refined structure enables excellent dye retention, leading to rich, lasting colors.
- Reduced pilling: Modal fibers resist pilling more effectively, helping fabrics maintain a fresh, new appearance over time.

Modal fibers are widely used in textiles such as underwear, sportswear, and bed linens, thanks to their optimal balance of softness, strength, and moisture management, providing lasting durability and comfort.

Properties of Lyocell (Tencel®) Fibers

Lyocell, a regenerated cellulose fiber, boasts strength comparable to polyester and surpasses cotton and other man-made cellulose fibers. It retains a high modulus in both dry and wet conditions, with minimal loss in strength when exposed to moisture. However, lyocell fibers are prone to fibrillation during wet abrasion, necessitating specialized finishing methods to ensure optimal performance.

Tencel®, a branded form of lyocell, can be processed with conventional machinery with minimal adjustments. Tencel® features a temporary crimp, high stiffness, and minimal fiber entanglement, producing yarns with superior tensile strength and fewer defects. It integrates seamlessly with other fibers, especially cellulose-based ones, improving the durability, functionality, and visual appeal of the finished fabric [51].

Lyocell's high crystallinity stems from its environmentally friendly solvent-spinning method, which dissolves cellulose in the non-toxic solvent N-methylmorpholine N-oxide (NMMO) and regenerates it in a coagulation bath. This structural quality grants lyocell several beneficial properties [44, 84, 124–126]:

- Strength and durability: lyocell fibers are exceptionally strong and durable, comparable to polyester, with wet strength on par with cotton, and reduced elongation compared to viscose.
- Drape and wrinkle resistance: excellent drape, providing an elegant look to fabrics. Better wrinkle resistance compared to viscose and modal.
- Comfort and softness: it combines strength with a soft, smooth texture, ensuring comfort in textile applications. Its fibrillar structure, with microfibrils oriented along the fiber axis, enables the formation of a fibrillated surface when subjected to abrasion.
- Moisture management: lyocell excels in moisture-wicking, keeping wearers dry and comfortable, ideal for activewear and intimate apparel.
- Biodegradability: made from natural cellulose with an eco-friendly process, lyocell is biodegradable.
- Sustainability: the manufacturing process is environmentally friendly, employing a closed-loop system that recovers and reuses the non-toxic solvent, minimizing waste and reducing ecological impact.
- Versatility: it's blend of durability, gentle texture, and effective moisture regulation makes it ideal for various uses, spanning fashion apparel to home furnishings.
- Impact of environmental: the closed-loop production system significantly reduces environmental impact, establishing lyocell as a high-quality and eco-friendly textile choice.

Properties of Cuprammonium Fibers (Cupro)

Cupro, created through the cuprammonium process, exhibits moderate crystallinity—greater than viscose rayon but less than natural cellulose fibers. This structural characteristic influences its properties, making it a unique material in textile applications. This structural balance grants Cupro fibers desirable traits, including a silky feel, superb drape, and a luxurious softness. The comparatively lower crystallinity improves dye absorption and results in a smoother texture, making Cupro a preferred choice for premium textiles and fashion applications.

Cupro fibers possess several distinctive qualities that make them a valuable choice in textiles [56, 127, 128]:

- Smoothness and softness: Cupro is highly valued for its delicate, silky feel and exceptionally smooth surface, making it a desirable choice in textiles.
- Moisture management: Cupro excels in moisture management, efficiently wicking away sweat while absorbing moisture to keep fabrics breathable and comfortable.
- Breathability: highly breathable.
- Drape: Cupro provides excellent drape and fluidity in fabrics.
- Strength: Cupro has lower strength compared to lyocell but surpasses certain varieties of viscose in durability, making it a balanced option among regenerated cellulose fibers.

Properties of Cellulose Acetate Fibers

Cellulose acetate fibers, often referred to as acetate fibers, exhibit lower crystallinity than other cellulose fibers such as cotton, modal, or lyocell. Unlike traditional regenerated cellulose fibers, acetate is derived from cellulose ester rather than pure cellulose. This distinction arises from the acetylation of cellulose with acetic anhydride during manufacturing, which disrupts the natural alignment of cellulose chains, leading to a more amorphous fiber structure.

The reduced crystallinity in acetate fibers influences their characteristics [56, 121, 124]:

- Softness and Drapability: Acetate's lower crystallinity results in a soft, silky texture with outstanding drapability, making it well-suited for lightweight, flowing fabrics. Its rich, lustrous finish enhances its fluidity and refined aesthetic.
- Sheen and Luster: Acetate fibers possess an inherent silky sheen and luster, imparting fabrics with a glossy, elegant appearance. This distinctive quality enhances their appeal in luxury textiles and fashion applications.
- Moisture Absorption: Acetate fibers exhibit moderate moisture absorption, providing a degree of comfort, though they are less absorbent than highly crystalline fibers such as cotton and lyocell. This property influences their breathability and suitability for specific textile applications.
- Color Retention: Acetate fibers have excellent dye absorption, allowing them to hold rich, vibrant colors—an asset in fashion and decorative textiles. Their ability to retain color enhances their aesthetic appeal and versatility.
- Resistance to Shrinkage and Wrinkling: Acetate's molecular composition reduces shrinkage and wrinkling, enhancing

durability and simplifying maintenance. These characteristics make it a practical choice for everyday and fashion textiles.
- Strength: Acetate fibers have lower tensile strength compared to highly crystalline cellulose fibers and are more vulnerable to heat and chemical exposure. Their composition emphasizes softness and fluid drapability rather than long-term durability.

14.6 Certain New Regenerated Cellulose Fibers Used in Textiles

Recent innovations in textile technology, driven by an increasing focus on sustainability, have led to the creation of advanced regenerated cellulose fibers. These developments aim to enhance fiber performance while reducing environmental impact. Some of the newest regenerated fibers are listed below and tier main mechanical characteristics are listed in Table 14.2. These advanced fibers are designed to enhance the qualities of traditional regenerated cellulose fibers like viscose, modal, and lyocell, while also incorporating more sustainable production methods. Innovations in processing techniques reduce environmental impact, improve durability, and refine fiber characteristics for diverse textile applications.

Ioncell fibers are a type of eco-friendly textile made using the novel regenerated cellulose fiber process of the Lyocell type—Ioncell process—ionic liquid solvent system. This approach not only makes textile production more sustainable but also supports recycling and the use of waste materials. Ioncell fibers characterizes high tensile strength (both wet and dry), smooth, soft, and a luxurious feel and are fully biodegradable.

Table 14.2 Properties of some certain new regenerated cellulose fibers [37, 44, 51, 120, 121, 124, 128–137].

Characteristic	Ioncell	Naia™	Spinnova	Refibra	Bast fiber-based
Density, g/cm³	1.5	1.48	1.45	1.5	1.5
Tenacity, cN/tex	8.9-13.3	8.0-10.6	8.9-11.6	8.9-11.6	10.6-16.0
Moisture regain, %	12-14	13-15	11.8	12-14	11-13
Elongation, %	18-22	20-25	12	15-20	15-18
Cristallinity, %	60	50	40-48	60	60-70

Ioncell fibers are versatile and can be used in various fashion and textile applications (i.e., apparel, home and technical textiles), offering a way to make high-quality fabrics with a smaller environmental impact [37, 129, 138].

Naia™ fibers are a type of sustainable textile material made from cellulose derived primarily from sustainably harvested wood pulp using a closed-loop process that minimizes waste and uses renewable resources. Naia™ fibers are notable for their environmental benefits and their smooth, soft texture, breathability, excellent moisture wicking, biodegradation and excellent drape. They making them suitable for a variety of textile products, including clothing, activewear, and home textiles, offering an eco-friendlier alternative to synthetic and conventional fibers [130, 139].

Spinnova fibers are made using a unique process that transforms cellulose from wood, agricultural residues, or other plant materials directly into fibers without the need for harmful chemicals or extensive water usage. The process uses mechanical and minimal chemical treatments to convert cellulose into fibers, making it both environmentally friendly and efficient. Spinnova fibers offer excellent strength, high durability, biodegradability, breathability, and comfort. These fibers are versatile and can be incorporated into a range of products, including clothing, sportswear, and home textiles [131, 140].

Refibra fibers are made from recycled cotton scraps and wood pulp. The process involves collecting cotton waste from textile production and combining it with sustainably sourced wood pulp to create new fibers. Refibra fibers are designed to be more sustainable compared to traditional textile fibers. The use of recycled cotton reduces textile waste and lowers the environmental impact of raw material extraction. Additionally, the wood pulp used is sourced from responsibly managed forests. Overall, Refibra fibers represent an important step toward creating a more sustainable and circular fashion industry by incorporating recycled materials into textile production. Refibra fibers are known for their softness, comfortable feeling, breathability, biodegradability, strength and durability, similar to lyocell. They offer similar performance characteristics to conventional fibers, making them suitable for a wide range of textile applications, including clothing, home textiles, and activewear [22, 132, 141].

Bast fiber-based regenerated cellulose are a category of sustainable textiles made from the fibers of plants that have bast (or outer) layers, such as hemp, flax, jute, and ramie. These fibers are processed and regenerated into cellulose fibers, offering an eco-friendly alternative to traditional textile materials. Bast fibers are first cleaned and processed to extract cellulose. The cellulose is then dissolved in a solution (such as ionic liquids or other solvents) to form a viscous liquid. The dissolved cellulose is spun into fibers through a process called regeneration, where the solution is extruded through spinnerets and solidified into fibers. This can be done using methods similar to those used for producing viscose or lyocell fibers. Using bast fibers is considered more sustainable compared to some synthetic fibers because these plants grow quickly, require fewer pesticides, and can be cultivated with less water. Regenerated cellulose fibers are generally biodegradable, reducing their environmental impact compared to synthetic alternatives. Bast fibers are valued for their strong tensile properties, long-lasting durability, and excellent breathability. Their effective moisture-wicking ability enhances comfort, making them well-suited for textile applications. They are used in clothing and accessories, and are also suitable for products like bed linens, curtains, and upholstery [23, 133, 142].

14.7 Applications of Regenerated Cellulose Fibers

The regenerated cellulose market was valued at approximately USD 18.0 billion in 2022 and is expected to grow to USD 27.3 billion by 2027, with a compound annual growth rate (CAGR) of 8.7%. This market includes both fibers and films made from regenerated cellulose. Fibers are expected to be the fastest-growing segment during the forecast period, driven by increasing demand for eco-friendly products that promote sustainability. Viscose remains the dominant product in the regenerated cellulose fiber market due to its widespread use in apparel and home textiles. However, the market for lyocell and other sustainable fibers is growing rapidly. Modal and lyocell are experiencing increased demand due to their higher performance and environmental benefits. Newer innovations like Ioncell, Naia™, and Spinnova are gaining traction as consumers and manufacturers seek more sustainable options [143].

Applications of regenerated cellulose fibers in textiles may be classified as follows:

- Fashion apparel: casual wear (t-shirts, skirts, pants), formal wear (dresses, suits, evening gowns); lingerie and underwear (bras panties, slips, nightgowns).
- Activewear and sportswear: gym clothing (leggings, sports bras, tank tops), outdoor gear (jackets, hiking pants, base layer); performance wear (running shirts, cycling shorts, yoga wear).

Table 14.3 Possible areas of applications for regenerated cellulose fibers in textiles [23, 51, 57, 134–137, 144, 145].

Fiber	Applications	Appropriate fiber characteristics
Viscose	Fashion apparel	Soft feel, good drape
	Home textiles	Aesthetic appeal, comfort
	Medical textiles as non-wovens	Absorbency, softness
Modal	Underwear and loungewear	Softness, moisture wicking
	Activewear	Strength, dimensional stability
	Home textiles	Durability, comfort
Lyocell, Tencel	High-fashion and casual wear	Smooth texture, high strength
	Activewear	Moisture management, breathability
	Home textiles	Expensive look, durability
	Medical textiles	Biocompatibility, absorbency
Cupro	High-end fashion	Silky texture, drape
	Lingerie	Softness, smoothness
Acetate	Formal wear	Shiny appearance, drape
	Linings	Smooth texture, ease of care
	Decorative fabrics	Luster, aesthetic attraction
Ioncell	High-quality apparel	Strength, biodegradability
	Home textiles	Durability, eco-friendly
	Technical textiles	Strong, sustainable
Naia™	Fashion apparel	Silky feel, drape
	Lingerie, loungewear	Softness, moisture wicking
	Home textiles	Comfort, sustainability

(Continued)

Table 14.3 Possible areas of applications for regenerated cellulose fibers in textiles [23, 51, 57, 134–137, 144, 145]. (*Continued*)

Fiber	Applications	Appropriate fiber characteristics
Spinnova	Casual and outdoor clothing	Durability, low environmental impact
	Home textiles	Sustainability
	Blends	Used to enhance performance and sustainability
Refibra™	Sustainable fashion	Sustainability
	Activewear	Strength, moisture management
	Home textiles	Sustainability, comfort
Bast fiber-based regenerated cellulose	Sustainable fashion	Natural origin, biodegradability
	Outdoor and technical textiles	Strength, moisture wicking
	Blends and composites	Used to enhance performance

- Home textiles: bedding (sheets, pillowcases, duvet covers); bath textiles (towels, bathrobes, washcloths); curtains and upholstery (drapes, cushion covers, furniture upholstery);
- Technical textiles: medical textiles (wound dressings, surgical gowns, hygiene products); industrial textiles (filters, conveyor belts, protective clothing); geotextiles (erosion control fabrics, soil stabilization mats).
- Luxury and high-end textiles: (silk alternatives (scarves, ties, luxury lining), designer fabrics (high-end fashion garments, bespoke clothing); eco-friendly luxury textiles (sustainable luxury garments and accessories).

Based on recent market research, the approximate consumption of regenerated cellulose fibers by application is as follows [143]: apparel, 60%; home textiles, 25%; technical textiles, 10%; medical textiles, 3%; others, 2%.

Regenerated cellulose fibers have revolutionized the textile industry by offering sustainable and high-performance alternatives to both natural and synthetic fibers. Their unique combination of properties and adaptability allows them to be used across various textile applications, spanning casual wear, high-end fashion, and specialized technical fabric (see Table 14.3).

Regenerated cellulose fibers offer a wide range of applications across the textile industry, from fashion and casual wear to home textiles and technical applications. The development of new fibers like Ioncell, Naia™, Spinnova, and Refibra™ reflects the industry's ongoing commitment to sustainability and high performance.

14.8 Recycling of Regenerated Cellulose Fibers

Recycling regenerated cellulose fibers is an important aspect of sustainable textile production. Viscose is made from wood pulp, primarily from eucalyptus, pine, and bamboo. The sustainability of this raw material depends on responsible forestry practices. Viscose rayon can be recycled using chemical or mechanical techniques. Chemical recycling involves dissolving the fibers in a solvent to extract cellulose, which is then regenerated into new fibers or products. While this method maintains fiber quality, it is complex and expensive. The traditional viscose process relies on hazardous chemicals like carbon disulfide, raising environmental and health concerns. In contrast, mechanical recycling shreds and re-spins the fibers into new textiles, though this approach often leads to diminished fiber strength, flexibility, and overall performance [146–150]. Innovations in recycling techniques for viscose rayon, such as enzyme-based processes, offer promising improvements in efficiency and effectiveness. Establishing closed-loop recycling systems, where viscose fibers are continuously reused within the same supply chain, could enhance sustainability while significantly reducing waste [147–149].

Lyocell fibers, derived from sustainably managed wood pulp, are well-suited for closed-loop recycling systems. They can be dissolved and re-spun into new fibers while largely retaining their original properties. This method significantly reduces pollution compared to viscose recycling and operates efficiently, with 99% of solvents being reclaimed, thus minimizing environmental impact. Lyocell can also be chemically recycled by dissolving and regenerating the cellulose or mechanically recycled by grinding the fibers into smaller pieces and re-spinning them. However, there are challenges in recycling lyocell fibers. The closed-loop process can be more expensive than chemical and mechanical recycling. Maintaining the quality of chemically recycled lyocell to match that of virgin fibers is another key challenge. Mechanical recycling, while less costly, can lead to a decrease in fiber quality [22, 47, 82, 150]. Advancements in recycling technologies, including enzyme-based approaches and optimized chemical processes, could enhance the efficiency and effectiveness of lyocell fiber recycling. Implementing closed-loop systems, where fibers are continuously repurposed within the same supply chain, would further support sustainability and minimize waste [150, 151].

Modal fibers are derived from beech wood, a renewable resource. Modal fibers production uses fewer toxic chemicals and is more environmentally friendly than viscose fibers due to reduced chemical waste and lower water consumption, despite still involving chemical processing. Modal fibers can be recycled through chemical or mechanical methods. Chemical recycling, akin to lyocell processing, dissolves and regenerates the fibers, preserving their quality but requiring a more intricate and expensive process. Mechanical recycling, in contrast, involves breaking down the fibers and re-spinning them into new materials like non-woven fabrics [64, 150, 152]. Advanced recycling methods, like enzyme-based processes and closed-loop systems, can enhance the effectiveness of modal fiber recycling by overcoming challenges associated with traditional techniques. Integrating circular economy practices, where modal fibers are continuously reused within the production cycle, can improve sustainability and significantly reduce waste [153].

Similar to other regenerated cellulose fibers, cupro can undergo mechanical or chemical recycling. Mechanical recycling, which breaks down and re-spins the fibers, often leads to a reduction in quality. Chemical recycling, while generally more efficient in preserving fiber integrity, is a more intricate process and may raise environmental concerns due to the use of specific chemicals [64, 154]. Advanced recycling methods, including solvent-based techniques and closed-loop systems, can enhance both the efficiency and ecological footprint of cupro fiber recycling. Implementing circular economy strategies, where cupro fibers are repeatedly reused within the supply chain, further promotes sustainability and reduces waste [155].

Mechanical recycling of acetate fibers, like other regenerated cellulose fibers, involves grinding and re-spinning, though this can lead to a reduction in fiber quality. Preserving key properties such as strength, flexibility, and moisture management remains a challenge. Refining chemical processes and adopting solvent-based recycling techniques, similar to cupro fibers, can enhance both efficiency and environmental sustainability. Implementing closed-loop systems, where acetate fibers are continuously repurposed within the supply chain, further supports waste reduction and sustainability efforts [150, 156, 157].

Ioncell fibers are produced from wood pulp and agricultural residues, making them a renewable resource. The recycling process for Ioncell fibers involves dissolving and regenerating the fibers, which can be efficiently reused multiple times. Ioncell fibers are produced using ionic liquids, which are recyclable and less harmful compared to traditional solvents. This makes Ioncell fibers easier to recycle and more sustainable [134, 138].

Naia™ fibers are made from sustainably sourced wood pulp. The recycling process of Naia™ fibers involves reclaiming the fibers, reprocessing them into new fibers, and integrating them into various textile products. Naia™ fibers are produced using a closed-loop process, and efforts are being made to enhance the recyclability of these fibers [135].

Spinnova fibers are processed through mechanical methods, allowing for efficient recycling and minimal environmental impact. Spinnova fibers are designed to be fully recyclable and are made from wood or agricultural waste. The recycling process is integrated into the production system, enhancing circularity [131, 136, 140].

The recycling process of Refibra™ fibers involves integrating cotton waste into the production of new fibers, creating a circular system. Refibra™ fibers are made from recycled cotton scraps combined with new cellulose. This approach reduces waste and improves the sustainability of textile production [22, 137, 141].

14.9 Future Perspectives

Regenerated cellulose fibers offer a sustainable and biodegradable option, serving as an eco-friendly substitute for synthetic materials like polyester and polyamide or nylon. Unlike synthetic fibers, which are heavily reliant on fossil-based resources, RCFs are derived from bio-based sources such as wood pulp and cotton linters. Their biodegradability offers a solution to the environmental challenges associated with textile waste and microplastic pollution, positioning RCFs as a more sustainable option in fiber production. RCFs play a significant role in advancing the circular economy within the textile industry. Their integration into circular fashion initiatives, along with the use of eco-friendly dyeing techniques, supports the industry's transition toward sustainability. The lyocell production process, for example, operates as an efficient closed-loop system, recovering over 99% of the solvent, thereby reducing both chemical waste and water consumption. Recent technological developments, such as the Ioncell and alkali/urea processes, further enhance the environmental sustainability of RCF production. These methods employ environmentally friendly solvents, improving the overall sustainability of fiber production. In particular, the Ioncell process produces fibers with superior mechanical properties, including increased strength and durability, making them suitable for applications such as high-performance fashion, automotive components, and technical fabrics. Similarly, fibers produced *via* the alkali/urea process show promise in applications requiring moisture management and comfort, such as wearable technology and smart textiles. Advancements in fiber formation techniques, including electrospinning, melt spinning of biopolymers, and microbial fiber production, have improved the efficiency of RCF production. These methods provide greater control over fiber structure and functionality, enabling the production of fibers with enhanced mechanical properties, such as improved durability, conductivity, and antimicrobial features. Despite these advantages, challenges remain in the production of RCFs, particularly regarding energy consumption and the use of chemical solvents, as seen in processes like viscose production. However, innovations in solvent systems, such as the use of ionic liquids and N-Methylmorpholine N-oxide (NMMO) in the lyocell process, are helping to mitigate these issues. The adoption of more energy-efficient production methods and enhanced solvent recycling systems can further reduce the environmental impact of RCF production. Recycling regenerated cellulose fibers is crucial for sustainable textile production. By implementing closed-loop systems, reducing the reliance on virgin materials, and integrating advanced recycling technologies, the industry can significantly reduce its environmental footprint. Technologies like Ioncell and solvent-based recycling systems hold significant potential for enabling a circular economy in textiles. As these technologies continue to evolve, the ability to recycle RCFs will become increasingly important for the industry's transition toward a more sustainable and resource-efficient future.

AI disclosure statement: The author acknowledges the use of AI-based tools for minor language editing or grammatical correction. However, all intellectual contributions, analyses, and conclusions presented in this chapter are entirely the author's own.

References

1. Sherif, F.S., A New Prospects to Enhance the Commercial and Economical Status in Textile Industry. *Int. Des. J.*, 6, 1, 141–148, 2016.
2. Schwarz, I. and Kovačević, S., Textile application: From need to imagination, in: *Textiles for advanced applications*, IntechOpen, London, UK, 2017.

3. Hossain, M.T., Shahid, M.A., Limon, M.G.M., Hossain, I., Mahmud, N., Techniques, applications, and challenges in textiles for sustainable future. *J. Open Innov. Technol. Mark. Complex*, 100230, 2024.
4. Uppal, N., *et al.*, Cellulosic fibres-based epoxy composites: From bioresources to a circular economy. *Ind. Crops Prod.*, 182, 114895, 2022.
5. Hassan, S.S., Williams, G.A., Jaiswal, A.K., Moving towards the second generation of lignocellulosic biorefineries in the EU: Drivers, challenges, and opportunities. *Renewable Sustainable Energy Rev.*, 101, 590–599, 2019.
6. Wang, S., Lu, A., Zhang, L., Recent advances in regenerated cellulose materials. *Prog. Polym. Sci.*, *53*, 169–206, 2016.
7. Liu, W., *et al.*, Eco-friendly post-consumer cotton waste recycling for regenerated cellulose fibers. *Carbohydr. Polym.*, 206, 141–148, 2019.
8. De Oliveira, N., Cardoso, G., *et al.*, Assessment of the Eco-Efficiency of the Circular Economy in the Recovery of Cellulose from the Shredding of Textile Waste. *Polymers*, 14.7, 1317, 2022.
9. Furferi, R., Volpe, Y., Mantellassi, F., Circular economy guidelines for the textile industry. *Sustainability*, 14.17, 11111, 2022.
10. Bao, H., Hong, Y., Yan, T., Xie, X., Zeng, X., A systematic review of biodegradable materials in the textile and apparel industry. *J. Text. Inst.*, 115, 7, 1173–1192, 2024.
11. Chang, C., Ginn, B., Livingston, N.K., Yao, Z., Slavin, B., King, M.W., Chung, S., Mao, H.Q., Medical fibers and biotextiles, in: *Biomaterials Science*, pp. 575–600, Academic Press, Amsterdam, NL, 2020.
12. Klemm, D., Heublein, B., Fink, H.P., Bohn, A., Cellulose: fascinating biopolymer and sustainable raw material. *Angew. Chem. Int. Ed.*, *44*, 22, 3358–3393, 2005.
13. Heinze, T., Cellulose: Structure and properties, in: *Cellulose chemistry and properties: fibers, nanocelluloses and advanced materials*, pp. 1–52, 2016.
14. Shaghaleh, H., Xu, X., Wang, S., Current progress in production of biopolymeric materials based on cellulose, cellulose nanofibers, and cellulose derivatives. *RSC Adv.*, 8.2, 825–842, 2018.
15. Sundarraj, A.A. and Ranganathan, T.V.A., review on cellulose and its utilization from agro-industrial waste. *Drug Invent. Today*, 10, 1, 89–94, 2018.
16. Tofanica, B.M., Belosinschi, D., Volf, I., Gels, Aerogels and Hydrogels: A Challenge for the Cellulose-Based Product Industries. *Gels*, 8, 8, 497, 2022.
17. Abdul Khalil, H.P.S., Adnan, A.S., Yahya, E.B., Olaiya, N.G., Safrida, S., Hossain, M.S., Balakrishnan, V., Gopakumar, D.A., Abdullah, C.K., Oyekanmi, A.A., Pasquini, D., A review on plant cellulose nanofibre-based aerogels for biomedical applications. *Polym.*, 12, 8, 1759, 2020.
18. Aziz, T., *et al.*, Developing multifunctional cellulose derivatives for environmental and biomedical applications: Insights into modification processes and advanced material properties. *Int. J. Biol. Macromol.*, 134695, 2024.
19. Kocić, A., Bizjak, M., Popović, D., Poparić, G.B., Stanković, S.B., UV protection afforded by textile fabrics made of natural and regenerated cellulose fibres. *J. Cleaner Prod.*, *228*, 1229–1237, 2019.

20. Ibrahim, N.A., Eid, B.M., Abd El-Ghany, N.A., Mabrouk, E.M., Polyfunctional cotton cellulose fabric using proper biopolymers and active ingredients. *J. Text. Inst.*, 2020.
21. Yue, Y., Han, G., Wu, Q., Transitional properties of cotton fibers from cellulose I to cellulose II structure. *BioResources*, 8, 4, 6460–6471, 2013.
22. Periyasamy, A.P. and Militky, J., Sustainability in regenerated textile fibers, in: *Sustainability in the Textile and Apparel Industries: Sourcing Synthetic and Novel Alternative Raw Materials*, pp. 63–95, 2020.
23. Ramamoorthy, S.K., Skrifvars, M., Persson, A.A., review of natural fibers used in biocomposites: Plant, animal and regenerated cellulose fibers. *Polym. Rev.*, 55, 1, 107–162, 2015.
24. Morris, H. and Murray, R., Medical textiles. *Text. Prog.*, 52, 1-2, 1–127, 2020.
25. Tu, H., Zhu, M., Duan, B., Zhang, L., Recent progress in high-strength and robust regenerated cellulose materials. *Adv. Mater.*, 33, 28, 2000682, 2021.
26. Mark, R.E., Viscose fiber production and its historical development. *J. Text. Inst.*, 92, 2, 68–75, 2001.
27. Hämmerle, F.M., The cellulose gap (the future of cellulose fibers). *Lenzinger Berichte.*, 89, 12–21, 2011.
28. Shaikh, T., Chaudhari, S., Varma, A., Viscose rayon: a legendary development in the manmade textile. *Int. J. Eng. Res.*, 2, 5, 675–680, 2012.
29. Vartiainen, J., Vähä-Nissi, M., Harlin, A., Biopolymer films and coatings in packaging applications—A review of recent developments. *J. Appl. Polym. Sci.*, 134, 15, 44367, 2017.
30. Kuen, T.H., Lau, T.Y., Ng, L.T.A., Study on the Sterilization of Regenerated Cellulose Fiber for Medical Textiles. *J. Ind. Text.*, 47, 8, 1803–1820, 2018.
31. Jiang, G., Huang, W., Li, L., Wang, X., Pang, F., Zhang, Y., Wang, H., Structure and properties of regenerated cellulose fibers from different technology processes. *Carbohydr. Polym.*, 87, 3, 2012–2018, 2012.
32. Fink, H.-P., Weigel, P., Purz, H.J., Ganster, J., Structure formation of regenerated cellulose materials from NMMO-solutions. *Prog. Polym. Sci.*, 26, 9, 1473–1524, 2001.
33. Kreze, T. and Malej, S., Structural characteristics of new and conventional regenerated cellulosic fibers. *Text. Res. J.*, 73, 8, 675–684, 2003.
34. Wang, H., Sixta, H., Lorenz, M., Properties and structure of viscose fibers spun from NMMO-cellulose solution. *Cellulose*, 11, 2, 163–172, 2004.
35. Kim, H.G., Bai, B.C., In, S.J., Lee, Y.S., Effects of an inorganic ammonium salt treatment on the flame-retardant performance of Lyocell fibers. *Carbon Lett.*, 17, 1, 74–78, 2016.
36. Liu, X.-H., Zhang, Y.-G., Cheng, B.-W., Ren, Y.-L., Zhang, Q.-Y., Ding, C., Peng, B., Preparation of durable and flame retardant Lyocell fibers by a one-pot chemical treatment. *Cellulose*, 25, 11, 6745–6758, 2018.
37. Sixta, H., Michud, A., Hauru, L., Asaadi, S., Ma, Y., King, A.W., Hummel, M., Ioncell-F: A high-strength regenerated cellulose fibre. *Nord. Pulp Pap. Res. J.*, *30*, 1, 43–57, 2015.

38. Shankar, S. and Rhim, J.W., Preparation of nanocellulose from microcrystalline cellulose: The effect on antimicrobial activity. *Carbohydr. Polym.*, 200, 163–172, 2018.
39. Sixta, H., *Handbook of Pulp*, Wiley-VCH, Weinheim, Germany2006.
40. Shen, L., Worrell, E., Patel, M.K., Environmental impact assessment of man-made cellulose fibers. *Resour. Conserv. Recy.*, 55, 2, 260–274, 2010.
41. Zhang, S., Chen, C., Duan, C., Hu, H., Li, H., Li, J., Liu, Y., Ma, X., Stavik, J., Ni, Y., Regenerated cellulose by the lyocell process, a brief review of the process and properties. *BioResources.*, 13, 2, 4577–4592, 2018.
42. Kim, H.S., Yun, C., Lee, D.S., Characterization of the mechanical properties of lyocell and viscose fibers. *J. Ind. Text.*, 45, 6, 875–890, 2016.
43. Müssig, J. and Haag, K., The tensile properties of lyocell made from wood pulp and from bamboo pulp fibers. *Compos. A Appl. Sci. Manuf.*, 42, 9, 1185–1192, 2011.
44. Rosenau, T., Potthast, A., Sixta, H., Kosma, P., The chemistry of side reactions and byproduct formation in the system NMMO/cellulose (Lyocell process). *Prog. Polym. Sci.*, 26, 9, 1763–1837, 2001.
45. Eichhorn, S.J., Baillie, C.A., Zafeiropoulos, N., Mwaikambo, L.Y., Ansell, M.P., Dufresne, A., Entwistle, K.M., Review: Current international research into cellulose nanofibres and nanocomposites. *J. Mater. Sci.*, 36, 9, 2107–2131, 2001.
46. Carrillo, F., Colom, X., Valldeperas, J., Evans, D., Huson, M., Church, J., Structural characterization and properties of lyocell fibers after fibrillation and enzymatic defibrillation finishing treatments. *Text. Res. J.*, 73, 11, 1024–1030, 2003.
47. Edgar, K.J. and Zhang, H., Antibacterial modification of Lyocell fiber: A review. *Carbohydr. Polym.*, 250, 116932, 2020.
48. Wang, X., Wang, S., Li, Y., Jin, X., Dong, C., Preparation and properties of low fibrillated antibacterial Lyocell fiber. *Arab. J. Chem*, 105658, 2024.
49. Parajuli, P., Acharya, S., Rumi, S.S., Hossain, M.T., Abidi, N., Regenerated cellulose in textiles: Rayon, lyocell, modal and other fibres, in: *Fundamentals of natural fibres and textiles*, pp. 87–110, Woodhead Publishing, Oxford, UK, 2021.
50. Lenzing, A.G., *Lenzing Modal® - The Fiber with a History of Innovation*, Lenzing AG, Lenzing, Austria, 2020, Available at: Lenzing Website.
51. Woodings, C., *Regenerated Cellulose Fibres*, Woodhead Publishing, Cambridge, UK, 2001.
52. Shen, H., Sun, T., Zhou, J., Recent progress in regenerated cellulose fibers by wet spinning. *Macromo. Mater. Eng.*, 308, 10, 2300089, 2023.
53. Kamide, K. and Nishiyama, K., Cuprammonium processes. *Regenerated Cellul. Fibers*, 88–155, 2001.
54. Seisl, S. and Hengstmann, R., Manmade Cellulosic Fibers (MMCF)—A Historical Introduction and Existing Solutions to a More Sustainable Production, in: *Sustainable Textile and Fashion Value Chains*, A. Matthes, K. Beyer, H. Cebulla, M.G. Arnold, A. Schumann (Eds.), Springer, Cham, 2021, https://doi.org/10.1007/978-3-030-22018-1_1.

55. Das, M., Man-made cellulose fibre reinforcements (MMCFR), in: *Biocomposites for High-Performance Applications*, pp. 23–55, Woodhead Publishing, Oxford, UK, 2017.
56. Hearle, J.W., Physical structure and fibre properties. *Regenerated Cellul. Fibres*, 18, 199, 2001.
57. Oral, O., Dirgar, E., İlleez, A.A., Cupro fibers, properties and usage areas, in: *3rd International Congress of Innovative Textiles ICONTEX 2022*, pp. 18–19, p. 26, May 2022.
58. Chen, J., Xu, J., Wang, K., Cao, X., Sun, R., Cellulose acetate fibers prepared from different raw materials with rapid synthesis method. *Carbohydr. Polym.*, 137, 685–692, 2016.
59. Fischer, S., Thümmler, K., Volkert, B., Hettrich, K., Schmidt, I., Fischer, K., Properties and applications of cellulose acetate, in: *Macromolecular symposia*, vol. 262, pp. 89–96, Wiley-VCH Verlag, Weinheim, 2008.
60. Wsoo, M.A., Shahir, S., Bohari, S.P.M., Nayan, N.H.M., Abd Razak, S.I., A review on the properties of electrospun cellulose acetate and its application in drug delivery systems: A new perspective. *Carbohyd. Res.*, 491, 107978, 2020.
61. https://textileexchange.org/app/uploads/2023/11/Materials-Market-Report-2023.pdf. Accessed 2024-08-27.
62. Kim, T., Kim, D., Park, Y., Recent progress in regenerated fibers for "green" textile products. *J. Clean. Prod.*, 376, 134226, 2022.
63. Shabbir, M. and Mohammad, F., Sustainable production of regenerated cellulosic fibres, in: *Sustainable Fibres and Textiles*, pp. 171–189, Woodhead Publishing, Oxford, UK, 2017.
64. Ma, Y., Nasri-Nasrabadi, B., You, X., Wang, X., Rainey, T.J., Byrne, N., Regenerated cellulose fibers wetspun from different waste cellulose types. *J. Nat. Fibers*, 18, 12, 2338–2350, 2021.
65. De Silva, R., Vongsanga, K., Wang, X., Byrne, N., Understanding key wet spinning parameters in an ionic liquid spun regenerated cellulosic fibre. *Cellulose*, 23, 4, 2741–51, 2016.
66. Hauru, L.K., Hummel, M., King, A.W., Kilpeläinen, I., Sixta, H., Role of solvent parameters in the regeneration of cellulose from ionic liquid solutions. *Biomacromolecules*, 13, 9, 2896–905, 2012.
67. Ma, Y., Rosson, L., Wang, X., Byrne, N., Upcycling of waste textiles into regenerated cellulose fibres: Impact of pretreatments. *J. Text. Inst.*, 111, 5, 630–638, 2020.
68. Felgueiras, C., Azoia, N.G., Gonçalves, C., Gama, M., Dourado, F., Trends on the cellulose-based textiles: raw materials and technologies. *Front. Bioeng. Biotechnol.*, 9, 608826, 2021.
69. Gorade, V., Chaudhary, B., Parmaj, O., Kale, R., Preparation and characterization of chitosan/viscose rayon filament biocomposite. *J. Nat. Fibers*, 19, 4, 1189–1200, 2022.

70. Eid, B.M. and Ibrahim, N.A., Recent developments in sustainable finishing of cellulosic textiles employing biotechnology. *J. Clean. Prod.*, 284, 124701, 2021.
71. Chen, J., Synthetic textile fibers: regenerated cellulose fibers, in: *Textiles and fashion*, pp. 79–95, Woodhead publishing, Oxford, UK, 2015.
72. Lavanya, D.K.P.K., Kulkarni, P.K., Dixit, M., Raavi, P.K., Krishna, L.N.V., Sources of cellulose and their applications—A review. *Int. J. Drug Formul. Res*, 2, 6, 19–38, 2011.
73. Wilkes, A.G., The viscose process. *Regenerated Cellul. Fibres*, 37–61, 2001.
74. Fechter, C., Fischer, S., Reimann, F., Brelid, H., Heinze, T., Influence of pulp characteristics on the properties of alkali cellulose. *Cellulose*, 27, 7227–7241, 2020.
75. Hon, D.N.S., Cellulose and its derivatives: structures, reactions, and medical uses, in: *Polysaccharides in medicinal applications*, pp. 87–105, Routledge, New York, NY, 2017.
76. Dyer, J. and Smith, F.R., New Methods for Studying Particles in Viscose and Their Effects on Viscose Rayon Production. Textile and Paper Chemistry and Technology, Chapter 1, in: *ACS Symposium Series*, vol. 49, pp. 3–19, 1977, published online 2009.
77. Ozipek, B. and Karakas, H., Wet spinning of synthetic polymer fibers, in: *Advances in filament yarn spinning of textiles and polymers*, pp. 174–186, Woodhead Publishing, Cambridge, UK, 2014.
78. El Seoud, O.A., Kostag, M., Jedvert, K., Malek, N.I., Cellulose regeneration and chemical recycling: closing the "cellulose gap" using environmentally benign solvents. *Macromol. Mater. Eng.*, 305, 4, 1900832, 2020.
79. Mondal, M.I.H., Ahmed, F., Rahman, M.H., Surface Modification of Viscose Fabric with Silane Coupling Agents-An Aspect of Comfort Properties. *ChemistrySelect*, 9, 17, e202400727, 2024.
80. Emam, H.E., El-Hawary, N.S., Ahmed, H.B., Green technology for durable finishing of viscose fibers *via* self-formation of AuNPs. *Int. J. Biol. Macromol.*, 96, 697–705, 2017.
81. Sixta, H., Iakovlev, M., Testova, L., Roselli, A., Hummel, M., Borrega, M., van Heiningen., A., Carmen, F., Schottenberger, H., Novel concepts of dissolving pulp production. *Cellul.*, 20, 1547–1561, 2013.
82. Jiang, X., Bai, Y., Chen, X., Liu, W.A., review on raw materials, commercial production and properties of lyocell fiber. *J. Bioresour. Bioprod.*, 5, 1, 16–25, 2020.
83. Perepelkin, K.E.E., Lyocell fibres based on direct dissolution of cellulose in N-methylmorpholine N-oxide: development and prospects. *Fibre Chem.*, 39, 2, 163–172, 2007.
84. Sayyed, A.J., Gupta, D., Deshmukh, N.A., Mohite, L.V., Pinjari, D.V., Influence of intensified cellulose dissolution process on spinning and properties of lyocell fibres. *Chem. Eng. Process.-Process Intensif.*, 155, 108063, 2020.

85. Jadhav, S., Lidhure, A., Thakre, S., Ganvir, V., Modified Lyocell process to improve dissolution of cellulosic pulp and pulp blends in NMMO solvent. *Cellulose*, 28, 973–990, 2021.
86. Foroughi, F., Rezvani Ghomi, E., Morshedi Dehaghi, F., Borayek, R., Ramakrishna, S., A review on the life cycle assessment of cellulose: From properties to the potential of making it a low carbon material. *Materials*, 14, 4, 714, 2021.
87. Röder, T., Moosbauer, J., Wöss, K., Schlader, S., Kraft, G., Man-made cellulose fibres–a comparison based on morphology and mechanical properties. *Lenzinger Ber.*, 91, 7–12, 2013.
88. Bali, G., Meng, X., Deneff, J., II, Sun, Q., Ragauskas, A.J., The effect of alkaline pretreatment methods on cellulose structure and accessibility. *ChemSusChem*, 8, 2, 275–279, 2015.
89. Colom, X. and Carrillo, F., Crystallinity changes in lyocell and viscose-type fibres by caustic treatment. *Eur. Polym. J.*, 38, 11, 2225–2230, 2002.
90. Singh, S.C. and Murthy, Z.V.P., Study of cellulosic fibres morphological features and their modifications using hemicelluloses. *Cellulose*, 24, 8, 3119–3130, 2017.
91. Wang, B., Nie, Y., Kang, Z., Liu, X., Effects of coagulating conditions on the crystallinity, orientation and mechanical properties of regenerated cellulose fibers. *Int. J. Biol. Macromol.*, 225, 1374–1383, 2023.
92. Isogai, A. and Bergström, L., Preparation of cellulose nanofibers using green and sustainable chemistry. *Curr. Opin. Green Sust. Chem.*, 12, 15–21, 2018.
93. Hasan, K.F., Champramary, S., Al Hasan, K.N., Indic, B., Ahmed, T., Pervez, M.N., Horvath, P.G., Bak, M., Sandor, B., Hofmann, T., Tolvak, L., Horvath, A., Koczan, Z., Sipos, G., Alpar, T., Bejó, L., Eco-friendly production of cellulosic fibers from Scots pine wood and sustainable nanosilver modification: A path toward sustainability. *Res. Eng*, 19, 101244, 2023.
94. Strnad, S., Šauperl, O., Fras-Zemljič, L., Cellulose fibres functionalised by chitosan: characterization and application. *Biopolymers*, 181–200, 2010.
95. Miyamoto, I., Matsuoka, Y., Matsui, T., Saito, M., Okajima, K., Studies on structure of cuprammonium cellulose III. Structure of regenerated cellulose treated by cuprammonium solution. *Polym. J.*, 28, 3, 276–281, 1996.
96. Karthik, T. and Gopalakrishnan, D., Environmental analysis of textile value chain: an overview, in: *Roadmap to sustainable textiles and clothing: Environmental and social aspects of textiles and clothing supply chain*, pp. 153–188, Springer, Singapore, 2014.
97. Ide, S., Filter made of cuprammonium regenerated cellulose for virus removal: A mini-review. *Cellulose*, 29, 5, 2779–2793, 2022.
98. Cheng, H.N., Dowd, M.K., Selling, G.W., Biswas, A., Synthesis of cellulose acetate from cotton byproducts. *Carbohyd. Polym.*, 80, 2, 449–452, 2010.
99. Homem, N.C. and Amorim, M.T.P., Synthesis of cellulose acetate using as raw material textile wastes. *Mater. Today Proc.*, 31, S315–S317, 2020.
100. Nawaz, H., He, A., Wu, Z., Wang, X., Jiang, Y., Ullah, A., Xu, F., Xie, F., Revisiting various mechanistic approaches for cellulose dissolution in

different solvent systems: A comprehensive review. *Int. J. Biolog. Macromol.*, 133012, 2024.
101. Swapnil, S.I., Datta, N., Mahmud, M.M., Jahan, R.A., Arafat, M.T., Morphology, mechanical, and physical properties of wet-spun cellulose acetate fiber in different solvent-coagulant systems and *in-situ* crosslinked environment. *J. Appl. Polym. Sci.*, 138, 18, 50358, 2021.
102. Slejko, E.A., Tuan, A., Scuor, N., From waste to value: Characterization of recycled cellulose acetate for sustainable waste management. *Waste Manag. Bull.*, 1, 4, 67–73, 2024.
103. Wolfs, J. and Meier, M.A., A more sustainable synthesis approach for cellulose acetate using the DBU/CO 2 switchable solvent system. *Green Chem.*, 23, 12, 4410–4420, 2021.
104. Kosan, B., Michels, C., Meister, F., Dissolution and forming of cellulose with ionic liquids. *Cellulose*, 15, 59–66, 2008.
105. Zhou, L., Kang, Z., Nie, Y., Li, L., Fabrication of regenerated cellulose fibers with good strength and biocompatibility from green spinning process of ionic liquid. *Macromol. Mat. Eng.*, 306, 4, 2000741, 2021.
106. Wang, C.G., Li, N., Wu, G., Lin, T.T., Lee, A.M.X., Yang, S.W., Li, Z., Luo, H.K., Carbon Dioxide Mediated Cellulose Dissolution and Derivatization to Cellulose Carbonates in a Low-pressure System. *Carbohydr. Polym. Technol. Appl.*, 3, 100186, 2022.
107. Zhang, J., Wu, J., Yu, J., Zhang, X., He, J., Zhang, J., Application of ionic liquids for dissolving cellulose and fabricating cellulose-based materials: State of the art and future trends. *Mater. Chem. Front.*, 1, 7, 1273–1290, 2017.
108. Xiong, B., Zhao, P., Hu, K., Zhang, L., Cheng, G., Dissolution of cellulose in aqueous NaOH/urea solution: role of urea. *Cellulose*, 21, 1183–1192, 2014.
109. Kayseri, GO., Bozdoğan, F., Hes, L., Performance properties of regenerated cellulose fibers. *Text. App.*, 20, 3, 208–212, 2010.
110. Maher, R.R. and Warsman, R.H., *The Chemistry of Textile Fibers*, 2nd Edition, pp. 127–128, Cambridge, UK: Royal Society of Chemistry, 2015, ISBN: 978-1-78262-023-5.
111. Abu-Rous, M., Cafuta, D., Schuster, K.C., The role of the enhanced moisture absorption of regenerated cellulose fibers (MMCF) in textile comfort and the thermoregulation of the body. *Lenzinger Berichte.*, 98, 5–13, 2023.
112. Kreze, T., Jeler, S., Strnad, S., Correlation between structure characteristics and adsorption properties of regenerated cellulose fibers. *Mater. Res. Innov.*, 5, 277–283, 2002.
113. Gizaw, M., Thompson, J., Faglie, A., Lee, S.Y., Neuenschwander, P., Chou, S.F., Electrospun fibers as a dressing material for drug and biological agent delivery in wound healing applications. *Bioeng.*, 5, 1, 9, 2018.
114. Zemljič, L.F., Kreže, T., Strnad, S., Šauperl, O., Vesel, A., Functional Vellulose Fibers for Hygienic and Medical Applications, in: *Textiles: Types, Uses and Production Methods*, pp. 467–487, Nova Science Publishers, Inc, Hauppauge, NY, 2012.

115. de Araújo Júnior, A.M., Braido, G., Saska, S., Barud, H.S., Franchi, L.P., Assunção, R.M., Scarel-Caminaga, R.M., Capote, T.S.O., Messaddeq, Y., Ribeiro, S.J., Regenerated cellulose scaffolds: Preparation, characterization and toxicological evaluation. *Carbohyd. Polym.*, 136, 892–898, 2016.
116. Suñol, J.J., Miralpeix, D., Saurina, J., Carrillo, F., Colom, X., Thermal behavior of cellulose fibers with enzymatic or Na 2 CO 3 treatment. *J. Therm. Anal. Calorim.*, 80, 117–121, 2005.
117. Nam, S., Hillyer, M.B., Condon, B.D., Method for identifying the triple transition (glass transition-dehydration-crystallization) of amorphous cellulose in cotton. *Carbohyd. Polym.*, 228, 115374, 2020.
118. Huson, M. and Denham, C., The Glass Transition of Cotton, in: *Cotton Fibres*, vol. 43, Nova Science Publishers, Hauppauge, NY, 2017.
119. Rebenfeld, L., Fibers and fibrous materials, in: *Textile Science and Technology*, vol. 13, pp. 199–232, Elsevier, Amsterdam, NL, 2002.
120. Mark, H.F., *Encyclopedia of polymer science and technology, concise*, John Wiley & Sons, Hoboken, NJ, 2013.
121. Kassab, Z., Bahloul, A., Hassani, F.-Z.S.A., Ablouh, E.-H., Bouhfid, R., Qaiss, A.E.K., Sehaqui, H., Achaby, M.E., Structure and Properties of Cellulose and Its Derivatives, in: *Annual Plant: Sources of Fibres, Nanocellulose and Cellulosic Derivatives. Composites Science and Technology*, R. Khiari, M. Jawaid, M.N. Belgacem (Eds.), Springer, Singapore, 2023.
122. Latif, W., Basit, A., Rehman, A., Ashraf, M., Iqbal, K., Jabbar, A., Baig, S.A., Maqsood, S., Study of mechanical and comfort properties of modal with cotton and regenerated fibers blended woven fabrics. *J. Nat. Fibers*, 16, 6, 836–845, 2019.
123. Kıvrak, N.M., Ozdıl, N., Mengüç, G.S., Characteristics of the yarns spun from regenerated cellulosic fibers. *Text. App.*, 28, 2, 107–117, 2018.
124. Lewin, M. (Ed.), *Handbook of Fiber Chemistry*, CRC Press, Boca Raton, FL, 2006.
125. Zhang, H., Shen, Y., Edgar, K.J., Yang, G., Shao, H., Influence of cross-section shape on structure and properties of Lyocell fibers. *Cellulose.*, 28, 1191–1201, 2021.
126. Udomkichdecha, W., Chiarakorn, S., Potiyaraj, P., Relationships between fibrillation behavior of lyocell fibers and their physical properties. *Text. Res. J.*, 72, 11, 939–943, 2002.
127. Iqbal, D., Zhao, R., Sarwar, M.I., Ning, X., Cellulose-Based Nanofibers Electrospun from Cuprammonium Solutions. https://doi.org/10.21203/rs.3.rs-3195822/v1. 2023.
128. Wakelyn, P.J., Bertoniere, N.R., French, A.D., Thibodeaux, D.P., Triplett, B.A., Rousselle, M.-A., Goynes Jr., W.R., Edwards, J.V., Hunter, L., McAlister, D.D., Gamble, G.R., *Cotton Fiber Chemistry and Technology*, 1st ed, CRC Press, Boca Raton, FL, 2006, https://doi.org/10.1201/9781420045888.
129. Michud, A., Tanttu, M., Asaadi, S., Ma, Y., Netti, E., Kääriainen, P., Sixta, H., Ioncell-F: ionic liquid-based cellulosic textile fibers as an alternative to viscose and Lyocell. *Text. Res. J.*, 86, 5, 543–552, 2016.

130. Taştan Özkan, E., Thermophysiological comfort and objective hand properties of modal, Naia™, cashmere and cotton included denim fabrics. *J. Eng. Fibers Fabr.*, 18, 15589250231219758, 2023.
131. Colmorgen, F. and Khawaja, C., *Small-scale technology options for regional bioeconomies*, BE-Rural, 2019.
132. Mathur, K., ITMA 2019-New developments in Fibers, Yarns and Fabrics. *J. Text. App. Technol. Manag.* Special Issue - ITMA Show, Barcelona, Spain, 1-6, 2019.
133. Zhu, S., Xie, J., Sun, Q., Zhang, Z., Wan, J., Zhou, Z., Lu, J., Chen, J., Xu, J., Chen, K., Fan, M., Recent advances on bast fiber composites: Engineering innovations, applications and perspectives. *Compos. B Eng*, 111738, 2024.
134. https://ioncell.fi/#:~:text=Ioncell%C2%AE%20in%20a%20nutshell%201%20The%20Ioncell%C2%AE%20process,could%20revolutionize%20the%20recycling%20of%20textile%20waste.%20. Accessed 2024-08-21.
135. Applications Naia Eastman. https://www.eastman.com/en/products/brands/naia/applications Accessed 2024-08-21.
136. Sustainable textile materials. | Spinnova. https://spinnova.com/fibre/#:~:text=Applications%20SPINNOVA%C2%AE%20fibre%20can%20be%20used%20in%20similar,brands%2C%20including%20T-shirts%2C%20trousers%2C%20jackets%2C%20and%20even%20skis. Accessed 2024-08-21.
137. TENCEL™ REFIBRA™ Technology: Transforming Textile Waste into New Fibers. https://www.tencel.com/b2b/technologies/refibra-technology. Accessed 2024-08-21.
138. Stepan, A.M., Michud, A., Hellstén, S., Hummel, M., Sixta, H., Ioncell-P&F: pulp fractionation and fiber spinning with ionic liquids. *Ind. Eng. Chem. Res.*, 55, 29, 8225–8233, 2016.
139. Şen, K., Kaya, A., Ozdemir, H.N., Seki, Y., Investigation on the fastness and dimensional stability of the knitted fabrics made of the new generation regenerated cellulose fibers. *Eur. J. Res. Develop*, 2, 4, 135–146, 2022.
140. Berlin, E. and Hamilton, M., Minimal footprint, maximum circularity—The new kind of smart, in: *Smart Clothes and Wearable Technology*, pp. 83–96, Woodhead Publishing, Cambridge, MA, 2023.
141. Celep, G., Tetik, G.D., Yilmaz, F., Limitations of textile recycling: the reason behind the development of alternative sustainable fibers, in: *Next-Generation Textiles*, IntechOpen, London, UK, 2022.
142. Björquist, S., Aronsson, J., Henriksson, G., Persson, A., Textile qualities of regenerated cellulose fibers from cotton waste pulp. *Text. Res. J.*, 88, 21, 2485–2492, 2018.
143. Global Regenerated Cellulose Market by Type (Fibers (Viscose, Lyocell, Modal), Films), Manufacturing Process, Source, End-user Industry (Fabrics, Automotive, Agriculture, Packaging) and Region (North America, Europe, APAC, Rest of the World) - Forecast to 2027. https://www.researchandmarkets.com/reports/5660071/global-regenerated-cellulose-market-by-type#src-pos-1. Accessed 2024-08-21.

144. Periyasamy, A.P., Processing and Application of Lyocell–An Overview. https://textilelearner.net/processing-and-application-of-lyocell/. 2021.
145. Gupta, A. and Dhakate, S.R., Cellulose Acetate-Based Nanofibers: Synthesis, Manufacturing, and Applications, in: *Emerging Technologies for Nanoparticle Manufacturing*, pp. 203–236, Springer International Publishing, Cham, 2021.
146. Mendes, I.S., Prates, A., Evtuguin, D.V., Production of rayon fibres from cellulosic pulps: state of the art and current developments. *Carbohyd. Polym.*, 273, 118466, 2021.
147. Forsberg, D.C.R., Bengtsson, J., Hollinger, N., Kaldéus, T., Towards Sustainable Viscose-to-Viscose Production: Strategies for Recycling of Viscose Fibres. *Sustainability*, 16, 10, 4127, 2024.
148. Li, Y., Peng, J., Liu, X., Song, D., Xu, W., Zhu, K., Dissolving waste viscose to spin cellulose fibers. *Polymer*, 237, 124349, 2021.
149. dos Santos, R.F., Oliveira, F.R., Rocha, M.R.D., Velez, R.A., Steffens, F., Reinforced cementitious composite using viscose rayon fiber from textile industry waste. *J. Eng. Fiber Fabr.*, 17, 15589250221115722, 2022.
150. Heikkilä, P., Määttänen, M., Jetsu, P., Kamppuri, T., Paunonen, S., Nonwovens from mechanically recycled fibres for medical applications. VTT Technical Research Centre of Finland. VTT Research Report No. VTT-R-00923-20, 2020.
151. Lakshmanan, S.O. and Raghavendran, G., Regenerated Sustainable Fibres, in: *Sustainable Innovations in Textile Fibres. Textile Science and Clothing Technology*, S. Muthu (Ed.), Springer, Singapore, 2018.
152. Yang, X., Fan, W., Wang, H., Shi, Y., Wang, S., Liew, R.K., Ge, S., Recycling of bast textile wastes into high value-added products: A review. Environment. *Chem. Lett.*, 20, 6, 3747–3763, 2022.
153. Liu, H., Fan, W., Miao, Y., Dou, H., Shi, Y., Wang, S., Zhang, X., Hou, L., Yu, X., Lam, S.S., Ge, S., Closed-loop recycling of colored regenerated cellulose fibers from the dyed cotton textile waste. *Cellulose*, 30, 4, 2597–2610, 2023.
154. Liu, X., Tian, Y., Wang, L., Chen, L., Jin, Z., Zhang, Q.A., Cost-Effective and Chemical-Recycling Approach for Facile Preparation of Regenerated Cellulose Materials. *Nano Lett.*, 24, 29, 9074–9081, 2024.
155. Wojciechowska, P., Fibres and textiles in the circular economy, in: *Fundamentals of natural fibres and textiles*, pp. 691–717, Woodhead Publishing, Oxford, UK, 2021.
156. Bracciale, M.P., de Caprariis, B., Musivand, S., Damizia, M., De Filippis, P., Chemical Recycling of Cellulose Acetate Eyewear Industry Waste by Hydrothermal Treatment. *Ind. Eng. Chem. Res.*, 63, 12, 5078–5088, 2024.
157. Rodrigues Filho, G., Monteiro, D.S., da Silva Meireles, C., de Assunção, R.M.N., Cerqueira, D.A., Barud, H.S., Ribeiro, S.J.L., Messadeq, Y., Synthesis and characterization of cellulose acetate produced from recycled newspaper. *Carbohyd. Polym.*, 73, 1, 74–82, 2008.

15

Durable and Sustainable Concrete with Spherical Nanocellulose: A Comprehensive Review

Mahapara Abbass[1], Syed Mohd. Arif[2*], Shalom Akhai[1], Rajni Verma[2] and Amandeep Singh Wadhwa[3]

[1]*Maharishi Markandeshwar Engineering College, Maharishi Markandeshwar (Deemed to be University), Mullana, Ambala, Haryana, India*
[2]*School of Engineering and Technology, CGC University, Jhanjeri, Mohali, Punjab, India*
[3]*University Institute of Engineering and Technology (UIET), Panjab University, Chandigarh, India*

Abstract
Challenges for the construction industry pertain more to developing sustainably long-lasting concrete solutions and mitigating impact on the environment as well as structurally ensuring durability. Several alternative materials are touting their alternatives, but one such emerging renewable biomass-derived material—spherical nanocellulose (SNC), whose properties may offer potential addition toward enhancing concrete performance—is researched in the following chapter as a review of advancement for incorporation into concrete. The properties of the concrete are improved by introducing SNC-based concrete with higher mechanical strength, sustainability, durability, and thermal insulation. The permeability can be reduced, chemical resistance is increased, and cracking in the material is minimized due to the increased lifespan of concrete. It also replaces traditional cement and aggregate materials while reducing greenhouse gas emissions and embodied energy. Furthermore, it possesses the ability of enhanced thermal resistance, minimizing energy consumption, and the maximum efficiency of buildings with respect to energy. Modification techniques for SNC are chemical functionalization, hybridization with other nanomaterials, and optimizing concentration and dispersion. Examples of applications include infrastructure development, building

**Corresponding author*: syed.aarif3@gmail.com

Mohd Shabbir (ed.) Regenerated Cellulose: A Renewable Alternative to Petroleum-Based Plastics, (463–490) © 2026 Scrivener Publishing LLC

insulation, and additive manufacturing through 3D printing. Challenges include scalability and cost effectiveness, standardization and regulation, long-term performance. Future research directions will continue to explore new applications, predict modeling, and complete life cycle assessments to enhance further understanding of the environment's impact. Such challenges include standards and guidelines for SNC-based concrete, deep investigation to long-term performance, as well as new applications in terms of marine structures and geopolymer concrete. This is with such great potential as a sustainable and performance-enhancing additive for concrete while research and development on-going efforts poised to overcome the challenges which are at hand, thus paving the way for a clear widespread adoption of SNC-based concrete solutions.

***Keywords*:** Spherical nanocellulose, sustainable concrete, nanomaterials in construction, green building materials, durability enhancement, bio-based construction materials

15.1 Background

The SNC initiative targets the environmental issues associated with the traditional concrete production process, specifically the carbon dioxide emissions released during cement production [1]. Using supplementary cementitious materials like Fly Ash (FA), Ground Granulated Blast Furnace Slag (GGBS), Silica Fume (SF), Metakaolin (MK), and Rice Husk Ash (RHA) produced from agricultural waste improves the sustainability and durability of concrete [2]. Concrete production is the largest source of CO_2 emissions worldwide, accounting for approximately 7% of total greenhouse gases; thus, using SCMs enables the production of more concrete with less cement, thereby reducing the impacts on the environment [3]. Sustainable concrete benefits include improved durability, economic efficiency, and reduced carbon footprint by incorporating waste materials and minimizing Portland cement usage. It will also ensure a shift toward concrete as more sustainable and longer-lasting in making environmentally friendly infrastructures [4]. SCMs together with performance-engineered approaches will also provide potential to reduce the ecological footprint of the construction industry as well as enhance the resilience and long-term life of concrete structures [5].

However, the construction industry faces several challenges that impact its efficiency and sustainability. These challenges include the environmental impacts of traditional concrete production, which contribute to carbon dioxide emissions and climate change; the increasing scarcity of natural resources such as high-quality aggregates and water; and the need for effective waste management strategies. Governments and regulating bodies are also driving the industry to greener practices and sustainability in the materials [6]. Hence, there is a great need for solutions that can be maintained sustainably and durably. Prominent features of this demand are the creation of greener materials in terms of fly ash, slag, and metakaolin; these could effectively reduce dependency on Portland cement and encourage recycling processes. Innovative mix designs can enhance their performance with fewer resources [7].

Concrete design should embrace the concept of lifecycle analysis in terms of estimating environmental impacts from production through disposal [8]. It would make use of materials that consume less footprint and serve longer. In durable long-lasting concrete, the aspect of longevity and minimal maintenance needs to be emphasized. As long as sustainability practices form a cornerstone in its business operations, the construction industry can conquer present challenges while establishing a stronger, greener future [9].

15.2 Properties and Benefits of SNC-Based Concrete

SNC-based concrete, therefore, is a sustainable solution based on supplementary cementitious materials and performance-engineered mixtures to enhance overall performance and sustainability [10]. The key advantages associated with SNC-based concrete include improved mechanical strength, durability, thermal insulation, low permeability, and minimization of cracking [11]. Therefore, SNC-based concrete reduces the reliance on Portland cement; consequently, carbon emissions are reduced in the manufacturing process of concrete, and also, it improves resistance toward environmental factors such as moisture, temperature fluctuations, and chemical attacks [12, 13]. The incorporation of specific SCMs provides a porous microstructure that increases thermal insulation, thus saving energy for heating and cooling in buildings and reducing energy efficiency [14]. SNC-based concrete is found to control shrinkage and tend to reduces thermal stresses during the curing process, which limits the cracking of the hardened material. Permeability reduction also minimizes corrosion in reinforcing steel and increases resistance to aggressive chemicals [15].

SNC-based concrete is found to control shrinkage and tend to reduces thermal stresses during the curing process, which limits the cracking of the hardened material; so the structures made of SNC concrete are found to have longer lifespan and require less maintenance over a period [16]. SNC-based concrete is thus structured to operate under adverse weather conditions [17]. Thus, the techniques used range from adjustment of mix designs up to the use of cooling in hot weather to reduce moisture evaporation [18]. In cold weather, heated water and insulating materials are used to maintain optimal conditions for curing [19].

The cost implications of SNC-based concrete on large-scale projects may be multiple, as material costs are often relatively high in initial stages, due to SCMs added and advanced mix designs [20]. However, these could be compensated by the longer-term savings in durability improvement and reduced maintenance costs. Extra reduction in carbon taxes or incentives using sustainable materials further improves economic viability [21]. Compared to other types of sustainable concrete, SNC-based concrete focuses more on the holistic integration of various SCMs such as fly ash, slag, and metakaolin to improve its mechanical strength while reducing environmental impacts [22].

Due to their durability and resistance to chemicals with reduced permeability, high-thermal insulations of the SNC-based concrete material make its applications viable for large-scale development works such as infrastructure facilities [16]. Its superior suitability spreads over bridges, highways, water tunnels, and notably infrastructure related to water conveyance systems such as dams and sewage systems [23]. The SNC-based concrete inclusion further significantly improves overall energy efficiency through regulation of internal temperature conditions with long-term maintenance significantly reduced [24].

Advanced monitoring technologies ensure full curing and strength development of concrete in extreme conditions of weather [25]. Real-time internal concrete temperature information is provided through wireless temperature sensors, and the present methods can change it and re-cure it when necessary [26]. In hot weather conditions, SNC-based concrete will overcome thermal cracking due to the prudent selection of material or a good mix design, which should make concrete workable with less heat generation while hydrating [27]. Cool aggregates and using ice in the mixture with shading or windbreaks aid in keeping it within a range of temperature.

Literature indicates that SNC-based concrete can be used successfully in cold climates, and projects equipped with wireless temperature sensors indicated that temperatures above freezing during early curing drastically reduce the risk of loss in strength and structural failure [28]. Generally, SNC-based concrete has higher strength than traditional concrete under extreme weather conditions because the mix design is optimized and there are effective moisture retention strategies in place, which makes it more resilient under fluctuating temperatures [29]. Table 15.1 provides a comprehensive overview of the properties, features, and advantages of SNC-based concrete solutions.

Table 15.1 Advantages of SNC-based concrete.

Properties and benefits	Description	Key features	Advantages
Improved Mechanical Strength	Enhanced compressive and flexural strength due to supplementary cementitious materials (SCMs) [30]	- Increased strength over time - Improved durability - Enhanced resistance to cracking	- Reduced risk of structural failure - Extended lifespan - Lower maintenance costs
Enhanced Sustainability and Durability	Reduced carbon emissions, increased resistance to environmental factors, and lower maintenance needs [31]	- Reduced reliance on Portland cement - Increased use of supplementary cementitious materials - Improved resistance to chemical attacks	- Reduced environmental impact - Lower greenhouse gas emissions - Improved public health and safety
Increased Thermal Insulation	Improved thermal performance due to porous microstructure, reducing energy consumption [16]	- Enhanced thermal resistance - Reduced heat transfer - Improved building energy efficiency	- Reduced energy consumption - Lower energy costs - Improved indoor air quality
Reduced Permeability and Improved Chemical Resistance	Minimized risk of corrosion, enhanced resistance to aggressive chemicals, and reduced water ingress [32]	- Reduced permeability - Improved chemical resistance - Enhanced durability	- Reduced risk of corrosion - Extended lifespan - Lower maintenance costs
Minimized Cracking and Extended Lifespan	Controlled shrinkage, reduced thermal stresses, and optimized mix designs for fewer cracks and extended lifespan [33]	- Controlled shrinkage - Reduced thermal stresses - Optimized mix designs	- Reduced risk of cracking - Extended lifespan - Lower maintenance costs

Overall, there are several advantages of SNC-based concretes. The benefits of SNC-based concrete include enhanced mechanical strength, sustainability, thermal insulation, reduced permeability, chemical resistance, and reduced cracking [34]. The integrated/blended cementitious materials improve the compressive and flexural strength of structure, thereby reducing the danger of structural failure and prolonging its lifespan [35]. SNC-based concrete also reduces carbon emissions, reduces harmful environmental effect, and maintenance expenses economically by reducing reliance on Portland cement and improving chemical resistance [36]. It also provides better thermal insulation, lessens energy usage, and betters the indoor air quality. It minimizes cracking and extends its life by managed shrinkage, reduced thermal stresses, and effective mix designs [37].

15.3 Environmental Benefits of SNC-Based Concrete

SNC-based concrete made by incorporating supplementary cementitious materials and performance-engineered mixtures, presents several environmental benefits that are significant to the construction industry's pursuit of sustainability [38]. As awareness of climate change and resource scarcity expands around the world, the pressure for environmentally conscious construction materials is raising [39]. This section discusses the three major most important/significant environmental advantages of SNC-based concrete that are - lower greenhouse gas emissions, lower embodied energy, and higher thermal resistance that reduces energy consumption. Table 15.2 presents these environmental advantages when Supplementary Cementitious Materials (SCMs) utilized.

15.3.1 Low Carbon Emissions

Among the biggest environmental benefits of SNC-based concrete is that it contributes less to greenhouse gas emissions caused by the production of concrete [31]. Cement production through traditional methods, for example, contributes 9% of CO_2 released worldwide; most of the emissions come from the process known as calcination where limestone is heated to make clinker [41, 42]. For every metric tonne of cement produced, around 900 kg of CO_2 is released [43]. By incorporating SCMs such as fly ash, ground granulated blast furnace slag (GGBS), and silica fume into the concrete mix, SNC-based concrete can substantially lower the carbon footprint of concrete production [44].

Table 15.2 Environmental benefits of using supplementary cementitious materials (SCMs).

Environmental benefit	Description	Key features	Advantages
Reduced Carbon Emissions	Lowered greenhouse gas emissions through the use of SCMs [60].	- 20% reduction in CO_2 emissions - Use of industrial by-products (fly ash, GGBS) - Reduced reliance on Portland cement [61]	- Mitigates climate change - Reduces environmental impact - Conserves natural resources
Lower Embodied Energy	Reduced energy consumption in production through the use of SCMs [62].	- 15-29% reduction in energy consumption - Lower operational costs	- Conserves energy resources - Reduces greenhouse gas emissions - Lowers operational costs
Enhanced Thermal Resistance	Improved thermal insulation properties through the use of SCMs [63].	- Porous microstructure of SCMs - Improved thermal resistance	- Energy efficiency - Reduces greenhouse gas emissions - Improves occupant comfort

According to research, use of SCMs will result in a reduction of more than 20% of CO_2 [45]. For example, it has been demonstrated that by replacing 40% of Portland cement with SCMs, the amount of CO_2 emitted is reduced from 398 kg to around 325 kg per cubic meter of concrete [47]. This decrease is significant as it directly reduces the amount of greenhouse gas released into the atmosphere and hence helps mitigate climate change.

Furthermore, SCMs consumption not only reduces emissions from production but also encourages recycling [48]. This is because these products are derived from industrial by-products, otherwise destined for landfills. Fly ash and GGBS are by-products from combustion of coal and steel manufacturing processes [48]. By reusing these by-products in concrete production, SNC-based concrete helps waste landfills and simultaneously improves mechanical properties and durability [49].

15.3.2 Lower Embodied Energy

Another critical environmental benefit of SNC-based concrete is its lower embodied energy compared to traditional concrete [50]. Embodied energy refers to the total energy consumed in the extraction, processing, transportation, and installation of building materials [51]. The production of Portland cement is highly energy-intensive, requiring significant amounts of fossil fuels and electricity [52].

SNC-based concrete reduces the overall energy consumption in its production by replacing a portion of Portland cement with SCMs [53]. The literature shows that SCMs reduce energy consumption between 15% and 29%, depending on the mix design used [54]. The reduction in embodied energy lowers the operational costs but also lessens the environmental impact associated with energy use in construction [55].

Besides, SNC-based concrete durability explains its lower life-cycle energy consumption. With structures using SNC-based concrete exhibiting longer service lives and much lower levels of maintenance requirement as opposed to structures built of conventional concrete, this ensures less resources usage in service for replacement or repair of portions throughout a building's lifetime hence consuming even lesser amounts of energy in consumption [57].

15.3.3 Enhanced Thermal Resistance and Minimized Energy Consumption

The incorporation of SCMs in SNC-based concrete improves its thermal resistance properties. It, therefore, enhances better insulation performance in buildings and reduces energy consumption for heating and cooling purposes. The porous microstructure developed by some SCMs gives improved thermal insulation compared to conventional concrete mixes [30].

The indoor temperature is stable for buildings made from SNC-based concrete. In that regard, these buildings demand less heating and cooling. As a result, the operational costs will be lower for the owners of such buildings. Besides that, it reduces greenhouse gases in the atmosphere generated during the production of energy [58]. Furthermore, the improved heat resistance enhances comfort for occupants through stabilizing the interior climate. This is particularly relevant in regions with large temperature differences, where maintaining comfortable interior conditions comes at a very high energy cost [59, 64].

Thus, SNC-based concrete generally offers strong environmental advantages for sustainable construction. Greenhouse gas emissions reduction from supplementary cementitious materials are fundamental steps toward combating the challenges posed by climate change [65]. With this, the associated embodied energy for SNC-based concrete is relatively lower. That implies that it conserves more resources and reduces adverse impacts on the environment through the life cycle. This improves the thermal resistance of SNC-based concrete, thus enhancing energy efficiency and improving occupant comfort in buildings [66]. The building industry still looks for answers that align with green goals. SNC-based concrete stands out as a material that matches well with caring for the environment and making money in the long run [67]. These advances in materials and building methods could help make the construction field greener and more open to needed changes in how we build. We see SNC-based concrete used more in big projects, which shows progress in cutting down on how much building hurts the planet. At the same time, it makes our buildings stronger and last longer.

15.4 Modification Techniques for SNC

SNC concrete are considered as unique among breakthroughs in construction materials. The innovation in its composition offers a set of very useful infrastructure properties. These features make structures perform better, become sustainable, and give them a longer lifespan [68]. It is possible to utilize SNC concrete in a wide variety of applications. This section considers three main pathways along which to modify SNC potency: modification of its surface modification, admixture with small other materials, and careful augmentation and dispersal.

15.4.1 Chemical Functionalization

The surface chemistry of nanomaterials is altered by chemical functionalization [69]. This procedure is intended to enhance nanomaterial interaction with the cement matrix and influence SNC behavior. This procedure may have significant effects on the mechanical properties and durability of concrete [70].

1. **Surface Modification:** Exposition of functional groups such as silanes or carboxylic acids bonded at the surface of nanomaterials produce good bonding between them and the cement pre-cursor. Enhancement of adhesion will in turn affect load transfer improvement leading to mechanical strength

enhancement [71]. For example, silane-functionalized nanoparticles improve interactions with the hydrating cement phases, generating denser microstructures [72].

2. **Compatibility Enhancement:** The dispersion behaviour of nanomaterials within the cement paste can be improved by alteration of surface charge or hydrophilicity/hydrophobicity. This improves their interaction with other constituents in the concrete mix. This spread helps to reduce the agglomeration, which would reduce the functionality of the SNC [73].
3. **Chemical Reactions:** Some functionalized nano–molecules target cement hydration products and undergo chemical reactions with them for an increased quantity of the binding phase that further improves strength and durability [74]. For instance, the use of functionalized silica nanoparticles may encourage pozzolanic reactions that lead to further strength gain with time [75].
4. **Improved Resistance to Environmental Factors:** Another way that SNC resists adverse environmental factors related to moisture and chemicals is through chemical functionalization. Making a hydrophobic surface on nanoparticle could significantly minimize water absorption and improve chemical resistance normalized to that of traditional concrete and thus, extending its service life [76].

15.4.2 Hybridization with Other Nanomaterials

Hybridization involves combining different types of nanomaterials to leverage their unique properties and synergistically enhance the performance of SNC. This technique allows for tailored solutions that address specific challenges in concrete applications [77].

1. **Combining Different Nanoparticles:** There are many possibilities which exist for researchers to couple hybrid nanoparticles such as CNT, GO, or nanosilica, with SCMs such as fly ash or slag, among others. The outcomes would be composites displaying significantly better mechanical properties compared to those made from each single type of nanomaterial. For instance, CNT fibers impart highly increased tensile strength and ductility to the silica fume-based mix, both of which are expected results of the exceptional mechanical properties of CNT [78].

2. **Synergistic Effects:** Integration of different nanomaterials yields synergistic effects with overall improvement in performance; for example, one nanoparticle can enhance compressive strength while the other improves workability or lowers permeability. The holistic approach enables the designing of mixes of SNC to meet specific performance criteria while maximizing environmental benefits [80].
3. **Optimizing Hybrid Ratios:** Hybridization and its further optimization proves its effectiveness by fine-tuning different nanomaterials ratios in concrete mixes [81]. Past studies show corelation of admixture ratios for obtaining fine balance between the nanoparticle distributions which is therefore vital to achieve new desirable properties of the composite without losing workability and other desirable characteristics [81].
4. **Increased Durability:** Hybrid SNCs show enhanced resistance to various environmental severities like in the form of freezing-thawing cycles, chemical attacks, or abrasion [82]. For instance, by addition of graphene oxide with conventional SCMs aids to improve resistance to penetration by chloride ions, and this is a critical parameter that is usually more dangerous for structures exposed to de-icing salts or seawater [83].

15.4.3 Optimizing Concentration and Dispersion

For achieving overall performance and resistance to failure, it is important that not only the concentration but also how evenly the nanomaterials are dispersed within the SNC is done in an optimized ratio/manner. This optimization allows nanomaterials utilization effectively without degrading workability or other mechanical properties [84].

1. **Determining Optimal Concentration:** The nanomaterials optimum concentration must yield all the expected output values while minimizing possible harmful effects caused by the consumption of the same. Excessive nanoparticles are detrimental but make it challenging to form the preparation properly [85]. According to literature available as of date, the range of concentration varies between 0.5 to 5% by weight of cement for many nanomaterials but depends on the material along with the application in case of specific ones.

2. **Dispersion Techniques:** Nanomaterials have to be homogeneously dispersed in the concrete matrix. Therefore, ultrasonication and high shear mixing methods are used that break agglomerates and distribute uniformly. At times, surfactants or dispersants may also be added to stabilize the nanoparticles during the mixing operation.
3. **Impact on Workability:** Added nanomaterials with concrete mixture modify workability changes associated with changes in viscosities and fluidities [86]. Concentrate levels needs thus to be optimized for satisfactory level of workability with good merit for improvement of mechanical property. All has to be designed with trade-offs of other mix design. Hence better design becomes easy because more improvements in performance take places.
4. **Monitoring Dispersion Quality:** The quality dispersion of nanomaterials in SNC mixes is examined using advanced characterization techniques, including DLS and SEM. The measurements provide size distributions along with information regarding agglomeration trends, thus opening up a chance for the modification of processing techniques to be applied [87].

SNC concrete modification techniques enhance performance and sustainability in interfacial bonding of nanomaterials to cement paste [88]. Chemical functionalization enhances compatibility and the exclusion of environmental factors [89]. Hybridization with other nano-materials allows for tailor-made exploitation of synergism mechanisms to achieve maximum mechanical properties and durability [90]. Optimization of nanoparticles concentration and dispersed patterns optimize the workability of concrete mixes [91]. It allows engineers to formulate SNC-based concrete solutions for immediate construction needs and greening challenges in traditional concrete manufacturing. As research moves forward, improvements will come into sustainable construction practices in the industry.

15.5 Applications of SNC-Based Concrete

SNC Concrete is a new material that, in contrast to its traditional counterpart, contains nanotechnology inside it. Therefore, in addition to enhanced mechanical properties and durability, SNC Concrete also supports environmental protection and sustainability. The areas of application of SNC-based concrete are very broad, including infrastructure development, building insulations, and 3D printing additive manufacture [92]. All of these are covered in this section.

15.5.1 Infrastructure Development

Infrastructure development is one of the primary applications of SNC-based concrete. The current need for durable, resilient, and sustainable materials in the construction of roads, bridges, tunnels, among other critical infrastructure has never been higher. Several advantages make it particularly suitable for infrastructure projects:

1. **Improved Strength:** The nanoparticles in SNC-based concrete include silica fume and carbon nanotubes, which enhance the mechanical properties of the concrete significantly. The compressive and tensile strengths increase and hence reduce cracking and structural failures. Such is important for the durability of infrastructure exposed to extreme environmental conditions such as freeze-thaw cycles, chemical attacks from de-icing salts, and abrasion from heavy traffic [93].
2. **Lower Maintenance:** The longer lifetime of the concrete-based using SNC material means fewer maintenance requirements. Structures, having used this material, withstand all the daily activities, including environmental stressors for a long time, reducing lifecycle cost. This makes critical infrastructure more favorable, especially when access to facilities can be challenging or costlier [16].
3. **Sustainability:** SNC-based concrete reduces the dependence of Portland cement, which causes a heavy contribution toward greenhouse emissions. Supplemental cementitious materials used in SNC-based concrete include fly ash and slag. The recycling of these waste materials can reduce both environmental impacts and promote resource savings [94].
4. **Self-Healing Properties:** Certain SNCs have self-healing capabilities through the use of self-healing technologies, which enable the concrete to heal itself upon cracking [30]. Microcapsules that contain healing agents can be incorporated into the concrete mix. Once cracks develop, these capsules break open and release the healing agent, thus filling in the cracks and restoring the structure. This is an added innovative feature that improves the durability of infrastructure [95].

5. **Application in Critical Structures:** SNC-based concrete is very ideal for critical infrastructure like bridges, tunnels, and high-rise buildings where structural integrity takes center stage. Its great performance characteristics make it a very suitable application for purposes requiring safety and longevity [96].

15.5.2 Building Insulation

The most important applications of SNC-based concrete are building insulation [97]. Over the last decade, energy efficiency has become one of the major considerations in modern construction practice, and SNC-based concrete presents unique advantages for sustainable building design:

1. **Improved Thermal Resistance:** Nanomaterials would ensure that the developed concrete would have better thermal resistance. That would mean improved insulation properties and, hence, can lead to stable indoor temperatures, particularly by reducing heat transfer through walls and ceilings. Buildings constructed from SNC-based concrete can, therefore, be made energy-efficient by reducing the need for heating and cooling systems [98].
2. **Energy Consumption:** SNC-based concrete reduces the energy consumption of the buildings because it offers good thermal insulation properties. Low energy consumption leads directly to lower operational heating and cooling costs and, eventually it cuts down on greenhouse effect-related emissions as a consequence of its energy production [97].
3. **Sustainable Building Practice:** The concrete material, being SNC-based, reflects the sustainable building practice that focuses on energy efficiency and resource savings. When this material is used to construct buildings, builders are able to produce structures that fulfill stringent energy codes while decreasing their environmental footprint [99].
4. **Chances for Green Certifications:** Concrete produced with SNC may even qualify for green building certifications such as LEED, Leadership in Energy and Environmental Design [100]. This may allow it to receive these qualifications if the buildings will contain environmentally friendly materials for its construction.

5. **Increased Comfort Sensitivity:** Increased thermal resistance contributes energy savings while improving the comfort level of occupants in any building [101]. Since it creates stable internal temperatures and excludes drafts, the SNC-based concrete promotes comfortable living or working conditions inside buildings.

15.5.3 Additive Manufacturing by 3D Printing

Additive manufacturing, also known as 3D printing, is a revolutionary construction approach that can fabricate structures layer by layer using various materials, including SNC-based concrete [102–105].

1. **Design Flexibility:** This is perhaps the greatest strength that SNC-based concrete adds to 3D printing, which is design freedom that architects and engineers now have in creating their products. Such complex geometries, which are quite challenging or even impossible to obtain using conventional construction techniques, can easily be achieved by additive manufacturing techniques [106–108].
2. **Material Efficiency:** Additive manufacturing eliminates material waste as it uses only a quantity of material that would be required for each successive layer of construction [109–114]. Material efficiency is especially useful when expensive or high-performance materials, such as SNC-based concrete, are involved.
3. **Quick Building Time:** 3D printing offers a much faster time of building than the conventional means. Structures can be built either on-site or in an off-site controlled environment with much less labor input and project duration [115].
4. **Customization:** The custom architectural feature or specific conditions of a site could simply be accommodated using additive manufacturing because it only permits adding customization of designs [103, 115–117].
5. **Smart Integration:** SNC concrete, in 3D printing with smart technologies, enables the tracking of properties and integrity of the material at all stages of building structure development. Such an integration can enhance the characteristics of performance further while making sure that the final product is within specifications.

6. **Sustainability Benefits:** SNC-based concrete additive manufacturing would be sustainable because it would minimize waste and promote resource efficiency at every stage of the construction process [118]. Applications of SNC concrete are extremely wide and diversified and involve such critical areas as infrastructure development, building insulation, and additive manufacturing by 3D printing [119]. All the applications prove that SNC-based concrete outperforms traditional materials in durability, with a lesser negative environmental impact, energy efficiency, and increased flexibility in design. The construction industry is changing to a more responsible and environmentally friendly process. SNC-based concrete offers a versatile solution capable of addressing the challenges in current practice: responsible resource use and environmental protection. Through the integration of nanotechnology breakthroughs and creative building methods, stakeholders will have developed sustainable structures that satisfy existing requirements but also pave the way for a more sustainable construction future.

15.6 Challenges and Future Research Directions

This immense potential that may be available for improving performance while sustaining the construction practices of SNC-based concrete is closely related to the complete realization of its full scope. However, still there are challenges to be met for realizing its full scope. The most crucial ones are scalability and cost-effectiveness, standardization and regulation, long-term performance and prediction modeling, life cycle assessments, environmental impact, and the development of standards and guidelines for specific materials [120, 121]. The major challenges include the production scalability of nanomaterials and cost-effectiveness [122].

15.6.1 Cost

The cost of nanomaterials can be so high that it alarms the users from employing such technologies. The main reason why nanomaterials are costly is due to the complexity of the process and the specialized equipment used, which may not be readily available in bulk. Therefore, the method of production must be scalable and mass-produce nanomaterials in large quantities without compromising on the quality for it to be cost-effective. This would make it impossible to make the current laboratory scale synthesis methods suitable for large-scale productions and limit supplies according to the demand. Long-term economic benefits include durability and reduced maintenance requirements. Life cycle cost analyses will explain to stakeholders the financial benefits of SNC-based concrete over a long period of time.

15.6.2 Standardization and Regulation

It challenges much concerning how concrete formed out of SNC would need fewer standardization and regulatory requirements in the construction field, and therefore standardized testing is required in order to develop consistency and reliability in an attempt to allow comparison amongst different formulations, thereby standardizing and comparing their values. Another key point relates to compliance with regulatory practices since the guidelines put should consider safety features about nanoparticles.

It would be necessary for that collaboration between researchers, industry professionals, the regulatory bodies, and organizations for standards development. Standards will depict best practice use of the SNC-based concrete only.

15.6.3 Long-Term Performance and Prediction Modeling

Long-term environmental conditioning of SNC-based concrete is essential for developing efficient projects. Long-term durability studies, therefore, need to be designed to understand the freeze-thaw cycles, chemical attack, and moisture-ingress performance of the material. The prediction modeling must, therefore, evolve with the long-term response of SNC-based concrete considering all aspects of environmental conditions along with loading scenarios and corresponding material properties [110]. Field testing in relevant site applications is also required due to valid predictive models. With time, the structures constructed using SNC will generate data on actual performance metrics that will be used to fine-tune the models.

15.6.4 Life Cycle Assessment and Environmental Impact

There should be the study of SNC-based concrete for comprehensive LCA by evaluating environmental impact in all the stages of its life cycle. Such evaluations will have to cover all the indicators about the environment, i.e., emission of green gas, energy consumption, and depletion of resources. With such production of SNC, the source should have sustainable raw materials taking either recycled or industrial by-products for reduced impacts on the environment. Knowledge of the behavior of SNC-based concrete at its end-of-life stage helps in determining appropriate recycling or waste management strategies, such as utilization of materials from demolished buildings in reusing or recycling material to minimize landfilling.

15.6.5 Development of Standards and Guidelines for SNC-Based Concrete

This has some promise, particularly in the aspects of a safer and more environment-friendly product. That is why its application in construction makes it so significant that it must have a proper set of guidelines and testing protocols. This covers material selection which may encompass compatibility with cementitious material, environmental stability, or potential health risks. Tests should standardize performance criteria for SNC-based concrete, including mechanical properties and durability as well as related environmental impacts. Such programs should be developed regarding training professionals on best practices of the use of nanotechnology in construction. Despite such a challenge it faces, SNC concrete can actually be integrated into practice and needs a best production process and long-term economic benefit through collaborative research, manufacturing, and industrial stakeholders. It will give standardized testing methods and regulatory frameworks that will ensure the safety and efficacy of SNC-based products. Continuous research will be done into long-term performance prediction modeling, improving material behavior under realistic conditions. Life cycle assessments would be conducted to provide insight into environmental impacts and sourcing practices. In doing this, the full realization of SNC technology in the resilience of infrastructure will be seen to meet modern demands and minimize ecological footprints.

15.7 Conclusion

SNC-based concrete is one of the major innovations in construction, responding to a very urgent demand for sustainable building materials that can deliver high performances. Nanotechnology integrated into concrete formulation improves SNC-based concrete: its mechanical strength, its durability, thermal insulation, and also environmental impact reduction. The use of supplementary cementitious materials such as Fly Ash, Ground Granulated Blast Furnace Slag, and Rice Husk Ash improves the sustainability through a reduction in emissions of greenhouse gases and embodied energy. SNC-based concrete demonstrates advanced characteristics, including very low permeability and strong chemical resistance, which is very suitable for essential applications of infrastructure such as bridges, tunnels, and water-related structures. The ability to control shrinkage, reduce thermal stresses when setting, and enhance the resilience of the structure against environmental factors has resulted in structures with increased lifetimes and less need for maintenance. The emerging application of 3D printing technique opens doors for customized, efficient, and sustainable construction practices.

However, there exists challenges seen with the widespread application of SNC-based concrete, despite its various advantages. Scalability, cost-effectiveness, standardization, and long-term performance are some of the key challenges that need to be addressed through improved modification techniques, such as chemical functionalization, hybridization with other nanomaterials, and optimization of concentration and dispersion.

Further studies on SNC-based concrete would include complete life cycle analyses, predictive modeling, and new applications to upgrade its potential. The approach taken by design and construction is crucial for reducing the ecological footprint of the industry to achieve global climate targets.

In conclusion, SNC-based concrete is a viable solution for durable, energy-efficient, and environmentally sustainable construction solutions. Through continuous innovation and collaboration among researchers, industry professionals, and regulatory bodies, SNC-based concrete has the potential to significantly transform the future of construction, aligning it with sustainability and resilience principles.

AI disclosure statement: The author confirms that no artificial intelligence (AI)-generated content or AI-assisted writing tools were used in the preparation of this chapter (including summaries and the abstract). All text, analysis, and conclusions are the result of the author's original work.

Bibliography

1. Asif, I., Hussain, M.U., Khan, A.A., Ashar, M., Usman, M., Shahid, Z., Utilization of Sawdust Ash as an additive of cement in concrete and study of its mechanical properties. *Memoria Investig. Ing.*, 26, 54–69, 2024.
2. Abbass, M., Singh, D., Singh, G., Properties of hybrid geopolymer concrete prepared using rice husk ash, fly ash and GGBS with coconut fiber. *Mater. Today Proc.*, *45*, 4964–4970, 2021.

3. Abbass, M. and Singh, G., Experimental investigation of engineered alkali-activated fibrous geopolymer concrete. *J. Build. Pathol. Rehabil.*, *8*, 1, 12, 2023.
4. D'Alessandro, A., Smart sustainable concrete materials and structures, in: *Sustainable Concrete Materials and Structures*, pp. 601–624, Woodhead Publishing, Cambridge, UK, 2024.
5. Younas, H., Yu, J., Leung, C.K., Mechanical and environmental performance of high-strength strain-hardening cementitious composites with high-dosage ternary supplementary cementitious materials: Fly ash, limestone, and calcined clay. *Constr. Build. Mater.*, *444*, 137856, 2024.
6. Bekele, A.A., Mahesh, G., Ingle, P.V., Enhancing SMCs' competitiveness through improving material supply chain management practice. *Int. J. Constr. Manag.*, *25*, 1, 77–88, 2025.
7. Mohammadi, A. and Ramezanianpour, A.M., Investigating the environmental and economic impacts of using supplementary cementitious materials (SCMs) using the life cycle approach. *J. Build. Eng.*, *79*, 107934, 2023.
8. Dhiman, S., Seema, S., Akhai, S., Performance evaluation of OPC 43 concrete incorporating micro silica and recycled ceramic tile waste as partial cement and coarse aggregate replacements. *Asian J. Civ. Eng.*, 1–19, 2025.
9. Harouache, A., Abkar, M.M.A., Gamil, Y., Al-Shameri, A.S.A., Gabir, A.A.M., Influence of supply chain management on the sustainable construction industry in Algeria. *Front. Built Environ.*, *10*, 1233266, 2024.
10. Hu, W., Zhang, D., Ftwi, E., Ellis, B.R., Li, V.C., Development of sustainable low carbon Engineered Cementitious Composites with waste polyethylene fiber, sisal fiber and carbonation curing. *Resour., Conserv. Recycl.*, *197*, 107096, 2023.
11. Ilyas, R.A., Asyraf, M.R.M., Rajeshkumar, L., Awais, H., Siddique, A., Shaker, K., Wahit, M.U., A review of bio-based nanocellulose epoxy composites. *J. Environ. Chem. Eng.*, 113835, 2024.
12. Liu, J. and Lv, C., Durability of cellulosic-fiber-reinforced geopolymers: a review. *Molecules*, *27*, 3, 796, 2022.
13. Kumari, A. and Yadav, S.K., Moving towards sustainable nanoengineered building materials with less energy consumption. *Energy Build.*, 114475, 2024.
14. Peng, T., Xiao, R., Rong, Z., Liu, H., Hu, Q., Wang, S., Zhang, J., Polymer nanocomposite-based coatings for corrosion protection. *Chem. Asian J.*, *15*, 23, 3915–3941, 2020.
15. Chen, L., Yang, M., Chen, Z., Xie, Z., Huang, L., Osman, A.I., Yap, P.S., Conversion of waste into sustainable construction materials: A review of recent developments and prospects. *Mater. Today Sustain.*, 100930, 2024.
16. Macías-Silva, M.A., Cedeño-Muñoz, J.S., Morales-Paredes, C.A., Tinizaray-Castillo, R., Perero-Espinoza, G.A., Rodríguez-Díaz, J.M., Jarre-Castro, C.M., Nanomaterials in construction industry: An overview of their properties and

contributions in building house. *Case Stud. Chem. Environ. Eng.*, 100863, 2024.

17. Alfarra, J., Mechanical behaviour of masonry assemblages built with mortar containing anti-freeze admixtures subject to sub-freezing curing temperatures, Doctoral dissertation, University of Saskatchewan, 2022.
18. Jay, O., Capon, A., Berry, P., Broderick, C., de Dear, R., Havenith, G., Ebi, K.L., Reducing the health effects of hot weather and heat extremes: from personal cooling strategies to green cities. *Lancet, 398*, 10301, 709–724, 2021.
19. Nilimaa, J. and Zhaka, V., An overview of smart materials and technologies for concrete construction in cold weather. *Eng.*, 4, 2, 1550–1580, 2023.
20. Reshmy, R., Philip, E., Thomas, D., Madhavan, A., Sindhu, R., Binod, P., Pandey, A., Bacterial nanocellulose: engineering, production, and applications. *Bioengineered, 12*, 2, 11463, 2021.
21. Popescu, C., Dissanayake, H., Mansi, E., Stancu, A., Eco Breakthroughs: Sustainable Materials Transforming the Future of Our Planet. *Sustainability, 16*, 23, 10790, 2024.
22. Amin, M., Elsakhawy, Y., Abu el-hassan, K., Abdelsalam, B.A., Behavior evaluation of sustainable high strength geopolymer concrete based on fly ash, metakaolin, and slag. *Case Stud. Constr. Mater.*, *16*, e00976, 2022.
23. Sikarskas, D., Antonovič, V., Malaiškienė, J., Boris, R., Stonys, R., Šahmenko, G., Effect of polyvinil alcohol fiber on the microstructure and properties of cement composite with expanded glass aggregates, 2020.
24. Marani, A., Zhang, L.V., Nehdi, M.L., Multiphysics study on cement-based composites incorporating green biobased shape-stabilized phase change materials for thermal energy storage. *J. Clean. Prod.*, *372*, 133826, 2022.
25. Taffese, W.Z. and Nigussie, E., Automated concrete curing and assessment of strength and durability using IoT system. *Mater. Today Proc.*, 2023.
26. Aleixandre, R.N., Shrikrishnan, S., Dalby, M.J., Salmeron-Sanchez, M., Engineered living biomaterials. *Nat. Rev. Mater.*, *6*, 12, 1175–1190, 2021.
27. Lu, J., Deng, F., Pei, W., Wan, X., You, Z., Zhang, Z., Mitigating frost heave and enhancing mechanical performance of silty clay with sisal fibre and geopolymer. *Constr. Build. Mater.*, *447*, 138120, 2024.
28. Zhu, M., Chen, H., Wang, L., Shi, Y., He, Y., Cui, T., Radiative cooling performance analysis of cooling wood with three delignification processes for building applications. *Energy Build.*, *312*, 114185, 2024.
29. Mei, X., Liu, C., Li, Z., Research progress on functional, structural and material design of plant-inspired green bionic buildings. *Energy Build.*, 114357, 2024.
30. Reza, T., Uddin, R., Tazwar, M.F., Hoque, M.E., Autul, Y.S., Mechanical, Thermal, Electrical, Optical, and Magnetic Properties of Organic Nanofillers, in: *Nanofillers*, pp. 71–113, CRC Press, Boca Raton, FL, USA, 2023.

31. Saleem, H., Zaidi, S.J., Alnuaimi, N.A., Recent advancements in the nanomaterial application in concrete and its ecological impact. *Materials, 14*, 21, 6387, 2021.
32. Barnat-Hunek, D., Omiotek, Z., Szafraniec, M., Dzierżak, R., An integrated texture analysis and machine learning approach for durability assessment of lightweight cement composites with hydrophobic coatings modified by nanocellulose. *Measurement, 179*, 109538, 2021.
33. Tushmanlo, A.S., Tushmanlo, H.S., Asadollahfardi, G., Cici, Y.M., Applications of micro-nanobubble and its influence on concrete properties: An in-depth review. *Nanotechnol. Rev., 13*, 1, 20240068, 2024.
34. Dhali, K., Ghasemlou, M., Daver, F., Cass, P., Adhikari, B., A review of nanocellulose as a new material towards environmental sustainability. *Sci. Total Environ., 775*, 145871, 2021.
35. Abbass, M. and Akhai, S., Exploring the sustainability potential of geopolymer concrete with coal bottom ash and basalt fibres. *Multiscale Multidiscip. Model. Exp. Des.*, 8, 3, 199, 2025.
36. Johnson, A.T., Production of Bio-Based Concrete Water Reducer from Crude Cellulose derived from Woody Biomass, Master's thesis, The University of Western Ontario, Canada, 2021.
37. Kumar, P. and Akhai, S., Effective energy management in smart buildings using VRV/VRF systems, in: *Additive Manufacturing in Industry 4.0*, pp. 27–35, CRC Press, Boca Raton, FL, USA, 2022.
38. Yu, K., Pan, J., Zhou, J., Zhou, Y., Zhou, X., Zhuge, Y. (Eds.), *Strain Hardening Cementitious Composites: material development, performance characterization, structural, and 3D printing applications*, Frontiers Media SA, Lausanne, Switzerland, 2022.
39. Walia, J., Sharma, R., Akhai, S., Stubble-reinforced materials for sustainable construction: Advancing SDG 9, 11, and 12 through waste-to-resource innovations, in: *Mitigating Unsustainable Practices in Construction and Architecture*, pp. 99–118, IGI Global, USA, 2025.
40. Lai, G., Liu, X., Li, S., Xu, Y., Zheng, Y., Guan, J., Cui, S., Development of chemical admixtures for green and environmentally friendly concrete: a review. *J. Cleaner Prod., 389*, 136116, 2023.
41. Nie, S., Zhou, J., Yang, F., Lan, M., Li, J., Zhang, Z., Sanjayan, J.G., Analysis of theoretical carbon dioxide emissions from cement production: Methodology and application. *J. Clean. Prod., 334*, 130270, 2022.
42. Krishnan, S., Emmanuel, A.C., Shah, V., Parashar, A., Mishra, G., Maity, S., Bishnoi, S., Industrial production of limestone calcined clay cement: experience and insights. *Green Mater., 7*, 1, 15–27, 2018.
43. Naqi, A. and Jang, J.G., Recent progress in green cement technology utilizing low-carbon emission fuels and raw materials: A review. *Sustainability, 11*, 2, 537, 2019.
44. Singh, R.P., Vanapalli, K.R., Cheela, V.R.S., Peddireddy, S.R., Sharma, H.B., Mohanty, B., Fly ash, GGBS, and silica fume based geopolymer concrete with

recycled aggregates: Properties and environmental impacts. *Constr. Build. Mater.*, *378*, 131168, 2023.

45. Abbass, M. and Akhai, S., Eco-friendly geopolymer bricks synthesized from alkali-activated sugarcane bagasse ash and rice husk ash. *Ind. Crops Prod.*, 233, 121410, 2025.
46. Keerio, M.A., Saand, A., Kumar, A., Bheel, N., Ali, K., Effect of local metakaolin developed from natural material soorh and coal bottom ash on fresh, hardened properties and embodied carbon of self-compacting concrete. *Environ. Sci. Pollut. Res.*, *28*, 60000–60018, 2021.
47. Snellings, R., Suraneni, P., Skibsted, J., Future and emerging supplementary cementitious materials. *Cem. Concr. Res.*, *171*, 107199, 2023.
48. Aslani, A., Hachem-Vermette, C., Zahedi, R., Environmental impact assessment and potentials of material efficiency using by-products and waste materials. *Constr. Build. Mater.*, *378*, 131197, 2023.
49. Naseem, S. and Rizwan, M., Sustainable construction and combo nanocellulose: A synergistic approach to greener building materials. *Energy Build.*, 115218, 2024.
50. Pathak, S., Sharma, H., Verma, H., Akhai, S., Arif, S.M., Singh, A., Enhancing flexible pavement performance through the incorporation of waste engine oil, crumb rubber, and polyethylene in bituminous mixes, in: *AIP Conf. Proc.*, 3261, 1, 020006, AIP Publishing LLC, June 2025.
51. Chen, W., Yang, S., Zhang, X., Jordan, N.D., Huang, J., Embodied energy and carbon emissions of building materials in China. *Build. Environ.*, *207*, 108434, 2022.
52. Fadayini, O.M., Madu, C., Oshin, T.T., Obisanya, A.A., Ajiboye, G.O., Ipaye, T.O., Kingsley, N.A., Energy and economic comparison of different fuels in cement production, in: *Cement Ind. Optim. Charact. Sustain. Appl*, 2021, 10.
53. Badawy, A.H., El-Feky, M.S., El-Tair, A.M., Kohail, M., Wood sawdust waste-derived nano-cellulose as a versatile reinforcing agent for nano silica cement composites: a systematic study on its characterization and performance. *Sci. Rep.*, *13*, 1, 12811, 2023.
54. Abu Shammeh, S., Influence of Sustainable Concrete Mix Designs on Costs and Global Warming Potential, Doctoral dissertation, Université d'Ottawa| University of Ottawa, 2024.
55. Gauch, H.L., Dunant, C.F., Hawkins, W., Serrenho, A.C., What really matters in multi-storey building design? A simultaneous sensitivity study of embodied carbon, construction cost, and operational energy. *Appl. Energy*, *333*, 120585, 2023.
56. Yadav, V., Singh, S., Garg, M.P. and Akhai, S., Effect of natural Sapindus mukorossi treatment process on bio-waste bagasse fibers for biocomposite fabrication and application purposes. *Ind. Crops Prod.*, 219, 119066, 2024.
57. Nguyen, M.T., Fernandez, C.A., Haider, M.M., Chu, K.H., Jian, G., Nassiri, S., Glezakou, V.A., Toward self-healing concrete infrastructure: review of

experiments and simulations across scales. *Chem. Rev.*, *123*, 18, 10838–10876, 2023.

58. Abbass, M., Akhai, S., Abbas, U., Jafri, R., Arif, S.M., AI-enabled sustainable urban planning and management, in: *Real-World Applications of AI Innovation*, pp. 233–260, IGI Global, USA, 2025.
59. Suresh, C., Hotta, T.K., Saha, S.K., Phase change material incorporation techniques in building envelopes for enhancing the building thermal Comfort-A review. *Energy Build.*, *268*, 112225, 2022.
60. Knight, K.A., Cunningham, P.R., Miller, S.A., Optimizing supplementary cementitious material replacement to minimize the environmental impacts of concrete. *Cem. Concr. Compos.*, *139*, 105049, 2023.
61. Ndahirwa, D., Zmamou, H., Lenormand, H., Leblanc, N., The role of supplementary cementitious materials in hydration, durability and shrinkage of cement-based materials, their environmental and economic benefits: A review. *Cleaner Mater.*, *5*, 100123, 2022.
62. Xing, W., Tam, V.W., Le, K.N., Hao, J.L., Wang, J., Life cycle assessment of sustainable concrete with recycled aggregate and supplementary cementitious materials. *Resour. Conserv. Recycl.*, *193*, 106947, 2023.
63. Ramzi, S. and Hajiloo, H., The effects of supplementary cementitious materials (SCMs) on the residual mechanical properties of concrete after exposure to high temperatures. *Buildings*, *13*, 1, 103, 2023.
64. Akhai, S., Abbass, M., Air conditioning for healthier living, in: *Managing the Health Risks of Climate Change*, pp. 1–20, IGI Global, USA, 2025.
65. Tayebani, B., Said, A., Memari, A., Less carbon producing sustainable concrete from environmental and performance perspectives: A review. *Constr. Build. Mater.*, *404*, 133234, 2023.
66. Akhai, S., Smart home environment for society: An overview of IoT-enabled air conditioners, in: *Modelling of Virtual Worlds Using the Internet of Things*, Vol. 1, pp. 74–87, CRC Press, Boca Raton, FL, USA, 2024.
67. Dhanasingh Sivalinga, V., Devarajan, P., Ramalingam, B., Soudagar, M.E.M., Mohanavel, V., Khan, T.Y., Cuce, E., Current trends and biotechnology infused cleaner production of biomaterials for the construction industry: A critical review. *Int. J. Low-Carbon Technol.*, *19*, 833–849, 2024.
68. Sanchez-Salvador, J.L., Xu, H., Balea, A., Negro, C., Blanco, A., Nanocellulose from a colloidal material perspective. *Front. Mater.*, *10*, 1231404, 2023.
69. Khoramian, R., Issakhov, M., Pourafshary, P., Gabdullin, M., Sharipova, A., Surface modification of nanoparticles for enhanced applicability of nanofluids in harsh reservoir conditions: A comprehensive review for improved oil recovery. *Adv. Colloid Interface Sci.*, 103296, 2024.
70. Zhang, J. and Zhang, X., Interfacial bonding characteristics of multi-walled carbon nanotube/ultralight foamed concrete. *Sci. Eng. Compos. Mater.*, *31*, 1, 20240028, 2024.

71. Ong, P.J., Wang, S., Suwardi, A., Cao, J., Wang, F., Yin, X., Loh, X.J., Functional Coatings for Built Environments Based on Nanotechnology. *World Sci. Annu. Rev. Funct. Mater.*, *2*, 2330004, 2024.
72. Huang, C.Y., Lin, Y.C., Chung, J.H., Chiu, H.Y., Yeh, N.L., Chang, S.J., Chen, G.Y., Enhancing cementitious composites with functionalized graphene oxide-based materials: Surface chemistry and mechanisms. *Int. J. Mol. Sci.*, *24*, 13, 10461, 2023.
73. Onthong, S., Edgar, A.O., Pongprayoon, T., Composite nanoarchitectonics by interfacial bonding for conductivity and strength development of grafted multiwall carbon nanotube/cement. *Constr. Build. Mater.*, *392*, 131940, 2023.
74. Meng, S., Shi, Z., Ouyang, X., Comparison of the effects of carbon-based and inorganic nanomaterials on early cement hydration. *Constr. Build. Mater.*, *421*, 135705, 2024.
75. Cacciatore, A., Smart Concrete for Urban Areas: Improved Performance and Sustainability of Functionalized Graphene Related Materials in Cementitious Products, 2023.
76. Szafraniec, M., Grabias-Blicharz, E., Barnat-Hunek, D., Landis, E.N., A critical review on modification methods of cement composites with nanocellulose and reaction conditions during nanocellulose production. *Materials*, *15*, 21, 7706, 2022.
77. Mekhzoum, M.E.M., El Bourakadi, K., Bouhfid, R., Cellulose nanocrystal/nanoparticles hybrid nanocomposites: From preparation to applications, in: *Cellulose Nanocrystal/Nanoparticles Hybrid Nanocomposites*, pp. 1–25, Woodhead Publishing, Cambridge, UK, 2021.
78. Janaki, A.M., Shafabakhsh, G., Hassani, A., Laboratory evaluation of alkali-activated slag concrete pavement containing silica fume and carbon nanotubes. *Eng. J.*, *25*, 5, 21–31, 2021.
79. Liu, C., Chen, F., Wu, Y., Zheng, Z., Yang, J., Yang, B., Luo, Y., Research progress on individual effect of graphene oxide in cement-based materials and its synergistic effect with other nanomaterials. *Nanotechnol. Rev.*, *10*, 1, 1208–1235, 2021.
80. Divya, S. and Praveenkumar, S., An integrated evaluation of graphene-based concrete mixture with copper slag and quarry dust using response surface methodology. *J. Build. Eng.*, *86*, 108876, 2024.
81. Klapiszewska, I., Ławniczak, L., Balicki, S., Gapiński, B., Wieczorowski, M., Wilk, K.A., Ślosarczyk, A., Influence of zinc oxide particles dispersion on the functional and antimicrobial properties of cementitious composites. *J. Mater. Res. Technol.*, *24*, 2239–2264, 2023.
82. Mahmud, A., Wasif, M.M., Roy, H., Mehnaz, F., Ahmed, T., Pervez, M.N., Islam, M.S., Aquatic microplastic pollution control strategies: sustainable degradation techniques, resource recovery, and recommendations for Bangladesh. *Water*, *14*, 23, 3968, 2022.

83. Castro-Borges, P., Isgor, B., Ogunsanya, I., Panesar, D., Suraneni, P., Weiss, J., Concrete Infrastructure: Recent Advancements and Needs with a Focus on North America. *RILEM Tech. Lett.*, *9*, 36–49, 2024.
84. Gorbunova, M., Grunin, L., Morris, R.H., Imamutdinova, A., Nanocellulose-based thermoplastic polyurethane biocomposites with shape memory effect. *J. Compos. Sci.*, *7*, 4, 168, 2023.
85. Khan, S., Naushad, M., Al-Gheethi, A., Iqbal, J., Engineered nanoparticles for removal of pollutants from wastewater: Current status and future prospects of nanotechnology for remediation strategies. *J. Environ. Chem. Eng.*, *9*, 5, 106160, 2021.
86. Huang, K., Xie, J., Wang, R., Feng, Y., Rao, R., Effects of the combined usage of nanomaterials and steel fibres on the workability, compressive strength, and microstructure of ultra-high performance concrete. *Nanotechnol. Rev.*, *10*, 1, 304–317, 2021.
87. Srivastava, S., Advancements in Nanocellulose Derived from Plant Waste: Strategies for Sustainable Innovations and Applications. *J. Solid Waste Technol. Manag.*, *50*, 3, 543–563, 2024.
88. Elhadary, R. and Bassuoni, M.T., Interfacial bonding between basalt fiber/polymer pellets and various nano-modified cementitious matrices. *J. Mater. Civ. Eng.*, *35*, 3, 04022471, 2023.
89. Yu, S., Pang, H., Huang, S., Tang, H., Wang, S., Qiu, M., Wang, X., Recent advances in metal-organic framework membranes for water treatment: A review. *Sci. Total Environ.*, *800*, 149662, 2021.
90. Sahoo, J., Sharma, R., Pachauri, V., Gandhi, S., Biomimetic/Bioderived Nanoengineered Interfaces for Biosensor Applications: A Review. *ACS Appl. Nano Mater.*, *7*, 17, 19854–19875, 2024.
91. Adithya, S.N., Sachin, K.C., Meda, U.S., Review on the correlation between mixing, microstructure and strength of cementitious products with nanoparticles. *IOP Conf. Ser.: Earth Environ. Sci.*, 822, 1, 012005, 2021, July. IOP Publishing.
92. She, W., Wu, Z., Yang, J., Pan, H., Du, F., Du, Z., Miao, C., Cement-based biomimetic metamaterials. *J. Build. Eng.*, 110050, 2024.
93. Alateah, A.H., Graphene concrete: Recent advances in production methods, performance properties, environmental impact and economic viability. *Case Stud. Constr. Mater.*, e02653, 2023.
94. Alkhaldi, S.K., Reducing the Carbon Footprint and Enhancing the Strength of Mortar (Concrete) Through the Incorporation of Date Seeds Ash, Master's thesis, Alfaisal University, Saudi Arabia, 2023.
95. Ma, E., Chen, X., Lai, J., Kong, X., Guo, C., Self-healing of microcapsule-based materials for highway construction: A review. *J. Traffic Transp. Eng. (Engl. Ed.)*, *10*, 3, 368–384, 2023.
96. Sepetcioglu, H., Lapčík, L., Lapčíková, B., Vašina, M., Hui, D., Ovsík, M., Zmeškal, O., Improved mechanical properties of graphene-modified basalt fibre–epoxy composites. *Nanotechnol. Rev.*, *13*, 1, 20240052, 2024.

97. Shen, Z., Qin, M., Xiong, F., Zou, R., Zhang, J., Nanocellulose-based composite phase change materials for thermal energy storage: status and challenges. *Energy Environ. Sci.*, *16*, 3, 830–861, 2023.
98. Bhardwaj, S., Singh, S., Gupta, P., Choudhary, N., Maji, P.K., Role of morphological arrangements in cellulose nanofiber-based aerogels for thermal insulation: A systematic review. *Int. J. Green Energy*, 1–22, 2023.
99. Azarmi, M., Berg, A., Geier, S., Gerhardt, E., Greimel, M., Groiß-Fuertner, D., Wacker, P., Scientific Background Paper, 2024.
100. Samaeen, S., Connecting Community Through Sports: A Design Approach for Community Sports Complex, Doctoral dissertation, Brac University, 2022.
101. Tushar, Q., Zhang, G., Bhuiyan, M.A., Giustozzi, F., Navaratnam, S., Hou, L., An optimized solution for retrofitting building façades: Energy efficiency and cost-benefit analysis from a life cycle perspective. *J. Cleaner Prod.*, *376*, 134257, 2022.
102. Altıparmak, S.C., Yardley, V.A., Shi, Z., Lin, J., Extrusion-based additive manufacturing technologies: State of the art and future perspectives. *J. Manuf. Processes*, *83*, 607–636, 2022.
103. Akhai, S. and Khang, A., Innovations in Medical Manufacturing: A Review of 3D Printing, Robotics, and Internet of Things (IoT), in: *The Quantum Evolution*, vol. 1, pp. 226–241, CRC Press, Elsevier, Amsterdam, Netherlands, 2024.
104. Khang, A. and Akhai, S., Green Intelligent and Sustainable Manufacturing: Key Advancements, Benefits, Challenges, and Applications for Transforming Industry, in: *Machine Vision and Industrial Robotics in Manufacturing*, pp. 405–417, 2024.
105. Akhai, S. and Wadhwa, A.S., Recent Advances in Bio-Tribology From Joint Lubrication to Medical Implants: A Review. *J. Mater.*, *2*, 2, 125–135, 2024.
106. Estakhrianhaghighi, E., Engineered Fiber-Reinforced Architected Cellular Composites: A Strategy to Develop Sustainable and Recyclable Advanced Materials, 2023.
107. Akhai, S. and Singh, H., Design optimization for modification of trough belt conveyor to reduce material spillage used in clinker transport in cement plant. *Int. J. Res.*, *1*, 27–37, 2013.
108. Wadhwa, S., A., M., A., A., S., K., Kumar, P., Integrating Taguchi Optimization for Multi-Criteria Decision Making in Engineering Applications, in: *Recent Theories and Applications for Multi-Criteria Decision-Making*, vol. 1, No. 1, pp. 125–150, IGI Global, Elsevier, Amsterdam, Netherlands, 2024.
109. Akhai, S., Bansal, S.A., Singh, S., A critical review of thermal insulators from natural materials for energy saving in buildings. *J. Crit. Rev.*, *7*, 19, 278–283, 2020.
110. Akhai, S., A Review on Optimizations in μ-EDM Machining of the Biomedical Material Ti6Al4V Using the Taguchi Method: Recent Advances Since 2020. *Latest Trends Eng. Technol.*, 395–402, 2024.

111. Verma, R., Akhai, S., Wadhwa, A.S., Use of Smart Materials in Physiotherapy, in: *Revolutionizing Healthcare Treatment With Sensor Technology*, pp. 300–319, IGI Global, Elsevier, Amsterdam, Netherlands, 2024.
112. Sharma, V. and Akhai, S., Trends in Utilization of Coal Fly Ash in India: A Review. *J. Eng. Design Analysis*, *2*, 1, 12–16, 2019.
113. Saleh Alghamdi, S., John, S., Roy Choudhury, N., Dutta, N.K., Additive manufacturing of polymer materials: Progress, promise and challenges. *Polymers*, *13*, 5, 753, 2021.
114. Siddiqui, S., Surananai, S., Sainath, K., Khan, M.Z., Kuppusamy, R.R.P., Suneetha, Y.K., Emerging trends in the development and application of 3D printed nanocomposite polymers for sustainable environmental solutions. *Eur. Polym. J.*, 112298, 2023.
115. Salman, N.M., Ma, G., Ijaz, N., Wang, L., Importance and potential of cellulosic materials and derivatives in extrusion-based 3D concrete printing (3DCP): Prospects and challenges. *Constr. Build. Mater.*, *291*, 123281, 2021.
116. Abbas, M., Akhai, S., Abbas, U., Jafri, R., Arif, S.M., AI-Enabled Sustainable Urban Planning and Management, in: *Real-World Applications of AI Innovation*, pp. 233–260, IGI Global Scientific Publishing, Elsevier, Amsterdam, Netherlands, 2025.
117. Abbass, M., Abbas, U., Jafri, R., Arif, S.M., Akhai, S., AI and Machine Learning Applications in Sustainable Smart Cities, in: *Sustainable Smart Cities and the Future of Urban Development*, pp. 1–32, IGI Global Scientific Publishing, Elsevier, Amsterdam, Netherlands, 2025.
118. Macías-Silva, M.A., Cedeno-Munoz, J.S., Morales-Paredes, C.A., Tinizaray-Castillo, R., Perero-Espinoza, G.A., Rodríguez-Díaz, J.M., Jarre-Castro, C.M., Case Studies in Chemical and Environmental Engineering.
119. Saleh, H.M. and Hassan, A.I., Synthesis and characterization of nanomaterials for application in cost-effective electrochemical devices. *Sustainability*, *15*, 14, 10891, 2023.
120. Kumar, H., Wadhwa, A.S., Akhai, S., Kaushik, A., Parametric analysis, modeling and optimization of the process parameters in electric discharge machining of aluminium metal matrix composite. *Eng. Res. Express*, *6*, 2, 025542, 2024.
121. Kumar, H., Wadhwa, A.S., Akhai, S., Kaushik, A., Parametric optimization of the machining performance of Al-SiCp composite using combination of response surface methodology and desirability function. *Eng. Res. Express*, *6*, 2, 025505, 2024.
122. Zhu, Y., Mohammed, K.J., Elsehrawy, M.G., Ali, H.E., AL Garalleh, H., Advancing building management with nano-enhanced carbon materials: a machine learning-driven business and economic analysis. *Carbon Lett.*, 1–22, 2024.

Index

Printed in the USA/Agawam, MA
September 18, 2026

910783.020